Model Optimization Methods for Efficient and Edge AI

IEEE Press
445 Hoes Lane
Piscataway, NJ 08854

Model Optimization Methods for Efficient and Edge AI

Federated Learning Architectures, Frameworks and Applications

Edited by

Pethuru Raj Chelliah
Chief Architect at the Edge AI division of Reliance Jio Platforms Ltd. (JPL), Bangalore, India

Amir Masoud Rahmani
Artificial intelligence faculty member at the National Yunlin University of Science and Technology, Taiwan

Robert Colby
Principal Engineer in IT Infrastructure responsible for Manufacturing Network Architecture and IoT Infrastructure at Intel Corporation

Gayathri Nagasubramanian
Assistant Professor, Department of Computer Science and Engineering, GITAM University, Bengaluru, India

Sunku Ranganath
Principal Product Manager for Edge Infrastructure Services at Equinix

IEEE Press

WILEY

Published by John Wiley & Sons, Inc., Hoboken, New Jersey.
Published simultaneously in Canada.

For general information on our other products and services or for technical support, please contact our Customer Care Department within the United States at (800) 762-2974, outside the United States at (317) 572-3993 or fax (317) 572-4002.

Wiley also publishes its books in a variety of electronic formats. Some content that appears in print may not be available in electronic formats. For more information about Wiley products, visit our web site at www.wiley.com.

Library of Congress Cataloging-in-Publication Applied for:

Hardback ISBN: 9781394219216

Cover Design: Wiley
Cover Image: © argus/Shutterstock

Set in 9.5/12.5pt STIXTwoText by Straive, Chennai, India

Contents

About the Editors

Pethuru Raj Chelliah, PhD, is the Chief Architect at the Edge AI division of Reliance Jio Platforms Ltd. (JPL) Bangalore, India. Previously, he worked at the IBM Global Cloud Centre of Excellence (CoE), Wipro Consulting Services (WCS), and Robert Bosch Corporate Research (CR). He has gained over 23 years of IT industry experience and 9 years of research experience. He completed his CSIR-sponsored PhD at Anna University, Chennai, and continued with UGC-sponsored postdoctoral research in the Department of Computer Science and Automation at the Indian Institute of Science (IISc), Bangalore. Following that, he was granted two international research fellowships (JSPS and JST) to work as a research scientist for 3.5 years in two leading Japanese universities.

He focuses on some of the emerging technologies like artificial intelligence (AI) model optimization techniques, Internet of Things use cases, cloud-native and edge-computing paradigms, reliability engineering practices, 6G communication and blockchain technologies, quantum cryptography algorithms, and multimodal generative AI methods.

Amir Masoud Rahmani, PhD, is an artificial intelligence faculty member at the National Yunlin University of Science and Technology, Taiwan. He has published over 240 journal papers (around 180 ISI-indexed papers, many in the best quartile journal, Q1) and more than 100 reference conference papers. His research interests are in the Internet of Things, cloud/fog computing, big data, and artificial intelligence.

Robert Colby is the Principal Engineer in IT Infrastructure, responsible for manufacturing network architecture, and IoT infrastructure at Intel Corporation. Rob joined Intel in 1999, and over the next 10 years drove a variety of network infrastructure and factory security architectures. In 2010, Rob shifted his focus to wireless sensor technologies (RFID, Bluetooth, 802.15.4, ultrasound, etc.) to standardize wireless telemetry and location sensor infrastructure. Over the past six years, Rob has been driving IoT infrastructure standardization, defining how market products land within Intel to solve business problems. His E2E IoT analytic architecture has enabled factories to accurately predict component failures weeks in advance. More recently, Rob has focused on connectivity modernization within Intel Factories and has shifted Intel from using “classic ethernet” to a software-defined networking strategy for industrial edge access. Rob has a long history of mentoring in repeatable innovation methodologies, running programs such as innovation “boot camps” and kick-starter projects. Rob has been an Intel Patent committee member for six years, is currently a voting member of the IoT and Fog Committee, and Rob now holds sixteen US patents.

Gayathri Nagasubramanian, PhD, is an assistant professor in the Department of Computer Science and Engineering at GITAM University, Bengaluru, India. She received her PhD in Information and Communication Engineering from Anna University, Chennai. She received her ME degree in Computer Science and Engineering and her, BTech degree in Information Technology from Thiagarajar College of Engineering, Madurai. She has several publications in journals and has edited many books. She holds 7 Indian patents, 3 Australian patents. Her research interests include Big Data Analytics, IoT, and networks, and she is actively involved as an IEEE WIE faculty advisor.

Sunku Ranganath is the Principal Product Manager for Edge Infrastructure Services, Equinix. He is a product management and solution leader in edge computing, private 5G, and IoT SaaS and PaaS services. He has expertise in defining technical product requirements, cross-functional requirements, product positioning, and developing Go-To and Get-To Market solutions with inhouse and public cloud-based (AWS, GCP) offerings. He has experience driving collaboration with CXOs and leadership teams in OEMs, SaaS/PaaS startups, ISVs, SIs, Telco vendors, and service providers, and establishing open-source alliances. He also has expertise in the architecture, design, integration, and performance analysis of IoT and edge services, edge infrastructure, private 5G, edge security, NFV, and microservices, as well as strong. His strong business acumen enables him to prioritize and map business requirements to solution requirements and software deliverables.

He has experience leading and working with architecture, engineering, marketing, sales, data science teams distributed across Europe, Asia, China, and Mexico, using agile and Scrum methodologies. He leads the solution architecture for building Edge and NFV reference architectures, PoCs, and demos with Kubernetes, Service Mesh, OpenStack, OVS, DPDK, Prometheus, Collectd, Influxdb, Kafka, etc.

He has demonstrated servant leadership as a maintainer and contributor to multiple open-source projects, leading cross-geographical teams of engineers and contract workers toward an established cadence of releases.

He is an avid innovator with multiple patent filings and he has presented at multiple industry and IEEE conferences. He has also authored multiple ETSI standards, industry publications, white papers, and solution briefs.

List of Contributors

Pavan Kumar Akkisetty
Lead AI Architect – LLM Pipelines & Cloud As as a Service Lead AI Architect
Intel Corporation
Hillsboro, Oregon
USA

S. Annamalai
School of Computer Science and Engineering
Jain (Deemed-to-be University)
Bangalore, Karnataka
India

Hanieh Mohammadi Arzanagh
Department of Computer Science and Engineering
Shahid Beheshti University
Tehran
Iran

M. Ashok Kumar
Department of Computer Science and Software Engineering
Skyline University Nigeria
Kano
Nigeria

Sundaravadivazhagan Balasubaramanian
Department of Information Technology
University of Technology and Applied Sciences
Al Mussanah
Sultanate of Oman

Balamurugan Baluswamy
Shiv Nadar University
Delhi-National Capital Region (NCR)
India

Neha Bhati
Department of Research & Development
AVN Innovations
Ajmer, Rajasthan
India

Likitha Chowdary Botta
School of Computer Science and Engineering
VIT-AP University
Inavolu
Andhra Pradesh
India

Madala Guru Brahmam
School of Computer Science Engineering & Information Systems
Vellore Institute of Technology
Vellore, Tamil Nadu
India

Sachin Chaudhary
School of Computer Science and Applications
IIMT University
Meerut, Uttar Pradesh
India

Balakrishnan Chinnaiyan
Department of Computer Science
CHRIST University
Bangalore, Karnataka
India

Avi Das
School of Computer Science and Engineering
VIT-AP University
Inavolu, Andhra Pradesh
India

Priyanka Gupta
Department of CSIT
Guru Ghasidas Central University
Bilaspur, Chhattisgarh
India

K. Hemanth Sai
Department of AI
Anurag University
Hyderabad, Telangana
India

Atefeh Hemmati
Department of Computer Engineering, Science and Research Branch
Islamic Azad University
Tehran
Iran

Anupriya Jain
School of Computer Applications MRIIRS
Faridabad
Faridabad, Haryana
India

Mahalakshmi Jeyabalu
Department of Computer Science
CHRIST University
Bangalore, Karnataka
India

Naresh Kumar Kar
GITAM Deemed to be University
Hyderabad, Telangana
India

A. Suresh Kumar
School of Computer Science and Engineering
Jain (Deemed-to-be University)
Bangalore, Karnataka
India

Bhupendra Kumar
School of Computer Science and Applications
IIMT University
Meerut, Uttar Pradesh
India

M. Kumaresan
School of Computer Science and Engineering
Jain (Deemed-to-be University)
Bangalore, Karnataka
India

Seyedeh Yasaman Hosseini Mirmahaleh
Department of Electrical Engineering, Science and Technology
Lille University
Lille
France

A. Mallikarjuna Reddy
Department of AI
Anurag University
Hyderabad, Telangana
India

Aliyu Mohammed
Department of Management
Skyline University Nigeria
Kano
Nigeria

Rakhi Mutha
Department of IT
Amity University Rajasthan
Jaipur, Rajasthan
India

Yamini Nimmagadda
Network and Edge, Intel
Portland, OR
USA

Manoj Kumar Pandey
Department of CSE
Pranveer Singh Institute of Technology
Kanpur, Uttar Pradesh
India

and

Department of CSE
Chandigarh University
Mohali
India

Rajneesh Panwar
School of Computer Science and Applications
IIMT University
Meerut, Uttar Pradesh
India

Jayalakshmi Periyasamy
School of Computer Science Engineering & Information Systems
Vellore Institute of Technology
Vellore, Tamil Nadu
India

Amir Masoud Rahmani
Future Technology Research Center
National Yunlin University of Science and Technology
Douliou
Taiwan

Vijay Anand Rajasekaran
School of Computer Science Engineering & Information Systems
Vellore Institute of Technology
Vellore, Tamil Nadu
India

U. Rakesh
Department of Computational Intelligence
MRCET
Secunderabad
India

Partha Pratim Ray
Department of Computer Applications
Sikkim University
Gangtok, Sikkim
India

Mure Sai Jaideep Reddy
School of Computer Science and Engineering
VIT-AP University
Inavolu, Andhra Pradesh
India

T. Rupa Rani
Department of CSE
Narsimha Reddy Engineering College
Hyderabad, Telangana
India

N. Sangeetha
Department of Artificial Intelligence and Machine Learning
City Engineering College
Bangalore, Karnataka
India

Kewal Krishan Sharma
School of Computer Science and Applications
IIMT University
Meerut, Uttar Pradesh
India

Rattan Sharma
Delhi School of Business
New Delhi, Delhi
India

Seema Sharma
School of Computer Applications MRIIRS
Faridabad
Faridabad, Haryana
India

Vikas Sharma
School of Computer Science and Applications
IIMT University
Meerut, Uttar Pradesh
India

M. Sirish Kumar
School of Computing
Mohan Babu University
Tirupati, Andhra Pradesh
India

V. Sivanantham
Department of Computer Science
Periyar University
Salem, Tamil Nadu
India

G. Sridhar Reddy
Department of AI
Anurag University
Hyderabad, Telangana
India

S. Sumanth
Department of Computer Science and Computer Applications
Government College for Women (Affiliated to Bengaluru North University)
Kolar, Karnataka
India

D. Sumathi
School of Computer Science and Engineering
VIT-AP University
Inavolu, Andhra Pradesh
India

G. Sunil Kumar
Department of CSE
Narsimha Reddy Engineering College
Telangana
India

Dyavarashetty Sunitha
Department of CSE
Narsimha Reddy Engineering College
Hyderabad, Telangana
India

Dommaraju Tejavarma
School of Computer Science and Engineering
Jain (Deemed-to-be University)
Bangalore, Karnataka
India

M. Trupthi
Department of AI
Anurag University
Hyderabad, Telangana
India

Gandhodi Harsha Vardhan
School of Computer Science and Engineering
Jain (Deemed-to-be University)
Bangalore, Karnataka
India

Tarun Kumar Vashishth
School of Computer Science and Applications
IIMT University
Meerut, Uttar Pradesh
India

G. Victor Daniel
Department of AI
Anurag University
Hyderabad, Telangana
India

Narayan Vyas
Department of Computer Science and Application
Vivekananda Global University
Jaipur, Rajasthan
India

Gayathry S. Warrier
Department of Computer Science
CHRIST University
Bangalore, Karnataka
India

Pawan Whig
Vivekananda Institute of Professional Studies-TC
New Delhi, Delhi, India

Nikhitha Yathiraju
University of The Cumberlands
Williamsburg, KY
USA

1

Fundamentals of Edge AI and Federated Learning

Atefeh Hemmati[1], Hanieh Mohammadi Arzanagh[2], and Amir Masoud Rahmani[3]

[1] *Department of Computer Engineering, Science and Research Branch, Islamic Azad University, Tehran, Iran*
[2] *Department of Computer Science and Engineering, Shahid Beheshti University, Tehran, Iran*
[3] *Future Technology Research Center, National Yunlin University of Science and Technology, Douliou, Taiwan*

1.1 Introduction

Machine Learning (ML) and Artificial Intelligence (AI) have advanced unparalleled in recent years. These developments have changed many industries, and interactions between people and technology have experienced significant changes. Data processing in cloud environments was a primary focus of traditional AI and ML techniques. The need for real-time processing has expanded due to the growth of the Internet of Things (IoT). Because of this, a new era of technology known as Edge AI and Federated Learning (FL) has appeared [1, 2].

Edge AI gives edge devices and servers the ability to assess data and carry out AI operations close to the data generated. Autonomous vehicles, remote monitoring, and customized user experiences are some of the applications of this technology that stand out. Because data is not centralized and is handled on local devices, edge AI can help with security and privacy issues [3–5].

At the same time, FL has become a crucial strategy for protecting data privacy and distributing the work of developing ML models among several devices. Devices train ML models using local data using this technique, sharing only the model's parameters with a centralized server. This ensures that consumers' privacy is protected and that sensitive data remains on the devices [2, 4].

With the assistance of edge AI and FL, we can create applications and systems that are smarter and safer. This is a new claim in the field of AI. With this novel approach, we can utilize the sophisticated capabilities of AI in a connected world while enhancing privacy, security, and efficiency [2, 6, 7].

The concepts and foundations of Edge AI and FL are entirely clarified in this chapter. This chapter also examines the advantages and challenges of Edge AI and FL. We explore the benefits, which include decreased latency, better bandwidth consumption, improved privacy, and increased robustness. We also present the challenges, such as resource limitations on edge devices, communication costs, and the requirement to manage uniformly distributed data distributions in FL. We also discuss real-world applications where Edge AI and FL have proven effective, opening the door for revolutionary solutions across numerous industries. We also discuss FL and Edge AI's challenges, future research directions, and open issues.

The remaining part of the chapter is organized as follows: In Section 1.2, the fundamental concepts of Edge AI are discussed, including the advantages of Edge AI and the challenges

Model Optimization Methods for Efficient and Edge AI: Federated Learning Architectures, Frameworks and Applications, First Edition. Edited by Pethuru Raj Chelliah, Amir Masoud Rahmani, Robert Colby, Gayathri Nagasubramanian, and Sunku Ranganath.

faced by Edge AI. Section 1.3 introduces the concepts and fundamentals of FL, including the advantages of FL and the challenges associated with FL. In Section 1.4, the chapter emphasizes the combined power of FL and Edge AI, highlighting the benefits of their integration. Section 1.5 offers background insights into the technological landscape and motivations behind exploring Edge AI and FL. Applications of Edge AI and FL are discussed in Section 1.6. Section 1.7 addresses challenges, future research directions, and potential solutions for integrating Edge AI and FL. Finally, Section 1.8 concludes the chapter.

1.2 Concepts and Fundamentals of Edge AI

In an era of data generation and IoT, Edge AI is emerging as a revolutionary paradigm that brings AI computing closer to the edge of the network, where data is generated. Unlike traditional AI architectures that rely on cloud servers for processing, Edge AI distributes computing tasks to local edge devices such as smartphones, smart sensors, and edge servers. This change in the deployment of AI brings significant benefits and exciting possibilities in various fields [3, 8].

1.2.1 Defining Edge AI

Figure 1.1 shows that Edge AI directly integrates AI and ML capabilities on edge devices. This allows devices to perform data processing and inference locally, reducing the need to transfer data to centralized cloud servers constantly. This decentralized operation reduces latency in terms of time and avoids dependence on a stable Internet connection [9, 10].

Edge devices play a vital role in the edge AI ecosystem. These devices are typically equipped with limited resources compared to powerful cloud servers but with sufficient computing capabilities to run lightweight AI models. In other words, these devices can perform AI operations with their limited resources [11]. There are many examples of edge devices, including smartphones, tablets, smart

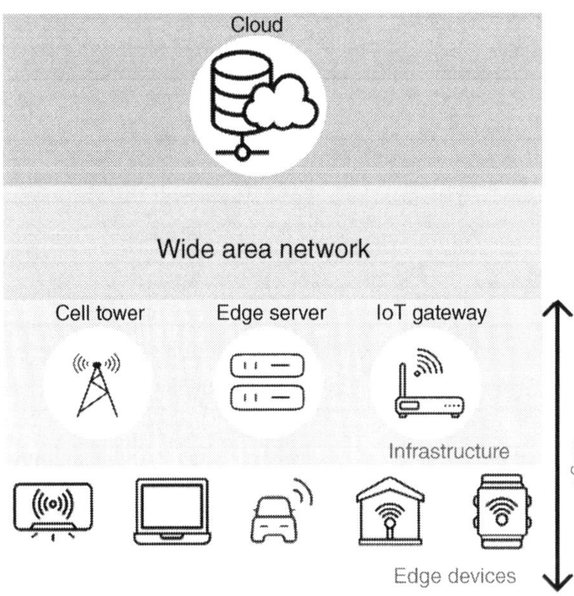

Figure 1.1 Edge AI architecture.

cameras, wearables, unmanned aerial vehicles (UAVs), and IoT sensors [12, 13]. These devices can collect, process, and interpret data without frequent communication with central servers and are ideally suited for situations where network bandwidth is limited, or data privacy is required. With these edge devices, the possibility of analyzing and using data locally increases to a certain extent, which leads to reduced response delays and improved user experience [5, 14, 15].

Edge AI has improved the process of directly deploying AI models to edge devices. This type of deployment results in significant latency reduction and increased privacy due to its local processing capabilities. Several different architectures have been used to implement edge AI, each of which covers its own specific needs and applications [16]:

Local Processing in IoT Devices: IoT devices mostly have limited computing resources, which makes direct deployment of complex AI models challenging. This type of architecture is suitable for applications that require fast processing, such as sensor data analysis, anomaly detection, and simple classification tasks [1, 17, 18]. In this instance, a unique local processing technique suggested by Bebortta et al. [19] supports an enhanced IoT platform structure for smart buildings. The proposed approach helps decrease bandwidth at the nodes' data collection level from a green computing standpoint. The researchers also discussed the most effective usage of sensors in wireless sensor networks (WSNs), who used the well-known queue model to calculate costs associated with non-Poisson and Poisson arrival of data packets at local processors. The experiments show that the suggested model successfully utilizes green computing standards. Therefore, this study offers a thing-centric, data-centric, and service-oriented IoT architecture within the framework.

Edge Servers and Gateways: Edge servers and gateways are more powerful computing devices closer to the edge devices. They act as intermediaries between edge devices and cloud servers, performing initial AI processing before sending data to cloud servers for further analysis. These devices can accommodate more resource-intensive models and are suitable for applications such as video analytics, natural language processing (NLP), and data preprocessing [2, 20]. As a real-world example, Rahmani et al. [21] utilized the ideal position of these gateways at the network's edge to provide several more advanced solutions, including storage locally, real-time local data processing, and integrated data mining, for giving an intelligent electronic health care gateway in the procedure. They then suggested creating a geo-distributed intermediate layer of awareness between sensor nodes and the cloud to use fog computing for medical IoT devices. Their fog-assisted design might handle obstacles in omnipresent medical facilities, such as mobility, energy consumption, flexibility, and dependability issues, by focusing on part of the duties of the sensor network and a distant medical center. Also, we can mention that Li et al. [22] suggested edge content-centric networking (ECCN). This enabling strategy combined Software-Defined Networking (SDN) with content-centric networking into a structured framework. To separate the data and control planes of ECC and CCN, SDN technology was included in the hierarchical framework. To manage data transmission, an SDN framework was created. To assess the effectiveness of the ECCN system, two apps were also deployed in the testbed. Thorough computations and results of experiments from the experimentation applications show that ECCN surpassed the original structures.

Edge Computing: The concept of edge computing expands the conceptual scope of edge computing by introducing a hierarchical architecture that spans from the edge to the cloud. In this setup, middle fog nodes are responsible for AI processing and decision-making, minimizing latency and network traffic. These nodes may be located in cellular base stations, access points, and local data centers. Edge computing is particularly suitable for applications that require real-time analytics in distributed environments such as smart cities and industrial IoT [23]. Chen et al. [24] addressed a distributed computing model called Mobile Edge Computing (MEC), which addresses energy and low latency requirements by integrating energy absorption technology into IoT devices. The research aimed to decrease system costs by proposing a hybrid power supply model for IoT devices and concurrently optimizing local computing, task transfer time, and edge computing decisions. Utilizing stochastic optimization theory, the study introduced an online dynamic algorithm named DTOME for task transfer in MEC with a hybrid energy supply. This algorithm facilitated task transfer decisions by balancing the system cost and queue stability. Simulation results affirmed the effectiveness of the DTOME approach, demonstrating significant improvements in reducing system costs compared to other comparative methods. Furthermore, Wu et al. [25] explored a multiuser MEC system within the context of the Internet of Vehicles, designed explicitly for handling computing tasks near vehicles. Compared to previous research that primarily focused on minimizing task transfer costs assuming complete channel estimation, this study addressed the issue of incomplete channel estimation caused by the dynamic movement of vehicles, which had been overlooked in prior work. The central objective was to decrease computation, communication delays, and energy consumption while considering the challenges of incomplete channel estimation. The study initially introduced a system cost metric that combined delay and energy consumption to achieve this goal. The optimization problem was then formulated to minimize this cost. The research further proposed an innovative approach combining deep reinforcement learning with Lagrange coefficients to reduce the overall system cost concurrently. Simulation results demonstrated the superiority of this approach over traditional methods in terms of performance.

Edge Multilayer Architecture: This architecture uses a multilayer approach with different levels of processing at other edge nodes. Low-level edge devices perform basic data preprocessing, feature extraction, and analysis tasks. Mid-level edge devices handle more complex tasks, including AI and model inference. Finally, the next level could be a local data center or cloud server that performs advanced analytics, model training, and long-term storage. This architecture optimizes resources and hides latency by distributing tasks at different levels [14, 26]. Robles-Enciso and Skarmeta [27] proposed an innovative multilayer enhancement to reinforcement learning (ML-RL) methods that enabled edge agents to reach an upper-level agent with supplementary information to improve efficiency in challenging and unpredictable scenarios. They initially proposed an RL approach to tackling the Task Assignment Problem (TAP) at the edge layer. Before task allocation, they assessed the potential balance between energy consumption and execution time. The scalability of each technique was subsequently validated through multiple simulations involving varying device counts.

Edge-Distributed AI: In this architecture, several edge devices work together to process and analyze data, and each one contributes to AI by executing a part of the model or performing specific tasks. Communication and device coordination are essential for efficient model integration and optimal decision-making. Distributed edge AI is particularly suitable for rapid data processing scenarios, such as autonomous vehicles and robotics [4, 28]. In this order, Mwase et al. [29] described the cloud-to-thing continuum and offered a framework to make AI possible in totally edge-based applications. They also provided ways to deal with the interaction difficulties brought on by totally edge-based situations' scattered nature. These methodologies' performance enhancements demonstrated in cutting-edge research were provided, along with suggestions for future research approaches. The information was offered to encourage comprehension and, as a result, the involvement of diverse researchers in tackling this difficulty. Moustafa [30] provided an innovative IoT experimentation design that could be utilized to assess safety features based on AI. To provide dynamic experimentation networks that enabled communication across edge, fog, and cloud layers, the platform named NSX vCloud NFV was used to ease the execution of Software-Defined Networks (SDNs), Network Function Virtualization (NFV), and Service Orchestration (SO). Real-world routine and assault scenarios were run while the architecture was deployed to gather datasets with labels.

Hybrid Cloud-Edge Architectures: Hybrid cloud-edge architectures combine the advantages of cloud and edge computing and use AI models on both cloud and edge devices. This approach depends on access to computing resources and data and allows for flexible processing based on different conditions. This architecture suits applications that balance real-time processing and more detailed analysis. For example, real-time monitoring may be done at the edge, while complex analysis of historical data is done in the cloud [31]. Some researchers use this architecture: Celesti et al. [32] focused on the problems that needed to be solved for the Cloud-to-Edge platform to provide privacy, reliability, genuineness, and nonrepudiation. Additionally, they examined a real-world case study while considering the Message-Oriented Middleware (MOM) architectural model. The overall efficiencies of the entire platform were unaffected by the additional safety capabilities, according to the experiments on a genuine testbed. Also, Lei et al. [33] considered that geographically dispersed edge servers had performance that varied over time and presented a dynamic offloading approach built on a framework of probabilistic evolutionary games. Researchers undertook practical case studies based on a real-world dataset of cloud-edge resource placements to evaluate their suggested framework. The results demonstrated that their proposed method outperformed conventional ones in terms of several measures.

Choosing the right Edge AI architecture depends on factors such as computing resources [2, 15, 17], latency constraints [28], communication bandwidth [14, 15], and application requirements [34]. As AI at the edge continues to evolve, various architectures have been introduced that enable AI models to be deployed closer to data sources. These choices enable improved performance, reduced network load, increased privacy, and the realization of scheduled AI applications across environments.

1.2.2 Advantages of Edge AI

Edge AI offers several advantages that will change AI. These advantages derive from the basic idea of processing data locally on edge devices and result in several benefits [35, 36]:

Reduction of Latency: Edge AI greatly decreases the time it takes to transport data to centralized servers by processing data locally on edge devices. This low-latency operation is vital for real-time applications like self-driving cars, video analytics, and augmented reality, where quick answers are required for a smooth user experience and safety. Edge AI, for example, analyses ambient data in augmented reality glasses to send timely digital information to the user's vision, resulting in a smooth and responsive augmented experience [28].

Advancing in Privacy and Security: Edge's AI solves privacy issues by decreasing the need for data sent to other servers. Personal data is kept on-site, lowering the risk of data breaches and illegal access. This benefit is especially essential when sensitive or secret data must be processed. Home security cameras, for example, may identify possible risks such as unlawful access or suspicious activity in the area using Edge AI, eliminating the need to transfer recordings containing sensitive information to external servers [37, 38].

Bandwidth Efficiency: Edge AI minimizes network bandwidth by processing data on edge devices, reducing the data transferred to the cloud for analysis. This is especially useful in areas with restricted connections or high data transmission costs. Field sensors collect information on soil moisture, temperature, and crop health in agriculture. Edge AI analyses this data locally to offer farmers precise irrigation and fertilization suggestions, conserving resources and lowering data transfer costs [14, 15, 34].

Network Reliability: Edge AI apps can function even if the Internet connection is unreliable or interrupted. Edge devices rely less on continual Internet connectivity since they can process data locally, ensuring ongoing performance and flexibility. Edge AI, for example, allows continuous monitoring and control of remote offshore wind farms by processing data locally on the turbines. It preserves dependability in the face of unreliable networks, optimizes data transfer, and boosts overall efficiency by recognizing abnormalities and taking prompt steps [39].

Real-Time Decision-Making: Edge AI allows for immediate decision-making without needing external servers. This is especially useful for applications requiring immediate reactions, such as predictive maintenance in industrial settings or healthcare monitoring. Edge AI, for example, analyses live cameras at junctions in an intelligent traffic management system to detect and respond to traffic infractions or emergencies in real time. This technology improves traffic flow efficiency and safety [17, 28].

Cost Efficiency: Edge AI can help cut the expenses of transporting and storing massive volumes of data on cloud servers. Organizations may optimize cloud utilization and decrease operating costs by processing data locally. Retail establishments, for example, employ Edge AI to manage real-time inventory, monitor stock inventories, and identify customer patterns. This gadget works locally, helps with refilling operations, and optimizes shelf space [26].

Scalability and Distributed Processing: Edge AI enables distributed processing across several edge devices, allowing more efficient data-intensive operations administration. This scalability is functional when centralized processing of vast data is impossible. Smart grids, for example, utilize edge AI to monitor and optimize energy use across an extensive network of connected devices. This technology enables more effective energy distribution across devices while avoiding overwhelming a centralized server, ensuring energy is transferred optimally and without needless stress [2].

1.2.3 Challenges of Edge AI

Although edge AI offers significant benefits, it also faces challenges that need to be carefully addressed:

Limitations and Resource Optimization: Edge devices frequently have limits owing to computer power, memory, and battery life. AI models must be carefully optimized to balance efficiency and effectiveness and assure more optimal resources throughout day-to-day operations [11, 40].

Complex Model Management: Managing sophisticated models on edge devices presents significant hurdles. To preserve the AI ecosystem's standards and accuracy, synchronous model updating, version control, and model change adaptation are required [15].

Navigation Heterogeneity: The increasing number of edge devices with varying hardware features, operating systems, and communication protocols have resulted in navigation heterogeneity. Imposing coherence and standards over this broad span creates unity, integration, and compatibility [2, 17].

Edge-Cloud Balancing: The ability to balance edge-driven AI with cloud computing is essential. Because not all jobs can be completed on edge devices, accurately determining which tasks should be performed on edge devices and which should be offloaded to cloud processing is critical. This complicated balance demands accurate task understanding and effective computing resource allocation while considering the dynamics of environment traversal and constant changes [7, 20].

Consequently, by utilizing edge devices' capabilities, edge AI will radically alter AI deployment and bring intelligence closer to the data source. The benefits provided by this technology in terms of latency reduction, privacy protection, and bandwidth efficiency make it a promising invention with broad applications in areas such as healthcare, manufacturing, transportation, and smart cities. It is critical to solve problems and optimize AI models.

1.2.4 Taxonomy of Edge AI Applications

Edge AI applications have affected various fields. In the context of IoT, smart devices have taken center stage, encompassing wearables like smartwatches, fitness trackers, and health monitors that empower individuals to monitor their well-being seamlessly. Home automation, featuring smart thermostats, cameras, and appliances, fosters a more connected and convenient lifestyle. Industrial sensors cater to manufacturing monitoring, enhancing operational oversight. The automotive industry leverages edge AI for self-driving cars and advanced driver assistance systems, ensuring safer and smarter mobility. Healthcare benefits from remote patient monitoring, medical imaging interpretation (MRIs, X-rays, CT scans), and personalized medicine. Agriculture embraces precision farming and livestock management for optimized resource allocation.

Computer vision thrives with applications like object detection and image classification. NLP drives speech recognition for voice commands and language translation, while sentiment analysis and named entity recognition refine textual insights.

Anomaly detection and predictive maintenance enhance sectors such as industrial equipment monitoring, infrastructure health assessment, healthcare tracking, energy management, and supply chain optimization. As edge AI continues to evolve, its transformative impact on diverse domains becomes increasingly evident. Figure 1.2 provides a taxonomy of Edge AI applications.

1.3 Concepts and Fundamentals of FL

FL networks have developed as a technique that enables collaborative learning in ML without requiring centralized data aggregation in an era where data privacy and security have become vital objectives. This decentralized learning paradigm enables numerous edge devices to train a shared model jointly while storing and keeping their data locally. FL offers a novel approach to utilizing the collective intelligence of dispersed devices while maintaining data privacy and control [11, 16, 41].

1.3.1 Defining FL

FL, or Federated ML, is a decentralized ML approach that allows edge devices to collectively train a shared model while concurrently keeping their data locally, as seen in Figure 1.3. Unlike classical

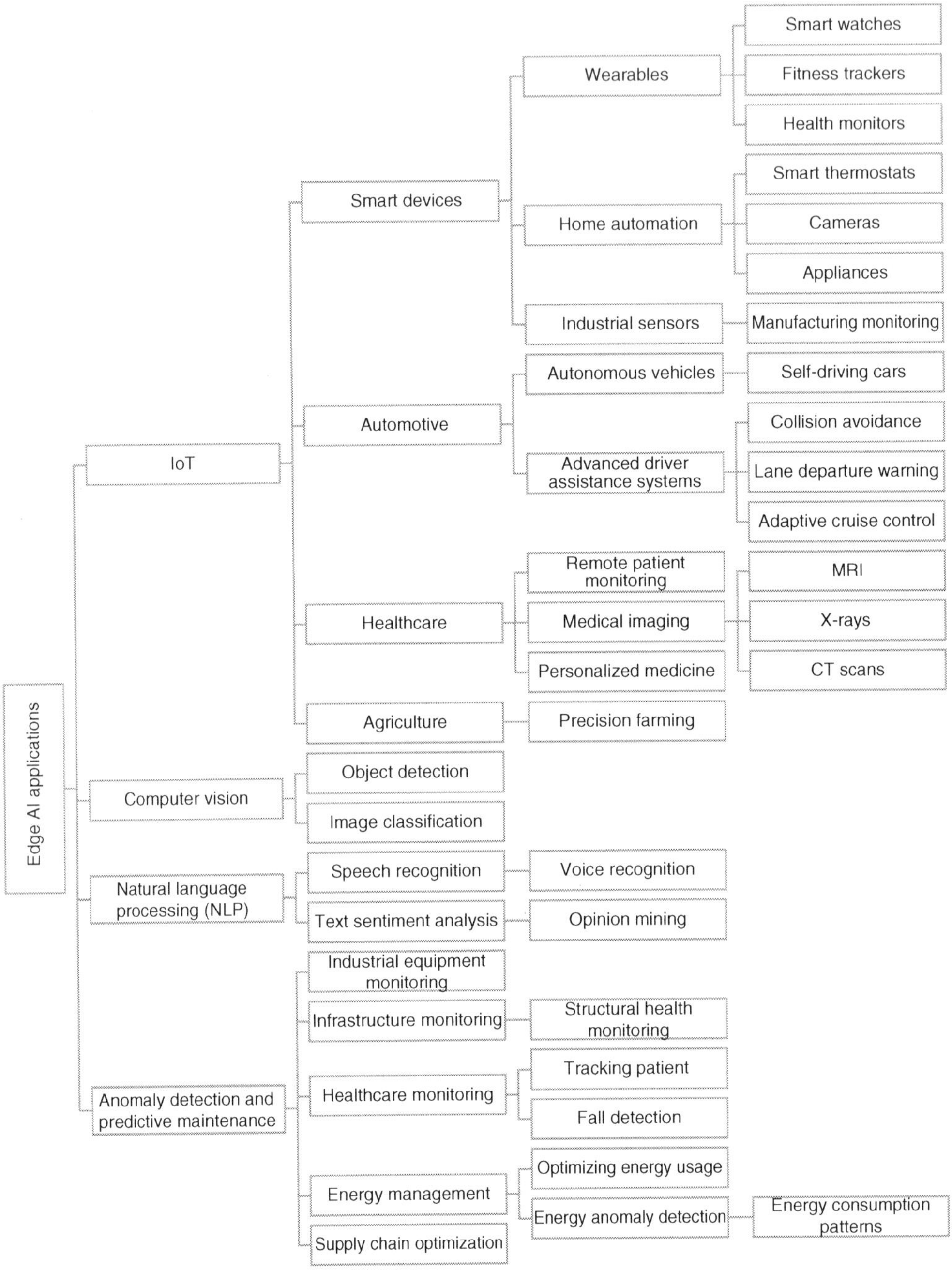

Figure 1.2 Taxonomy of edge AI applications.

ML, which collects and processes data on a centralized cloud server, FL models are trained directly on edge devices. Instead of delivering raw data, model updates are also sent to a central server or peer devices [41, 42].

The key steps involved in the FL process are as follows [16, 43]:

1) **Initialization**: An initial global model is generated randomly or by pretrained weights on a central server.

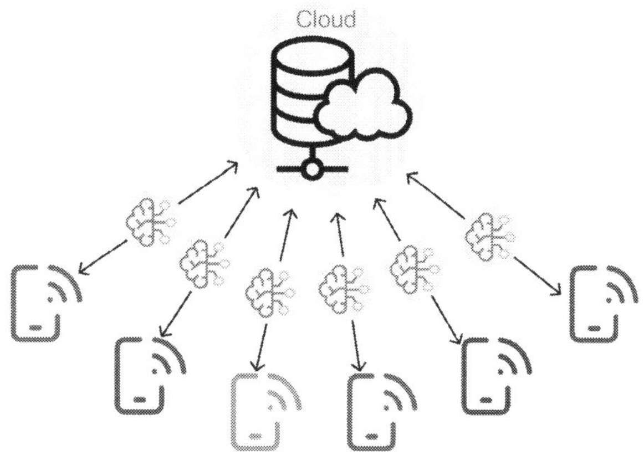

Figure 1.3 FL architecture.

2) **Distribution**: The initial global model is delivered to participating edge devices, forming a dispersed network of devices.
3) **Local Training**: The global model is trained locally at each edge device without exchanging raw data with the central server or other devices.
4) **Model Update**: Following local training, each edge device provides a model update incorporating training-related modifications into the global model.
5) **Aggregation**: All device model changes are aggregated to a central server using various approaches, such as averaging or weighted averaging.
6) **Global Model Update**: The central server aggregates the bulk model updates to provide a more accurate global model.
7) **Iteration**: This procedure is done numerous times, allowing edge devices to build the global model collectively utilizing their different datasets.

1.3.2 Advantages of FL

FL offers a range of advantages, including [41]:

Data Privacy and Security: FL solves privacy concerns since raw data is never sent beyond the edge devices, lowering the danger of exposing sensitive data. This method keeps data decentralized and confidential, enhancing confidence and compliance with data protection requirements. In a medical research context, for example, FL allows hospitals to collaboratively train a model to forecast illnesses without revealing individual patient data, allowing them to comply with privacy requirements [37, 42].

Decentralization: FL utilizes the computational capacity of several devices by spreading the training process among edge devices, making it scalable and appropriate for big data sets. FL enables multiple urban sensors in a smart city setting to train a model for optimal traffic flow. The approach is effortlessly scalable to suit the intricacies of an extensive and complex metropolitan environment by utilizing the processing capacity of these dispersed devices [5, 16].

Low Communication Cost: FL minimizes communication costs over standard centralized ML systems requiring raw data delivered to a central server since only model changes are communicated. Consider a retail chain that employs FL to enhance customer referrals. Instead of sending vast volumes of client purchase data to a central server, only refined model updates are communicated, lowering communication costs while ensuring data privacy [26].

Adaptation to Heterogeneous Data: FL is appropriate for instances where edge devices have heterogeneous data distributions, allowing the model to be trained on various data sources.

FL allows several fields with diverse soil and climate variables to prepare a crop yield prediction model in agriculture jointly. This enables the model to account for various data distributions, resulting in more accurate and adaptable predictions [2, 44].

Disturbance Resistance: FL is more resistant to network disruptions and connectivity interruptions due to model training on edge devices. This capability is beneficial in unstable circumstances or with intermittent Internet connections. FL's toughness shows through in industrial automation. FL is used in manufacturing facilities to train quality control models on edge devices, ensuring that production is not hampered by network outages, which is crucial for maintaining constant product quality [43].

Local Improvement: Each edge device can improve locally by training the global model on its device. This enables network operators to update their models depending on their needs, resulting in a more ideal user experience. FL lets each cell tower in a telecommunications network enhance its coverage prediction model locally. This personalization improves network performance in specific geographic areas, assuring continuous connectivity and customer pleasure [17].

Model Flexibility: FL enables network operators to construct customized models depending on the demands and interactions of edge devices. This allows the models to adjust and respond to real-world changes dynamically. FL reacts to changing situations, such as environmental monitoring. Weather stations work together to train a model that predicts storms, update and fine-tune parameters based on real-time data from numerous sensors, and produce accurate and fast weather predictions [34].

Reduction of Training Time: Because of local data usage in edge devices, FL can cut model training time. This enables operators to train more up-to-date models in less time and enhance performance more quickly. FL speeds fraud detection model changes in the banking industry. Banks collaborate to train the model utilizing data from dispersed branches, allowing faster adaptability to developing fraud trends and reducing possible losses [4, 43].

1.3.3 Challenges of FL

Despite the FL process's vast commitment, it also presents problems that must be carefully considered:

Communication Efficiency: Ensuring efficient communication between edge devices and the central server is critical. Managing possible communication delays and faults is essential for maintaining optimal FL performance. For example, if a cellular network with restricted capacity is used for FL, the communication overhead may be significant, resulting in delays and inefficiencies [14, 23].

Privacy and Security: Protecting user privacy is a significant priority in FL. Sharing model updates may expose sensitive information about individual participants' data. For example, while designing a tailored recommendation system, user activity data from numerous devices may be collected. Updates to the sharing mechanism may mistakenly reveal users' preferences and behaviors [4].

Heterogeneity of Models: Edge devices' computational capabilities necessitate adaptive model architectures and adaptive learning rates for various hardware restrictions.

Quality and Accuracy of Models: Because of the decentralized structure and diversity of data, model changes may result in a decline in model quality and accuracy. To increase the accuracy of models in FL, quality standards and techniques must be established.

Using Local Resources: The emphasis on local training at the edge devices may result in insufficient use of aggregated data across the FL network. Balancing local training and global model updates is complex and requires careful consideration.

Management Complexity: Managing numerous devices, executing updates, and balancing local training and update collection can be complex and challenging in a more extensive FL network.

Optimization of Energy Consumption: The most efficient use of energy resources in edge devices for training models and collecting updates necessitates consideration of the energy elements of this process.

1.3.4 Taxonomy of FL Applications

FL is an innovative form of ML that uses several data sources while protecting the privacy and security of that data. Device-centric and server-centric data sources can be divided into two categories. Data originating from mobile devices such as smartphones, tablets, and IoT gadgets; wearables such as smartwatches and fitness trackers; and edge devices such as cameras, sensors, and edge servers, are all included under the device-centric category. Data centers, including centralized data centers, cloud servers, distributed servers, and edge servers, are the data sources for the server-centric type.

Many different industries can benefit from this method, including healthcare and medical for clinical data analysis, medical imaging diagnosis, and drug discovery; finance and banking for fraud detection, credit scoring, and risk assessment; smart cities and IoT for traffic management, energy consumption prediction, and environmental monitoring; manufacturing and industrial for quality control, predictive maintenance, and process optimization; and retail and e-commerce.

With a focus on methods like secure aggregation, Differential Privacy, and Homomorphic Encryption, the methodology also emphasizes security and privacy. It navigates ethical issues like bias reduction, fairness, responsibility, and openness to ensure responsible and effective implementation. Figure 1.4 presents an FL taxonomy.

1.4 Combining FL and Edge AI

The combination of FL and Edge AI technologies is a novel and sophisticated strategy in the field of ML and AI that dramatically increases the performance, security, and privacy of AI devices and IoT. ML models are trained on local devices and regularly synced to each other utilizing FL approaches in this combination. The local models are updated, and their training results are communicated to the central hub at the end of each cycle. Model combination approaches are then used to turn local models into a new combined model with improved accuracy and predictive ability [2, 5, 7]. These two approaches offer complementary benefits that can help achieve unique real-world benefits:

Using Local Data for Global Insight: The use of federated training approaches is one solution for ensuring privacy and maximizing the usage of AI models in networks of dispersed edge devices. This method enables AI models to be trained without relying on raw data. This method defines the models' concentration on data processing in the immediate vicinity or at the data source. This improves data privacy while also addressing privacy concerns and legal mandates. Edge AI, conversely, is concerned with processing data at or near the place to reduce latency and enable real-time decision-making. Organizations may use the large quantity of local data accessible on edge devices by integrating FL with edge AI. This information may be utilized to

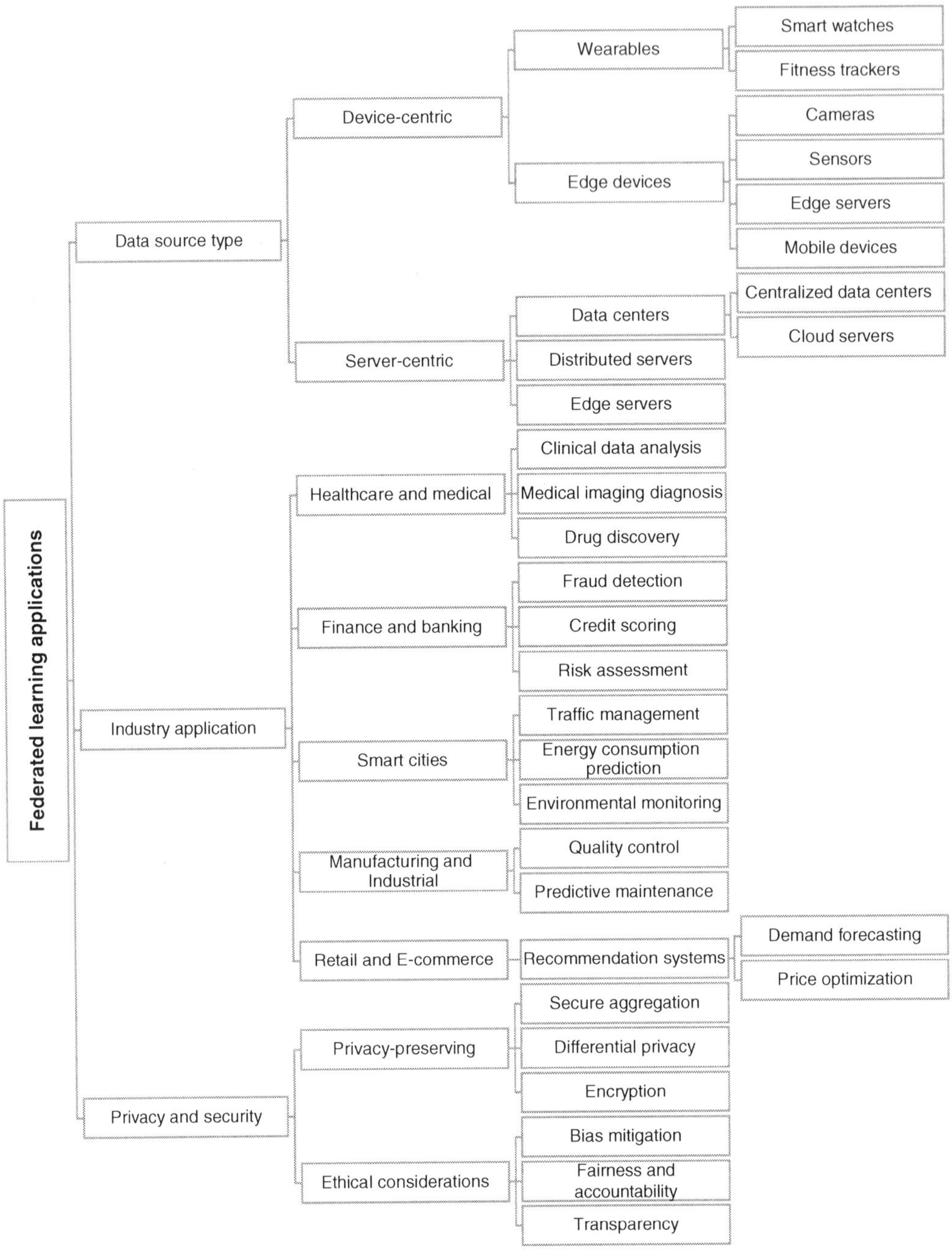

Figure 1.4 Taxonomy of FL applications.

train and develop appropriate AI models for each device and environment. This method protects sensitive local information and allows devices to make intelligent real-time judgments based on local data. This combination provides for developing customized, adaptable, and efficient AI systems [45, 46].

Collaborative Learning Across Edge Devices: FL focuses on teaching collaborative paradigms across various edge devices. A global and representative representation of the whole network is produced by spreading data amongst devices. This combined feature complements Edge's AI. Devices in various areas generate local knowledge, improving the system's accuracy and dependability. In a smart city, for example, edge devices of multiple regions may train models for traffic control, energy optimization, and public safety while sharing their local expertise. This cooperative learning strategy enhances model accuracy and generalization while making them more adaptable to dynamic situations [47].

Reducing Communication Costs: Managing the communication burden is one of the issues in FL and edge AI, especially in resource-constrained contexts. Federated training avoids sending raw data to servers and keeps models updated as little as possible during aggregation. On the other hand, Edge AI is concerned with locally processing data to lessen dependency on cloud communications. When these two strategies are combined, data sharing becomes more efficient. Edge devices process initial input and train models locally, transmitting only necessary updates to the central server. This method considerably minimizes the communication burden, saves bandwidth, and speeds up the model update process. AI systems gain from faster convergence, lower latency, and edge resource optimization in this manner [26, 40].

Enhanced Privacy and Security: In FL and Edge AI, privacy is a significant problem. Federated Training protects sensitive data by anonymizing and localizing it. Edge AI, on the other hand, minimizes privacy issues by eliminating the need to transport data to cloud servers and places sensitive data in less danger. By combining these two strategies, enterprises may improve data privacy and security. Edge devices preserve data control and FL guarantees that model changes are captured securely without disclosing individual data points. This essential combination improves privacy and data security while providing a solid framework for launching AI applications in industries as diverse as healthcare, finance, and manufacturing [7, 37].

Integrating FL with Edge AI makes it feasible to use local data without transferring it and increases model performance in AI devices and Internet-connected gadgets. This combination is presented as a strong solution for AI and IoT's future. Here are some examples of practical uses:

- **Health**: In smart medical devices, the combination of "FL" and "Edge AI" can be used to diagnose diseases, predict health status, and even prevent the occurrence of diseases [16, 48].
- **Smart Cities**: In smart cities, this technology can help improve traffic, waste and energy management, and public resource management and strengthen smart city infrastructure [8].
- **Smart Cars**: This combination improves navigation services, prevents accidents, and optimizes fuel consumption [13].

The combination of Edge AI and FL is a valuable combo that will fundamentally alter the landscape of AI and ML. Reduced latency, better privacy, efficient communication, and distributed learning have substantial consequences across various sectors and applications, laying the groundwork for ethical, safer, and smarter AI. Furthermore, as these technologies evolve, the potential for creating creative and revolutionary applications in health, industrial automation, smart cities, and other sectors becomes more apparent [10, 28].

Combining FL and Edge AI offers us significant advantages. However, in this context, there are several challenges and essential points that we should pay attention to:

- **Synchronization of Models**: Ensuring efficient and timely updating of models from edge devices to the central server and back to other devices requires well-designed synchronization protocols.
- **Heterogeneity and Non-i.i.d.**: Managing heterogeneity among edge devices and i.i.d. data distributions while maintaining models' performance and fairness is a complex challenge.
- **Resource Constraints**: Managing resource constraints in edge devices and optimizing the size and complexity of models for placement in resource-constrained devices is very necessary.
- **Data Quality and Security**: Ensuring data quality and protecting against security risks may be particularly important, as these are fundamental to maintaining accuracy and confidence in models.

As a result, the combination of FL and Edge AI has the potential to usher in a new age of decentralized, privacy-conscious, and intelligent AI systems. The benefits of greater privacy, real-time decision-making, and communication efficiency combine to make this convergence an appealing and adaptable option for a wide range of applications. Overcoming obstacles and strengthening the convergence of these two paradigms will drive the future of AI and form intelligent systems that leverage the potential of FL and Edge AI to create a smart and connected environment [1, 15, 20].

1.5 Background

Training ML models in resource-constrained edge node contexts is a significant problem in IoT and edge computing. Traditional FL solutions typically result in longer training durations and more significant resource costs, making them unsuitable for edge nodes like base stations and access points. The authors in [14] address the challenges of training ML models at resource-constrained edge nodes in the context of FL. The article proposes a novel asynchronous FL mechanism (CE-AFL) that aims to improve training efficiency by aggregating updates from a subset of edge nodes in an order-based manner. They introduced the CE-AFL approach and provided a detailed analysis of its performance. They proved the convergence of the proposed algorithm and examined its efficiency in two scenarios of single and multiple-task learning under bandwidth limitations. Simulations confirm the ability of CE-AFL to reduce training time and improve the accuracy of models in resource-constrained environments. Also, the authors acknowledge the importance of continued research, especially in dynamic scenarios, demonstrating their commitment to advancing efficient communicative FL in edge computing environments. Compared with existing methods, CE-AFL optimizes updating models by collecting only a fraction of local updates from edge nodes based on the order in which they are received in each training session. This approach responds to challenges such as bandwidth limitations and uncertainty in edge environments, and in addition to improving training efficiency, it also secures users' privacy. The article suggested that the problem of handling dynamic scenarios will be the focus of future studies. This implies that the authors plan to extend their research to address scenarios where edge nodes dynamically join or leave the network. As a weak point of their presented approach, the scalability of the proposed mechanism to handle a larger number of tasks or edge nodes could be a potential concern.

The authors in [15] address an essential challenge of the effectiveness of training models on edge devices due to limited computing resources and bandwidth. In contrast to the traditional methods that mainly focus on improving communication, this paper introduces an innovative federated training framework focusing on model pruning. This framework aims to simultaneously deal with the limitations of computing and communication resources. The original approach replaces the dynamic selection of neurons or convolutional kernels before global model propagation. This step performs optimal subnet pruning, and the updated compressed model is distributed to each client for training. A new expansion scheme is generalized to increase the parameter aggregation and updating. This scheme provides training opportunities for global model parameters, and by regenerating and reusing parameters, it preserves the integrity of the model structure and minimizes pruning errors. Extensive experiments emphasize the effectiveness of this framework on independent and smooth (i.i.d. data) and non-i.i.d datasets. This framework performs better by reducing upstream and downstream communication, maintaining global model accuracy, and reducing client computational costs. Traditional approaches face communication challenges and privacy concerns, which have led to the proposal of edge training based on FL. This approach uses separate edge devices to jointly train models using local data and perform global model updating. However, the resource limitation of edge devices poses challenges due to limited computing capacities and memory. Extensive experiments confirm the effectiveness of this approach in reducing communication, maintaining model accuracy, and facilitating client computational costs. The authors will focus on integrating model pruning and reinforcement learning in future efforts. This integration will play an influential management role in collecting participating learning devices, emphasizing dynamic and unpredictable network environments.

In the traditional cloud computing paradigm, data processing involves uploading data to centralized cloud servers, which results in data transfer delays and response times. The concept of edge computing has been born according to the new needs. Edge computing involves data processing on edge nodes closer to data sources, thereby reducing transmission delays and improving response speed. Recently, the incorporation of AI into edge networks has attracted much attention. However, the data available at a single edge node is often insufficient for effective ML. For this reason, performing ML in edge networks while maintaining data privacy has become a significant challenge. To preserve privacy in the edge network computing process, the authors of [10] introduced a privacy-preserving asynchronous FL mechanism for edge network computing (PAFLM). PAFLM allows multiple edge nodes to participate in an efficient FL process without sharing private data. Compared with traditional distributed learning methods, PAFLM can compress the communication between nodes and the parameter server during training without losing accuracy. Also, PAFLM allows nodes to join or leave the learning process at any time, which is suitable for scenarios involving mobile edge devices. PAFLM has two main aspects: self-adaptive threshold gradient compression and asynchronous FL. The self-adaptive threshold gradient compression algorithm adjusts the compression threshold based on the gradient change during the model training process, reducing the possibility of privacy leakage. Asynchronous FL addresses the challenges posed by unbalanced learning instances and varying learning progress due to the mobility of edge nodes.

In the rapidly evolving technological landscape, researchers devote their time and effort to envisioning and advancing the capabilities of 6G networks, leading to superior capabilities over 5G. Concepts such as edge computing and applications such as self-driving cars have improved in the context of 6G services, creating the ability to perform various tasks successfully. These programs generate a large amount of data that has the potential to be helpful in the fields of AI and ML. However, data privacy concerns have led to data collection and use changes. To address these concerns, FL has emerged as a critical solution that combines privacy and technology by enabling learning from decentralized data sets. In traditional ML methods, models are trained based on centralized data. But recently, data privacy has become a critical concern and thus has become an important issue in data collection. Federated training allows devices such as mobiles, IoT-connected objects, and vehicles to train ML models based on their data without sharing their raw data. In this method, different devices jointly participate in learning, and each device locally trains its model and then shares its updated weights with a central server for aggregation. But while many methods improve convergence speed and accuracy, neglecting real-time deployment and selection of clients can be a problem, especially in dynamic and resource-constrained environments. For this reason, researchers introduce a new approach in [17] in 6G networks. This approach increases the ability to deploy clients in dynamic and changing environments by proposing an on-demand client deployment mechanism in federated training. This approach leverages containerization technologies such as Docker and the Kubernetes tool Kubeadm to allow each client device to participate in the FL process, even in dynamic environments. This approach will enable customers to participate whenever and wherever needed, increasing the volume and variety of data used in the learning process. This approach allows efficient organization and deployment of services and ML models on new client devices. The experiments show that this approach has proven effective in reducing the number of rounds, increasing the volume and variety of data, and deploying requested clients. In summary, this paper emphasizes the potential of this approach and highlights the importance of adapting to technological developments and user demands.

In the field of IoT, the vast amount of data collected from various devices requires intelligent and practical analysis. Recently, a trend called "from cloud-based services to edge devices" has grown in the field of IoT. This means bringing computation closer to where the data is collected. One of the critical solutions for training AI models in this decentralized scenario is FL. This approach involves exchanging training data from different devices in a privacy-preserving environment. In this scenario, data migration to edge servers enables semi-trained computing from the cloud to edge devices. In FL, edge servers receive semi-trained models from the cloud, complete them using their local data, and send the refined models. Finally, these models are aggregated by the cloud node to produce a global model [20]. The authors investigate the quantization of neural networks in FL in the IoT environment. This approach uses quantization techniques in FL to improve the performance of data exchange between edge servers and cloud nodes while having a negligible impact on the final performance of the model. One of this research focuses on training Recurrent Neural Network (RNN) models using data from edge producers, which are commonly used in time series forecasting. By combining quantization techniques with FL, this approach improves the efficiency of data exchange and minimizes its impact on model performance. This research shows that combining FL with quantization can effectively tackle the challenges of AI and data analysis in the IoT ecosystem.

Combining AI and MEC convergence and implementing FL in this context helps transfer information and machine models between devices in newer and dynamic wireless networks. This combination enables the use of distributed computing resources in MEC systems and allows users to maintain the privacy of their data. However, challenges related to information transmission in wireless networks and resource limitations in MEC systems can affect the performance and efficiency of FL. One of the proposed solutions to these challenges is the optimization of edge-based communication that the researchers in [9] presented. This approach fits well with MEC goals and offers the advantage of preserving local data privacy. By optimizing edge-based communication for distributed FL, the overall training efficiency can be improved, and resource constraints in MEC systems can be used more optimally. This paper proposes an optimization framework for FL in MEC systems by addressing the challenges of model accuracy, training efficiency, and resource constraints. This approach significantly improves the accuracy and cost of training models. The integration of federated training and MEC holds great promise for advancing advanced AI networks and the evolution of future wireless systems. This combination means increasing the capabilities and efficiency of wireless networks, improving the speed and efficiency of learning models, and preserving users' privacy in distributed and wireless environments.

1.6 Applications of Edge AI and FL

Edge AI and FL have opened opportunities in various sectors and application cases. With dispersed information and privacy in ML, these technologies have the potential to revolutionize the way AI is employed in real-world settings. Some of the most common applications of Edge AI and FL are listed below [2, 41]:

- **Healthcare**
  - **Remote Patient Monitoring**: Edge AI allows continuous patient health monitoring via wearable devices and sensors. Real-time data analytics at the edge enables healthcare workers to respond to emergencies promptly and make sound judgments.
  - **Personalized Medicine**: FL enables medical institutions to participate in generating prediction models without disclosing sensitive patient data. These algorithms can deliver individualized treatment regimens based on individual health information and enhance patient outcomes.
  - **Analytics that Protect Patient Privacy**: FL facilitates collaborative research on medical datasets from many institutions while protecting patient privacy. This method increases medical science without compromising data security.
  - **Smart Wearable Devices**: Wearables with edge AI, such as fitness trackers and smartwatches, may continually monitor a person's vital signs and health parameters. FL allows diverse user data to be collected and collated to develop tailored health insights while respecting people's privacy. These gadgets can monitor a patient's health in real time and spot problems early on, improving patient outcomes.
  - **Distributed Medical Research**: FL in medical research allows healthcare facilities to use patient data to train prediction models. This technology enables universities to collaborate on research and data aggregation while keeping sensitive information on a secure network. Edge devices in hospitals or clinics can engage in model training and aid in illness diagnosis, treatment planning, and medical research advancements.
- **IoT**
  - **Smart Home Devices**: Edge AI extends its capabilities by allowing them to execute local processing for voice recognition, facial recognition, and NLP. This shortens reaction time and decreases reliance on cloud services.
  - **Industrial IoT**: Edge AI is used in industrial settings to evaluate data from sensors and equipment, which aids in predictive maintenance, anomaly identification, and process optimization. This local analysis eliminates the need to connect to central servers [23, 43] regularly.
  - **Smart Cities**: FL facilitates the collaborative analysis of sensor data from many IoT devices in smart cities. This leads to more efficient and sustainable urban life by improving traffic management, trash management, and resource allocation.
    - **Traffic Management and Optimization**: Using Edge AI cameras and sensors is critical for traffic patterns and urban traffic management. These devices can assess traffic flow and movement patterns to optimize traffic light timing. Data from multiple junctions and locations is collected through federated training, and more accurate traffic prediction models are developed. Combining edge AI with federated training improves traffic flow and decreases travel time throughout the city.

  - **Environmental Monitoring**: Edge sensors implanted in urban surroundings may monitor air quality, noise levels, pollution, and other environmental issues. Data from many sources is analyzed and predicted using the federated training approach. These contribute to more accurate and up-to-date environmental models, allowing cities and residents to make better decisions on sustainable urban growth and pollution management. This edge AI and FL combination enables cities to increase environmental stewardship and contribute to sustainable development.
- **Self-driving Vehicles**
  - **Real-time Decision-making**: In self-driving vehicles, Edge AI interprets sensor data directly, providing real-time decision-making for navigation, object identification, and collision avoidance. This decreases reliance on cloud computing, even in places with weak network connections.
  - **Map and Traffic Updates**: FL lets cars anonymously transmit information on road conditions, traffic patterns, and possible risks. This shared information provides better navigation and overall traffic safety.
- **Agriculture**
  - **Precision Agriculture**: AI and IoT sensors on agricultural equipment allow real-time analysis of soil conditions, crop health, and environmental variables. This data-driven strategy boosts product performance and resource usage.
  - **Animal Monitoring**: AI-enabled edge cameras and sensors monitor cattle health and activity, allowing for early illness identification and improved treatment.
- **Disasters and Emergency Response**
  - **Real-time Monitoring**: Edge AI and IoT devices help monitor natural disasters like earthquakes, floods, and fires. Data analytics at the edge helps emergency responders make quick decisions and allocate optimal resources [31].
  - **Flexible Communication**: FL ensures communication and cooperation between emergency responders in emergencies, even in areas with impaired communication networks.
- **Finance and Fraud Detection**
  - **Local Fraud Detection**: The application of edge AI in identifying fraud in financial transaction data is critical. This technique processes real-time transaction data on individual devices to discover suspicious trends and fraudulent transactions. This approach is used with federated training to collect cross-device insight data from transactions and develop more effective fraud detection models. This hybrid strategy assists the financial network in improving the overall quality of fraud detection and increasing financial security.
  - **Risk Assessment**: AI-based risk assessment algorithms analyze loan applications, insurance claims, and investment decisions using edge devices. These gadgets analyze risk based on local insights and data in their contexts, allowing for better-informed decision-making. The accuracy of risk assessment models is increased by combining this method with FL and using data from numerous sources. This combination allows for better and more accurate judgments in scenarios involving financial hazards, and it assists businesses in improving their financial decisions.

These technologies allow devices to process data locally, reducing latency and reliance on cloud services. Furthermore, FL addresses data privacy problems and facilitates remote entity collaboration without jeopardizing individual data privacy. More inventive applications that combine Edge AI and FL capabilities will drive the future of intelligent, decentralized AI systems as research and development develop.

Table 1.1 FL and Edge AI challenges, future research directions, and solutions.

Challenge	Future Research Direction	Solutions
Data Heterogeneity [2, 17]	Develop algorithms that can handle diverse data types and distributions across edge devices.	• **Transfer Learning**: Employ transfer learning techniques to adapt models from one device to another. • **Data Synthesis**: Generate synthetic data to bridge the gap between different data distributions.
Communication Efficiency [34, 40]	Investigate techniques for compressing and quantizing models to reduce communication overhead.	• **Localized Updates**: Perform local model updates before sending aggregated updates to reduce communication frequency. • **Sparsification**: Send only important model updates using sparsification techniques.
Model Heterogeneity [2, 17]	Investigate multitask learning approaches to handle diverse models across edge devices.	• **Model Distillation**: Distill complex models into smaller, simpler models for deployment. • **Dynamic Model Loading**: Load appropriate model components based on device capabilities.
Resource Constraints [11, 40, 46]	Develop algorithms that adapt based on edge device resources.	• **Model Pruning**: Prune less important model parameters to reduce resource requirements. • **Knowledge Distillation**: Train edge models with guidance from a central server.
Scalability and System Efficiency [2]	Explore hierarchical structures for managing large-scale federated systems.	• **Distributed Computing**: Utilize distributed computing paradigms to improve scalability. • **Adaptive Resource Allocation**: Dynamically allocate resources based on demand and availability.
Adversarial Attacks [31, 37]	Investigate defenses against adversarial attacks on edge devices.	• **Adversarial Training**: Train models with adversarial examples to enhance robustness. • **Secure Aggregation**: Employ secure aggregation methods to protect against poisoned updates.
Lack of Centralized Data [16, 38]	Research ways to generate shared synthetic data for model training.	• **Data Federations**: Collaborate to create federated datasets with representative samples from each edge device. • **Data Augmentation**: Apply augmentation techniques to expand limited data on edge devices.
Energy Efficiency [4]	Develop energy-efficient algorithms for resource-constrained devices.	• **Quantization and Pruning**: Apply quantization and model pruning to reduce computational and energy requirements. • **Adaptive Computation**: Adjust model complexity based on available power.
Dynamic Network Topology [11, 15]	Research ways to handle changing device participation and connectivity.	• **Topology Adaptation**: Develop methods for models to adapt to changing network structures. • **Neighbor Selection**: Optimize the selection of neighboring devices for communication.
Privacy-Preserving Aggregation [37]	Explore advanced cryptographic techniques for privacy-preserving aggregation.	• **Secure Multiparty Computation**: Implement secure computation protocols for privacy-preserving aggregation. • **Federated Averaging with Differential Privacy**: Combine privacy-preserving techniques with federated averaging.
Communication Reliability [15]	Develop protocols that ensure reliable communication between edge devices and the central server.	• **Error Correction**: Integrate error correction codes to recover lost communication. • **Retry Mechanisms**: Implement retry mechanisms for failed communications.
Edge-to-Cloud Collaboration [7, 34]	Research hybrid models that leverage both edge and cloud resources.	• **Dynamic Task Allocation**: Dynamically allocate tasks between edge devices and the cloud based on resource availability. • **Model Partitioning**: Partition models to optimize execution between edge and cloud.

1.7 Challenges, Future Research Directions, and Solutions

The FL and Edge AI domains cannot overstate the importance of addressing challenges, setting future research directions, and devising practical solutions. Table 1.1 presents a comprehensive overview of the challenge, future research direction, and FL and Edge AI solutions.

1.8 Conclusion

With its decentralized approach, Edge AI pushes AI computation closer to the network's edge, allowing for real-time processing and lowering latency. Edge AI delivers benefits such as better privacy, less dependency on cloud connectivity, and improved bandwidth efficiency by exploiting the computational capabilities of edge devices. However, difficulties relating to edge device resource restrictions and data heterogeneity must be overcome to realize edge AI's potential. FL, on the other hand, presents a privacy-preserving and collaborative ML paradigm that lets numerous edge devices learn from their local data collaboratively without sharing raw data. This distributed learning technique tackles data privacy issues and promotes collaboration across many groups while keeping data decentralized and safe. However, some continuing research issues are managing non-i.i.d.data distribution, communication efficiency, and model fairness. Edge AI and FL work well together because they combine the benefits of real-time local processing with privacy and decentralized learning. Reduced latency, better privacy, efficient communication, and adaptive learning offer promise for applications in various fields, including healthcare, finance, autonomous cars, and smart cities. Further advancements in model optimization, privacy, and secure multiparty computation will impact the progress of Edge AI and FL's future progress. Innovative edge devices, hybrid cloud-edge architectures, and standardization initiatives are assisting in their widespread acceptance and smooth integration into various applications. Continued research and development in these areas is required to solve existing obstacles, maintain data privacy, and boost the scalability and resilience of edge-based AI systems. Edge AI and FL will power a smarter, more connected, and privacy-aware society as they become more prevalent, transforming how AI is applied and enhancing the lives of individuals and industries alike. This chapter delves into the principles, advantages, problems, and future directions of Edge AI and FL, two technologies reshaping the AI and ML ecosystem.

References

1 Yu, B., Mao, W., Lv, Y. et al. (2022). A survey on FL in data mining. *WIREs Data Mining and Knowledge Discovery* 12 (1): e1443. https://doi.org/10.1002/widm.1443.

2 Brecko, A., Kajati, E., Koziorek, J., and Zolotova, I. FL for edge computing: a survey. *Applied Sciences* 12 (18). https://doi.org/10.3390/app12189124.

3 Singh, R. and Gill, S.S. (2023). Edge AI: a survey. *Internet of Things and Cyber-Physical Systems* 3: 71–92. https://doi.org/10.1016/j.iotcps.2023.02.004.

4 Hosseinzadeh, M., Hemmati, A., and Rahmani, A.M. (2022). FL-based IoT: a systematic literature review. *International Journal of Communication Systems* 35 (11): e5185. https://doi.org/10.1002/dac.5185.

5 Alam, T. and Gupta, R. FL and its role in the privacy preservation of IoT devices. *Future Internet* 14 (9). https://doi.org/10.3390/fi14090246.

6 Lim, W.Y.B., Luong, N.C., Hoang, D.T. et al. (2020). FL in mobile edge networks: a comprehensive survey. *IEEE Communications Surveys & Tutorials* 22 (3): 2031–2063. https://doi.org/10.1109/COMST.2020.2986024.

7 Doku, R. and Rawat, D.B. IFLBC: on the edge intelligence using FL blockchain network. *2020 IEEE 6th Intl Conference on Big Data Security on Cloud (BigDataSecurity), IEEE Intl Conference on High Performance and Smart Computing, (HPSC) and IEEE Intl Conference on Intelligent Data and Security (IDS)* (25–27 May 2020), Baltimore, MD, USA, pp. 221–226. https://doi.org/10.1109/BigDataSecurity-HPSC-IDS49724.2020.00047.

8 Hemmati, A. and Rahmani, A.M. (2022). The internet of autonomous things applications: a taxonomy, technologies, and future directions. *Internet of Things* 20: 100635. https://doi.org/10.1016/j.iot.2022.100635.

9 Wang, T., Liu, Y., Zheng, X. et al. (2021). Edge-based communication optimization for distributed FL. *IEEE Transactions on Network Science and Engineering* 9. https://doi.org/10.1109/TNSE.2021.3083263.

10 Lu, X., Liao, Y., Lio, P., and Hui, P. (2020). Privacy-preserving asynchronous FL mechanism for edge network computing. *IEEE Access* 1–1. https://doi.org/10.1109/ACCESS.2020.2978082.

11 Lim, W.Y.B., Ng, J.S., Xiong, Z. et al. (2022). Decentralized edge intelligence: a dynamic resource allocation framework for hierarchical FL. *IEEE Transactions on Parallel and Distributed Systems* 33 (3): 536–550. https://doi.org/10.1109/TPDS.2021.3096076.

12 Alsamhi, S.H., Almalki, F.A., Afghah, F. et al. (2022). Drones' edge intelligence over smart environments in B5G: blockchain and FL synergy. *IEEE Transactions on Green Communications and Networking* 6 (1): 295–312. https://doi.org/10.1109/TGCN.2021.3132561.

13 Hemmati, A., Zarei, M., and Souri, A. (2023). UAV-based Internet of Vehicles: a systematic literature review. *Intelligent Systems with Applications* 18: 200226. https://doi.org/10.1016/j.iswa.2023.200226.

14 Liu, J., Xu, H., Xu, Y. et al. (2021). Communication-efficient asynchronous FL in resource-constrained edge computing. *Computer Networks* 199: 108429. https://doi.org/10.1016/j.comnet.2021.108429.

15 Wu, T., Song, C., and Zeng, P. (2023). Efficient FL on resource-constrained edge devices based on model pruning. *Complex & Intelligent Systems*. https://doi.org/10.1007/s40747-023-01120-5.

16 Banabilah, S., Aloqaily, M., Alsayed, E. et al. (2022). FL review: fundamentals, enabling technologies, and future applications. *Information Processing & Management* 59 (6): 103061. https://doi.org/10.1016/j.ipm.2022.103061.

17 Chahoud, M., Otoum, S., and Mourad, A. (2023). On the feasibility of FL towards on-demand client deployment at the edge. *Information Processing & Management* 60 (1): 103150. https://doi.org/10.1016/j.ipm.2022.103150.

18 Cai, Y., Genovese, A., Piuri, V. et al. IoT-based architectures for sensing and local data processing in ambient intelligence: research and industrial trends. *2019 IEEE International Instrumentation and Measurement Technology Conference (I2MTC)* (20–23 May 2019), Auckland, New Zealand, pp. 1–6. https://doi.org/10.1109/I2MTC.2019.8827110.

19 Bebortta, S., Singh, A.K., Pati, B., and Senapati, D. (2020). A robust energy optimization and data reduction scheme for IoT based indoor environments using local processing framework. *Journal of Network and Systems Management* 29 (1): 6. https://doi.org/10.1007/s10922-020-09572-7.

20 Tonellotto, N., Gotta, A., Nardini, F.M. et al. (2021). Neural network quantization in FL at the edge. *Information Sciences* 575: 417–436. https://doi.org/10.1016/j.ins.2021.06.039.

21 Rahmani, A.M., Gia, T.N., Negash, B. et al. (2018). Exploiting smart e-health gateways at the edge of healthcare internet-of-things: a fog computing approach. *Future Generation Computer Systems* 78: 641–658. https://doi.org/10.1016/j.future.2017.02.014.

22 Li, H., Ota, K., and Dong, M. (2018). ECCN: orchestration of edge-centric computing and content-centric networking in the 5G radio access network. *IEEE Wireless Communications* 25 (3): 88–93. https://doi.org/10.1109/MWC.2018.1700315.

23 Lu, Y., Huang, X., Zhang, K. et al. (2021). Communication-efficient FL for digital twin edge networks in industrial IoT. *IEEE Transactions on Industrial Informatics* 17 (8): 5709–5718. https://doi.org/10.1109/TII.2020.3010798.

24 Chen, Y., Zhao, F., Lu, Y., and Chen, X. (2023). Dynamic task offloading for mobile edge computing with hybrid energy supply. *Tsinghua Science and Technology* 28 (3): 421–432. https://doi.org/10.26599/TST.2021.9010050.

25 Wu, Y., Xia, J., Gao, C. et al. (2022). Task offloading for vehicular edge computing with imperfect CSI: a deep reinforcement approach. *Physical Communication* 55: 101867. https://doi.org/10.1016/j.phycom.2022.101867.

26 Zhou, Z., Yang, S., Pu, L., and Yu, S. (2020). CEFL: online admission control, data scheduling, and accuracy tuning for cost-efficient FL across edge nodes. *IEEE Internet of Things Journal* 7 (10): 9341–9356. https://doi.org/10.1109/JIOT.2020.2984332.

27 Robles-Enciso, A. and Skarmeta, A.F. (2023). A multi-layer guided reinforcement learning-based tasks offloading in edge computing. *Computer Networks* 220: 109476. https://doi.org/10.1016/j.comnet.2022.109476.

28 Xianjia, Y., Queralta, J.P., Heikkonen, J., and Westerlund, T. (2021). FL in robotic and autonomous systems. *Procedia Computer Science* 191: 135–142. https://doi.org/10.1016/j.procs.2021.07.041.

29 Mwase, C., Jin, Y., Westerlund, T. et al. (2022). Communication-efficient distributed AI strategies for the IoT edge. *Future Generation Computer Systems* 131: 292–308. https://doi.org/10.1016/j.future.2022.01.013.

30 Moustafa, N. (2021). A new distributed architecture for evaluating AI-based security systems at the edge: network TON_IoT datasets. *Sustainable Cities and Society* 72: 102994. https://doi.org/10.1016/j.scs.2021.102994.

31 Nair, A.K., Sahoo, J., and Raj, E.D. (2023). Privacy-preserving FL framework for IoMT based big data analysis using edge computing. *Computer Standards & Interfaces* 86: 103720. https://doi.org/10.1016/j.csi.2023.103720.

32 Celesti, A., Fazio, M., Galletta, A. et al. (2019). An approach for the secure management of hybrid cloud–edge environments. *Future Generation Computer Systems* 90: 1–19. https://doi.org/10.1016/j.future.2018.06.043.

33 Lei, Y., Zheng, W., Ma, Y. et al. (2021). A novel probabilistic-performance-aware and evolutionary game-theoretic approach to task offloading in the hybrid cloud-edge environment. In: *Collaborative Computing: Networking, Applications and Worksharing* (ed. H. Cham, X. Gao, M. Wang, et al.), 255–270. Springer International Publishing.

34 Zhang, X., Zhu, X., Wang, J. et al. (2020). FL with adaptive communication compression under dynamic bandwidth and unreliable networks. *Information Sciences* 540: 242–262. https://doi.org/10.1016/j.ins.2020.05.137.

35 Ding, A.Y., Peltonen, E., Meuser, T. et al. (2022). Roadmap for edge AI: a Dagstuhl perspective. *ACM SIGCOMM Computer Communication Review* 52 (1): 28–33. https://doi.org/10.1145/3523230.3523235.

36 Zhou, Z., Chen, X., Li, E. et al. (2019). Edge intelligence: paving the last mile of AI with edge computing. *Proceedings of the IEEE* 107 (8): 1738–1762. https://doi.org/10.1109/JPROC.2019.2918951.

37 Wan, Y., Qu, Y., Gao, L., and Xiang, Y. (2022). Privacy-preserving blockchain-enabled FL for B5G-Driven edge computing. *Computer Networks* 204: 108671. https://doi.org/10.1016/j.comnet.2021.108671.

38 Hemmati, A., Zarei, M., and Souri, A. (2023). Blockchain-based internet of vehicles (BIoV): a systematic review of surveys and reviews. *Security and Privacy* e317. https://doi.org/10.1002/spy2.317.

39 Hosseinzadeh, M., Hemmati, A., and Rahmani, A.M. (2022). 6G-enabled internet of things: vision, techniques, and open issues. *Computer Modeling in Engineering and Sciences* 133 (3).https://doi.org/10.32604/cmes.2022.021094.

40 Khan, L.U., Pandey, S.R., Tran, N.H. et al. (2020). FL for edge networks: resource optimization and incentive mechanism. *IEEE Communications Magazine* 58 (10): 88–93. https://doi.org/10.1109/MCOM.001.1900649.

41 Rahman, K.M.J., Ahmed, F., Akhter, N. et al. (2021). Challenges, applications and design aspects of FL: a survey. *IEEE Access* 9: 124682–124700. https://doi.org/10.1109/ACCESS.2021.3111118.

42 Yang, Z., Shi, Y., Zhou, Y. et al. (2023). Trustworthy FL via blockchain. *IEEE Internet of Things Journal* 10 (1): 92–109. https://doi.org/10.1109/JIOT.2022.3201117.

43 Hiessl, T., Rezapour Lakani, S., Kemnitz, J. et al. (2022). Cohort-based FL services for industrial collaboration on the edge. *Journal of Parallel and Distributed Computing* 167: 64–76. https://doi.org/10.1016/j.jpdc.2022.04.021.

44 Nishio, T. and Yonetani, R. Client selection for FL with heterogeneous resources in mobile edge. *ICC 2019—2019 IEEE International Conference on Communications (ICC)* (20–24 May 2019), Shanghai, China, pp. 1–7. https://doi.org/10.1109/ICC.2019.8761315.

45 Qi, Q. and Chen, X. (2022). Robust design of FL for edge-intelligent networks. *IEEE Transactions on Communications* 70 (7): 4469–4481. https://doi.org/10.1109/TCOMM.2022.3175921.

46 Yu, S., Chen, X., Zhou, Z. et al. (2021). When deep reinforcement learning meets FL: intelligent multitimescale resource management for multiaccess edge computing in 5G ultradense network. *IEEE Internet of Things Journal* 8 (4): 2238–2251. https://doi.org/10.1109/JIOT.2020.3026589.

47 Lin, Z., Liu, H., and Zhang, Y.J.A. (2022). Relay-Assisted Cooperative FL. *IEEE Transactions on Wireless Communications* 21 (9): 7148–7164. https://doi.org/10.1109/TWC.2022.3155596.

48 Hemmati, A. and Rahmani, A.M. Internet of medical things in the COVID-19 era: a systematic literature review. *Sustainability* 14 (19).https://doi.org/10.3390/su141912637.

2

AI Applications – Computer Vision and Natural Language Processing

Balakrishnan Chinnaiyan[1], *Sundaravadivazhagan Balasubaramanian*[2], *Mahalakshmi Jeyabalu*[1], *and Gayathry S. Warrier*[1]

[1] *Department of Computer Science, CHRIST University, Bangalore, Karnataka, India*
[2] *Department of Information Technology, University of Technology and Applied Sciences, Al Mussanah, Sultanate of Oman*

2.1 Introduction

Natural language processing (NLP) and computer vision have been used to broaden the field of artificial intelligence (AI) and open up novel opportunities for technological advancement. Many AI applications that are changing industries, transforming human–computer interaction, and extending the limits of what robots can accomplish have emerged from this dynamic connection of fields. We examine the exciting potential of these two crucial disciplines, computer vision and NLP, as we travel across the world of AI applications in this chapter. The research area of computer vision, which is concerned with the interpretation of visual data, has advanced considerably over time. Computers can now comprehend, evaluate, and draw conclusions from photos and videos with previously unheard-of accuracy thanks to AI.

Major developments in object identification, image recognition, and the creation of autonomous systems have been made possible by this achievement. In the automotive sector, computer vision determines autonomous vehicles, enhancing safety and efficiency of transportation. In the healthcare sector, it helps with early health identification through medical imaging. It improves public safety in the security domain by providing strong surveillance and facial recognition capabilities [1]. Simultaneously, a notable shift has occurred in the discipline of NLP [2]. NLP deals with the use of human language as a communication channel between computers and people. The ability of computers to comprehend, produce, and interpret text has been made possible by AI-driven language models, revolutionizing human communication.

The whole world has become smaller as well as more linked via language translation services, sentiment analysis tools that help businesses evaluate public opinion, and chatbots that increasingly offer automated customer assistance. The true magic comes at the intersection of NLP and computer vision. Imagine a system that is able to comprehend and react to human language in addition to being able to interpret the visual environment. Innovative uses like image captioning, which lets computers explain the contents of images in natural language, are the result of this synergy. Additionally, it makes it possible to create augmented reality (AR) apps that smoothly

Model Optimization Methods for Efficient and Edge AI: Federated Learning Architectures, Frameworks and Applications, First Edition. Edited by Pethuru Raj Chelliah, Amir Masoud Rahmani, Robert Colby, Gayathri Nagasubramanian, and Sunku Ranganath.

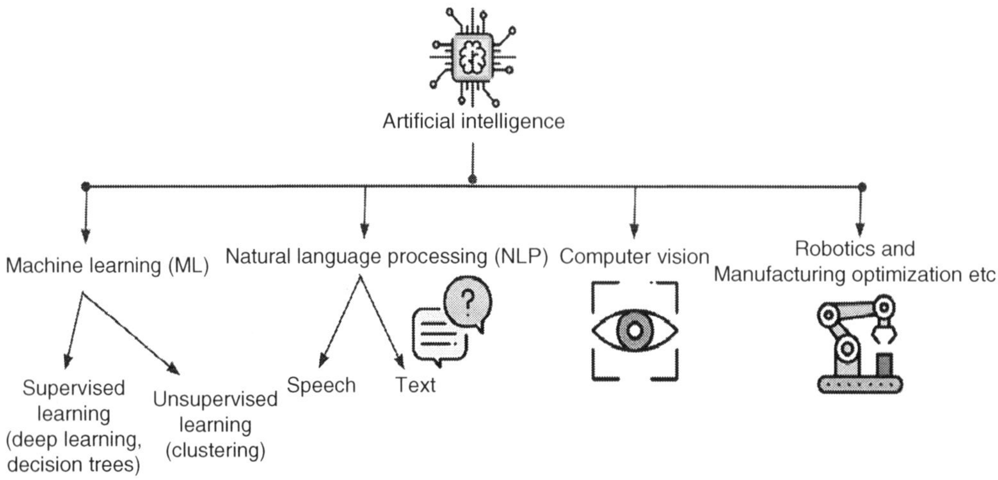

Figure 2.1 Overview of AI—computer vision and natural language processing.

combine the virtual and real worlds. Furthermore, it creates fascinating opportunities in domains like robotics [3], where human-robot cooperation is made possible by robots' ability to observe and comprehend human orders.

But as AI technologies develop further, a number of challenges also follow them. We need to devote special attention to issues like data privacy, discrimination in AI applications, and ethical concerns. Ensuring AI's power is used wisely and for the benefit of society is crucial. Our goal as we dive into this chapter of AI Applications: Computer Vision and NLP is to offer an understanding of the major patterns, advancements, and obstacles that define this multidisciplinary area. We will explore the moral implications and talk about using AI in a responsible manner [4]. In addition, we stress the continued necessity of research and development to tackle practical issues and realize AI's full potential in these fields as shown in Figure 2.1.

2.1.1 What Is It? Everything to Know About Artificial Intelligence

Artificial intelligence, or AI, is the process of imitating human intellect in machines that are built to process information and think like people. The goal of this multidisciplinary area of computer science is to develop technologies and systems that can carry out operations that normally call for human intellect. AI systems are capable of information processing, pattern recognition, decision-making, and performance improvement over time. There are several varieties of AI that are now widely available in daily life. Two excellent instances of AI are the speakers with AI on our mantle that come along with Google or Alexa speech assistants. Popular AI chatbots like Google Bard, ChatGPT, and the recently released Bing Chat are excellent examples. Figure 2.2 illustrates the three broad classifications of AI: super AI, general AI, and narrow AI.

2.1.1.1 Super AI

For voice assistants like Siri, Alexa, and Google Assistant, Artificial Narrow Intelligence (ANI) is essential. This group comprises intelligent systems that, although not specifically intended to do so, have been trained or developed to perform particular functions or address particular issues. Although ANI is sometimes referred to as weak AI because it lacks general intelligence, image recognition systems, tools that flag illicit material online, and the voice assistants mentioned above

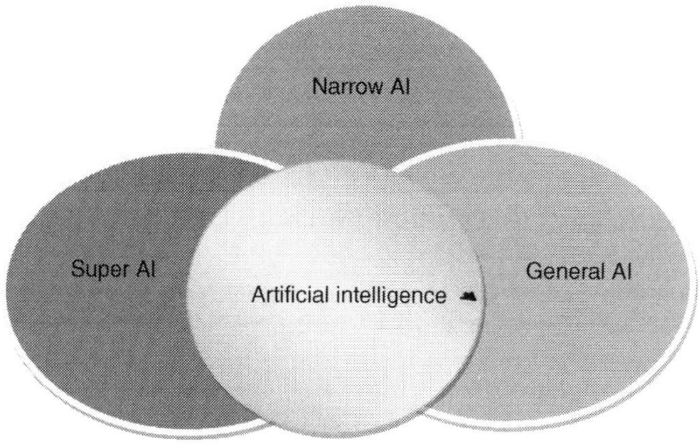

Figure 2.2 Three subcategories of artificial intelligence.

are just a few examples of the power of narrow AI. Other examples consist of technologies that can react to basic customer service requests and image recognition. An example of an artificial neural network (ANI) is ChatGPT, which is designed to respond to commands by producing text messages.

2.1.1.2 General AI

Artificial General Intelligence (AGI) could be able to reason abstractly, comprehend any intellectual endeavor, acquire information from its observations, and utilize that knowledge to tackle new challenges, much like a human. Basically, what we're discussing is a computer or system that is capable of common sense, which is now unattainable with any kind of AI that is currently on the market [5]. Developing a sentient machine is the primary objective of AI research; however, it is still potentially a long way off.

2.1.1.3 Narrow AI

A system known as Artificial Super Intelligence (ASI) has the potential to wipe out humanity in addition to upending it completely. If it seems like something from a science fiction book, that's because ASI is a system where machine intelligence exceeds all human intellect in all domains and performs better than humans in all capacities. It is currently only a theoretical idea to create an intelligent system that is capable of self-improvement through learning. However, this framework has the ability to lead to incredible breakthroughs in science, health, and other sectors if it is applied sensibly and responsibly [6].

2.1.2 Key Ideas and Elements in the Study of Artificial Intelligence

- **Machine Learning**: This area of AI is concerned with teaching algorithms how to gain knowledge from data and provide conclusions or forecasts without the need for explicit programming. Applications such as recommendation systems and predictive analytics make extensive use of it [7].
- **Deep Learning**: Deep learning is a type of machine learning that makes use of multiple-layered neural networks, often known as deep neural networks. Advancements in image and speech recognition have been made possible by this technology [8].
- **Neural Networks**: The architecture and operations of the human brain are imitated in these computer models. Artificial neurons are used for tasks like voice and image recognition. They are made up of interconnected nodes that process information.

- **Natural Language Processing**: Natural language processing, or NLP, aims to enable computers to understand, translate, and generate human language. It is essential to virtual assistants, chatbots, and language translation services.
- **Computer Vision**: Used in applications such as object identification and facial recognition, computer vision allows machines to analyze and comprehend visual information from photos and videos.
- **Robotics**: The development of autonomous robots and drones, which are used in many industries including industry, healthcare, and agriculture, is greatly aided by AI.
- **Reinforcement Learning**: Through interaction with the environment and feedback through the use of incentives or penalties, an agent can develop the ability to make decisions through Reinforcement Learning, a type of machine learning.
- **Expert Systems**: AI programs created to simulate a human expert's decision-making process in a particular field, like finance or medicine, are called expert systems.
- **AI Ethics**: As AI is used more often, moral questions like bias, accountability, transparency, and privacy have come to light.

Healthcare, banking, transportation, education, entertainment, and other industries are just a few of the industries and applications where AI is used [9]. Its powers might revolutionize whole sectors, boost productivity, and provide solutions for challenging issues. However, ethical and responsible developments are necessary to guarantee that AI maximizes advantages to society while limiting.

2.1.3 Today's Modern Role of Artificial Intelligence

Society is being redefined by AI in ways that we never could have predicted. Technology has an impact on all aspects of our lives, from using mobile devices for going through our daily lives to buying online, intelligent dashboards in cars, self-driving robots, and so on. Even though the concept of AI was first proposed at the beginning of the 1950s and formed the basis for numerous computer learning programs and complex decision-making systems, it has only been more recently that this area of technology has started to gain traction in scenarios requiring the processing of large volumes of data. The amount of data produced nowadays—by both humans and machines—far exceeds our capacity to comprehend, analyze, and draw conclusions from such data.

All computer learning is based on AI, which is also the foundation for all complicated decision-making in the future. For instance, most people can learn how to win at tic tac toe (noughts and crosses) despite the fact that the game has 255,168 possible plays, 46,080 of which result in a tie. With almost 500×10^{24}, or 500 quintillions, possible moves, many fewer people would be regarded as great champions of checkers. In order to determine the optimal choice, computers can calculate these permutations and combinations incredibly quickly. Deep learning and AI, which are the natural progression of machine learning, will be the cornerstones of commercial decision-making in the future.

2.1.3.1 In the AI Basket, What Is There?

AI is a science and a subject of study, not a technology. It is a constellation of pre- and post-analytical approaches for working with both structured and unstructured data, as well as several statistical calculation methods. It is an interesting effort to replicate and increase human intelligence through machines and deep learning platforms, natural language production, virtual agents, text-voice-image recognition, AI-optimized hardware, automated robotic operations, intelligent engine systems, and other techniques [10]. Using all available technologies, it seeks to create intelligent machines.

2.1.3.2 AI's Growth in India

AI is being experimented with as a tool for innovations in nearly all Indian industries, like healthcare, education, agriculture, financing, automobiles energy, retailing, manufacturing, and scientific study with autonomous discoveries. AI-based study and creation of goods is being carried out by several Indian firms, including Walmart, Google, Microsoft, Amazon, and Samsung. However, there is still more space for our country to progress in its research on this innovative technology. Most of our governmental, commercial, and academic institutions fund and promote AI inventors, researchers, and startups. With the goal of encouraging AI-based startups and develop modern technological innovations, the government is supporting the private sector and offering many channels for it to do so, namely DST, Niti Aayog, IndiaAI, and many more. Bangalore, Hyderabad, Ahmedabad, Mumbai, and Delhi are the cities with the highest concentration of startups for AI-related industries.

2.1.3.3 Importance of AI in the Modern Era

AI has been gradually scaling up as a proof of concept in response to the world's growing digitization. AI is famous because of its large-scale data processing, task automation, intelligent goods, smart services—content, enhanced accuracy and decision-making assistance, and end-to-end decoding efficiency of sophisticated procedures. Its rapid expansion is evident in the fields of e-commerce, science, medicine, and pharmaceuticals [11]. Google's interactive apps, DeepMind's Alpha Fold, BenevolentAI, chatbots like Clara and Zini, Aryoga Setu, Co-Win, Amazon, Zomato, and Swiggy are just a few of the companies that are emerging as our epidemic digital saviors.

2.1.3.4 AI's Effects on Business

Over time, businesses have changed from being little corner stores to the rapidly expanding Internet marketplaces. These novel approaches not only make people's lives easier, but they also help organizations accomplish their goals by enhancing customer experience, sales forecasting, and automated decision-making. It functions well when machinery, technology, and people collaborate for the benefit of the company. The only technologies used by business now are AI, Cloud, and Big Data. The two main applications of these technologies are e- and m-commerce, which have a huge impact on global trade. In tandem with global advances in automation and innovation, India has experienced a digital revolution throughout the last 20 years. Today, technology is developing faster than expected; the pandemic was a major factor in the technology's rapid acceptance and change.

2.1.3.5 How Can One Search for AI Jobs?

In the competitive IT sector, AI is a profitable field that promises employment growth today. According to four out of five C-suite executives, in order to stay in business, companies must accelerate automation and data processing. Thus, recruiters want candidates with substantial practical experience and superior technical abilities, AI capabilities stand among the job descriptions that have expanded the fastest in recent years. Among the well-known job titles are computer vision experts, AI-deep learning–machine learning engineers, business intelligence developers, data scientists, data analysts, hackers and experts, big data engineers, and related research jobs [12].

How can an AI professional get a good job? In summary, there are various avenues to connect with experts, such as technical blogs, job fairs, LinkedIn, and company employment pages. We will gain insight into the real employment demands and objectives by attending tech speeches provided by corporations at universities and conclaves that host academic, government, and industrial organizations. When we are first starting our postsecondary education, make it a point to explore opportunities in government or industry-sponsored research facilities. This will help us maximize the development of our skill set. To improve our technical profile, participate in stack overflow and open-source projects. Technical competitions like hackathons, ideathons, and makeathons can help develop essential life skills and sharpen your creative thinking.

We are currently facing a difficult global scenario. Everyone is working on the overhaul strategy to balance the economy after COVID-19, regardless of the industries. AI will work to resurrect industrial development and profitability. We anticipate the emergence of novel and sophisticated prospects. The new normal will have a bright future driven by AI. It will serve as the primary force behind new and developing technologies. So, to improve your abilities in an area that is always changing, adopt an interdisciplinary approach. Begin small, think large, and move quickly.

2.2 Artificial Intelligence: 2023's Top 18 Applications

2.2.1 AI Application in E-Commerce

- **Powered Personalized Shopping**: Recommendation engines are made possible by AI technology, which helps us to interact more effectively with our clients. Their browsing history, preferences, and interests are taken into consideration while making these suggestions. It enhances the bond between our clients and increases their fidelity to our business.
- **AI-Renowned Helpers**: Chatbots and virtual shopping assistants both enhance the online shopping experience for users. In order to make the interaction sound as genuine and intimate as possible, NLP is applied. Additionally, these helpers are able to interact with your clients in real time.
- **Fraud Prevention**: Two of the main issues that e-commerce companies deal with are credit card theft and fake reviews. By considering usage trends, AI can reduce the probability of credit card theft. Many consumers decide which product or service to buy based on comments from previous customers. AI can identify and handle fraudulent reviews.

2.2.2 Artificial Intelligence Applications in Education

Even though it affects people's lives the most, AI is progressively finding its way into the educational field. The progressive deployment of AI has enhanced faculty productivity even in the education sector, freeing them up to concentrate more on teaching rather than office or administrative work. A handful of applications in this sector include:

- **Automated Administrative Processes to Support Educators**: Back-office duties like organizing and facilitating parent-guardian communications, organizing and enabling enrollment, courses, and HR-related matters are just a few of the non educational tasks that AI can help educators with. It can also help with task-related duties like facilitating and streamlining personalized messages to students.
- **Creating Smart Content**: AI may be used to create digital versions of textbook aids, conferences, and video courses. With modification, we may use various interfaces, such as animations and educational materials, for pupils in various grade levels. AI generates and provides audio and video summaries and comprehensive lesson plans, which contribute to the creation of a rich learning environment.
- **Voice Assistants**: With Voice Assistants, students may obtain additional learning resources or support without ever having to interact directly with lecturers or teachers. This reduces the

expense of publishing temporary handbooks and makes it simple to get answers to frequently asked queries.

- **Customized Education**: Advanced AI technologies enable the comprehensive monitoring of students' data through hyper personalization techniques. Furthermore, it is simple to create routines, lesson plans, reminders, study tools, flash notes, frequency of review, etc.

2.2.3 Artificial Intelligence Applications in Lifestyle

AI greatly impacts our way of life. Let's talk about a couple of them.

- **Autonomous Vehicles**: Automakers like Toyota, Audi, Volvo, and Tesla are using machine learning to train computers to think and react like people when it comes to driving in any type of circumstance and seeing things to avoid collisions.
- **Spam Detectors**: AI in the email we use on a daily basis filters out junk emails, moving them to trash or spam folders. So, we only view the material that has been screened. Gmail, a well-known email service, has achieved a filtering capacity of around 99.9%.
- **Facial Recognition**: Facial recognition algorithms are used by our favorite gadgets, such as mobile phones, laptops, and PCs, to recognize and detect faces with the goal of granting safe access. Facial recognition is a popular AI use outside of personal utilize, even in high-security settings across a variety of sectors.
- **Recommendation System**: Many of the websites and e-commerce sites we use on a daily basis, such as social networking sites, entertainment websites, and video-sharing sites like YouTube, employ recommendation systems to collect user data and present users with tailored recommendations in an attempt to increase interaction. This is one of the most widely used uses of AI in almost every business.

2.2.4 Navigational Applications of Artificial Intelligence

An MIT study claims that GPS technology can provide customers with accurate, timely, and thorough information to improve safety. This method makes life easier for users by automatically recognizing the number of lanes and types of roads behind obstructions on the roadways using convolutional and graph neural networks. Uber and many other logistics businesses utilize AI extensively to enhance operational efficiency, assess traffic, and optimize routes.

2.2.5 Applications of Artificial Intelligence in Robotics

Robotics is an additional discipline in which AI techniques are widely used. Robots driven by AI use real-time updates to identify obstacles in their way and instantly plan their journey. It can be used for:

- The transportation of supplies in industries, hospitals, and warehouses
- Cleaning huge machinery and offices
- Inventory control

2.2.6 Artificial Intelligence Applications in Human Resources

AI aids in the practice of blind hiring. Machine learning software allows for analyzing applications according to particular criteria. AI-powered systems have the ability to analyze resumes and job prospects' profiles to provide recruiters with a better idea of the talent pool from which to select [13].

2.2.7 Artificial Intelligence Applications in Healthcare

There are several uses for AI in the healthcare industry. AI is being applied in the healthcare industry to create intelligent computers that can recognize cancer cells and diagnose illnesses. To guarantee an early diagnosis, AI can assist in the analysis of lab and other medical data related to chronic illnesses. AI finds novel medications by combining medical knowledge with historical data.

2.2.8 Applications of Artificial Intelligence in Agriculture

AI is used to identify nutrient deficiencies and soil defects. AI is applied to computer vision, machine learning, and robotics to examine the growth patterns of weeds. AI bots can help with agricultural harvesting more rapidly and in larger quantities than human laborers can [14].

2.2.9 Artificial Intelligence Applications in Gaming

The gaming industry is one more area where AI technologies have gained popularity. NPCs that are intelligent and human-like can be created using AI and communicate with gamers. Additionally, it may be used to forecast human behavior, which can help with testing and game design. AI is used by the 2014 game Alien Isolation to follow the player around. The game makes use of two AI systems: the "Alien AI," which is powered by sensors and behaviors and constantly hunts the player, and the "Director AI," which usually knows where the players are.

2.2.10 Artificial Intelligence Applications in Automobiles

Autonomous cars are built with the help of AI. AI may be utilized to drive a vehicle when coupled with its camera, radar, GPS, cloud services, and control signals. In along with offering extra features like driver-assist steering, blind-spot surveillance, and brake assistance, AI may enhance the in-car experience.

2.2.11 Applications of Artificial Intelligence in Social Media

- **Instagram**: This platform employs AI to select posts for our Explore tab by analyzing our likes and the accounts we follow.
- **Facebook**: AI is utilized in conjunction with a program known as DeepText. Facebook is able to better understand discussions with the help of this technology. It may be used for automatically translating postings between several languages.
- **Twitter**: Uses AI to detect fraud and eliminate offensive and propaganda material. Twitter also makes use of AI to suggest tweets to users depending on the kinds of tweets they interact with.

2.2.12 Applications of Artificial Intelligence in Marketing

AI applications are also widespread in the marketing sector. With the use of AI, marketers may offer extremely targeted and tailored ads by identifying patterns in machine learning, behavioral analysis, etc. It also helps with retargeting visitors at the right times to improve results and reduce feelings of mistrust and irritation. AI can help content marketing in a way that maintains the brand and aesthetic of the company. It can handle a variety of routine tasks, including performance management and campaign reporting. Chatbots that are driven by NLP, AI, natural language understanding, and natural language production are able to comprehend user language and respond appropriately. AI can adjust and improve marketing efforts to meet the demands of a local market in addition to providing consumers with real-time personalizations depending on their activity.

2.2.13 Artificial Intelligence Applications for Chatbots

Chatbots with AI can comprehend natural language and respond to consumers utilizing the "live chat" feature that many companies provide for customer service. AI chatbots are effective when they make use of machine learning and may be integrated into a range of websites and applications. In addition to extracting data from a pre-existing collection of integrated responses, AI chatbots have the capacity to eventually create a database of replies. These chatbots can efficiently handle client concerns, reply to basic questions, enhance customer care, and offer round-the-clock assistance as AI advances. All things considered, these AI chatbots can raise client satisfaction levels.

2.2.14 Artificial Intelligence Applications in Finance

According to reports, 80% of banks are aware of the advantages AI may offer. AI offers highly advanced technology that could significantly enhance a wide range of financial services, whether they are personal, business, or consumer financial services. For instance, clients seeking assistance with wealth management solutions may quickly obtain the information they want via online chat or SMS text messaging, both of which are driven by AI. AI may also spot possible red flags, like as changes in transaction patterns, which people may overlook but which could indicate fraud. By doing so, it can save firms and individuals from suffering large losses. AI is capable of better predicting and evaluating loan risks in addition to work automation and fraud detection.

2.2.15 AI in Astronomical Science

Without a doubt, in this wonderful world of technology, AI is the idea that has everyone completely captivated. Over the years, AI has found many uses in industries such as healthcare, robotics, eCommerce, and even finance. In contrast, astronomy is a subject that has not yet been thoroughly studied but is equally fascinating and exhilarating as the others. The analysis of data is one of the most challenging issues in astronomy. Because of this, astronomers are using AI and machine learning to develop new tools. Nevertheless, take into account how AI has changed astronomy and is satisfying astronomers' needs [15].

Recently, scientists have established that galaxy mergers are the major drivers behind starbursts through the application of AI in a galactic merger inquiry. Because of the volume of data, the researchers created a deep learning system that taught itself to identify merging galaxies. AI, according to one of the astronomers, has the advantage of increasing the study's repeatability [16]. This is due to the fact that the algorithm's definitions of a merger are always the same. The sky is changing, and it's one of the most incredible things I've ever seen. The goal of this endeavor seeks to explore the entire night sky every night, gathering about 80 terabytes of data in one go, to investigate how stars and galaxies in the universe evolve over time.

One of the most important things an astronomer has to do is find map. According to the theory, part of the light is visible to humans when an exoplanet passes in the direction of its parent star but is blocked in other places. This site is used by astronomers to examine the orbit of exoplanets and create a map of the light dips. The planet's numerous characteristics are then determined, including its size, mass, and separation from its star, to name a few. But in this instance, AI turns out to be more than a rescuer. It is possible to evaluate information as an ordered sequence to recognize planetary signals with up to 96% accuracy by utilizing AI's time-series analysis skills.

Astronomers must identify the signatures of the most catastrophic events in the cosmos. Time dilation occurs when exoplanets collide with one another. Weak signals on Earth can be monitored to further identify them. The Ligo and Virgo collaborations on gravitational-wave detectors have done a fantastic job in this area. They were both proficient in applying machine learning to identify signals. Now that they get notifications, astronomers can aim their telescopes appropriately.

2.2.16 AI for Data Security

AI is seen by many as the technology industry's future and today. AI is used by many business executives for a number of objectives, including offering valuable services and future-proofing their organizations [17]. One of the most common and crucial uses of AI is data security, which is among the most valuable resources for any tech-focused company. Because private information, including organizational secrets and consumer data (like credit card information), is maintained online, data security is crucial for any institution to meet its operational and legal requirements. These days keeping data out of the wrong hands is a critical and difficult responsibility for many businesses, which is why many of them adopt AI-based security solutions. The use of AI in business is vital since the modern world is more intelligent and connected than ever. Security teams will need to rely on AI technologies to keep control over systems and data since cyberattacks are expected to become more persistent over time [18].

- **Identifies Unknown Threats**: It's possible that a human cannot identify every risk that a company faces. For a multitude of causes, hundreds of millions of attacks are launched by hackers annually. Severe network damage can be caused by unknown attacks. Even worse, they can already be in effect while you don't realize, identify, or stop them. Modern solutions should be employed to prevent attacks as they evolve, ranging from malware assaults to more complex malware assaults. AI has shown to be one of the greatest security techniques for mapping out and preventing unforeseen dangers from wreaking havoc on a business.
- **Identification of Flaws**: AI helps identify buffer overflows of data. Buffer overflow is the term used when programs use more data than they should. Apart from the error resulting from human triggers destroying important data, AI can see these errors as well, and it can identify them in real time to avert further risks. With machine learning, AI can accurately identify flaws, vulnerabilities, and other issues related to cybersecurity. AI is also aided by machine learning in spotting dubious data supplied by any application. Programming language bugs allow criminals to employ malware or viruses to infiltrate systems and steal data.
- **Threat Detection**: New AI technologies are continually being developed by cybersecurity vendors. AI has evolved to the point where it can now identify updates or system problems. It would immediately turn off anyone trying to take advantage of such problems. AI would be a fantastic tool for averting any hazard. In addition to installing more firewalls, it could fix dangerous code errors.
- **Addressing Dangers**: It is a consequence of the danger having infiltrated the system. AI is utilized, as previously said, to identify anomalous activity and generate a profile of malware or viruses. Right now, AI is combating malware and viruses as necessary. Primarily, the response is eliminating the infection, fixing the issue, and addressing the damage caused. Last, AI ensures that a similar occurrence won't occur again and implements the necessary preventative measures.

- **Identify Uncharacteristic Action**: AI enables us to identify odd system behavior. It can identify anomalous activity by continuously monitoring a system and collecting a sufficient quantity of information. AI also recognizes unauthorized access. AI uses certain components to assess if detected odd activity is a real threat or a manufactured alert. AI uses machine learning to distinguish between abnormal and non abnormal behavior. As machine learning advances over time, AI will be able to recognize even little irregularities. AI would therefore highlight any issues with the system.

2.2.17 AI in Transportation and Travel

In recent years, intelligent technology has permeated every aspect of our existence. Furthermore, new applications of AI—most notably in transportation—are becoming commonplace as society adopts more advanced technology [19]. This has created a new market for companies and entrepreneurs to create innovative ways to raise the standard of public transportation's accessibility, comfort, and safety. Intelligent transportation systems might emerge as one of the most significant means of enhancing the standard of living for individuals worldwide. There are several examples of comparable systems in use across different industries.

- **Transportation of Heavy Goods**: For example, truck platooning, which arranges HGVs (heavy goods trucks), can be quite useful when moving cars or other big items. A human driver drives the lead truck in a platoon of trucks; the other human drivers drive the other vehicles in a passive way, only taking over the wheel in exceptionally dangerous or difficult situations. The platoon's trucks move in formation and simultaneously initiate the activities performed by the lead vehicle's human driver since they are all connected via a network. Therefore, all of the cars that are following the lead driver halt completely as well.
- **Traffic Control**: Globally, one of the main obstacles to urban mobility is congested city streets. Cities all across the world have built bridges, expanded highways, and added more transit options like trains, yet the traffic issue still exists. The application of AI to traffic management, however, offers a real chance to improve the issue. Road safety may be increased and traffic laws can be enforced with the help of intelligent traffic management. For instance, Alibaba's City Brain project in China tracks road networks in real time and eases traffic using AI technologies including big data analysis, visual search engines, and predictive analysis. An effective transformation system is necessary for the construction of a city, and next-generation smart cities are being powered by AI-based traffic management technology [20].
- **Ride-sharing**: By bringing drivers and passengers together, enhancing communications and user interaction, and streamlining decision-making, ride-sharing platforms such as Uber and Ola employ AI to enhance user experiences. For instance, Uber's in-house, patent-pending Michelangelo machine-learning platform can forecast demand and supply, and spot anomalies in travel, such as accidents, and project arrival times.
- **Route Planning**: Predictive analytics combined with AI-enabled route planning may benefit individuals and companies alike. This is already accomplished by ride-sharing services through the analysis of multiple real-world characteristics for optimal route planning. Businesses, especially those in the shipping and logistics sectors, may build a more effective supply network by utilizing AI-enabled route planning, which anticipates road conditions and optimizes vehicle routes. In route planning, predictive analytics refers to the intelligent assessment of road usage data by a machine, including traffic patterns, customer preferences, road limits, and congestion levels. Cargo logistics companies, as well as other general-purpose logistics firms, may employ

this technology to reduce delivery costs, shorten delivery times, and enhance asset and operation management.

2.2.18 AI in the Automobile Industry

The notion that robots will be able to comprehend, carry out complex computations, and provide workable answers to pressing issues was mostly reserved for science fiction authors a century ago [21]. Even so, as the twenty-first century draws to a close, it is hard for us to imagine living without the innovations made possible by advances in machine learning and AI, such as virtual travel agents, factory robots, intelligent assistance, and bots for stock trading and marketing. It is impossible to exaggerate the significance of AI and machine learning in the automobile industry. An increasing number of firms are choosing to integrate machine-learning models and AI into their operations as a result of the increasing uses of AI in the automobile industry.

- **Production**: By integrating AI into the manufacturing process, automakers can take advantage of smarter factories that increase efficiency and save expenses. AI may be used in a variety of manufacturing processes, including supply chain optimization, robot employment on the assembly line, sensor-based performance enhancement, automotive design, and post-production tasks.
- **Supply Chain**: In 2021 and 2022, supply chain disruptions and difficulties impacted the automotive industry. AI is also useful in this context. AI assists businesses in anticipating obstacles and restocking supply networks as necessary. AI may also help with volume projections, routing issues, and other issues.
- **Driver and Passenger Experience**: Everyone wants to travel in their cars in comfort. In this regard, AI is also helpful. AI has the potential to help drivers stay focused while driving by reducing distractions, studying driving habits, and improving the overall customer experience. Due to AI, passengers can take advantage of in-car delivery services and tailored accessibility.
- **Inspections**: Renting an automobile, getting insurance, or simply having a garage to inspect it is a highly subjective and manual process. AI has the potential to digitize auto inspections as current technology can evaluate an automobile, locate its defects, and generate an extensive status report.
- **Quality Assurance**: Everybody desires a luxury vehicle and lifestyle. Wouldn't it be preferable to find out about a problem with your car before it breaks down? This program makes use of AI to offer capabilities like predictive monitoring and extremely accurate fracture diagnosis.

2.3 AI in Computer Visions

When machine learning and AI are combined, computer vision becomes capable of comprehending and interpreting visual data from photos and videos. This is known as AI in computer vision. In computer vision, this is referred to as AI. AI has made strides in computer vision and the ability of computers to do a wide range of tasks that typically require human visual perception and interpretation to be conceivable [22]. The following are important elements of AI in computer vision:

- **Object Detection and Recognition**: AI-powered computer vision systems are capable of identifying and detecting things in pictures and movies. Applications for this include industrial quality control, security, and driverless cars.

- **Image Classification**: AI systems are capable of classifying photos into pre-established groups. Applications like content filtering, medical diagnostics, and e-commerce all make extensive use of this.
- **Image Segmentation**: AI makes it possible to precisely identify and separate items inside a picture using image segmentation. It is employed in scene comprehension, industrial automation, and medical imaging.
- **Facial Recognition**: Access control, social media tagging, and security are among the uses for AI-driven facial recognition technology.
- **Anomaly Detection**: AI is useful in applications like fraud detection and defect inspection because it can recognize odd or abnormal patterns in photos and movies.
- **Text Recognition (OCR)**: AI-powered Optical Character Recognition (OCR) enables the extraction of text from pictures, which is helpful for data input automation and document digitization.
- **Image captioning**: AI systems are able to provide natural language descriptions for images, increasing accessibility and helping those with vision impairments.
- **Visual Search**: AI in computer vision allows users to utilize visual searches to find items, information, or related pictures.
- **Augmented Reality (AR) and Virtual Reality (VR)**: AI improves AR and VR experiences by monitoring motion and identifying real-world objects, resulting in more realistic interactions.
- **Medical Image Analysis**: AI-powered computer vision systems help with medical image analysis, which helps with X-ray and MRI interpretation as well as early illness identification.
- **Autonomous Vehicles**: AI is essential to autonomous vehicles because it enables them to sense and comprehend their environment via cameras and other sensors.
- **Manufacturing Quality Control**: AI-powered computer vision systems are used to verify items and guarantee quality minimize faults, and increase productivity.
- **Environmental Monitoring**: AI can process visual data, such as observing wildlife populations and examining climatic trends, to assess and track changes in the environment.
- **Entertainment and Gaming**: To create immersive gaming experiences and recognize gestures, the gaming industry leverages AI in computer vision.
- **Robotics**: AI-powered computer vision is used by robots to navigate, recognize objects, and interact with their environment.

AI and computer vision together have the power to completely transform a range of fields and applications, improving decision-making, streamlining operations, and opening up new avenues for human–computer connection. The accuracy, speed, and variety of uses for computer vision are expected to increase as AI technologies develop.

2.4 AI in Natural Language Processing

In NLP, AI is the application of AI technology to the comprehension, production, and interaction with human language [23]. Within the field of AI, NLP examines the ways in which computers and natural language interact to allow machines to comprehend, interpret, and generate text and speech. Here's a summary of AI in NLP:

- **Language Understanding**: NLP AI systems are able to extract information and perspectives from written or spoken language by analyzing and comprehending the structure and meaning of text. This has several uses in fields like virtual assistants, chatbots, and sentiment analysis.

- **Sentiment Analysis**: Whether a text is a product review, a social media post, or customer feedback, AI-driven NLP systems may ascertain its sentiment or emotional tone. For customer service, brand management, and market research, this is priceless.
- **Chatbots and Virtual Assistants**: NLP is a tool that AI chatbots and virtual assistants use to communicate with users, answer questions, and provide assistance. They are employed in interactive applications such as online support and customer care.
- **Language Translation**: Accurate and instantaneous language translation is now possible because of AI's substantial improvements to machine translation systems. This is utilized in worldwide commerce and communication, as well as in technologies like Google Translate.
- **Information Extraction**: AI in NLP has the ability to automatically extract structured data from unstructured text, which is useful for data mining, news aggregation, and content curation, among other activities [24].
- **Speech Recognition**: AI-driven NLP systems can translate spoken words into text, enabling voice prompts and dictation. Voice assistants like Siri and Alexa are common examples of this technology in use.
- **Language Generation**: AI systems are capable of producing text that is human-like for chatbot answers, automated report generation, or content creation. This has uses in journalism, content marketing, and other fields.
- **Question Answering**: Natural language inquiries may be comprehended and responded to by AI-driven NLP systems. Both knowledge bases and search engines employ them.
- **Named Entity Recognition**: AI assists in locating and categorizing named entities in text, such as names of individuals, locations, and organizations. This is helpful for organizing material and retrieving information.
- **Summarizing**: AI can produce clear, logical summaries of lengthy texts, which helps with knowledge management and content summarizing.
- **Speech Synthesis**: NLP and AI work together to produce realistic-sounding synthesized speech, which powers voice assistants and improves accessibility for those with visual impairments.
- **Language Models**: By comprehending context and producing contextually appropriate replies, sophisticated AI language models like GPT-3 and BERT have transformed NLP. They are employed in many different contexts, such as content production and chatbots.
- **Conversational AI**: AI is being used to power chatbots and conversational agents, which will provide more conversational and natural interactions in customer service, healthcare, and education.
- **Conversational AI**: Chatbots and conversational agents, allowing for more conversational and natural connections in customer service, healthcare, and education, are powered by AI.
- **Healthcare Language Understanding**: NLP is assisting with patient data and medical record analysis, enhancing clinical decision support and medical transcribing services.

AI in NLP is a quickly developing topic that has a big influence on many different sectors. As NLP continues to progress, human–computer interactions will become increasingly intuitive and effective. Natural language understanding and generation offer opportunities to enhance communication, automation, and decision-making processes.

2.5 Conclusion

We have entered the world of AI Applications, namely Computer Vision and NLP, where we will explore the revolutionary potential of these two crucial fields that are reshaping human interaction and technology in the present and the future. With its amazing developments in object identification, picture interpretation, and autonomous systems, computer vision has opened the door for ground-breaking discoveries in autonomous cars, security, healthcare, and other fields. It is amazing how NLP has made it possible for computers to understand, speak, and comprehend human language. As a result, sentiment analysis, chatbots, and language translation services have all been developed. The combination of computer vision and NLP has opened up new creative possibilities. This synergy has ushered in a future in which technology is even more interwoven with our daily lives and businesses by enabling robots to comprehend and respond to human language in addition to perceiving the visual environment.

It has gotten us one step closer to realizing robotics applications, AR applications, and the merging of digital and physical experiences. But even as we welcome these technical wonders, it's critical to acknowledge the moral issues raised by this quick advancement. We need to maintain openness in AI applications, protect data privacy, and be on the lookout for biases. Responsible deployment is crucial to maximizing the use of AI in computer vision and NLP. As we get to the end of our investigation, it is clear that these technologies have advanced to the point where they are essential parts of several businesses and not just tools, providing game-changing solutions that alter how we connect with the outside world and with one another. AI applications at the intersection of NLP and computer vision have a promising future, provided that ethical considerations have been taken into account and these advances can be developed ethically. There is countless potential for innovation and advancement in a future where human collaboration and AI are set to grow increasingly more dynamic.

2.6 AI's Opportunities for Computer Vision and Natural Language Processing

The future lies in the inventive and dynamic realm of AI applications situated at the intersection of computer vision and NLP. We should expect a plethora of fascinating innovations that will transform industries, enhance human skills, and improve our daily lives as technology advances and our understanding of AI expands. The smooth integration of NLP with computer vision is one of the most exciting prospects. As a result of this convergence, intelligent systems will be developed that can perceive and react to human language more sophisticatedly and human-like, in addition to interpreting visual input. The end product will be a virtual environment that interacts with and reflects the physical world in previously unthinkable ways. For example, AR apps will use these latest innovations to provide immersive and interactive experiences for learning, gaming, and remote collaboration.

Robots will enter a new age in the future when they will be even more skilled at communicating and comprehending human behavior. These AI-driven robots will do sophisticated jobs in the manufacturing, healthcare, and service sectors. They will help surgeons assemble complicated parts as well as offer excellent companionship and customer service. Automation and efficiency will significantly increase as a result of the interaction between AI and robots. AI in Computer Vision and NLP will revolutionize diagnosis, treatment suggestions, and patient care, causing a major upheaval in the healthcare industry. Processes related to healthcare will be streamlined and health outcomes will be improved with more precise illness detection, tailored medicines, and seamless administration of medical information becoming standard.

Ethical issues and appropriate AI use will become more and more important as these technologies develop. As we work to mitigate possible biases and hazards and assure the good effect of AI, maintaining fairness, transparency, and data privacy will remain critical. High-quality, human-like text and media content will be produced by AI systems, leading to an increasingly automated process for content generation and curation. This will give new opportunities for narrative and innovation, redefining journalism, content marketing, and the creative industries. AI-driven chatbots and voice assistants will provide conversational and responsive user interfaces, facilitating more intuitive and natural AR interaction. Personalized information and recommendations that are catered to user preferences will be the driving force behind a wide range of apps.

Language models will advance to become more contextually aware, which will increase their usefulness in chatbot development, customer support, and content production. Transportation, logistics, and delivery services will continue to change as a result of autonomous systems like drones and self-driving cars. AI applications in environmental monitoring will support biodiversity preservation, climate change tracking, and conservation initiatives. Language translation services will close the gap in worldwide communication and promote mutual understanding and cooperation. AI in NLP and Computer Vision will be crucial in improving accessibility for people with impairments by guaranteeing equitable access to opportunities and more inclusive information and services. AI-driven security and privacy solutions will progress, defending sensitive data and thwarting changing cybersecurity threats.

Furthermore, a new era of unheard-of computational power will be ushered in by the combination of AI with quantum computing, enabling more sophisticated and quick processing in these fields. All things considered, computer vision and NLP applications of AI have a promising future. Technology has the ability to expand human creativity and open up previously uncharted territories. These technologies will impact a world where the distinctions between the digital and physical are progressively dissolved and the potential of AI is fulfilled in more significant and meaningful ways than before, provided they are researched, developed ethically, and deployed responsibly.

References

1 Li, C., Zhang, Y., Weng, Y. et al. (2023). Natural language processing applications for computer-aided diagnosis in oncology. *Diagnostics* 13 (2): 286.

2 Fanni, S. C., Febi, M., Aghakhanyan, G., & Neri, E. (2023). Natural language processing. In Peter Van Oijen, Erik R. Ranschaert, Annalisa Trianni, Michail E. Klontzas. *Introduction to Artificial Intelligence* (pp. 87-99). Cham: Springer International Publishing.

3 El-Komy, A., Shahin, O.R., Abd El-Aziz, R.M., and Taloba, A.I. (2022). Integration of computer vision and natural language processing in multimedia robotics application. *Information Sciences* 7 (6): 1.

4 Devi, V.A. and Naved, M. (2021). Dive in deep learning: computer vision, natural language processing, and signal processing. In: *Machine Learning in Signal Processing* (ed. S. Tanwar, A. Nayyar, and R. Rameshwar), 97–126.

5 Xu, F., Uszkoreit, H., Du, Y. et al. (2019). Explainable AI: a brief survey on history, research areas, approaches and challenges. *Natural Language Processing and Chinese Computing: 8th CCF International Conference, NLPCC 2019*, Dunhuang, China (9–14 October 2019), Proceedings, Part II 8 (pp. 563–574). Springer International Publishing.

6 Sennott, S.C., Akagi, L., Lee, M., and Rhodes, A. (2019). AAC and artificial intelligence (AI). *Topics in Language Disorders* 39 (4): 389.

7 Goyal, A.A. (2023). The role of machine learning in natural language processing and computer vision. *Iconic Research and Engineering Journals* 6 (11): 185–195.

8 Chai, J., Zeng, H., Li, A., and Ngai, E.W. (2021). Deep learning in computer vision: a critical review of emerging techniques and application scenarios. *Machine Learning with Applications* 6: 100134.

9 Lund, B.D., Wang, T., Mannuru, N.R. et al. (2023). ChatGPT and a new academic reality: artificial intelligence-written research papers and the ethics of the large language models in scholarly publishing. *Journal of the Association for Information Science and Technology* 74 (5): 570–581.

10 Malhotra, Y. (2018). MIT Computer Science & AI Lab: AI-Machine Learning-Deep Learning-NLP-RPA: Executive Guide: including Deep Learning, Natural Language Processing, Autonomous Cars, Robotic Process Automation.

11 Le Glaz, A., Haralambous, Y., Kim-Dufor, D.H. et al. (2021). Machine learning and natural language processing in mental health: systematic review. *Journal of Medical Internet Research* 23 (5): e15708.

12 Xu, F., Wang, L., and Gao, J. (2020, August). Thoughts on application of artificial intelligence in teaching of different disciplines. *2020 15th International Conference on Computer Science & Education (ICCSE)*, Organized by National Research Council for Computer Education in Colleges & Universities, China, Hosted by Delft University of Technology, Netherlands, 18–22 August 2020, pp. 703–707. IEEE.

13 Manovich, L. (2021). Computer vision, human senses, and language of art. *AI & SOCIETY* 36: 1145–1152.

14 Dhanya, V.G., Subeesh, A., Kushwaha, N.L. et al. (2022). Deep learning based computer vision approaches for smart agricultural applications. *Artificial Intelligence in Agriculture* 6: 211.

15 Galassi, A., Lippi, M., and Torroni, P. (2020). Attention in natural language processing. *IEEE Transactions on Neural Networks and Learning Systems* 32 (10): 4291–4308.

16 Mahadevkar, S.V., Khemani, B., Patil, S. et al. (2022). *A Review on Machine Learning Styles in Computer Vision-Techniques and Future Directions*. IEEE Access.

17 Selvaraj, C., Elakkiya, E., Prabhu, P. et al. (2023). Chapter 9—advances in QSAR through artificial intelligence and machine learning methods. In: *QSAR in Safety Evaluation and Risk Assessment* (ed. H. Hong), 101–116. Academic Press. ISBN 9780443153396.

18 Tran, B.X., Vu, G.T., Ha, G.H. et al. (2019). Global evolution of research in artificial intelligence in health and medicine: a bibliometric study. *Journal of Clinical Medicine* 8 (3): 360.

19 Abioye, S.O., Oyedele, L.O., Akanbi, L. et al. (2021). Artificial intelligence in the construction industry: a review of present status, opportunities and future challenges. *Journal of Building Engineering* 44: 103299.

20 Chung, S., Moon, S., Kim, J. et al. (2023). Comparing natural language processing (NLP) applications in construction and computer science using preferred reporting items for systematic reviews (PRISMA). *Automation in Construction* 154: 105020.

21 Szegedy, C. (2020). A promising path towards autoformalization and general artificial intelligence. *Intelligent Computer Mathematics: 13th International Conference, CICM 2020*, Bertinoro, Italy (26–31 July 2020), Proceedings 13 (pp. 3–20). Springer International Publishing.

22 Iyer, L.S. (2021). AI enabled applications towards intelligent transportation. *Transportation Engineering* 5: 100083.

23 Hwang, G.J., Xie, H., Wah, B.W., and Gašević, D. (2020). Vision, challenges, roles and research issues of artificial intelligence in education. *Computers and Education: Artificial Intelligence* 1: 100001.

24 De-Lima-Santos, M.F. and Ceron, W. (2021). Artificial intelligence in news media: current perceptions and future outlook. *Journalism and Media* 3 (1): 13–26.

3

An Overview of AI Platforms, Frameworks, Libraries, and Processors

Pavan Kumar Akkisetty

Lead AI Architect – LLM Pipelines & Cloud As as a Service Lead AI Architect, Intel Corporation, Hillsboro, Oregon, USA

3.1 Introduction: Artificial Intelligence Platforms, Frameworks, Libraries, and Processors: The Building Blocks of AI Development

Artificial intelligence (AI) platforms, frameworks, libraries, and processors are the essential building blocks of AI development. They provide the tools and infrastructure that AI developers need to build and deploy AI applications. Deploying AI applications in the cloud vs. edge needs to be a well-thought-through process as each brings its respective challenges and requirements.

3.2 Edge AI

Edge AI is the combination of edge computing and AI, and it is a critical piece of the software-defined business. It is powered by AI applications running at the edge, and it can be used to transform businesses with intelligence. Edge AI is the combination of edge computing and AI. Edge computing is the process of performing computation at the edge of the network, close to where the data is being generated. AI is the ability of machines to learn and make decisions without being explicitly programmed. Edge AI can be used to transform businesses by providing real-time insights and automating tasks. Figure 3.1 gives an overview of the layers of Edge AI.

For example, Edge AI can be used to:

- Improve customer service by identifying and resolving issues in real time.
- Optimize manufacturing processes by automating tasks such as quality control.
- Improve traffic management by detecting and responding to traffic congestion.
- Provide personalized recommendations for products and services.
- Detect and prevent fraud.
- Improve health care by providing early diagnosis and treatment of diseases.

In the next sections, let us discuss the various modules that are needed to make Edge AI possible.

3.2.1 AI Platforms

AI platforms provide a comprehensive set of tools and services, the full AIOps pipeline that includes a runtime environment, a set of libraries, development, deployment, orchestration, monitoring,

Model Optimization Methods for Efficient and Edge AI: Federated Learning Architectures, Frameworks and Applications, First Edition. Edited by Pethuru Raj Chelliah, Amir Masoud Rahmani, Robert Colby, Gayathri Nagasubramanian, and Sunku Ranganath.

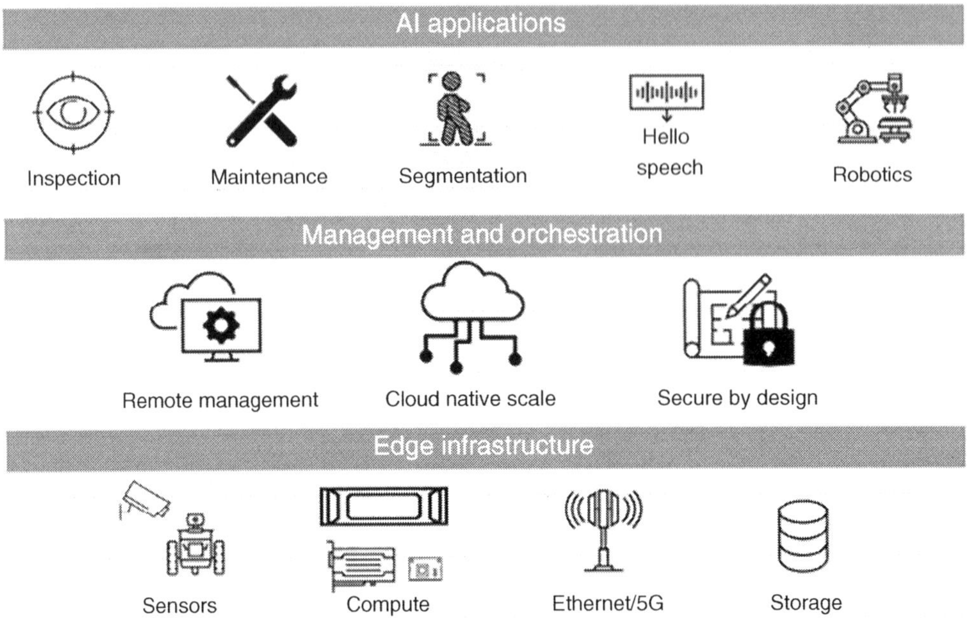

Figure 3.1 Overview of the layers of Edge AI [1]. Source: https://developer.nvidia.com/blog/steps-to-getting-started-with-edge-ai//with\ignorespacespermission\ignorespacesof\ignorespacesNVIDIA/Last accessed on May 05.

telemetry environments, etc. The runtime environment provides the infrastructure for running AI models, the libraries provide the functionality for common AI tasks, and the development environment provides the tools for developing and debugging AI applications.

The deployment environment provides the infrastructure for deploying AI models into production. The monitoring environment provides the tools for monitoring the performance of AI models in production. The telemetry environment provides the tools for collecting and analyzing data about the performance of AI models in production. The deployment environment provides the infrastructure for deploying AI models into production. The monitoring environment provides the tools for monitoring the performance of AI models in production. The telemetry environment provides the tools for collecting and analyzing data about the performance of AI models in production. Orchestration is the process of managing and coordinating the deployment, monitoring, and telemetry of AI models in production. It ensures that AI models are deployed correctly, that they are performing as expected, and that data about their performance is collected and analyzed effectively.

Orchestration is important because it helps ensure that AI models are deployed and managed effectively in production. It can help improve the performance of AI models, identify and fix problems, and make sure that AI models are used in a safe and responsible way.

These environments are essential for building, deploying, and managing AI applications in production. They help ensure that AI applications are reliable, scalable, and secure.

Examples of AI platforms:

- Google Cloud AI Platform[2]
- Amazon Web Services (AWS) AI Platform[3]

- Microsoft Azure AI Platform[4]
- IBM Watson Platform[5]
- NVIDIA AI Platform[6]
- Other platforms like RunAI, LatentAI, Qualcomm AI, etc.

3.2.1.1 AIOps

AIOps [7] is a combination of DevOps, DataOps, MLOps, and AI modules. DevOps is the combination of software development and IT operations. It is a set of practices that automate and streamline the process of software development and delivery. DataOps is the application of DevOps to data. It is a set of practices that automate and streamline the process of data management and analytics. MLOps is the application of DevOps to machine learning. It is a set of practices that automate and streamline the process of machine learning model development and deployment. AIOps is the use of AI for a variety of IT operations. It is a set of practices that use AI to automate and improve IT operations tasks such as in health care technologies to monitor patients' health, retail domain to improve logistics and customer support, industrial domain to improve quality and safety, etc.

These AIOps are all available in several AI platforms we referred to in the above section. They offer a variety of services that can be used to automate and streamline the software development, data management, machine learning, and IT operations processes.

3.2.2 AI Frameworks

AI frameworks offer a set of tools and APIs for developing AI models. They typically include a compiler, a debugger, and a set of libraries for common AI tasks. The compiler converts the AI model into a format that can be executed by the AI platform, the debugger helps find and fix errors in the AI model, and the libraries provide the functionality for common AI tasks.

Some popular AI frameworks and their inference solutions are in Table 3.1:

3.2.3 AI Libraries

AI libraries are a collection of tools and resources that can be used to build and deploy AI applications quickly and easily. They provide pretrained models and code for common AI tasks, such as image recognition, natural language processing, and machine translation.

AI libraries can also be used for preprocessing and postprocessing data. Preprocessing is the process of preparing data for training and deploying an AI model. This may involve cleaning the data, removing outliers, and converting the data into a format that is compatible with the AI model. Postprocessing is the process of cleaning up the output of an AI model. This may involve filtering the output to remove false positives, or converting the output into a format that is more readable or useful for humans.

Some of the Most Popular AI Libraries Include

- Intel offers a variety of AI libraries, including:
  - oneAPI [23] is a cross-platform programming model that provides a unified API for developing AI applications on Intel CPUs, GPUs, and FPGAs. Intel® oneAPI Math Kernel Library (oneMKL) is a high-performance math library that provides optimized implementations of common math functions for AI workloads. Intel oneAPI Deep Neural Network Library (oneDNN) is a high-performance deep learning library that provides optimized implementations of common deep learning operations for Intel CPUs and GPUs. Intel oneAPI

Table 3.1 AI frameworks and their inference solutions.

Framework	Library	Description	Hardware and architecture
TensorFlow	TensorFlow Serving [8]	A production-ready inference server for TensorFlow models.	CPU, GPU, FPGA
	TensorFlow Lite [9]	A framework for running TensorFlow models on mobile and embedded devices.	ARM, x86, RISC-V
PyTorch	TorchServe [10]	A production-ready inference server for PyTorch models.	CPU, GPU, FPGA
	TorchScript [11]	A compiler for PyTorch models that converts them to a portable and efficient format.	CPU, GPU, FPGA
	TorchMobile [12]	A framework for running PyTorch models on mobile devices.	ARM, x86, RISC-V
ONNX Runtime	ONNX Runtime [13]	An open-source inference engine that supports a variety of frameworks, including TensorFlow, PyTorch, and Core ML.	CPU, GPU, FPGA
Chainer	Chainer [14]	A deep learning framework developed by Preferred Networks.	CPU, GPU
Keras	Keras [15]	A high-level API for building and training deep learning models with TensorFlow, PyTorch, and other frameworks.	CPU, GPU, FPGA
MXNet	MXNet Serving [16]	A production-ready inference server for MXNet models.	CPU, GPU, FPGA
	MXNet Gluon Inference [16]	A lightweight library for running MXNet models on mobile and embedded devices.	ARM, x86, RISC-V
Caffe2	Caffe2 Serving [17]	A production-ready inference server for Caffe2 models.	CPU, GPU, FPGA
	Caffe2 Mobile [17]	A framework for running Caffe2 models on mobile devices.	ARM, x86, RISC-V
Intel OpenVINO	OpenVINO Runtime [18]	An open-source inference engine that supports a variety of Intel hardware platforms, including CPUs, GPUs, FPGAs, and the Intel Neural Compute Stick 2 (NCS2).	Intel CPU, Intel GPU, Intel FPGA
	OpenVINO Model Optimizer [18]	A tool for converting trained models from popular frameworks (e.g., TensorFlow, PyTorch, Caffe2) to the OpenVINO Intermediate Representation (IR) format.	Intel CPU, Intel GPU, Intel FPGA
	OpenVINO Deployment Tools [18]	A set of tools for deploying OpenVINO models to a variety of platforms, including edge devices and 5G networks.	Intel CPU, Intel GPU, Intel FPGA
NVIDIA TensorRT	TensorRT [19]	A high-performance inference optimizer and runtime for deep learning models.	NVIDIA GPU
	TensorRT Inference Server [19]	A production-ready inference server for TensorRT models.	NVIDIA GPU
	TensorRT Converter [19]	A tool for converting trained models from popular frameworks (e.g., TensorFlow, PyTorch, Caffe2) to the TensorRT Engine format.	NVIDIA GPU

Table 3.1 (Continued)

Framework	Library	Description	Hardware and architecture
AMD ROCm ML	ROCm ML Runtime [20]	An open-source inference engine that supports a variety of AMD hardware platforms, including GPUs and FPGAs.	AMD GPU, AMD FPGA
	ROCm ML Converter [20]	A tool for converting trained models from popular frameworks (e.g., TensorFlow, PyTorch, Caffe2) to the ROCm ML Engine format.	AMD GPU, AMD FPGA
Apple Core ML	Core ML Model [21]	A framework for running machine learning models on Apple devices.	Apple A-series CPU, Apple M-series CPU, Apple Neural Engine
	Core ML Tools [21]	A set of tools for converting and deploying machine learning models to Apple devices.	Apple A-series CPU, Apple M-series CPU, Apple Neural Engine
ARM NN	ARM NN SDK [22]	Machine learning (ML) inference engine for Android and Linux	Arm Cortex-A CPUs and Arm Mali GPUs

Data Analytics Library (oneDAL) is a high-performance data analytics library that provides optimized implementations of common data analytics operations for Intel CPUs and GPUs.

- NVIDIA offers a variety of AI libraries, including:
  - CUDA [24] is a parallel computing platform and programming model that enables developers to accelerate AI applications on NVIDIA GPUs. cuDNN is a high-performance deep learning library for NVIDIA GPUs. TensorRT is an inference engine that optimizes TensorFlow, PyTorch, and ONNX models for deployment on NVIDIA GPUs. RAPIDS is a collection of GPU-accelerated libraries for data science tasks such as data preparation, machine learning, and graph analytics.
- AMD offers a variety of AI libraries, including:
  - ROCm [20] is an open-source software platform for parallel computing on AMD GPUs. MIOpen is a high-performance deep learning library for AMD GPUs. HIP is a programming model that enables developers to port CUDA applications to AMD GPUs. RadeonML is a software development kit that provides tools and libraries for developing AI applications on AMD GPUs.
- Scikit-learn [25] is a free, open-source Python library for machine learning. It provides a wide range of algorithms for classification, regression, clustering, and other machine-learning tasks. Scikit-learn is known for its ease of use and its well-documented code. It is also one of the most popular AI libraries in the world, with over 1 million users.
- NumPy [26] is a free, open-source Python library for scientific computing. It provides a variety of functions for working with multidimensional arrays. NumPy is used in a wide range of scientific and engineering applications, including machine learning. NumPy is known for its speed and efficiency, and it is widely used in AI libraries such as scikit-learn and TensorFlow.

- Pandas [27] is a free, open-source Python library for data analysis and manipulation. It provides a variety of functions for working with data frames, which are two-dimensional data structures with labeled columns. Pandas is known for its ease of use and its powerful data manipulation capabilities. It is widely used in AI libraries such as scikit-learn and TensorFlow for data preprocessing and feature engineering.
- OpenCV [28] is a free, open-source library for computer vision and image processing. It provides a variety of functions for working with images and videos, including image processing, feature detection, and object tracking. OpenCV is widely used in AI libraries such as TensorFlow for computer vision tasks such as image classification and object detection.
- TensorFlow Hub [29] is a repository of pretrained TensorFlow models. These models can be used for a variety of AI tasks such as image recognition, natural language processing, and machine translation. TensorFlow Hub is known for its wide range of pretrained models, which can be used to build AI applications quickly and easily.

Here are some examples of how these libraries can be used together to build AI applications:

- A developer could use NumPy to load and preprocess a dataset of images.
- Then, the developer could use scikit-learn to train a machine learning model to classify the images.
- Finally, the developer could use TensorFlow Hub to deploy the trained model to a web service or mobile app.

This is just one example of how AI libraries can be used together to build AI applications. AI libraries are a powerful tool that can be used to build a wide variety of AI applications, and they are essential for any developer who wants to work with machine learning.

3.2.3.1 MLIR in the AI Domain

A special mention is needed to MLIR [30, 31] as its importance is growing in the AI compiler library development domain. MLIR (Multi-Level Intermediate Representation) is a compiler infrastructure project that provides a modern toolkit for building domain-specific compilers. It is highly influenced by LLVM and can be used to optimize performance and memory consumption of AI models. It is well-suited for the AI domain, as it can be used to represent and transform.

3.2.3.1.1 AI Models at Multiple Levels of Abstraction One of the key benefits of using MLIR for AI is that it allows for more efficient compilation. MLIR's compiler passes can be used to optimize AI models for specific hardware platforms, considering the unique capabilities and constraints of each platform. This can lead to significant performance improvements, especially for complex AI models that are deployed on resource-constrained devices.

Another benefit of MLIR is that it makes it easier to develop new AI compilers and frameworks. MLIR's extensible design allows developers to create custom dialects that represent specific AI models or frameworks. This makes it possible to build compilers and frameworks that are tailored to the specific needs of different AI applications. See Figure 3.2 for how MLIR is used to lower the models from the frameworks to the hardware.

3.2.3.1.2 Use Cases for MLIR in AI MLIR is being used in a variety of ways in the AI domain. Some examples include:

- **Compiling AI Models for Heterogeneous Hardware**: MLIR can be used to compile AI models for heterogeneous hardware platforms, such as CPUs, GPUs, and TPUs. This can help improve the performance and efficiency of AI applications.

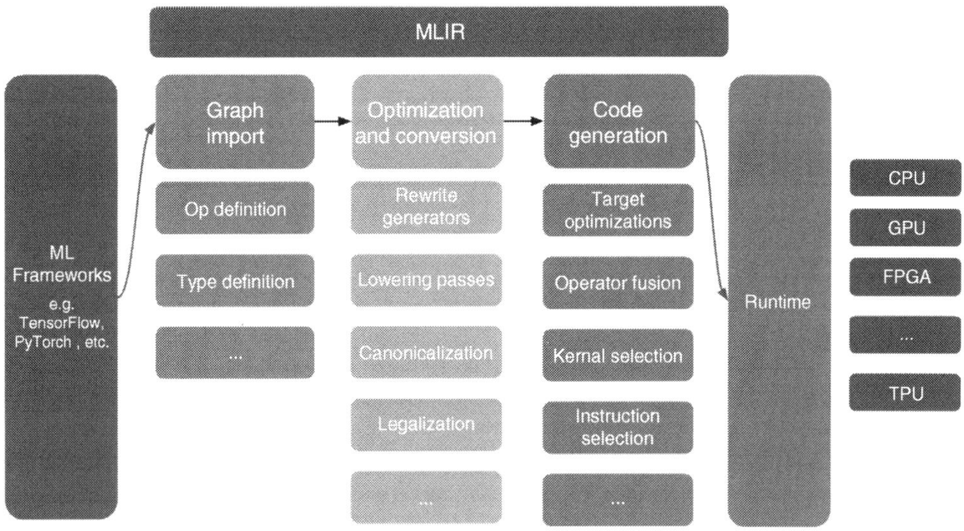

Figure 3.2 MLIR workflow from frameworks to hardware [31]. Source: https://www.tensorflow.org/mlir/overview/TensorFlow/CC-BY-4.0/Last accessed on May 05.

- **Developing New AI Compilers and Frameworks**: MLIR can be used to develop new AI compilers and frameworks that are tailored to the specific needs of different AI applications.
- **Optimizing AI Models for Specific Hardware Platforms**: MLIR compiler passes can be used to optimize AI models for specific hardware platforms, taking into account the unique capabilities and constraints of each platform.
- **Improving the Performance and Efficiency of AI Applications**: MLIR can be used to improve the performance and efficiency of AI applications by compiling them to efficient machine code and optimizing them for specific hardware platforms.

3.2.4 AI Processors

AI processors are specialized hardware designed to accelerate AI workloads. They typically include multiple cores and a large amount of memory in the cloud segment and the smaller variants on the edge. The cores are used to execute the AI model, and the memory is used to store the data and the model parameters. AI processors can significantly improve the performance of AI applications.

Examples of AI processors include:

- NVIDIA GPUs
- Intel and AMD CPUs, GPUs
- Google TPUs
- Examples of low-power AI processors used by mobile device manufacturers and 5G Network providers include:
  - **Apple**: Neural Engine
  - **Qualcomm**: Hexagon Neural Engine DSP
  - **MediaTek**: NPU
  - **HiSilicon**: Kirin NPU
  - **Broadcom**: Jericho3-AI

Example use cases:

Here are some specific examples of how AI platforms, frameworks, libraries, and processors are being used in the industry:

- **Retail**: AI platforms are being used to develop chatbots that can provide customer service and support. AI frameworks are being used to develop personalized recommendations, predict demand, and optimize inventory levels. AI libraries are being used to develop fraud detection systems and product image recognition systems. AI processors are being used to accelerate the performance of these AI applications.
- **Health Care**: AI platforms are being used to develop diagnostic tools, treatment plans, and virtual assistants. AI frameworks are being used to develop risk assessment models and to identify patterns in patient data. AI libraries are being used to develop medical imaging analysis systems and to develop natural language processing systems for processing medical records. AI processors are being used to accelerate the performance of these AI applications.
- **Industrial**: AI platforms are being used to develop predictive maintenance systems, quality control systems, and supply chain management systems. AI frameworks are being used to develop robotic systems and to automate complex tasks. AI libraries are being used to develop machine vision systems and natural language processing systems for processing industrial data. AI processors are being used to accelerate the performance of these AI applications.
- **Safety Monitoring**: AI can be used to monitor video footage from security cameras to detect potential safety hazards. This can help prevent accidents and injuries.

3.2.4.1 Conclusion

AI platforms, frameworks, libraries, and processors are essential for developing AI applications. They provide the tools and infrastructure that are necessary to build and deploy AI applications in a wide range of industries.

3.2.4.2 Specialty ASIC Hardware for AI Training and Inference Workloads

Application-specific integrated circuits (ASICs) are chips that are designed for a specific purpose. ASICs for AI training and inference are becoming increasingly popular because they can offer significantly higher performance and efficiency than other types of hardware.

Here are some examples of specialty ASIC hardware for AI training and inference. Note that this is a fast-changing landscape and there are several similar products that are being developed and designed besides the ones listed below.

3.2.4.2.1 Google TPUs TPUs [32] stand for tensor processing units and are custom-designed ASICs developed by Google specifically for machine learning and AI.

3.2.4.2.2 Specs

- **TPU v4 Pod**: 4096 TPU v4 chips, 1024GB HBM2e memory
- **TPU v4**: 2048 TPU v4 cores, 64GB HBM2e memory
- **TPU v3 Pod**: 4096 TPU v3 chips, 64GB HBM2 memory
- **TPU v3**: 2048 TPU v3 cores, 64GB HBM2 memory
- **Software Stack**: TensorFlow, PyTorch, Jax, JAXLib
- **AI Frameworks**: TensorFlow, PyTorch, Chainer, Sonnet

Product Line-up: Google TPUs are available in a variety of forms, including:

- **TPU Pods**: TPU Pods are the largest and most powerful TPU systems, and they are typically used by large enterprises and research organizations.
- **TPU v4**: The TPU v4 is the latest generation of TPU chips, and it is available as a standalone chip or as part of a TPU Pod.
- **TPU v3**: The TPU v3 is the previous generation of TPU chips, and it is still available as a standalone chip or as part of a TPU Pod.

3.2.4.2.3 Intel Habana Gaudi Intel Habana Gaudi [33] is an AI ASIC that is designed for training and deploying large language models. It is available to the public in two variants, Gaudi and Gaidi2.

Specifications:

- Intel Gaudi2 is built with 7nm process technology, Heterogeneous compute, 24 Tensor Processor Cores, Dual matrix multiplication engines, 24 100 Gigabit Ethernet integrated on chip, 96 GB HBM2E memory on board, 48 MB SRAM, Integrated Media Control
- It is designed to accelerate TensorFlow and PyTorch, and it can provide up to 40x performance improvement over CPUs for TensorFlow and PyTorch workloads.

Software Stack:

- Intel Gaudi is supported by the TensorFlow and PyTorch frameworks. Supports its SynapseAI SW stack.
- Intel Gaudi can also be used with other frameworks, such as MXNet, through third-party libraries.

AI Frameworks Supported: TensorFlow, PyTorch, MXNet, and others.

3.2.4.2.4 Cerebras The Cerebras [34] CS-2 is a wafer-scale AI accelerator that is designed for training and deploying large language models. It has a massive 40 million transistors, 400,000 cores and 18 gigabytes of on-chip memory. It supports TensorFlow, PyTorch and has their homebrewed Cerebas SDK.

3.2.4.2.5 Graphcore IPU The Graphcore [35] IPU is a parallel processor that is designed for training and deploying complex AI models. It has a unique architecture that allows it to efficiently process large amounts of data in parallel. One of their products has 1,472 cores with 768 gigabytes of on-chip memory, making it one of the most popular AI hardware in the market. Similar to Cerebras, Graphcore supports TensorFlow and PyTorch and also has their homebrewed Poplar SDK.

3.2.4.2.6 Sambanova SN10 The Sambanova[36] SN10 is a cloud-scale AI accelerator that is designed for training and deploying large language models. It has a massive 1,024 cores and 128 gigabytes of on-chip memory. Sambanova HW supports Sambanova SDK and also popular AI frameworks like TensorFlow and PyTorch.

3.2.4.2.7 Benefits of Specialty ASIC Hardware for AI Specialty ASIC hardware for AI offers several benefits over other types of hardware, including:

- **Performance**: ASICs can offer significantly higher performance than other types of hardware, such as GPUs and CPUs. This is because ASICs are designed specifically for AI workloads.

- **Efficiency**: ASICs can also be more efficient than other types of hardware, meaning that they can consume less power and generate less heat.
- **Scalability**: ASICs can be scaled to meet the needs of large-scale AI workloads.

3.2.4.2.8 Use Cases for Specialty ASIC Hardware for AI Specialty ASIC hardware for AI is used in a variety of applications, including:

- **Training and Deployment of Large Language Models**: ASICs are ideal for training and deploying large language models, such as GPT-3 and LaMDA. These models require a massive number of computational resources, and ASICs can provide the performance and efficiency that are needed to train and deploy these models.
- **Computer Vision**: ASICs are also used for computer vision applications, such as image recognition and object detection. ASICs can provide the performance and efficiency that are needed to process large amounts of image and video data in real time.
- **Natural Language Processing**: ASICs are also used for natural language processing applications, such as machine translation and text summarization. ASICs can provide the performance and efficiency that are needed to process large amounts of text data in real time.

3.2.4.2.9 Conclusion Specialty ASIC hardware for AI is a rapidly growing market. As AI models become more complex and data-intensive, ASICs are becoming increasingly important for training and deploying these models.

If you are considering using ASIC hardware for AI, it is important to carefully evaluate your specific needs and requirements. There is a wide range of ASIC hardware available, and the best hardware for you will depend on your specific application.

3.2.4.3 Hardware Selection for AI Solutions for Cloud and Edge Use Cases

The hardware selection for AI solutions for cloud and edge use cases depends on a number of factors, including the type of AI algorithm being used, the amount of data being processed, and the latency requirements of the application. For example, if you are using a deep learning algorithm to process large amounts of video data, you will need a powerful GPU-accelerated server. If you are developing an edge application that needs to respond to real-time data, you will need a low-latency system with a small form factor.

Here are some of the key factors to consider when selecting hardware for AI solutions:

- **Type of AI Algorithm**: The type of AI algorithm you are using will dictate the type of hardware you need. For example, deep learning algorithms require GPUs, while natural language processing algorithms can be run on CPUs.
- **Amount of Data**: The amount of data you need to process will also affect your hardware selection. If you are working with large datasets, you will need a system with a lot of storage and processing power.
- **Latency Requirements**: If you are developing an application that needs to respond to real-time data, you will need a system with low latency. This is especially important for applications such as self-driving cars and medical imaging.
- **Cost**: The cost of hardware is also a factor to consider. AI solutions can be expensive, so you need to make sure you select the right hardware for your budget.

Once you have considered these factors, you can start to narrow down your choices. There are a number of different vendors that offer hardware for AI solutions, so you should do your research to find the best option for your needs (Tables 3.2 and 3.3).

Table 3.2 GPU Specifications and their importance.

GPU Specification [37]	Why Is It Important?
Memory	This determines how much data the GPU can store in memory. AI models can be very large, so a high memory size is important for ensuring that the GPU can store the entire model in memory.
Compute	This determines how fast the GPU can perform calculations on AI models. A higher compute performance will result in faster training and inference times.
Memory bandwidth	This determines how fast the GPU can access data from memory. AI models can be very large, so a high memory bandwidth is important for ensuring that the GPU can access the data it needs quickly.
Interconnects	GPUs can be interconnected to each other to increase the overall compute performance and memory bandwidth of the system. It is important to choose an interconnect that is appropriate for the specific requirements of the AI application.
TDP	Thermal design power (TDP) is the maximum amount of heat that the GPU can dissipate. It is important to choose a GPU with a TDP that is appropriate for the cooling system of the system in which it will be installed.
Precision	GPUs can perform calculations using different precision levels, such as FP32, FP16, and INT8. It is important to choose a precision level that is appropriate for the specific requirements of the AI application.
Decoders	GPUs can decode video and image formats. This can be useful for AI applications that need to process video or image data.
Form factors	GPUs come in a variety of form factors, such as PCIe cards and embedded modules. It is important to choose a form factor that is compatible with the system in which the GPU will be installed.
Boost clock	The boost clock is the maximum clock speed that the GPU can achieve. It is important to note that the GPU may not always operate at the boost clock, depending on the workload and the cooling system.
CUDA cores	CUDA cores are the processing units in the GPU that are used to perform calculations on AI models. A higher number of CUDA cores will result in faster training and inference times.
Tensor cores	Tensor cores are specialized processing units in the GPU that are designed to accelerate matrix multiplication operations. This can significantly speed up training and inference for AI models that use matrix multiplication extensively.
Support for AI frameworks	It is important to choose a GPU that is supported by the AI frameworks that you will be using.
TOPS/W	Metrics are used to measure the performance of AI hardware. It stands for Tera Operations per Second per Watt, and it measures how many trillion operations per second an AI accelerator can perform per watt of power.

Table 3.3 CPU Specifications and their importance.

CPU Specification [38]	Why Is It Important?
Core count	This determines the number of threads that the CPU can execute simultaneously. A higher core count will result in faster training and inference times for AI applications that can be parallelized.
Clock speed	This determines how fast the CPU can perform calculations. A higher clock speed will result in faster training and inference times for AI applications.
Cache size	The cache is a high-speed memory that stores frequently accessed data. A larger cache can improve the performance of AI applications by reducing the number of times that the CPU needs to access memory.
Instruction set architecture (ISA)	The ISA is the set of instructions that the CPU can understand. It is important to choose a CPU with an ISA that is supported by the AI frameworks that you will be using.
Integrated graphics	Integrated graphics are a type of graphics processor that is built into the CPU. Integrated graphics can be used to accelerate AI applications that use computer vision or machine learning operations.
Power consumption	This determines how much power the CPU consumes. A lower power consumption is important for edge devices, where power is limited.

References

1 Saunders, A. (2022). Steps to getting started with Edge AI. https://developer.nvidia.com/blog/steps-to-getting-started-with-edge-ai/
2 Google (2024). AI and Machine Learning Products. https://cloud.google.com/products/ai.
3 Amazon AWS (2024). Easily add intelligence to your applications. https://aws.amazon.com/ai/services/
4 Microsoft Azure (2024). Welcome to the age of AI transformation. https://www.microsoft.com/en-us/ai.
5 IBM (2024). watsonx.ai. https://www.ibm.com/products/watsonx-ai.
6 NVIDIA (2024). The most advanced AI, Ready for Enterprise. https://www.nvidia.com/en-us/ai-data-science/
7 IBM (2024). All the Ops: DevOps, DataOps, MLOps, and AIOps. https://developer.ibm.com/articles/all-the-ops-devops-dataops-mlops-and-aiops/
8 Tensorflow (2024). Serving Models. https://www.tensorflow.org/tfx/guide/serving.
9 Tensorflow (2024). For Mobile and Edge. https://www.tensorflow.org/lite.
10 Pytorch (2024). TorchServe. https://pytorch.org/serve/
11 Pytorch (2024). TorchScript. https://pytorch.org/docs/stable/jit.html.
12 Pytorch (2024). Pytorch Mobile. https://pytorch.org/mobile/home/
13 OnnxRuntime (2024). Accelerated Machine Learning. https://onnxruntime.ai/
14 Chainer (maintenance mode) (2024). A Powerful, Flexible, and Intuitive Framework for Neural Networks. https://chainer.org/
15 Keras (2024). https://keras.io/
16 MXNet (Retired) (2024). https://mxnet.apache.org/versions/1.9.1/
17 Caffe2 (Deprecated) (2024). https://caffe2.ai/
18 OpenVino (2024). https://docs.openvino.ai/
19 NVIDIA (2024). TensorRT. https://developer.nvidia.com/tensorrt.

20 AMD RocM (2024). AMDRocM Documentation. https://rocm.docs.amd.com/en/latest/
21 Apple CoreML (2024). CoreML Documentation. https://developer.apple.com/documentation/coreml.
22 ARMNN (2024). ArmNN Github. https://github.com/ARM-software/armnn.
23 Intel (2024). OneAPI Documentation. https://www.intel.com/content/www/us/en/developer/articles/technical/oneapi-what-is-it.html.
24 NVIDIA (2024). Cuda toolkit. https://developer.nvidia.com/cuda-toolkit.
25 Scikit-learn (2024). Scikit-learn documentation. https://scikit-learn.org/stable/
26 Numpy (2024). Numpy documentation. https://numpy.org/
27 Pandas (2024). Pandas documentation. https://pandas.pydata.org/
28 OpenCV (2024). OpenCV Computer vision documentation. https://opencv.org/
29 Tensorflow (2024). TensorFlow Hub is a repository of trained machine learning models. https://www.tensorflow.org/hub.
30 Lattner, C. et al. (2020). MLIR: A Compiler Infrastructure for the End of Moore's Law. https://arxiv.org/abs/2002.11054.
31 Tensorflow MLIR (2024). https://www.tensorflow.org/mlir/overview.
32 Google, TPU v4 architecture (2024). https://cloud.google.com/tpu/docs/v4#system_architecture.
33 Intel Gaudi (2024). Deep Learning inference and training accelerator. https://habana.ai/products/gaudi/
34 Cerebras (2024). Revolutionary compute for Generative AI. https://www.cerebras.net/
35 Graphcore (2024). https://www.graphcore.ai/
36 Sambanova (2024). https://sambanova.ai/
37 NVIDIA (2024). GPU Performance user guide. https://docs.nvidia.com/deeplearning/performance/dl-performance-gpu-background/index.html.
38 David Mulnix, Intel (2024). Intel® Xeon® Processor Scalable Family Technical Overview. https://www.intel.com/content/www/us/en/developer/articles/technical/xeon-processor-scalable-family-technical-overview.html.

4

Model Optimization Techniques for Edge Devices

Yamini Nimmagadda

Network and Edge, Intel, Portland, OR, USA

Edge AI is a distributed computing paradigm in which machine learning models are deployed on edge devices, such as smartphones, smartwatches, and IoT devices. This allows for real-time inference and decision-making at the source of data, without the need to send data to a central cloud server. However, edge devices often have limited processing power and memory, that can make it challenging to deploy and run complex machine learning models. Model optimization [1, 2] is the process of reducing the size and complexity of a machine learning model without sacrificing its accuracy. This is essential for deploying machine learning models on edge.

4.1 Overview of Model Optimization Techniques

Model optimization techniques can be categorized into three buckets:

- **Predeployment**: Predeployment model optimization techniques are applied to an AI model before it is deployed to production. These techniques can be used to improve the performance, efficiency, and accuracy of the model.
- **Deployment-Time**: Deployment-time model techniques are applied to a machine learning model after it has been deployed to production. These techniques can be used to improve the performance and efficiency of the model on the target hardware platform.
- **Postdeployment**: Postdeployment model optimization techniques are applied to a machine learning model after it has been deployed to production. These techniques can be used to improve the performance and efficiency of the model and to ensure that it continues to meet the needs of the business.

4.1.1 Predeployment Optimization Techniques

Model optimization techniques are presented in Figure 4.1.

Predeployment optimization techniques are applied to the model before it is deployed to production. These techniques can be used to:

- **Model Architecture Selection**: Start with the right model architecture. Not all model architectures are created equal. Some model architectures are more efficient and easier to optimize than others. When choosing a model architecture, it is important to consider the specific requirements of the edge device [3].

Model Optimization Methods for Efficient and Edge AI: Federated Learning Architectures, Frameworks and Applications, First Edition. Edited by Pethuru Raj Chelliah, Amir Masoud Rahmani, Robert Colby, Gayathri Nagasubramanian, and Sunku Ranganath.

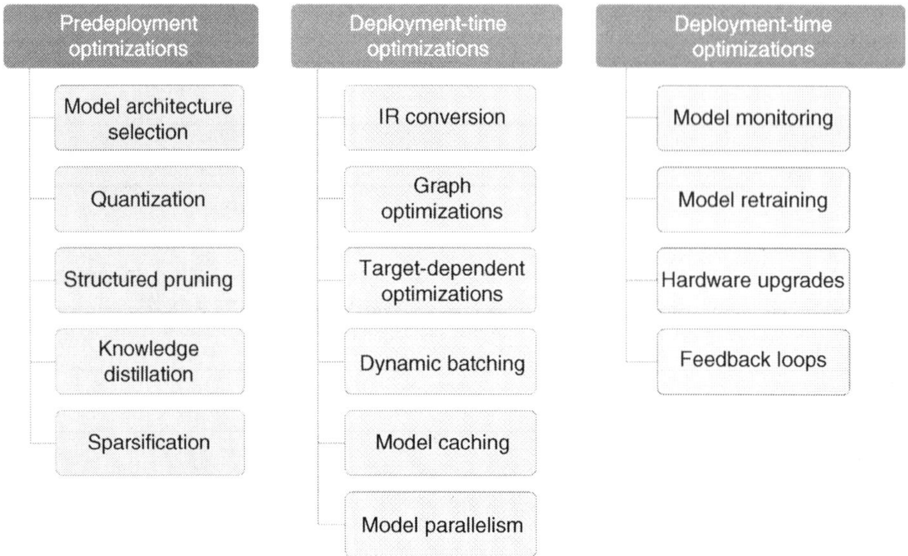

Figure 4.1 Model optimization techniques classified into predeployment, deployment-time, and postdeployment phases.

- **Quantization**: Quantization reduces the size of the model by reducing the precision of the model's weights and activations. This can be done without significantly impacting the accuracy of the model [4].
- **Structured Pruning**: Pruning removes unnecessary weights and connections from the model. This can reduce the size and complexity of the model and improve the inference speed [5].
- **Knowledge Distillation**: Knowledge distillation (KD) is a technique that can be used to transfer knowledge from a large and complex model to a smaller and simpler model. This can be used to reduce the size and complexity of the model while still maintaining a high level of accuracy [6].
- **Sparsification**: Sparsification is the process of making machine learning models sparser by removing or zeroing out certain model parameters, such as weights and neurons [7]. The goal of sparsification is to reduce the model's size and computational complexity while maintaining or even improving its performance.

4.1.2 Deployment-Time Optimization Techniques

Deployment-time optimization techniques are applied to the model while it is being deployed to production and during runtime. These techniques can be used to:

- **IR Conversion**: Intermediate representation (IR) is a format for representing machine learning models that is independent of the specific machine learning framework that the model was trained in. This makes it easier to optimize the model for different hardware platforms and deployment environments [8].
- **Graph Optimizations**: Graph optimizations include techniques like node fusion, redundant node removal, and constant folding to reduce the overall number of operations that need to be performed during inference, and improve the inference speed [9].
- **Target-Dependent Optimizations**: The model can be optimized for the specific hardware platform on which it is running. This can be done by using specialized instructions and libraries that are available for the target hardware [2].

- **Dynamic Batching**: Dynamic batching groups input data into batches of different sizes, depending on the availability of resources. This can help improve the performance of the model on hardware platforms that have limited resources [10].
- **Model Caching**: Model caching in AI inference for edge reduces inference latency and improves performance by storing pretrained machine learning models on edge devices for quick access and use [11].
- **Model Parallelism**: Model parallelism is a technique that can be used to distribute the workload of inference across multiple devices. This can be used to improve the performance of the model on hardware platforms that have multiple processors or hardware accelerators [12].

4.1.3 Postdeployment Optimization Techniques

Postdeployment model optimization techniques can work together to help improve the performance, reliability, and efficiency of edge AI deployments.

- **Model Monitoring**: Model monitoring helps identify model performance degradation and drift over time. It enables proactive measures to be taken to maintain model accuracy and reliability [13].
- **Model Retraining**: Model retraining can be used to update and improve deployed models with new data or to adapt them to changing environmental conditions. It can help maintain model accuracy and performance over time [14].
- **Hardware Upgrades**: In some cases, the hardware on which the model is deployed may be upgraded or replaced to enhance performance. For example, moving to more powerful GPUs or dedicated accelerators can significantly speed up inference.
- **Feedback Loops**: Collect feedback from end-users and stakeholders to identify areas where the model can be improved. This feedback can provide insights into user satisfaction and help uncover issues that might not be apparent through quantitative metrics alone.

4.1.4 Examples

The following examples illustrate how model optimization techniques can be used to improve the performance and efficiency of machine learning models on edge devices:

Example 4.1 A company is developing a mobile application that uses a machine learning model to classify images. The model is trained on a large dataset of images, and is deployed to the mobile application.

To reduce the size of the model and improve the inference speed, the company uses a predeployment optimization technique called quantization. Quantization reduces the size of the model by reducing the precision of the model's weights and activations. This can be done without significantly impacting the accuracy of the model.

After quantization, the model is deployed to the mobile application. To further improve the performance of the model on the mobile device, the company uses a runtime optimization technique called dynamic batching. Dynamic batching groups input data into batches of different sizes, depending on the availability of resources. This helps improve the performance of the model on hardware platforms that have limited resources.

As a result of these optimizations, the company is able to reduce the size of the model by 50% and improve the inference speed by 30%. This makes the model more suitable for deployment on mobile devices.

Example 4.2 A company is using a machine learning model to detect fraud in financial transactions. The model is trained on a large dataset of historical transactions and is deployed to production to detect fraud in real time.

To improve the performance of the model on edge devices, the company uses a predeployment optimization technique called pruning. Pruning removes unnecessary weights and connections from the model. This can reduce the size and complexity of the model, and improve the inference speed.

After pruning, the model is deployed to edge devices. To further improve the performance of the model on edge devices, the company uses a runtime optimization technique called model compression. Model compression techniques can be used to reduce the size of the model without significantly impacting the accuracy of the model.

As a result of these optimizations, the company is able to reduce the size of the model by 40% and improve the inference speed by 20%. This makes the model more suitable for deployment on edge devices.

4.2 Deep Dive of Predeployment Model Optimization Techniques

4.2.1 Model Architecture Selection

When selecting a model architecture for edge devices, there are a number of factors to consider, including:

- **Hardware Constraints**: Edge devices often have limited processing power and memory, which can constrain the types of models that can be deployed. It is important to choose a model architecture that is compatible with the hardware constraints of the target edge device.
- **Accuracy Requirements**: The accuracy requirements of the application will also dictate the choice of model architecture. More complex models tend to be more accurate, but they also require more resources to run. It is important to choose a model architecture that provides the required level of accuracy while meeting the hardware constraints of the target edge device.
- **Latency Requirements**: For some applications, latency is a critical factor. For example, real-time applications require models that can produce predictions very quickly. It is important to choose a model architecture that can meet the latency requirements of the application.
- **Model Size**: The size of the model is also a factor to consider, especially for edge devices with limited storage space. It is important to choose a model architecture that is small enough to fit on the target edge device.

In addition to these factors, it is also important to consider the specific requirements of the application. For example, if the application needs to be able to handle a wide variety of inputs, then a more complex model may be required.

Here are some general guidelines for selecting a model architecture for edge devices:

- **Start with a Simple Architecture**: Complex models are often more accurate, but they also require more resources to run. It is best to start with a simple architecture and then add complexity as needed.
- **Use Pretrained Models**: There are a number of pretrained models available for a variety of tasks. Using a pretrained model can save time and effort, and it can also lead to better performance.

- **Use Transfer Learning**: Transfer learning is a technique that allows you to use a pretrained model as a starting point for training your own model. This can be a good way to get good performance on edge devices without having to train a model from scratch.

Here are some specific model architectures that are well-suited for edge devices:

4.2.1.1 Vision Model Architectures

- **MobileNets**: MobileNets [15] are a family of lightweight convolutional neural networks that are designed for mobile devices. MobileNets are very efficient and can be deployed on edge devices with limited resources.
- **SqueezeNets**: SqueezeNets [16] are another family of lightweight convolutional neural networks that are designed for mobile devices. SqueezeNets are even more efficient than MobileNets, but they may sacrifice some accuracy.
- **ShuffleNets**: ShuffleNets [17] are a family of convolutional neural networks that are designed for mobile devices. ShuffleNets are a good compromise between efficiency and accuracy.
- **EfficientNets**: EfficientNets [18] are a family of convolutional neural networks that are designed to be both efficient and accurate. EfficientNets are a good choice for edge devices that have more resources available.

4.2.1.2 Lightweight Transformers

Lightweight transformers are a type of transformer model that is designed to be efficient and run on edge devices. This makes them suitable for a variety of applications, such as real-time image classification, object detection, and natural language processing.

One of the key challenges in developing lightweight transformers is that the self-attention mechanism, which is the core component of transformers, is computationally expensive. However, there have been a number of advances in recent years that have made it possible to develop lightweight transformers that are still able to achieve good performance.

Some of the techniques that have been used to develop lightweight transformers include:

- **Reducing the Number of Self-Attention Heads**: Self-attention heads are used to learn different representations of the input. Reducing the number of self-attention heads can reduce the computational cost of self-attention without sacrificing too much accuracy.
- **Using Efficient Self-Attention Algorithms**: There are a number of efficient self-attention algorithms that have been developed in recent years. These algorithms can significantly reduce the computational cost of self-attention without sacrificing accuracy.
- **Using Hierarchical Self-Attention**: Hierarchical self-attention is a technique that allows you to focus on different levels of detail in the input. This can reduce the computational cost of self-attention while still allowing the model to learn long-range dependencies.

Here are some examples of lightweight transformers that are suitable for edge devices:

- **Swin Transformer**: Swin Transformer [19] is a hierarchical transformer model that is designed for image classification. Swin Transformer has achieved state-of-the-art results on a number of image classification benchmarks while being very efficient.
- **MobileViT**: MobileViT [20] is a lightweight transformer model that is designed for mobile devices. MobileViT has achieved state-of-the-art results on a number of mobile benchmarks while being very efficient.
- **TinyViT**: TinyViT [21] is a very lightweight transformer model that is designed for edge devices with limited resources. TinyViT has achieved good results on a number of image classification and natural language processing benchmarks while being very efficient.

- **DistilBERT**: DistilBERT [22] is a lightweight version of the BERT LLM from Google AI. DistilBERT is trained to be efficient and run on edge devices, and it can be used for a variety of tasks, such as text classification, natural language inference, and question answering.
- **TinyBERT**: TinyBERT [23] is a very lightweight version of the BERT LLM from Google AI. TinyBERT is designed to be efficient and run on edge devices with limited resources, and it can be used for a variety of tasks, such as text classification, question answering, and summarization.

4.2.1.3 Lightweight LLMs

Here are some examples of lightweight large language models (LLMs) that may be suitable for edge devices:

- **TinyChat**: TinyChat [24] is an efficient and lightweight system for LLMs on the edge. It seamlessly supports state-of-the-art LLMs on resource-constrained NVIDIA Jetson Orin, and runs the popular LLaMA-2-7B model from Meta at 30 tokens/sec on the edge.
- **TinyLLAMA**: TinyLLAMA [25] is a generative AI model based on Llama model that is designed to be efficient and run on edge devices. TinyLLAMA is available as a 1.1B parameter model that is suitable for edge devices.
- **Gecko**: Gecko [26] is a lightweight version of the PaLM 2 LLM from Google AI. Gecko is designed to be efficient and run on mobile devices, and it can be used for a variety of tasks, such as text generation, translation, and code generation.

Ultimately, the best model architecture for a particular edge device will depend on the specific requirements of the application. However, the guidelines and model architectures provided in this chapter can help you select a good starting point.

4.2.2 Quantization

Quantization is a technique used to reduce the size and improve the performance of machine learning models by converting the model's weights and activations to lower-precision values, such as integers. This can be done without significantly impacting the accuracy of the model. Quantization is particularly important for edge devices, which have limited memory and computational resources. By quantizing a model, developers can deploy it to edge devices without sacrificing performance or accuracy. Benefits of quantization for edge include:

- **Reduced Model Size**: Quantized models are typically much smaller than floating-point models. This can be important for edge devices, which have limited storage space.
- **Improved Inference Speed**: Quantized models can be inferred much faster than floating-point models. This is because integer operations are faster than floating-point operations on most hardware platforms.
- **Reduced Power Consumption**: Quantized models consume less power than floating-point models. This is because integer operations are more energy-efficient than floating-point operations.

Quantization results in the below challenges that need to be overcome for successful deployments:

- **Accuracy Degradation**: Quantization can lead to accuracy degradation if it is not done carefully. It is important to choose the right quantization technique and parameters to minimize the impact on accuracy.

- **Hardware Support**: Quantization is not supported by all hardware platforms. It is important to check that the target hardware platform supports quantization before deploying a quantized model.

There are two main approaches to quantizing a model for edge: post-training quantization (PTQ) and quantization-aware training (QAT).

- PTQ quantizes a model after it has been trained. The weights and activations of the model are first converted to integer values, and then the model is fine-tuned to improve its accuracy.
- QAT trains a model with quantization in mind. During training, the model's weights and activations are quantized and then dequantized to floating-point values before being used in the forward and backward passes. This allows the model to be trained to retain its accuracy even after quantization.

PTQ is a quantization technique where a model is quantized after it has been trained. PTQ is relatively simple to implement, but it can lead to a decrease in model accuracy, especially for models with high precision requirements (Figure 4.2).

QAT is a quantization technique where a model is trained from the start using quantized weights and activations. QAT is more complex to implement than PTQ, but it can provide better accuracy an PTQ, especially for models with high precision requirements.

Table 4.1 summarizes the key differences between PTQ and QAT:

The best quantization approach to choose depends on the specific requirements of the application. If accuracy is the primary concern, then QAT is the better option. However, if performance is the primary concern and accuracy is not as critical, then PTQ may be a good option.

Other factors to consider when choosing a quantization approach include:

- **Model Complexity**: QAT is generally more efficient for complex models.
- **Inference Latency**: QAT can add latency to inference.
- **Hardware Support**: Not all hardware platforms support QAT.

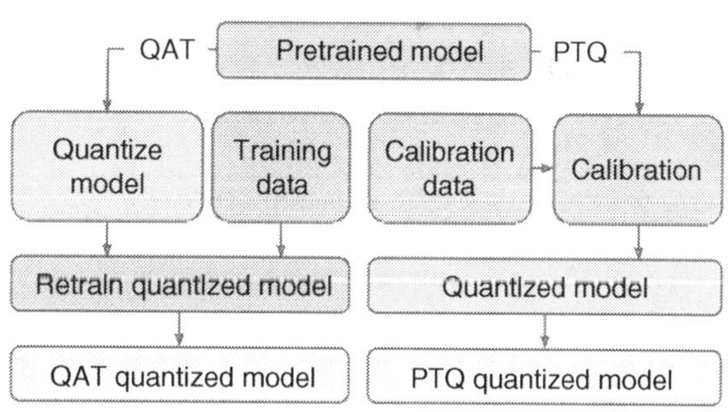

Figure 4.2 Quantization-aware training and post-training quantization overview. Source: Weng et al. [27]/arXiv/CC-BY-4.0.

Table 4.1 Comparison overview of post-training quantization and quantization-aware training. Source: Zhao et al. [28].

Characteristic	PTQ	QAT
When quantization is applied	After training	During training
Implementation complexity	Easier	More difficult
Accuracy impact	Can lead to a decrease in accuracy	Can provide better accuracy
Best suited for	Models with low precision requirements	Models with high precision requirements

PTQ and QAT are both effective techniques for optimizing the performance and memory footprint of deep learning models. The best approach to choose depends on the specific requirements of the application [28].

Quantization can use the below approaches:

- **Static Quantization**: Static quantization is a quantization technique where the weights and activations of a model are quantized offline, before deployment. This is done by first calibrating the model on a representative dataset to determine the optimal quantization parameters. Once the model is calibrated, the weights and activations are quantized using these parameters. Static quantization is relatively simple to implement and can provide significant performance and memory savings. However, it is important to note that static quantization can lead to a decrease in model accuracy, especially for models with high precision requirements.
- **Dynamic Quantization**: Dynamic quantization is a quantization technique where the activations of a model are quantized on-the-fly, during inference. This is done by first calculating the min and max values of the activations for each layer of the model. The activations are then quantized using these min and max values. Dynamic quantization is more complex to implement than static quantization, but it can provide better accuracy than static quantization, especially for models with high precision requirements. However, dynamic quantization can also lead to a decrease in performance, due to the overhead of quantizing the activations on-the-fly.
- **Hybrid Quantization**: Hybrid quantization combines elements of both static and dynamic quantization. For example, the weights of a model can be quantized statically, while the activations are quantized dynamically.

The best quantization approach to choose depends on the specific requirements of the application. If performance is the primary concern, then static quantization is a good option. However, if accuracy is the primary concern, then dynamic quantization is a better option. Hybrid quantization can provide a good balance between performance, accuracy, and latency. However, it is more complex to implement than either static or dynamic quantization. Static quantization quantizes a model once before it is deployed. This is the simplest type of quantization, but it can lead to accuracy degradation if the model is deployed on hardware with different characteristics than the hardware used for quantization. Dynamic quantization quantizes a model at runtime, based on the characteristics of the target hardware. This can help improve the accuracy of the model on different hardware platforms, but it can also add complexity to the model.

Below are some example tools that provide quantization support:

- **TensorFlow Quantization**: TensorFlow Quantization [29] is a library that provides tools for quantizing TensorFlow models. It supports both PTQ and QAT quantization.
- **PyTorch Quantization**: PyTorch Quantization [30] is a library that provides tools for quantizing PyTorch models. It supports both PTQ and QAT quantization.
- **OpenVINO NNCF**: OpenVINO NNCF [31] is a compiler that can optimize models for Intel hardware. It supports quantization, as well as other optimizations.
- **ONNX Runtime**: ONNX Runtime is a cross-platform inference engine that supports a variety of machine learning frameworks. It supports quantization for TensorFlow and PyTorch models [32].

Example quantization techniques

- **Quantization-Aware Training (QAT)**: QAT is a powerful quantization technique that can achieve high accuracy with minimal performance loss. However, it can be difficult to implement and tune.

- **Post-Training Quantization (PTQ)**: PTQ is a simpler quantization technique that is easier to implement and tune. However, it may not achieve the same level of accuracy as QAT.
- **Symmetric Quantization**: Symmetric quantization represents both positive and negative values using a single set of quantization levels. This is the simplest type of quantization, but it can lead to accuracy degradation for models with a wide range of values.
- **Asymmetric Quantization**: Asymmetric quantization uses two sets of quantization levels, one for positive values and one for negative values. This can improve the accuracy of quantized models, but it can also add complexity to the model.

4.2.3 Structured Pruning

Structured pruning [33] is a technique for pruning neural networks by removing entire neurons, channels, or layers at a time, rather than pruning individual weights. This approach can be more effective in reducing the size and complexity of neural networks while still maintaining a high level of accuracy.

Structured pruning is particularly well-suited for edge devices, which have limited memory and computational resources. By pruning a neural network, developers can reduce the size of the model and improve the inference speed on edge devices.

Examples of structured pruning techniques

- **Channel Pruning**: Channel pruning removes entire channels from a convolutional neural network. This can be done by pruning channels that are less important for the accuracy of the model.
- **Filter Pruning**: Filter pruning removes individual filters from a convolutional neural network. This can be done by pruning filters that are less important for the accuracy of the model.
- **Neuron Pruning**: Neuron pruning removes entire neurons from a neural network. This can be done by pruning neurons that are less important for the accuracy of the model.
- **Layer Pruning**: Layer pruning removes entire layers from a neural network. This can be done by pruning layers that are less important for the accuracy of the model.

Examples of industry tool chains

There are a number of industry tool chains that support structured pruning for edge devices. Some popular examples include:

- **TensorFlow Model Optimization**: TensorFlow Model Optimization is a library that provides tools for optimizing TensorFlow models for edge devices. It supports structured pruning, as well as other optimizations.
- **PyTorch Model Optimization**: PyTorch Model Optimization is a library that provides tools for optimizing PyTorch models for edge devices. It supports structured pruning as well as other optimizations.
- **OpenVino Model Optimization Toolkit**: The OpenVino Model Optimization Toolkit is a tool that provides tools for optimizing models for Intel hardware. It supports structured pruning as well as other optimizations.
- **NVIDIA TensorRT**: NVIDIA TensorRT is an inference engine that provides tools for optimizing and deploying models on NVIDIA GPUs. It supports structured pruning as well as other optimizations.

Structured pruning is a powerful technique for reducing the size and improving the performance of neural networks on edge devices. There are a number of industry tool chains that support structured pruning for edge devices.

There are a number of benefits of using structured pruning for edge devices:

- **Reduced Model Size**: Structured pruning can significantly reduce the size of neural networks. This is important for edge devices, which have limited storage space.
- **Improved Inference Speed**: Structured pruning can improve the inference speed of neural networks. This is important for edge devices, which need to respond to requests in real time.
- **Reduced Power Consumption**: Structured pruning can reduce the power consumption of neural networks. This is important for edge devices, which need to conserve battery power.

There are a few challenges to using structured pruning for edge devices:

- **Accuracy Loss**: Structured pruning can lead to accuracy loss. It is important to choose the right pruning technique and parameters to minimize the impact on accuracy.
- **Hardware Support**: Structured pruning is not supported by all hardware platforms. It is important to check that the target hardware platform supports structured pruning before deploying a pruned model.

Overall, structured pruning is a powerful technique for improving the performance and efficiency of neural networks on edge devices. By carefully choosing the right pruning technique and parameters, developers can deploy their models to edge devices without sacrificing performance or accuracy.

4.2.4 Knowledge Distillation

KD [34, 35] is a model compression technique that can be used to train a smaller, faster, and more efficient model, called the student model, to mimic the behavior of a larger, more accurate model, called the teacher model. KD is particularly well-suited for edge computing applications where resources are constrained and latency is critical (Figure 4.3).

KD works by training the student model to predict the soft labels of the teacher model. Soft labels are probability distributions over the possible output classes, rather than hard labels, which are

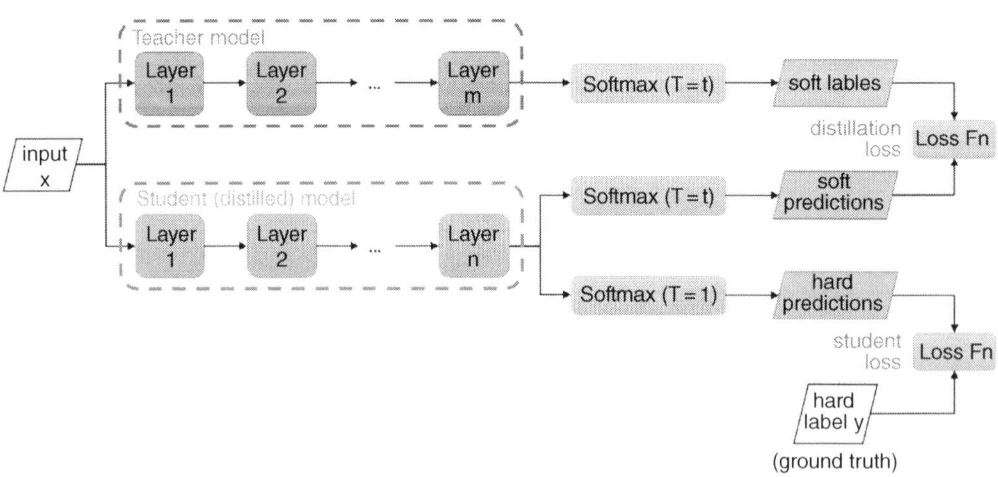

Figure 4.3 Knowledge Distillation Overview [36]. Source: https://intellabs.github.io/distiller/knowledge_distillation.html/with permission of GitHub, Inc.

simply the most likely output class. KD is typically implemented using a two-stage process:

1) Train the teacher model: The teacher model is trained on a large dataset to achieve high accuracy.
2) Train the student model: The student model is trained on a smaller dataset, using the soft labels of the teacher model as its training targets.

During the training of the student model, the following loss function is typically used:

```
loss = L_soft(student_output, teacher_output) + L_hard(student_output,
ground_truth)
```

where:

- L_soft() is the soft loss function, which measures the difference between the soft labels of the student and teacher models.
- L_hard() is the hard loss function, which measures the difference between the hard labels of the student model and the ground truth labels.
- student_output is the output of the student model.
- teacher_output is the output of the teacher model.
- ground_truth is the ground truth label.

KD offers a number of benefits for edge computing applications including:

- **Model Compression**: KD can be used to significantly compress the size of machine learning models, making them more suitable for deployment on edge devices.
- **Efficiency Improvements**: KD can also lead to efficiency improvements in terms of inference latency and power consumption.
- **Improved Robustness**: KD can also improve the robustness of machine learning models to noise and adversarial attacks.

Despite its many benefits, KD also faces some challenges for edge computing applications, including:

- **Training Complexity**: Training a student model using KD can be more complex than training a model from scratch, as it requires the training of a teacher model.
- **Accuracy Loss**: KD can also lead to a decrease in model accuracy, especially if the student model is too small or the hyperparameters are not tuned correctly.

In recent years, there has been a significant amount of research on improving KD for edge computing applications. Some of the notable advances in this area include:

- **Model-agnostic KD**: Model-agnostic KD methods can be used to distill knowledge from any teacher model to any student model, regardless of their architecture. This makes them more flexible and adaptable than traditional KD methods.
- **Progressive KD**: Progressive KD methods train the student model in stages, gradually increasing the difficulty of the training data. This can help improve the accuracy and robustness of the student model.
- **Quantization-aware KD**: Quantization-aware KD methods train the student model to use quantized weights and activations. This can further reduce the size and improve the efficiency of the student model.

KD is a powerful technique for training smaller, faster, and more efficient machine learning models without sacrificing too much accuracy. It is particularly well-suited for edge computing applications, where resources are constrained and latency is critical.

4.2.5 Sparsification

Sparsification [37, 38] for edge deployment is a technique used to optimize and deploy machine learning models on resource-constrained edge devices such as smartphones, IoT devices, and edge servers. The goal is to make these models more compact and computationally efficient while preserving their accuracy and functionality. Sparsification reduces the number of nonzero parameters in a machine learning model. This can be done by pruning out small and insignificant parameters, or by quantizing parameters to lower precision values, or by a combination of various techniques described below.

- **Pruning**: A variety of pruning techniques can be applied to make the model sparse. Weight pruning involves removing or zeroing out model parameters (weights) with low magnitudes, typically applied to neural network layers. Structured Pruning removes entire neurons, channels, or layers from the model, rather than individual weights, for more aggressive model size reduction. Magnitude-based Pruning prunes weights with magnitudes below a certain threshold, often applied iteratively to achieve the desired level of sparsity. Iterative Pruning is used to finetune the model after each round of pruning to recover any loss in performance, and this process can be repeated for higher sparsity.
- **Dynamic Sparsity**: Dynamic sparsity is a technique used in machine learning and deep learning to adapt the sparsity level of a neural network or model dynamically based on the input data. In dynamic sparsity, the model changes its level of sparsity on the fly during inference, adjusting to the specific characteristics of the data being processed. This approach aims to reduce computational complexity and resource usage when it is feasible while maintaining model accuracy. Dynamic sparsity is data-dependent, meaning it takes into account the nature of the input data. Dynamic sparsity allows the model to become sparser or denser depending on the input. When the data allows for it, the model can be made sparser to reduce computational demands. For instance, in an image recognition task, if the input image contains a simple background with no complex features, dynamic sparsity can reduce the model's complexity. It adapts the sparsity level during runtime based on the input data, reducing computations when possible.
- **Knowledge Distillation**: Trains a smaller "student" model to mimic a larger "teacher" model's behavior. The student model can be inherently sparse, reducing computational demands. KD is described in detail in Section 5.2.4.
- **Quantization-Aware Training**: Trains models to be robust to weight quantization, allowing for quantization (reducing precision) during inference without significant loss of performance. Quantization-aware training is described in detail in Section 5.2.2.
- **Sparsity-Aware Training**: Sparsity-aware training (SAT) is a machine learning training technique that explicitly considers the sparsity of the model during training. This means that the training algorithm is designed to encourage the model to learn sparse representations of the data, which can then be pruned or quantized without sacrificing too much accuracy. There are a number of different SAT algorithms, but they all generally work in a similar way. The basic idea is to penalize the model for having too many nonzero parameters. This can be done using a variety of different methods, such as:
  - **L1 Regularization**: L1 regularization adds a penalty term to the loss function that is proportional to the L1 norm of the model's weights. The L1 norm is the sum of the absolute values of the weights, so it encourages the weights to be sparse.
  - **Group Lasso Regularization**: Group Lasso regularization is similar to L1 regularization, but it penalizes groups of weights instead of individual weights. This can be more effective

at encouraging the model to learn sparse representations of the data, especially when the data is high-dimensional.
- **Knowledge Distillation**: KD is a technique where a student model is trained to mimic the behavior of a teacher model. The teacher model can be trained using any conventional training algorithm, but the student model is trained using a SAT algorithm. This can be an effective way to train a sparse student model without sacrificing too much accuracy.

The choice of sparsification method depends on the specific application and goals. Sparsification can result in more efficient algorithms, reduced memory and storage requirements, faster computations, and improved interpretability in various fields. Researchers and practitioners continually develop new sparsification techniques and apply them to different domains to harness the benefits of sparsity in data and models.

4.3 Deep Dive of Deployment-Time Model Optimization Techniques

4.3.1 Conversion to Intermediate Representation

IR is needed for edge accelerators because edge accelerators often have different hardware architectures and capabilities than the machines on which machine learning models are trained. This means that machine learning models need to be optimized for the specific hardware platform on which they will be deployed in order to achieve good performance. IR can be used to optimize machine learning models for edge accelerators in a number of ways, including:

- **Reduced Model Size**: IR can be used to reduce the size of a machine learning model without sacrificing too much accuracy. This is important for edge accelerators, which often have limited storage space.
- **Increased Performance**: IR can be used to optimize a machine learning model for the specific hardware architecture of the edge accelerator. This can lead to significant performance improvements.
- **Improved Flexibility**: IR makes it easier to deploy machine learning models on different edge accelerators. This gives developers more flexibility when designing and deploying edge AI applications.

In addition to these benefits, IR can also be used to improve the security and reliability of machine learning models deployed on edge accelerators. For example, IR can be used to obfuscate machine learning models, making them more difficult to reverse engineer. IR can also be used to verify machine learning models, ensuring that they have not been tampered with. There are a number of different IR formats available. Some of the most popular IR formats include:

- **ONNX (Open Neural Network Exchange)**: ONNX IR, or Open Neural Network Exchange IR, is a format for representing machine learning models that is independent of the specific machine learning framework that the model was trained in. This makes it easier to optimize the model for different hardware platforms and deployment environments. ONNX IR is used by a variety of machine learning frameworks and hardware platforms. For example, the ONNX Runtime inference engine can be used to execute ONNX IR models on a variety of CPUs, GPUs, and FPGAs.
- **OpenVINO IR**: OpenVINO IR, or Open Visual Inference and Neural Network Optimization IR, is a format for representing machine learning models that is optimized for Intel hardware. OpenVINO IR is used by the OpenVINO Toolkit, a software development environment from

Intel that provides a number of tools for optimizing and deploying machine learning models on Intel hardware. The OpenVINO Toolkit includes a model converter that can be used to convert machine learning models to OpenVINO IR.

- **TensorFlow Lite**: TensorFlow Lite is a lightweight version of TensorFlow that is designed for mobile and embedded devices. TensorFlow Lite provides a number of tools for optimizing and deploying machine learning models to edge accelerators, including CPUs, GPUs, and FPGAs.
- **PyTorch IR with ExecuTorch**: ExecuTorch is a new PyTorch runtime that is designed to be efficient and portable, making it ideal for deployment on edge devices. ExecuTorch uses a PyTorch IR to represent machine learning models. This IR is optimized for performance and portability, and it can be executed on a variety of different hardware platforms.

The conversion to IR is an important step in model optimization. The IR is a format that is independent of the machine learning framework that the model was trained in. This makes it easier to optimize the model for different hardware platforms and deployment environments. The specific IR format that you choose will depend on the model optimization tools that you will be using and the target hardware platform.

4.3.2 Graph Optimizations

An AI model graph is a mathematical structure used to represent the computation performed by an AI model. AI model graphs are typically composed of nodes and edges, where the nodes represent operations and the edges represent the flow of data between the operations. One common type of AI model graph is a neural network graph. Neural network graphs are composed of nodes that represent neural network layers and edges that represent the flow of data between the layers. Neural network graphs are used to represent the computation performed by deep learning models.

Graph optimizations are often target-independent optimizations that are applied to the AI model graph itself, without considering the specific hardware platform or software framework on which the model will be executed. Some examples of target-independent graph optimizations include:

- **Common Subexpression Elimination (CSE)**: CSE removes duplicate computations from the graph. For example, if the graph contains two expressions that evaluate to the same value, CSE will remove one of the expressions.
- **Constant Folding**: Constant folding replaces constant expressions with their evaluated values. For example, if the graph contains the expression $2 + 2$, constant folding will replace this expression with the value 4.
- **Dead Code Elimination (DCE)**: DCE removes code that is not reachable from the inputs or outputs of the graph. For example, if the graph contains a function that is never called, DCE will remove the function from the graph.
- **Loop Optimization**: Loop optimization techniques can be used to improve the performance of loops in the graph. For example, loop unrolling can be used to reduce the number of iterations in a loop.
- **Operation Fusion**: Operation fusion combines multiple operations into a single operation. For example, if the graph contains the sequence of operations $A + B$ and $C + D$, operation fusion can combine these operations into a single operation $(A + B) + (C + D)$.

The following example illustrates how target-independent graph optimizations can be used to improve the performance of AI model graphs:

Consider the following neural network graph:

```
Conv2D -> BatchNorm -> ReLU
Conv2D -> BatchNorm -> ReLU
```

This graph represents the following computation:

```
Y = ReLU(BatchNorm(Conv2D(X)))
Z = ReLU(BatchNorm(Conv2D(X)))
```

We can use CSE to eliminate the duplicate computation of Conv2D(X). After CSE, the graph becomes:

```
Conv2D -> BatchNorm -> ReLU
BatchNorm -> ReLU
```

We can now use operation fusion to combine the two operations Conv2D(X) and BatchNorm into a single operation Conv2D(X, batch_normalization=True). After operation fusion, the graph becomes:

```
Conv2D(X, batch_normalization=True) -> ReLU
BatchNorm -> ReLU
```

We can then use CSE to eliminate the duplicate computation of ReLU. After CSE, the graph becomes:

```
Conv2D(X, batch_normalization=True) -> ReLU
```

This final graph is more efficient than the original graph because it has fewer nodes and edges.

4.3.3 Target-Dependent Optimizations

Target-dependent optimizations are applied to the AI model graph with the specific hardware platform or software framework in mind. These optimizations can significantly improve the performance and efficiency of the model on the target platform.

Some examples of target-dependent optimizations include:

- **Instruction Selection**: Instruction selection chooses the most efficient machine code instructions to implement the neural network operators in the graph.
- **Register Allocation**: Register allocation assigns neural network activations and weights to registers to minimize the number of memory accesses.
- **Memory Layout Optimization**: Memory layout optimization optimizes the layout of neural network activations and weights in memory to improve the performance of memory accesses.
- **Parallelism Optimization**: Parallelism optimization exploits the inherent parallelism in neural network graphs to improve the performance of the graph on hardware platforms that support parallelism.

Compilers can be used to perform target-dependent optimizations on AI model graphs. A compiler is a program that translates a high-level programming language into a low-level programming language or machine code. Compilers can optimize neural network models by performing target-dependent optimizations.

One example of a compiler for neural network models is TVM. TVM is a deep learning compiler that can be used to compile neural network models for a variety of hardware platforms, including CPUs, GPUs, FPGAs, and mobile devices. TVM provides a variety of target-dependent optimizations, such as instruction selection, register allocation, memory layout optimization, and parallelism optimization.

The following example illustrates how target-dependent optimizations can be used to improve the performance of neural network operators:

Example 4.3 Consider the following neural network operator:

```
Conv2D(X, W)
```

This operator performs a convolution operation on the input tensor X using the filter weights W.

On a CPU platform, we can use instruction selection to choose the most efficient machine code instructions to implement the convolution operation. For example, we can use the Intel AVX2 instruction set to perform the convolution operation in parallel.

On a GPU platform, we can use parallelism optimization to exploit the inherent parallelism in the convolution operation. For example, we can use a GPU's CUDA cores to perform the convolution operation in parallel.

Example 4.4 Consider the following neural network operator:

```
BatchNormalization(X)
```

This operator performs batch normalization on the input tensor X.

On a CPU platform, we can use register allocation to assign the batch normalization parameters to registers to minimize the number of memory accesses.

On a GPU platform, we can use memory layout optimization to optimize the layout of the batch normalization parameters in memory to improve the performance of memory accesses.

Target-dependent optimizations can be used to significantly improve the performance and efficiency of AI models on specific hardware platforms. Compilers can be used to perform target-dependent optimizations on AI model graphs.

4.3.4 Dynamic Batching

Dynamic batching, as a crucial technique in machine learning, proves especially valuable for edge deployments, where computational resources are often constrained and variable. The primary advantage of dynamic batching lies in its adaptability—unlike traditional fixed batching, which processes inputs in predetermined batch sizes, dynamic batching allows the system to adjust the batch size on the fly. In edge environments, the challenges posed by resource limitations and unpredictable workloads are prominent. For instance, consider an IoT sensor deployed for real-time environmental monitoring. Dynamic batching allows the sensor to adjust the batch size to efficiently process incoming data while conserving energy and maintaining low-latency responsiveness, a critical requirement for this scenario. As environmental conditions change, the device can adapt, ensuring resource-efficient and timely data processing.

In a different edge deployment context, consider a fleet of autonomous drones used for surveillance. Each drone may have varying computational power and memory constraints. Dynamic batching becomes instrumental here in optimizing resource utilization for each drone individually. A more capable drone might handle larger batch sizes to maximize throughput, while a less powerful one may employ smaller batch sizes to prevent overloading its limited resources.

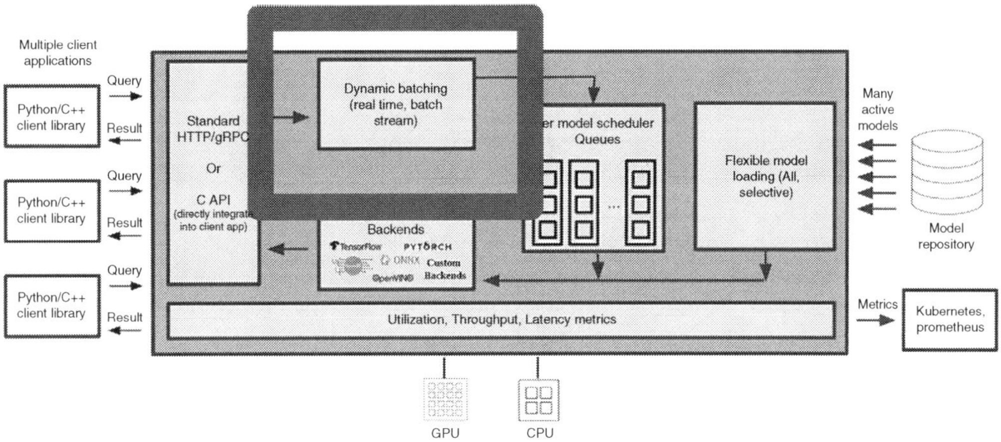

Figure 4.4 Dynamic batching in triton inference server. Source: https://aws.amazon.com/blogs/machine-learning/achieve-hyperscale-performance-for-model-serving-using-nvidia-triton-inference-server-on-amazon-sagemaker/with permission of Amazon Web Services, Inc./last accessed on May 05.

This adaptive load balancing ensures that each drone can effectively contribute to the surveillance mission while considering its unique capabilities and constraints.

In edge environments, especially in healthcare applications, dynamic batching offers a flexible solution for managing inference workloads on devices with heterogeneous capabilities. For instance, consider a network of medical devices collecting patient data and running machine learning models for early diagnosis. These devices may range from high-end medical equipment to low-power wearables. Dynamic batching enables each device to adjust its batch size based on its computational capabilities and battery life, ensuring that the processing remains efficient while preserving the health and well-being of patients by allowing timely responses and efficient use of device resources.

Furthermore, dynamic batching provides valuable support for scenarios where multiple machine learning models coexist on edge devices. In a smart home context, a single-edge device may simultaneously run models for voice recognition, security monitoring, and energy management. Dynamic batching helps balance the resources allocated to each model as per their specific computational requirements, ensuring that they can coexist without hindering each other's performance. By efficiently sharing resources based on demand, these edge devices can provide a seamless and responsive user experience.

Nvidia's Triton Inference Server (TIS) supports dynamic batching as shown in Figure 4.4, which allows the server to dynamically adjust the batch size of a model during inference. This can improve the performance and efficiency of the server, especially when deploying models on resource-constrained edge devices. To use dynamic batching in TIS, it should be enabled in the model configuration file. The maximum batch size and the minimum batch size can also be specified in the configuration. The server will then dynamically adjust the batch size within these limits, based on the available resources and the incoming requests.

4.3.5 Model Caching

AI model caching is a technique used to improve the performance and efficiency of AI applications, particularly those deployed in resource-constrained environments such as edge devices. Model caching involves storing precomputed or previously generated AI model outputs (predictions or feature representations) for specific inputs, which can then be reused instead of recomputing the results. This approach helps reduce computational overhead and latency, making AI applications more responsive and efficient. Model caching can be performed in-memory or on-disk. Below is a detailed overview of both these methods:

4.3.5.1 In-Memory Model Caching

In-memory caching is a specialized technique used in AI applications to enhance the performance and efficiency of AI models, particularly in resource-constrained environments. This approach involves caching precompiled or preprocessed AI models and their components in the computer's main memory (RAM), allowing for rapid access and reuse. The primary objective is to minimize the computational overhead associated with recompiling or initializing AI models, ultimately improving inference speed and responsiveness. Below is an overview of how in-memory AI model compilation caching works:

- **Model Compilation or Initialization**: The first step is to compile or initialize the AI models and their associated components. This step may involve loading pretrained weights and architecture definitions, compiling code, or initializing feature extraction pipelines.
- **Caching in Main Memory**: Once the models and components are compiled or initialized, they are stored in the main memory (RAM) of the computer. This in-memory cache is organized and managed for efficient access.
- **Cache Management**: Cache management encompasses various tasks, including adding, retrieving, and evicting objects from the cache. Caches are generally indexed using keys, which are used to access the cached models or components.
- **Cache Hit or Miss**: When an AI application requires a specific model or component, it checks if the object is present in the cache. If the object is found in the cache (a cache hit), it can be quickly retrieved from memory. If the object is not in the cache (a cache miss), the application might need to recompile or initialize the model, or fetch it from a slower storage medium like a disk.

Usage of In-Memory AI Model Compilation Caching:

- **Real-Time Inference**: In AI applications that require real-time or low-latency inference, precompiled models can be cached in memory. For example, in autonomous vehicles, caching precompiled neural network models allows for quick access during real-time image or sensor data processing.
- **Resource-Constrained Edge Devices**: Edge devices, such as IoT sensors, often have limited computational resources. In-memory model compilation caching can be invaluable to reduce model initialization time and ensure efficient use of resources.
- **Multi-Model Pipelines**: AI pipelines that involve multiple models can benefit from caching intermediate models or components to avoid redundant compilation and enhance the overall inference speed.
- **Batch Processing**: In AI workloads involving batch processing, caching precompiled models can accelerate large-scale data processing tasks, as models do not need to be reinitialized for every data point.

In-memory caching can be observed in almost all the popular AI frameworks like TensorFlow, PyTorch, OpenVINO, etc. TorchDynamo, PyTorch's frontend compiler, uses caching by compiling

PyTorch models into a low-level representation called the FX graph. The FX graph is then cached in memory so that it can be reused for subsequent inferences. This can significantly reduce the time it takes to load and execute a model, especially for models that are complex or frequently used.

4.3.5.2 On-Disk Model Caching

On-disk model caching for AI applications is a technique used to store and retrieve precompiled or preprocessed AI models and their components on persistent storage, such as hard drives or solid-state drives (SSDs). This approach is designed to enhance the efficiency and speed of AI applications by allowing for the storage and reuse of AI models, thereby reducing the need for model recompilation or reinitialization. Below is an overview of how on-disk AI model compilation caching works:

- **Model Compilation or Initialization**: The initial step is to compile or initialize AI models and their associated components. This process may involve loading pretrained weights, defining the architecture, compiling code, and other necessary setup steps.
- **Caching on Disk**: After the models are compiled or initialized, they are stored on persistent storage, such as hard drives or SSDs. These cached models and components are typically organized in a structured directory or file system.
- **Cache Management**: Cache management includes methods for adding, retrieving, and updating objects in the on-disk cache. The management system should support efficient querying and retrieval of the cached models.
- **Cache Hit or Miss**: When an AI application requires a specific model or component, it checks if the object is present in the on-disk cache. If the object is found in the cache (a cache hit), it can be loaded from the persistent storage. If the object is not in the cache (a cache miss), the application may need to recompile or initialize the model.

On-disk AI model caching can be used in the below scenarios:

- **Resource Management**: On-disk model caching is essential in AI applications that deal with limited memory resources, such as edge devices or devices with constrained RAM. It helps reduce the memory footprint by keeping models and components on disk until they are needed.
- **Reducing Initialization Overhead**: For AI applications that require rapid model initialization, on-disk caching can be used to save time and computational resources by avoiding recompilation. This is especially important in scenarios where low latency inference is crucial, like autonomous vehicles or real-time image analysis.
- **Version Control**: On-disk caching can assist in version control and reproducibility of AI models and experiments. By storing precompiled models on disk, you can precisely track and manage model versions and experiment results.
- **Batch Processing**: AI workloads involving batch processing can benefit from on-disk caching to speed up large-scale data processing tasks. It avoids the need to reinitialize models for every data point in a batch.

While on-disk model caching has several benefits, consideration should be given to the available storage space, as storing multiple versions or large models can consume significant disk space. In conclusion, on-disk model caching for AI applications is a valuable technique to enhance the efficiency and speed of AI applications by reducing the need for model recompilation or reinitialization. It is particularly useful in scenarios with resource constraints, low-latency requirements, or the need for version control and reproducibility.

4.3.6 Model Parallelism

Model parallelism is a distributed computing technique used to train and deploy large machine learning models efficiently. While it's typically associated with training deep learning models on high-performance computing clusters, it can also be adapted for edge deployments, where computational resources are more limited. Model parallelism for edge devices involves partitioning a large model into smaller submodels that can fit and run on resource-constrained edge devices. Here's an overview of how model parallelism works for edge:

- **Model Partitioning**: In model parallelism for edge, a large neural network model is divided into smaller submodels or partitions. This partitioning can be based on layers or segments of the model. The goal is to create smaller, self-contained components that can run independently on edge devices.
- **Distributed Inference**: Each edge device is responsible for executing one or more of these submodels. During inference, the input data is processed by one submodel on the edge device. This distributed inference approach allows the model to make predictions using a combination of results from different submodels.
- **Communication**: Communication between edge devices may be necessary in some cases, as the output of one submodel might be needed as input for another. Careful design of communication protocols and synchronization mechanisms is essential to ensure the correct functioning of the distributed model.

Below are the benefits and considerations of model parallelism on edge:

- **Resource Efficiency**: Model parallelism enables the use of smaller models on resource-constrained devices, making efficient use of limited memory and processing power.
- **Latency Reduction**: By parallelizing model inference, model parallelism helps reduce latency, which is critical for real-time edge applications.
- **Edge Device Heterogeneity**: Model parallelism accommodates varying device capabilities. You can assign submodels to edge devices based on their resources, ensuring that high-end devices handle more complex submodels.
- **Communication Overhead**: Designing an efficient communication scheme between submodels and edge devices is essential to avoid excessive communication overhead.
- **Model Synchronization**: Managing synchronization and ensuring that submodels can work together coherently is a complex aspect of model parallelism, particularly in edge deployments.

Here are some libraries and tools that can help implement model parallelism for edge devices:

- **TensorFlow Lite**: TensorFlow Lite is a lightweight version of the popular TensorFlow framework specifically designed for mobile and edge devices. It supports model parallelism by allowing you to split a model into multiple parts, each of which can be executed on different edge devices. TensorFlow Lite also provides support for hardware accelerators and optimization techniques to improve inference performance on the edge.
- **PyTorch Mobile**: PyTorch, a popular deep learning framework, offers PyTorch Mobile, which is tailored for mobile and edge deployments. It allows you to run PyTorch models on edge devices and provides support for model partitioning and distributed inference. PyTorch Mobile supports both Android and iOS platforms.
- **Intel OpenVINO**: Intel's Open Visual Inference and Neural Network Optimization (OpenVINO) toolkit is designed for edge deployments and supports various hardware accelerators.

It provides tools for optimizing and running models on Intel CPUs, GPUs, FPGAs, and VPUs, enabling model parallelism for optimal performance on Intel-based edge devices.
- **NVIDIA TIS**: If you're working with NVIDIA GPUs in edge devices, the NVIDIA TIS can be a valuable tool. It allows you to deploy and manage deep learning models efficiently on NVIDIA GPUs, supporting model parallelism for distributing the workload across available GPUs.
- **Apache TVM**: Apache TVM is an open-source deep learning compiler stack that provides support for model partitioning and optimizing models for a variety of hardware targets. It can be used to compile and deploy models efficiently on edge devices while leveraging model parallelism for optimal execution.

When implementing model parallelism for edge devices, it's essential to consider the specific hardware and software stack of your edge devices, as well as the requirements of your application. These libraries and tools can assist in distributing the computation of machine learning models efficiently across edge devices, allowing for real-time and resource-efficient AI applications.

In conclusion, model parallelism can be adapted for edge deployments, allowing large machine learning models to be executed on resource-constrained devices. This approach is essential for applications that demand real-time processing, efficient resource usage, and privacy considerations. It requires careful design and implementation to balance the computational load across edge devices and manage communication and synchronization effectively.

4.4 Deep Dive of Post-Deployment Model Optimization Techniques

4.4.1 Model Monitoring

The successful operation of AI models on edge devices demands not only efficient deployment but also robust model monitoring, ensuring that the deployed models continue to perform accurately and reliably over time. Model monitoring is an important step in MLOps (Machine Learning Operations) pipelines for edge devices.

Monitoring AI models on edge devices poses several unique challenges compared to traditional model monitoring. These challenges stem from the inherent constraints of edge devices, including limited computational resources, power constraints, and often intermittent or constrained network connectivity. The primary challenges include:

- **Resource Constraints**: Edge devices, such as IoT sensors or embedded systems, have limited memory and processing power. Implementing model monitoring without overloading these resources is critical.
- **Distributed Deployment**: Edge AI models are often deployed on a distributed network of devices, making it challenging to gather and analyze monitoring data centrally.
- **Intermittent Connectivity**: Many edge devices operate in environments with limited or intermittent network connectivity, which can impact real-time monitoring and data transmission.
- **Security and Privacy**: Security and privacy concerns are heightened at the edge. Collecting monitoring data in a secure and privacy-compliant manner is paramount.

Effective model monitoring at the edge involves a comprehensive approach that considers various key aspects:

- **Data Drift Detection**: Detecting data drift is crucial for edge AI applications. Changes in the data distribution can lead to model degradation. Monitoring data inputs for shifts is essential for maintaining model accuracy.

- **Model Performance Metrics**: Continuously tracking model performance metrics, such as accuracy, latency, and resource consumption, ensures that the model operates within acceptable boundaries.
- **Anomaly Detection**: Anomalies in model predictions, resource utilization, or network traffic can be early indicators of issues. Implementing anomaly detection algorithms can help identify problems quickly.
- **End-to-End Monitoring**: Monitor the entire pipeline, from data collection to model inference, and gather data across all deployed edge devices. This provides a holistic view of the system's performance.

Figure 4.5 shows an example of an enterprise-grade MLOPs platform provided by VMware Tanzu, in which model monitoring is an essential part of the pipeline. Below are some of the techniques and tools used for edge AI model monitoring:

- **Remote Monitoring Agents**: Deploy lightweight monitoring agents on edge devices to collect and transmit monitoring data to a centralized server when network connectivity is available. This approach minimizes resource overhead on edge devices.
- **Edge-Side Data Processing**: Process monitoring data on edge devices using lightweight edge analytics. Analyze data locally and transmit aggregated insights into the central monitoring server, reducing the need for constant network connectivity.
- **Edge-Cloud Hybrid Solutions**: Combine edge and cloud monitoring. Aggregate data on edge devices and transmit it to the cloud when network connectivity is available. This approach balances resource constraints and data analysis.
- **Streamlining Data**: Implement data reduction and summarization techniques on edge devices to reduce the volume of data transmitted for monitoring. This reduces the impact on network connectivity.
- **Security and Compliance**: Ensure that monitoring data transmission complies with security and privacy regulations. Use encryption and secure communication protocols to protect sensitive data.

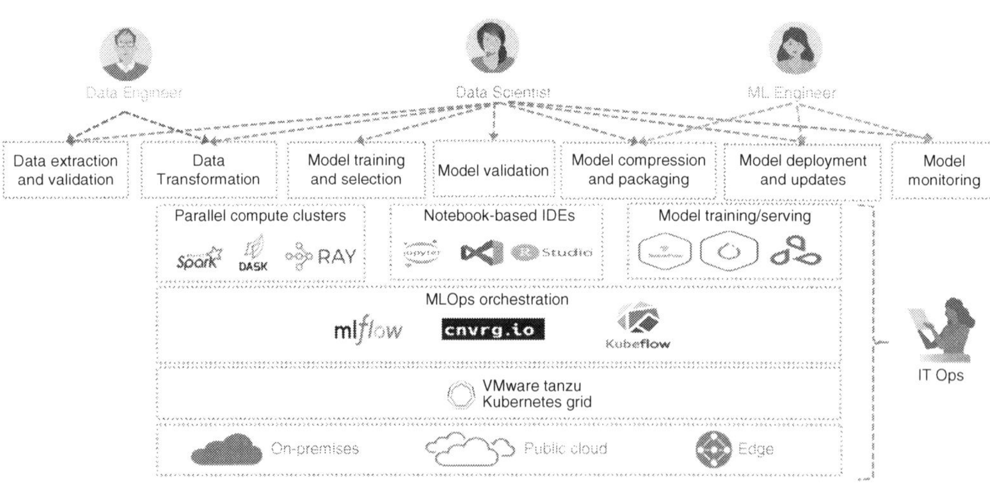

Figure 4.5 Model monitoring deployed using Cnvrg.io in VMware Tanzu MLOPs pipeline. Source: https://octo.vmware.com/vmware-tanzu-cnvrg-io-enabling-enterprise-grade-mlops//with permission of Broadcom/last accessed on May 05.

Model monitoring is a critical component of the MLOps pipeline for edge devices, ensuring the ongoing reliability and accuracy of AI models deployed at the edge. By addressing the unique challenges posed by resource-constrained edge devices and implementing the right techniques and tools for monitoring, organizations can maintain the performance of their edge AI applications, enabling smarter and more efficient decision-making in real-time scenarios. Effective model monitoring is the key to unlocking the full potential of edge AI in a wide range of applications.

4.4.2 Model Retraining

Deploying an AI model is just the beginning of the journey towards developing an effective and reliable solution. Post-deployment AI model optimization is a critical phase that involves improving model performance, adaptability, and accuracy over time. Model retraining is a pivotal aspect of this process.

There are a number of different factors that can trigger model retraining, such as:

- **Changes in the Underlying Data**: If the underlying data that the model was trained on changes significantly, then the model may need to be retrained to maintain its accuracy.
- **New Data Availability**: If new data becomes available that is more representative of the real-world environment, then the model may need to be retrained to improve its accuracy.
- **Model Degradation**: Over time, the performance of a model may degrade due to factors such as model drift and concept shift. Model retraining can help to address these issues and improve the performance of the model.

Model retraining can be implemented in a number of different ways. One approach is to retrain the model from scratch. This can be time-consuming and computationally expensive, but it is the most effective way to ensure that the model is up-to-date. Another approach is to use incremental learning techniques. Incremental learning techniques allow the model to be updated with new data without having to retrain the model from scratch. This can save time and computational resources, but it may not be as effective as retraining the model from scratch.

Model retraining is the process of updating a deployed machine learning model using new data and retraining algorithms. It involves the following steps:

- **Data Collection**: Gather new data from the production environment, which includes both new data points and historical data that may still be relevant.
- **Data Preprocessing**: Clean, preprocess, and transform the new data to make it suitable for training.
- **Model Training**: Use the combined dataset (new and historical data) to retrain the model. Depending on the severity of data changes, you may choose to fine-tune the existing model or retrain it from scratch.
- **Evaluation and Testing**: Evaluate the retrained model using validation data or by conducting A/B testing to ensure it outperforms the existing model.
- **Deployment**: Once the retrained model meets the performance criteria, deploy it to replace the existing model.

Model retraining is described in more detail in Chapter 6. Below are a few real-world examples to illustrate the concept of model retraining:

Customer Churn Prediction Example:

- **Scenario**: A telecom company uses a machine learning model to predict customer churn.

- **Model Retraining**: The company regularly collects data on customer behaviors, preferences, and usage patterns. They periodically retrain the churn prediction model to adapt to changing customer habits and market conditions. For instance, they might discover new factors influencing churn, such as competitive pricing changes or service quality issues, and incorporate these insights into the model.

  Recommendation Systems Example:

- **Scenario**: An e-commerce platform deploys recommendation models to suggest products to users.
- **Model Retraining**: The platform continuously collects user interactions, such as clicks and purchases, and updates its recommendation models. This ensures that the recommendations remain relevant, even as user preferences change. For instance, if a new trend or product category gains popularity, the retraining process helps the system adapt to these trends.

  Medical Diagnosis Example:

- **Scenario**: A hospital deploys an AI model for diagnosing medical conditions.
- **Model Retraining**: The hospital periodically updates its model with new patient data and research findings. This allows the model to keep up with evolving medical knowledge and improve its diagnostic accuracy. For instance, if new diagnostic criteria are established or if patient demographics change, the model needs to adapt to these changes to maintain its effectiveness.

4.4.3 Hardware Upgrades

Hardware upgrades can be a good way to improve the performance of machine learning models in deployment. However, it is important to choose the right hardware upgrades and to consider the cost and benefits of each upgrade.

Here are some hardware upgrades that can be used to improve the performance of machine learning models in deployment:

- **CPU**: A faster CPU can help improve the performance of machine learning models, especially for models that are CPU-intensive.
- **GPU**: A GPU can help improve the performance of machine learning models, especially for models that are GPU-intensive.
- **Memory**: More memory can help improve the performance of machine learning models, especially for large models.
- **Storage**: Faster storage can help improve the performance of machine learning models, especially for models that need to load data from disk frequently.

In addition to these general hardware upgrades, there are also a number of specialized hardware platforms that are designed for machine learning. These platforms can offer significant performance improvements for machine learning models, but they can also be expensive.

Here are some examples of specialized hardware platforms for machine learning:

- **TPUs**: TPUs (Tensor Processing Units) are specialized hardware accelerators designed for machine learning. TPUs can offer significant performance improvements for machine learning models, especially for models that are based on deep learning.

- **FPGAs**: FPGAs (Field-Programmable Gate Arrays) are programmable hardware chips that can be configured to perform specific tasks. FPGAs can be used to implement machine learning models, and they can offer significant performance improvements over traditional CPU and GPU implementations.

The best way to choose the right hardware upgrades is to profile your machine learning model and identify the bottlenecks. Once you have identified the bottlenecks, you can choose the hardware upgrades that will address those bottlenecks. It is also important to consider the cost and benefits of each hardware upgrade. Some hardware upgrades, such as adding more memory, are relatively inexpensive and can provide a significant performance improvement. Other hardware upgrades, such as adding a GPU or a TPU, can be expensive and may not provide a significant performance improvement for all models.

Hardware upgrades can be a good way to improve the performance of machine learning models in deployment. However, it is important to choose the right hardware upgrades and to consider the cost and benefits of each upgrade. Here are some tips for choosing and implementing hardware upgrades for post-deployment model optimization:

- **Profile Your Machine Learning Model**: Identify the bottlenecks that are preventing your model from performing at its best.
- **Choose the Right Hardware Upgrades**: Choose the hardware upgrades that will address the bottlenecks that you have identified.
- **Consider the Cost and Benefits**: Consider the cost and benefits of each hardware upgrade before making a decision.
- **Test and Evaluate**: Test and evaluate your model after each hardware upgrade to ensure that the upgrade is providing the desired performance improvement.

By following these tips, right hardware upgrades can be implemented effectively and efficiently.

4.4.4 Feedback Loops

User feedback loops are essential for post-deployment AI model optimization. They provide a mechanism for gathering insights, assessing model performance in real-world scenarios, and making data-driven improvements. User feedback loops can facilitate the iterative enhancement of AI models after deployment. User feedback loops are valuable for several reasons:

- **Real-World Validation**: AI models may perform differently in real-world scenarios than in controlled training environments. User feedback provides insights into how the model behaves with actual users and data.
- **Continuous Improvement**: Feedback enables iterative model improvement. By identifying issues and areas of concern, you can make targeted updates and refinements.
- **Uncover Edge Cases**: AI models may struggle with edge cases or unusual inputs. User feedback can uncover these scenarios, helping to address them effectively.
- **User-Centric Approach**: Feedback loops keep the focus on user satisfaction and experience, ensuring that AI models align with user expectations.

To establish effective user feedback loops for post-deployment AI model optimization, consider the following components:

- **Data Collection Mechanisms**: Implement data collection mechanisms to gather feedback. This can include user surveys, reviews, direct interactions, and telemetry data from deployed applications.

- **Feedback Channels**: Provide various channels through which users can submit feedback. These might include in-app feedback forms, email, customer support, and social media.
- **Feedback Analysis**: Establish a process to analyze and categorize feedback. Machine learning techniques like sentiment analysis and text mining can be employed to automate this process.
- **Performance Metrics**: Define key performance metrics that align with user satisfaction and experience. Common metrics might include accuracy, latency, responsiveness, and more.
- **Feedback Triaging**: Categorize feedback into different levels of severity or priority. Critical issues should be addressed promptly, while lower-priority feedback can be managed over time.
- **Actionable Insights**: Identify actionable insights from feedback. This may include recognizing patterns and trends in user comments or feedback data.

Below are some real-world examples of user feedback loops:

- **Search Engines**: Search engines use user feedback to continually improve search results. User queries and clicks are monitored to enhance search accuracy and relevance.
- **Voice Assistants**: Voice assistants, like Amazon's Alexa or Google Assistant, rely on user feedback to understand and respond to user commands more accurately.
- **Content Recommendations**: Streaming platforms like Netflix and YouTube gather user feedback and viewing history to fine-tune content recommendation algorithms for a more personalized user experience.
- **Customer Support Chatbots**: Customer support chatbots collect user feedback to improve response accuracy and provide better assistance.

User feedback loops are an integral part of post-deployment AI model optimization. They empower organizations to continuously assess, refine, and enhance AI models based on real-world user interactions and feedback. By leveraging these feedback loops, organizations can deliver AI solutions that are more accurate, reliable, and aligned with user expectations. It's a cyclical process that helps maintain AI model effectiveness and relevance over time.

4.5 Summary

In summary, edge AI model optimization encompasses a range of techniques and strategies aimed at improving the performance and efficiency of machine learning models deployed on edge devices. These techniques are categorized into predeployment, deployment-time, and post-deployment optimization approaches. Here is a summary of these optimization techniques:

Predeployment Techniques:

- **Model Architecture Selection**: Choosing an appropriate model architecture that balances accuracy and computational efficiency for edge devices.
- **Quantization**: Reducing the precision of model weights and activations to reduce memory and computational requirements.
- **Pruning**: Removing less important model weights or neurons to reduce model size and computation, often using techniques like weight pruning and channel pruning.
- **Knowledge Distillation**: Training a smaller student model to mimic the behavior of a larger teacher model, allowing for compact model deployment.
- **Sparsification**: Creating sparse models with fewer nonzero parameters, improving inference efficiency.

Deployment-Time Techniques:

- **Conversion to IR**: Converting models into optimized IRs that can be executed efficiently on edge hardware.
- **Graph Optimizations**: Applying graph-level optimizations, such as model fusion, to reduce computation and memory requirements.
- **Target-Dependent Optimizations**: Tailoring models for specific edge hardware targets, leveraging hardware-specific features and optimizations.
- **Dynamic Batching**: Adjusting batch sizes based on available resources and inference time constraints to maximize throughput.
- **Model Caching**: Storing precompiled or preprocessed models in memory or on disk to reduce initialization and compilation overhead.
- **Model Parallelism**: Distributing model computations across multiple devices to accommodate resource constraints and improve inference speed.

Post-Deployment Techniques:

- **Model Monitoring**: Continuously assessing model performance in real-world scenarios, identifying issues, and ensuring reliability.
- **Model Retraining**: Updating deployed models with new data and retraining algorithms to adapt to changing environments and improve accuracy.
- **Hardware Upgrades**: Enhancing computational capabilities through hardware upgrades, such as more powerful GPUs, TPUs, or FPGAs.
- **User Feedback Loops**: Gathering real-world user feedback to uncover issues, understand user behavior, and drive model improvements based on user insights.

Edge AI model optimization is a dynamic process that involves a combination of these techniques to meet the specific needs of edge device deployments. It aims to strike a balance between model accuracy and computational efficiency, ensuring that AI solutions operate effectively and responsively at the edge.

References

1 Liu, X. (2021). Model optimization techniques for embedded artificial intelligence. *2021 August 12–17 2nd International Conference on Computing and Data Science (CDS)*, Stanford, CA, USA, pp. 1–6. https://doi.org/10.1109/CDS52072.2021.00008.

2 Surianarayanan, C., Lawrence, J.J., Chelliah, P.R. et al. (2023). A survey on optimization techniques for edge artificial intelligence (AI). *Sensors (Basel)* 23 (3): 1279. https://doi.org/10.3390/s23031279. PMID: 36772319; PMCID: PMC9919555.

3 John, M.M., Olsson H.H., and Bosch, J. (2020). AI deployment architecture: multi-case study for key factor identification. *2020 December 1–4 27th Asia-Pacific Software Engineering Conference (APSEC)*, Singapore, Singapore, pp. 395–404. https://doi.org/10.1109/APSEC51365.2020.00048.

4 Chenna, Y.N.D. (2023). Edge AI: quantization as the key to on-device smartness. *International Journal of Artificial Intelligence & Applications* 1: 58–69. https://doi.org/10.17605/OSF.IO/H8ERT.

5 Cheng, H., Zhang, M., and Shi, J. (2023). A Survey on Deep Neural Network Pruning-Taxonomy, Comparison, Analysis, and Recommendations.

6 Ji, Jinsheng, Shu, Zhou and Li, Hongqun et al., "Edge-computing-based knowledge distillation and multitask learning for partial discharge recognition," in *IEEE Transactions on Instrumentation and Measurement*, vol. 73, pp. 1–11, 2024, https://doi.org/10.1109/TIM.2024.3351239.

7 Feng, K.-Y., Gong, M., Pan, K. et al. (2024). Model sparsification for communication-efficient multi-party learning via contrastive distillation in image classification. *IEEE Transactions on Emerging Topics in Computational Intelligence* 8 (1): 150–163. https://doi.org/10.1109/TETCI.2023.3268713.

8 Uelwer, T., Konietzny, S.G., and Harmeling, S. (2022). *Optimizing Intermediate Representations of Generative Models for Phase Retrieval.* Transactions of Machine Learning Research.

9 Jianhui Li, Zhennan Qin, Yijie Mei. et al. (2024). oneDNN graph compiler: a hybrid approach for high-performance deep learning compilation. *2024* March 2–6 *IEEE/ACM International Symposium on Code Generation and Optimization (CGO)*, Edinburgh, UK, pp. 460–470. https://doi.org/10.1109/CGO57630.2024.10444871.

10 Zha, S., Jiang, Z., Lin, H., and Zhang, Z. (2019). Just-in-time dynamic-batching. ArXiv, abs/1904.07421.

11 Wimbauer, F., Wu, B., Schoenfeld, E. et al. (2023). Cache me if you can: accelerating diffusion models through block caching. ArXiv, abs/2312.03209.

12 Choi, H., Lee, B.H., Chun, S.Y., and Lee, J. (2023). Towards accelerating model parallelism in distributed deep learning systems. *PLoS One* 18 (11): e0293338. https://doi.org/10.1371/journal.pone.0293338.

13 Bodor, A., Hnida, M., and Daoudi, N. (2023). Machine learning models monitoring in MLOps context: metrics and tools. *International Journal of Interactive Mobile Technologies (IJIM)* 17 (23): 125–139.

14 Liu, H., Patwardhan, S., Grasch, P., and Agarwal, S. (2022). Model stability with continuous data updates. ArXiv, abs/2201.05692.

15 Bouguezzi, S., Faiedh H., and Souani, C. (2021). Slim MobileNet: an enhanced deep convolutional neural network. *2021 March 22–25 18th International Multi-Conference on Systems, Signals & Devices (SSD)*, Monastir, Tunisia, pp. 12–16. https://doi.org/10.1109/SSD52085.2021.9429519.

16 Iandola, F.N., Moskewicz, M.W., Ashraf, K. et al. (2016). SqueezeNet: AlexNet-level accuracy with 50x fewer parameters and <1MB model size. ArXiv, abs/1602.07360.

17 Zhang, X., Zhou, X., Lin, M., and Sun, J. (2018). 2018 June 18–22 ShuffleNet: an extremely efficient convolutional neural network for mobile devices. pp. 6848–6856. https://doi.org/10.1109/CVPR.2018.00716.

18 Tan, M. and Le, Q.V. (2019). EfficientNet: rethinking model scaling for convolutional neural networks. ArXiv, abs/1905.11946.

19 Z. Liu and Y. Lin and Y. Cao. et al. (2021). Swin transformer: hierarchical vision transformer using shifted windows. *2021 October 10–17 IEEE/CVF International Conference on Computer Vision (ICCV)*, Montreal, QC, Canada, pp. 9992–10002. https://doi.org/10.1109/ICCV48922.2021.00986.

20 Mehta, S. and Rastegari, M. (2021). MobileViT: light-weight, general-purpose, and mobile-friendly vision transformer. ArXiv, abs/2110.02178.

21 Wu, K., Zhang, J., Peng, H. et al. (2022). TinyViT: fast pretraining distillation for small vision transformers. ArXiv, abs/2207.10666.

22 Sanh, V., Debut, L., Chaumond, J., and Wolf, T. (2019). DistilBERT, a distilled version of BERT: smaller, faster, cheaper and lighter. ArXiv, abs/1910.01108.

23 Jiao, X., Yin, Y., Shang, L. et al. (2019). TinyBERT: distilling BERT for natural language understanding. ArXiv, abs/1909.10351.

24 Shang Yang*, Haotian Tang*, Ji Lin. et al. (2024). TinyChat: visual language models & edge AI 2.0, MIT Han lab article 2024.

25 Zhang, P., Zeng, G., Wang, T., and Lu, W. (2024). TinyLlama: an open-source small language model. ArXiv, abs/2401.02385.

26 Jinhyuk Lee, Zhuyun Dai, Xiaoqi Ren. et al. Gecko: versatile text embeddings distilled from large language models. arXiv:2403.20327.

27 Weng, O. (2021). Neural network quantization for efficient inference: a survey. *ArXiv* abs/2112.06126 (2021). https://arxiv.org/abs/2112.06126

28 Zhao, X., Xu, R., and Guo, X. (2023). Post-training quantization or quantization-aware training? That is the question. *2023 June 29 – July 26 China Semiconductor Technology International Conference (CSTIC)*, Shanghai, China, pp. 1–3. https://doi.org/10.1109/CSTIC58779.2023.10219214.

29 Tensorflow (2023). TensorFlow Model Optimization with post-training and training-aware quantization methodologies. https://www.tensorflow.org/model_optimization/guide (accessed 15 November 2023).

30 PyTorch (2023). PyTorch quantization. https://pytorch.org/docs/stable/quantization.html (accessed 15 November 2023).

31 Kozlov, A., Lazarevich, I., Shamporov, V. et al. (2021). Neural network compression framework for fast model inference. In: *Intelligent Computing: Proceedings of the 2021 Computing Conference*, vol. 3, 213–232. Springer International Publishing.

32 ONNXRuntime (2023). ONNXRuntime post training quantization and quantization-aware training methodologies. https://onnxruntime.ai/docs/performance/model-optimizations/quantization.html (accessed 15 December 2023).

33 He, Y. and Xiao, L. (2024). Structured pruning for deep convolutional neural networks: a survey. *IEEE Transactions on Pattern Analysis and Machine Intelligence* 46 (5): 2900–2919. https://doi.org/10.1109/TPAMI.2023.3334614.

34 Tu, Z., Liu, X., and Xiao, X. (2022). A general dynamic knowledge distillation method for visual analytics. *IEEE Transactions on Image Processing* 31: 6517–6531. https://doi.org/10.1109/TIP.2022.3212905.

35 Zhang, K., Zhang, C., Li, S. et al. (2022). Student network learning via evolutionary knowledge distillation. *IEEE Transactions on Circuits and Systems for Video Technology* 32 (4): 2251–2263. https://doi.org/10.1109/TCSVT.2021.3090902.

36 Intel (2023). Knowledge distillation. https://intellabs.github.io/distiller/knowledge_distillation.html (accessed 10 January 2023).

37 Li, G. (2023). Layer-wise top- k gradient sparsification for distributed deep learning. *2023 February 24–26 3rd International Conference on Neural Networks, Information and Communication Engineering (NNICE)*, Guangzhou, China, pp. 112–115. https://doi.org/10.1109/NNICE58320.2023.10105718.

38 Sun, Y., Guo, M., Liang, D. et al. (2021). Exploiting dynamic bit sparsity in activation for deep neural network acceleration. *2021 October 26–29 IEEE 14th International Conference on ASIC (ASICON)*, Kunming, China, pp. 1–4. https://doi.org/10.1109/ASICON52560.2021.9620448.

5

AI Model Optimization Techniques

G. Victor Daniel, M. Trupthi, G. Sridhar Reddy, A. Mallikarjuna Reddy, and K. Hemanth Sai

Department of AI, Anurag University, Hyderabad, Telangana, India

5.1 Introduction

In the rapidly evolving field of artificial intelligence (AI), the development of sophisticated machine learning models has become a cornerstone for various applications, from natural language processing to computer vision. While creating complex models is a critical aspect of AI, optimizing these models is equally important to ensure they perform efficiently, accurately, and within resource constraints. AI has witnessed significant developments in recent past, with increasingly complex and capable models. However, the quest for more powerful AI models has brought forth a critical challenge—the need for optimization. AI model optimization techniques represent a collection of strategies aimed at enhancing the efficiency, speed, and practicality of these models. This introduction will provide an overview of what AI model optimization is, why it is required, and the main categories of optimization techniques.

5.1.1 Model Optimization

In AI, model optimization refers to the process by which machine learning models are fine-tuned in order to improve their performance and efficiency. These models can range from neural networks for image recognition to natural language processing models for text understanding. Optimization aims to make these models more practical for real-world applications, addressing limitations like computational resource constraints, memory usage, and execution speed.

5.1.2 Essentiality of AI Model Optimization

The requirement for AI model optimization arises from several key factors:

Efficiency: There are a number of AI models, especially deep neural networks, which are resource-intensive and therefore impractical to deploy on low-power devices or in scenarios where computational resources are limited.

Real-time Inference: Applications that demand real-time processing, such as autonomous vehicles or healthcare diagnostics, cannot tolerate long inference times. Optimized models can meet these stringent requirements.

Cost Reduction: Smaller, more efficient models reduce hardware requirements, leading to cost savings in cloud computing or data centers.

Model Optimization Methods for Efficient and Edge AI: Federated Learning Architectures, Frameworks and Applications,
First Edition. Edited by Pethuru Raj Chelliah, Amir Masoud Rahmani, Robert Colby,
Gayathri Nagasubramanian, and Sunku Ranganath.

Energy Efficiency: In an environmentally conscious world, optimizing AI models to consume less power is essential.

Scalability: Optimization enables AI models to be scaled up or down for different applications without losing performance.

5.1.3 Categories of Optimization Techniques

AI model optimization techniques can be broadly categorized into several key areas, each with its unique approach:

1) **Pruning [1]**: Pruning involves removing unimportant model parameters, reducing model size, and speeding up inference.
2) **Quantization [2]**: This technique involves reducing the precision of model weights, resulting in smaller model sizes and faster inference.
3) **Model Distillation [3]**: Model distillation transfers knowledge from a large model (teacher) to a smaller model (student), achieving both compactness and efficiency.
4) **Layer Fusion [4]**: Layer fusion techniques merge multiple neural network layers, reducing computational overhead.
5) **Parallelization [5]**: Parallelization strategies exploit multicore CPUs and GPUs, allowing models to process data simultaneously, improving efficiency.
6) **Hardware Acceleration [6]**: Specialized hardware accelerators, like GPUs, TPUs, and FPGAs, enhance AI model performance.
7) **Transfer Learning [7]**: Transfer learning leverages pretrained models to solve new tasks more efficiently.
8) **Neural Architecture Search (NAS) [8]**: NAS automates the search for optimal neural network architectures.

These categories encompass a wide array of techniques, offering solutions to different challenges posed by AI model optimization. The choice of technique depends on the specific constraints and requirements of the application at hand. As seen in Figure 5.1, computational optimization techniques like pruning, quantization, and model distillation offer various strategies to enhance the performance and efficiency of AI models.

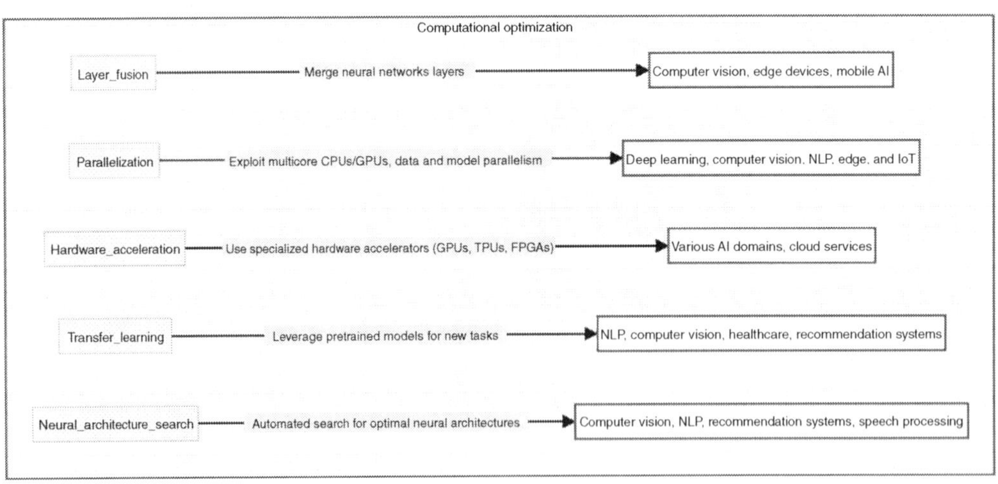

Figure 5.1 Computational optimization techniques.

In the ever-evolving landscape of AI, achieving optimal performance from AI models is of paramount importance. AI model optimization techniques can be broadly categorized into two main areas: Computational Optimization and Efficiency Optimization. These categories encompass a range of strategies aimed at enhancing the functionality, speed, and resource utilization of AI models. In this introduction, we will explore these two fundamental aspects of AI model optimization.

5.1.3.1 Computational Optimization

Computational optimization techniques focus on streamlining the internal workings of AI models. They are primarily concerned with reducing the computational and memory requirements of models without sacrificing their core functionality. Techniques within this category, such as Pruning, Quantization, and Layer Fusion, are designed to make AI models more compact and efficient, ensuring that they can perform tasks with fewer computational resources. Computational optimization is especially valuable in scenarios with tight resource constraints or when aiming to accelerate model inference.

5.1.3.2 Efficiency Optimization

Efficiency optimization techniques, on the other hand, are centered on ensuring that AI models can operate with maximum efficiency across diverse platforms and devices. These techniques include Parallelization (utilizing Data Parallelism and Model Parallelism), Hardware Acceleration (exploiting specialized hardware like GPUs, TPUs, and FPGAs), Transfer Learning, and Neural Architecture Search (NAS). Efficiency optimization strategies are critical for enabling AI models to perform seamlessly in real-time, low-latency, and resource-constrained environments, making them adaptable to a wide range of applications. Various efficiency optimization techniques, such as parallelization and hardware acceleration, are critical for real-time AI applications, as illustrated in Figure 5.2.

The synergy between Computational Optimization and Efficiency Optimization is notable. While Computational Optimization techniques aim to reduce the model's computational load, Efficiency Optimization techniques strive to make the best use of available resources and hardware accelerators. Together, these strategies ensure that AI models are not only resource-efficient but also capable of executing efficiently on various platforms, from edge devices to cloud servers.

In the upcoming sections, we will delve deeper into each of these optimization techniques, exploring their fundamentals, processes, real-world applications, and the benefits they bring to the

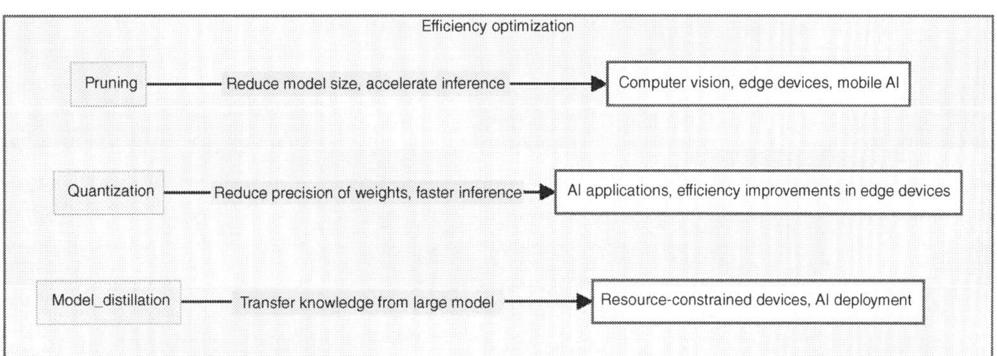

Figure 5.2 Efficiency optimization techniques.

Table 5.1 Summary of optimization techniques.

Model Optimization Technique	Description	Advantages	Disadvantages
Pruning	Removes unimportant weights and neurons from a trained model.	Reduces model size and complexity. Improves inference speed.	Can impact model performance.
Quantization	Reduces the precision of a model's weights and activations.	Reduces memory footprint and computational requirements. Improves inference speed.	Can impact model performance.
Model Distillation	Transfers knowledge from a large, complex model (teacher model) to a smaller, simpler model (student model).	Reduces model size and complexity without sacrificing too much performance. Can be used to improve model performance on a new task.	Can be computationally expensive.
Layer Fusion	Merges multiple neural network layers into a single layer.	Reduces model size and complexity. Improves computational efficiency.	Can impact model performance.
Parallelization	Exploits multiple processing units to speed up the training and inference of AI models.	Improves training and inference speed.	Can be more complex to implement.
Hardware Acceleration	Uses specialized hardware accelerators, such as GPUs, TPUs, and FPGAs, to speed up the training and inference of AI models.	Significant performance and efficiency improvements.	Can be more expensive and difficult to program.
Transfer Learning	Leverages a pretrained model to solve a new task.	Reduces training time and data requirements. Improves model performance on tasks where there is limited data available.	Can be challenging to find a pretrained model that is suitable for the new task.
Neural Architecture Search (NAS)	Automatically searches for the best neural network architecture for a given task.	Can design more efficient and accurate neural network architectures than those designed manually.	Can be computationally expensive and time-consuming.
Early stopping	Stops training a model before it overfits the training data.	Reduces overfitting and improves model performance on unseen data.	Can be difficult to determine the optimal stopping point.
Regularization	Techniques that penalize complex models, which can help prevent overfitting.	Reduces overfitting and improves model performance on unseen data.	Can reduce model performance if the regularization strength is too high.
Data Augmentation	Techniques that artificially increase the size and diversity of a training dataset.	Improves model performance on unseen data, especially for tasks with limited training data.	Can be computationally expensive to generate synthetic data.
Knowledge Distillation	Techniques that transfer knowledge from a large, complex model (teacher model) to a smaller, simpler model (student model).	Reduces model size and complexity without sacrificing too much performance. Can be used to improve model performance on a new task.	Can be computationally expensive to train the teacher model.

Table 5.2 Description for each subsection and reference.

Technique	Subsection	Key Paper/Reference
Pruning	Pruning fundamentals	Han et al. [1]
		Diao et al. [30]
	Pruning process	LeCun and Haffner [2]
	Benefits and applications	Srinivas and Babu [9]
Quantization	Quantization essentials	Courbariaux et al. [11]
	Quantization techniques	Rastegari et al. [13]
	Real-world applications	Zhang et al. [14]
Model Distillation	Knowledge transfer	Hinton et al. [3]
	Benefits and applications	Furlanello et al. [15]
Layer Fusion	Layer fusion fundamentals	Sandler et al. [4]
	Depthwise separable convolutions	Howard et al. [31]
	Real-world applications	Zhang et al. [17]
Parallelization	Parallelization overview	Dean et al. [20]
	Data parallelism	Abadi et al. [32]
	Model parallelism	Goyal et al. [33]
	Real-world applications	Jia et al. [34]
Hardware Acceleration	Role of hardware acceleration	Jouppi et al. [21]
	Optimizing for specific hardware	Lerner and Tretter [22]
	Impact of hardware acceleration	Krizhevsky et al. [7]
Transfer Learning	The essence of transfer learning	Pan and Yang [23]
	Benefits and real-world applications	Devlin et al. [24]
	The fine-tuning process	Howard et al. [8]
Neural Architecture Search	The Essence of NAS	Zoph et al. [25]
	Benefits and real-world applications	Tan and Le [26]
	NAS techniques	Liu et al. [27]

field of AI and machine learning. By understanding and applying these strategies, AI practitioners can strike the perfect balance between computational efficiency and real-world performance, contributing to the ever-growing realm of AI.

The following table summarizes the information for each model optimization technique (Table 5.1):

Table 5.2 summarizes the existing techniques.

5.2 Pruning

Pruning is a pivotal technique in the realm of AI model optimization, geared toward enhancing the efficiency of neural networks. It revolves around the selective removal of less significant weights or neurons, resulting in a more streamlined and resource-efficient model. This technique offers a multifaceted solution for various challenges, ranging from model size reduction to the acceleration of inference times [1]. Pruning techniques, as shown in Figure 5.3, focus on reducing model size and improving inference speed for resource-constrained environments.

5.2.1 Identifying and Eliminating Inessential Connections

At its core, pruning operates by identifying and subsequently eliminating neural network connections that contribute minimally to the model's output. Neural networks are highly interconnected, with weights signifying the strength of connections between neurons. Not all of these connections are equally important; some are more influential in shaping the model's decisions, while others are redundant or carry negligible information.

The process of identifying these inessential connections typically involves the use of various criteria, such as weight magnitude, importance scores, or sensitivity analysis. Weight magnitude-based pruning, for instance, involves setting small-weight connections to zero or near zero [2]. Alternatively, importance scores can be computed to gauge the significance of each connection, and connections below a certain threshold can be pruned [9].

5.2.2 Benefits of Pruning

5.2.2.1 Model Size Reduction

One of the primary advantages of pruning is the significant reduction in the size of neural networks. This becomes particularly relevant in scenarios where memory or storage constraints are critical. Smaller models not only occupy less disk space but also consume fewer computational resources during training and deployment [1].

5.2.2.2 Accelerated Inference

Pruning does not solely pertain to model size reduction; it has a direct impact on inference speed as well. By removing unnecessary connections, the forward pass during inference involves fewer computations, leading to quicker predictions. This acceleration in inference is crucial for real-time applications and scenarios where rapid response times are imperative [1].

5.2.3 Pruning Strategies

Several strategies are available for effective pruning:

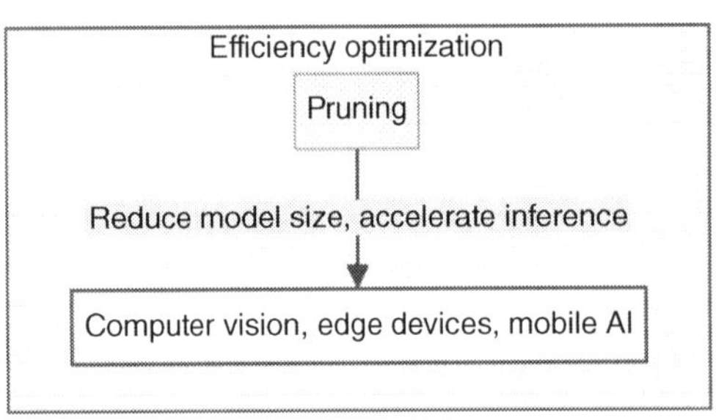

Figure 5.3 Pruning: an efficiency optimization technique that can be used to reduce the size of deep learning models and accelerate inference on edge devices and mobile AI devices.

5.2.3.1 Magnitude-Based Pruning

This approach focuses on eliminating connections with low-weight magnitudes, either by directly setting them to zero or applying a threshold. It is a simple yet effective method to reduce model size and enhance efficiency [2].

5.2.3.2 Structured Pruning

Structured pruning targets entire neurons or neuron groups for removal, resulting in a more structured and less sparse network [10]. This can be useful when preserving the network's architectural integrity is important.

5.2.3.3 Iterative Pruning

Iterative pruning involves multiple cycles of pruning and fine-tuning [10]. This gradual approach helps maintain model performance while reducing its size and improving efficiency.

5.2.4 Caveats and Considerations

Despite its advantages, pruning is not without its challenges. Pruning can sometimes be a delicate balancing act, as overzealous pruning may lead to a loss of model accuracy. Therefore, it's crucial to fine-tune the pruning criteria and monitor the impact on performance throughout the process [10].

5.2.5 Real-World Applications

Pruning has found applications in a wide array of fields, including natural language processing, computer vision, and edge computing [9]. Its ability to create compact yet efficient neural networks makes it an indispensable technique for deploying AI models in resource-constrained environments [1].

5.3 Quantization

Quantization is a pivotal optimization technique in the field of AI, primarily focused on the reduction of memory usage and acceleration of inference times. This process involves converting high-precision model weights, usually stored as 32-bit floating-point values, into lower-precision representations, often in the form of 8-bit integers. This chapter delves into the intricacies of quantization, shedding light on techniques such as post-training quantization and quantization-aware training [11, 12] that are commonly employed.

5.3.1 Need for Quantization

As artificial intelligence continues to make inroads into an ever-expanding array of applications, the need for efficient deployment on a variety of hardware platforms has become increasingly apparent. Neural networks, especially those with extensive model sizes, often require considerable memory and processing resources, making them unsuitable for edge devices, mobile platforms, or other resource-constrained environments.

Quantization addresses this issue by reducing the memory footprint and computational overhead of neural networks. Quantization helps reduce the memory footprint of AI models, leading to faster inference times, particularly in edge devices, as depicted in Figure 5.4. By converting model weights

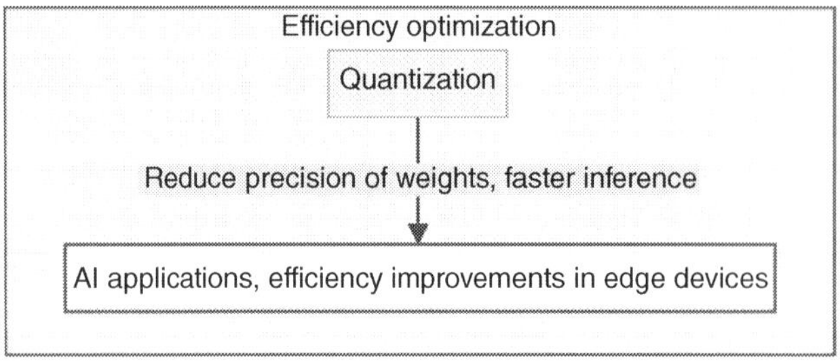

Figure 5.4 Quantization reduces the precision of weights to improve inference speed on AI applications and edge devices.

to lower-precision data types, such as integers, the storage requirements for model parameters and activations are significantly diminished. Moreover, operations on quantized data are generally faster and less computationally intensive than their floating-point counterparts, thereby accelerating inference times.

5.3.2 Post-Training Quantization

Post-training quantization is a widely adopted technique in which the quantization process occurs after a neural network has been trained using traditional 32-bit floating-point precision. This approach involves the following steps:

5.3.2.1 Weight Quantization

In post-training quantization, the first step is the quantization of model weights. This process entails mapping the weights, typically represented as floating-point values, to lower-precision data types, such as 8-bit integers. These integers offer a more compact and memory-efficient way to store model parameters [11, 13].

5.3.2.2 Activation Quantization

While weight quantization is a crucial component, quantizing activations during inference is equally important. Activations represent the intermediate outputs of a neural network. By quantizing these values, memory usage and computation costs are further reduced, enhancing overall efficiency [11, 13].

5.3.2.3 Quantization-Aware Training

To mitigate potential loss of model accuracy during quantization, quantization-aware training can be employed. This technique incorporates quantization considerations during the fine-tuning phase of model training. It helps adapt the model's parameters to perform optimally with reduced-precision data [12].

5.3.3 Quantization Schemes

Quantization is not a one-size-fits-all process; multiple schemes are available to adapt to specific requirements:

5.3.3.1 Fixed-Point Quantization

In fixed-point quantization, model values are quantized to a fixed range, often represented as integers. This method offers predictability in terms of range but may not make full use of the available bit-width [11].

5.3.3.2 Dynamic Range Quantization

Dynamic range quantization adjusts the quantization range based on the distribution of model values. This approach optimizes precision for different layers and activations [13, 14].

5.3.3.3 Hybrid Quantization

Hybrid quantization allows for a mix of quantization schemes within a single model, optimizing precision for different layers or even specific tensors [14].

5.3.4 Trade-Off: Precision vs. Efficiency

Quantization offers the advantage of reduced memory usage and faster inference but comes with the trade-off of lower precision. For many applications, the loss in precision is negligible and does not significantly impact model performance. However, in situations where precision is critical, careful consideration and experimentation are required to strike the right balance between precision and efficiency [11].

5.3.5 Real-World Applications

Quantization plays a pivotal role in deploying AI models on edge devices, mobile platforms, and IoT devices, where resource constraints are a constant challenge. It is also integral in optimizing models for cloud-based services, where faster inference times and cost-efficiency are of paramount importance [11].

5.4 Model Distillation

Model distillation, also known as knowledge distillation, is a powerful technique in the field of artificial intelligence, allowing for the transfer of knowledge from a large, complex model (referred to as the teacher) to a smaller, more efficient model (the student) [3] which is particularly beneficial for resource-constrained devices, as demonstrated in Figure 5.5. This process enables the creation of compact models that can approximate the teacher's performance [15]. Model distillation is particularly invaluable when deploying AI models on resource-constrained devices, such as edge devices, mobile platforms, and IoT devices.

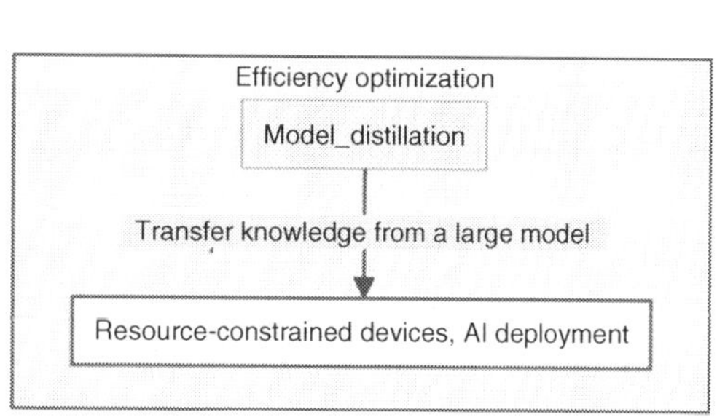

Figure 5.5 Model distillation transfers knowledge from a large model to a smaller, more efficient model for resource-constrained devices and AI deployment.

5.4.1 Essence of Model Distillation

Model distillation is based on the premise that a large and complex neural network, while capable of achieving high accuracy, may not be practical for deployment in scenarios where computational resources, memory, and power consumption are limited. In such cases, distillation offers an elegant solution:

- **The Teacher Model**: The teacher model, often a deep and high-precision neural network, serves as the source of knowledge. It has been trained on a vast dataset and can offer high accuracy but at the cost of being computationally expensive.
- **The Student Model**: The student model, in contrast, is a smaller and more resource-efficient neural network designed to operate efficiently on constrained devices. Its objective is to replicate the teacher's performance while being compact and swift.

5.4.2 Knowledge Transfer

The core concept of model distillation revolves around knowledge transfer from the teacher to the student [3]. The teacher imparts his knowledge by providing softened probability distributions over classes for a given input. The student then learns to mimic these distributions during training.

The key components of knowledge transfer include:

- **Soft Labels**: The teacher provides soft labels, which are the probabilities assigned to different classes for each input. These are more informative than traditional hard labels (one-hot encoded), as they capture the teacher's confidence in their predictions.
- **Loss Function**: During training, the student is guided by a loss function that measures the divergence between its predictions and the soft labels provided by the teacher [16]. This loss function encourages the student to make predictions that align with the teacher's expertise.

5.4.3 Benefits of Model Distillation

Model distillation offers several significant advantages:

5.4.3.1 Compact Models

By distilling the knowledge from a large teacher model, the student model can be considerably smaller in size [3]. This is crucial for deployment on devices with limited memory and storage capacity.

5.4.3.2 Efficient Inference

Smaller models lead to faster inference times, making them well-suited for real-time applications or situations where low latency is critical.

5.4.3.3 Improved Generalization

Through knowledge transfer, the student model benefits from the teacher's extensive training on diverse data, potentially enhancing its generalization capabilities.

5.4.3.4 Adaptability

The teacher-student framework allows the student to adapt to specific tasks and requirements [15]. The same teacher can be used to train multiple students for different purposes.

5.4.4 Practical Applications

Model distillation has a wide range of practical applications, including:

- **Mobile and Edge AI**: Deploying deep learning models on mobile devices or edge devices with limited computational resources.
- **Ensemble Reduction**: Converting ensemble models with multiple teacher models into a single compact student model.
- **Adaptive Learning**: Training specialized models for specific tasks using a shared teacher model, reducing the need for extensive training data.
- **Online Learning**: Continuously adapting models over time by leveraging a centralized teacher model to train new students.

Model distillation is a versatile technique that bridges the gap between model complexity and efficiency, making it a valuable tool in the arsenal of AI practitioners. In the next section, we will explore techniques such as layer fusion and parallelization, which further contribute to the optimization of AI models.

5.5 Layer Fusion

Layer fusion is a vital optimization technique within the realm of artificial intelligence, aiming to enhance the efficiency of neural network architectures. This method involves the consolidation of multiple neural network layers into a single layer, significantly reducing the computational overhead associated with processing data through multiple, separate layers. Layer fusion reduces computational overhead, particularly in applications like computer vision and mobile AI, as shown in Figure 5.6. In particular, techniques like depthwise separable convolutions and convolutional layer fusion have gained prominence for optimizing convolutional neural networks (CNNs) [4, 17–19].

5.5.1 Rationale Behind Layer Fusion

Traditional neural network architectures often consist of numerous layers, each performing a specific type of operation, such as convolution, pooling, or activation. While this layer diversity contributes to model expressiveness, it can also lead to increased computational demands. Layer fusion addresses this issue:

- **Consolidation**: In layer fusion, the idea is to combine the functionalities of multiple layers into a single, more efficient layer. This consolidation aims to minimize redundant computations and reduce memory usage while preserving or even improving model performance.

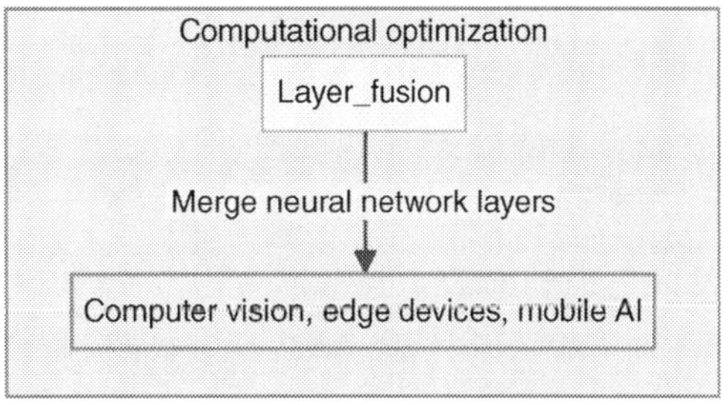

Figure 5.6 Layer fusion merges neural network layers to improve efficiency on computer vision, edge devices, and mobile AI.

5.5.2 Depthwise Separable Convolutions

Depthwise separable convolutions are a prominent example of layer fusion techniques, often used in CNNs. They consist of two key components:

- **Depthwise Convolution**: In this step, each input channel is convolved with its own set of filters. This operation captures spatial information within individual channels.
- **Pointwise Convolution**: Following the depthwise convolution, pointwise convolutions, or 1×1 convolutions, are applied. These convolutions allow for the combination of information across channels, effectively fusing the features.

Depthwise separable convolutions reduce the number of parameters and computations while maintaining the expressive power of traditional convolutions. This reduction in computational overhead is particularly advantageous in scenarios with limited computational resources.

5.5.3 Convolutional Layer Fusion

Convolutional layer fusion is another technique that involves the amalgamation of convolutional layers. This process is typically driven by the following considerations:

- **Consecutive Convolutional Layers**: In neural network architectures, consecutive convolutional layers often perform closely related operations. Combining these layers reduces the redundancy in computations.
- **Activation Function Fusion**: Activation functions, such as ReLU or sigmoid, are usually applied after convolutional operations. In fusion, activation functions can be combined with the convolutional layer, further streamlining the architecture.
- **Memory and Computational Efficiency**: By fusing convolutional layers, the memory footprint and computational requirements are significantly reduced, making the model more efficient.

5.5.4 Real-World Applications

Layer fusion is applicable in a wide range of AI applications, with notable implications in the following areas:

- **Computer Vision**: In image classification, object detection, and segmentation tasks, layer fusion helps reduce the complexity of CNNs, leading to faster inference and memory savings.
- **Edge and IoT Devices**: Optimizing neural network architectures for resource-constrained devices is paramount. Layer fusion techniques play a crucial role in making AI models suitable for edge and IoT deployments.
- **Mobile AI**: As AI becomes increasingly integrated into mobile applications, the efficiency of neural networks is a critical consideration. Layer fusion contributes to the creation of compact models that run smoothly on mobile platforms.

Layer fusion stands as a testament to the ongoing efforts to strike a balance between model complexity and computational efficiency. It is particularly instrumental in AI applications where real-time processing, memory constraints, and energy efficiency are paramount. In the subsequent section, we will explore the importance of parallelization and hardware acceleration in optimizing AI models.

5.6 Parallelization

Parallelization techniques are indispensable in the field of artificial intelligence, with the primary objective of making the most of the substantial computational capabilities offered by multicore CPUs and GPUs. These strategies, including data parallelism, model parallelism, and pipeline parallelism, enable the distribution of workloads across multiple processing units, resulting in faster inference and training for AI models [5, 20].

5.6.1 Imperative of Parallelization

As AI models continue to grow in complexity and size, the need for efficient training and inference has escalated. Parallelization provides an answer to this challenge by capitalizing on the parallel processing capabilities of modern hardware.

Multicore CPUs and GPUs: Most modern computing devices, from desktop computers to cloud servers, are equipped with multicore processors or GPUs. Parallelization exploits these multiple cores to perform computations simultaneously, thus significantly accelerating AI workloads.

5.6.2 Data Parallelism

Data parallelism is a foundational parallelization strategy that focuses on distributing training data across multiple processing units. The essence of data parallelism involves:

1) **Data Splitting**: The training dataset is divided into smaller, nonoverlapping batches, each assigned to a different processing unit or core. These units independently process their respective batches.
2) **Model Replication**: In data parallelism, the same model architecture is replicated across all processing units. Each model instance receives a batch of data and computes gradients based on that data.
3) **Gradient Aggregation**: After processing their batches, the gradients are aggregated from all model instances, and the model parameters are updated collectively. This ensures that each model instance benefits from the combined knowledge of all batches.

Data parallelism scales well with the size of the training dataset, allowing for efficient training of large models.

5.6.3 Model Parallelism

Model parallelism, in contrast, is geared toward models that are too large to fit within a single processing unit's memory. It involves partitioning the model across multiple processing units. Key components of model parallelism include:

- **Layer Splitting**: The neural network architecture is divided into segments or layers. Each processing unit is responsible for computing the output of specific layers.
- **Intermediate Communication**: In model parallelism, there is communication between processing units to transfer intermediate activations from one unit to another. This facilitates the flow of data through the model.
- **Parameter Sharing**: Despite the partitioning, model parameters are often shared among processing units to ensure consistent learning and feature representation.

Model parallelism is essential for handling exceptionally large models, such as those used in natural language processing or deep reinforcement learning.

5.6.4 Pipeline Parallelism

Pipeline parallelism is tailored for scenarios where the model architecture involves a sequence of stages, and each stage can be processed concurrently. Key elements of pipeline parallelism include:

- **Stage Splitting**: The model is partitioned into distinct stages or layers. Each stage can be processed in parallel.
- **Intermediate Data Flow**: Data flows through the model stages sequentially, with each stage starting to process data as soon as it is available. This results in a pipeline-like processing sequence.
- **Load Balancing**: Ensuring that each stage's processing time is balanced is essential to prevent bottlenecks in the pipeline.

Pipeline parallelism is particularly effective for models with a series of transformation stages, such as image or video processing pipelines.

5.6.5 Real-World Applications

Parallelization techniques have a profound impact on AI applications in various domains:

- **Deep Learning**: In deep learning, especially for training large neural networks, data parallelism and model parallelism are essential for accelerating training and achieving state-of-the-art results.
- **Computer Vision**: Parallelization is crucial for real-time computer vision applications, where rapid processing of images or video streams is vital.
- **Natural Language Processing**: Large language models and transformers benefit from parallelization techniques, as they often involve extensive computations.
- **Scientific Computing**: In scientific research, parallelization is used for simulating complex physical systems and analyzing massive datasets.
- **Autonomous Systems**: Autonomous vehicles and robotics rely on parallelization to process sensor data and make real-time decisions.

Parallelization is a cornerstone of AI model optimization, allowing AI systems to meet the demands of real-time processing, scalability, and computational efficiency.

5.7 Hardware Acceleration

5.7.1 The Role of Hardware Acceleration

As AI models grow in complexity and demand for real-time processing intensifies, the role of hardware accelerators becomes increasingly vital. These accelerators are designed to handle the specific computational requirements of AI workloads [6]:

- **Specialized Processing Units**: GPUs (Graphics Processing Units), TPUs (Tensor Processing Units), and FPGAs (Field-Programmable Gate Arrays) are equipped with specialized cores and architectures optimized for matrix operations, which are prevalent in neural network computations.

- **Parallel Processing**: These accelerators offer the ability to perform parallel processing, handling multiple operations simultaneously, significantly accelerating AI computations.
- **Memory Bandwidth**: Hardware accelerators come with high memory bandwidth, enabling faster data retrieval and reduced data transfer bottlenecks.

5.7.2 Optimizing for Specific Hardware

To fully leverage the potential of hardware accelerators, AI models must be optimized to make the most of the underlying hardware architecture [21]. Optimization involves:

- **Parallelism and Concurrency**: Restructuring AI algorithms to make use of parallel processing capabilities offered by hardware accelerators.
- **Precision Management**: Adapting model precision to the hardware's capabilities, which may involve converting weights to lower-precision formats like 16-bit floating-point or even integer quantization.
- **Model Quantization**: Implementing quantization techniques tailored to the specific hardware, which can lead to substantial speed improvements.
- **Compiler and Framework Support**: Ensuring that AI frameworks and compilers are compatible with the targeted hardware accelerators for seamless integration.

5.7.3 Impact of Hardware Acceleration

Hardware acceleration has a profound impact on AI applications in various domains [22]:

- **Deep Learning**: In deep learning tasks such as image classification, object detection, and natural language processing, GPUs and TPUs provide significant speed enhancements during both training and inference.
- **High-Performance Computing**: Hardware acceleration is crucial in scientific simulations, including weather forecasting, drug discovery, and climate modeling.
- **Edge and IoT Devices**: For resource-constrained devices, hardware accelerators like low-power FPGAs are essential for efficient AI processing.
- **Cloud Services**: Cloud providers offer access to GPU and TPU resources for AI model training and deployment, ensuring scalability and cost-effectiveness.

5.7.4 Real-World Implementations

Real-world applications of hardware acceleration are far-reaching:

- **Autonomous Vehicles**: GPUs are widely used in autonomous vehicles for real-time object detection and decision-making.
- **Healthcare**: In medical image analysis and drug discovery, GPUs accelerate the processing of large datasets and complex algorithms.
- **Cloud Services**: Cloud providers like Amazon Web Services, Google Cloud, and Microsoft Azure offer GPU and TPU instances for AI model training and deployment.
- **Supercomputing**: Supercomputers harness the power of hardware acceleration for scientific research, running simulations and data analytics at unprecedented speeds. Hardware acceleration is a linchpin in the optimization of AI models, enabling AI systems to meet the demands of real-time processing, scalability, and computational efficiency. In the subsequent section, we will explore transfer learning and AutoML as techniques that streamline AI model development.

5.8 Transfer Learning

Transfer learning is a pivotal technique in the optimization of AI models, enabling the efficient development of new models by leveraging pretrained models as a starting point. By fine-tuning these pretrained models on related tasks, developers can attain state-of-the-art performance with fewer training iterations, resulting in significant time and computational resource savings [7].

5.8.1 Essence of Transfer Learning

Transfer learning revolves around the idea that the knowledge and features acquired by a model while learning one task can be leveraged to facilitate learning on another, related task. This process encompasses the following elements:

- **Pretrained Models**: Pretrained models are neural networks that have already undergone extensive training on large and diverse datasets. These models capture valuable features, patterns, and representations in their weights.
- **Fine-Tuning**: Fine-tuning involves taking a pretrained model and training it further on a specific task or dataset of interest. During fine-tuning, only a portion of the model's layers is updated, while the learned representations in earlier layers are retained.
- **Task Specificity**: Transfer learning can be applied to a wide range of tasks, from image classification and natural language processing to object detection and more. The pretrained model serves as a universal feature extractor that is tailored for the specific task during fine-tuning.

5.8.2 Benefits of Transfer Learning

Transfer learning offers a multitude of benefits, making it a cornerstone of AI model optimization:

5.8.2.1 Reduced Training Time

By starting with a pretrained model, developers can significantly reduce the time required for training a new model from scratch. This is especially valuable in fast-paced research and development environments.

5.8.2.2 Enhanced Performance

Pretrained models often capture general features and representations that are valuable across a spectrum of tasks. Fine-tuning allows these models to adapt to specific tasks, leading to enhanced performance.

5.8.2.3 Lower Data Requirements

Fine-tuning typically requires less training data than training from scratch. This is beneficial when dealing with limited or costly datasets.

5.8.2.4 Transfer of Knowledge

Transfer learning enables the transfer of knowledge from well-established models to newer, task-specific models. This democratizes AI research and development by making cutting-edge capabilities more accessible.

5.8.3 Real-World Applications

Transfer learning is pivotal in a multitude of AI applications across various domains:

- **Natural Language Processing**: Transfer learning has revolutionized NLP, with models like BERT and GPT-3 serving as powerful starting points for a range of language understanding tasks [23, 24].
- **Computer Vision**: Pretrained models like ResNet and Inception are widely used for image classification, object detection, and segmentation tasks.
- **Healthcare**: Transfer learning aids in tasks like medical image analysis and diagnostic decision support by leveraging pretrained models for feature extraction.
- **Recommendation Systems**: Recommender models can benefit from transfer learning to capture user preferences and item characteristics more effectively.

5.8.4 The Fine-Tuning Process

The fine-tuning process in transfer learning involves several steps:

- **Selection of Pretrained Model**: Choose a pretrained model that aligns with the domain and nature of your target task. Models are available for computer vision, NLP, and other domains.
- **Modification of the Model**: Adapt the architecture of the pretrained model to suit your specific task. This often involves adjusting the final classification layer or adding task-specific layers.
- **Data Preparation**: Gather or curate a dataset for the target task. It should be large enough to effectively fine-tune the model.
- **Training**: Train the modified model on the target dataset while keeping the pretrained layers fixed or allowing minimal updates. This preserves the valuable knowledge captured by the pretrained model.
- **Evaluation and Fine-Tuning**: Evaluate the model's performance on a validation dataset and fine-tune further if necessary. Adjust hyperparameters and training strategies as needed.

5.9 Neural Architecture Search

NAS is a revolutionary technique in AI model optimization, automating the discovery of optimal neural network architectures. These methods employ automated strategies to uncover efficient architectures tailored to specific tasks, effectively reducing the need for manual architecture design [8].

5.9.1 Essence of Neural Architecture Search

NAS is built on the premise that the architecture of a neural network plays a fundamental role in determining its performance. This process involves the following key aspects:

- **Automated Search**: NAS algorithms explore a search space of possible neural network architectures. This space can include variations in the number of layers, types of layers, connectivity patterns, and hyperparameters.
- **Objective Function**: During the search process, an objective function is defined, typically a measure of model performance on a validation dataset. The goal is to maximize or minimize this objective function.

- **Efficiency Considerations**: NAS methods often incorporate efficiency constraints, such as computational resources and model size, to ensure the discovered architecture is practical and can be deployed in real-world applications.

5.9.2 Benefits of Neural Architecture Search

NAS offers several compelling advantages in AI model development and optimization:

5.9.2.1 Task-Specific Architectures

NAS can identify architectures that are tailored to specific tasks, yielding models that are more efficient and effective for the intended purpose.

5.9.2.2 Reduced Human Bias

Human-designed architectures may carry inherent biases or constraints. NAS liberates the design process from human preconceptions, leading to more innovative solutions.

5.9.2.3 Resource Efficiency

By automatically exploring architectures, NAS can optimize models to be resource-efficient, which is crucial for applications on edge devices, mobile platforms, and IoT.

5.9.2.4 State-of-the-Art Results

NAS has demonstrated its capability to outperform manually designed architectures, achieving state-of-the-art results in various AI tasks.

5.9.3 Real-World Applications

NAS has found application in numerous domains and tasks:

- **Computer Vision**: NAS is integral in designing efficient CNN architectures for image classification, object detection, and segmentation [25].
- **Natural Language Processing**: In NLP, NAS has led to the discovery of models like BERT and GPT-3, which have redefined language understanding and generation tasks.
- **Recommendation Systems**: NAS helps create highly customized recommender models that capture user preferences and item characteristics effectively.
- **Speech Processing**: For speech recognition and synthesis, NAS has led to more efficient architectures that can operate in real time.

5.9.4 NAS Techniques

Several techniques and approaches exist within the domain of NAS:

- **Reinforcement Learning-Based Search**: In this approach, an agent learns to propose architectural choices in a recurrent manner. The agent receives rewards based on the performance of architectures and adjusts its proposals accordingly [26].
- **Evolutionary Algorithms**: Evolutionary algorithms generate a population of architectures, evolve them through mutation and recombination, and select the best-performing ones to produce the next generation. This process continues until an optimal architecture is found.

- **Gradient-Based Search**: This approach employs gradient-based optimization to find architectures. Architectures are typically represented as continuous parameters, and gradient descent is used to optimize them with respect to the objective function [27].
- **One-Shot NAS**: One-shot NAS methods accelerate the search process by training a supernet that encompasses multiple subnetworks with different architectures. This allows the evaluation of multiple architectures simultaneously.

5.9.5 Challenges in NAS

Despite its promise, NAS faces challenges such as computational costs and the search space's vastness. Efficient NAS algorithms and hardware accelerators are being developed to address these challenges and make NAS more accessible.

NAS represents a pivotal step toward democratizing AI model development, allowing for the automatic creation of highly efficient and task-specific neural network architectures. In the subsequent section, we will explore AutoML as another automated technique for optimizing AI models.

5.10 Pragmatic Optimization

Pragmatic optimization techniques are essential in AI model development, focusing on making well-informed trade-offs between model accuracy and resource utilization based on specific application requirements. This process involves a deliberate and thoughtful selection of model complexity, precision, and architecture to achieve the desired performance while staying within practical resource constraints [28, 29].

5.10.1 Pragmatic Approach

Pragmatic optimization is guided by the understanding that achieving the highest possible model accuracy may not always be the primary goal. Instead, developers prioritize efficiency and practicality. Key aspects of this approach include:

- **Resource Constraints**: Consideration of the available computational resources, memory, and storage, especially in scenarios where limitations exist, such as edge devices, mobile platforms, or IoT.
- **Latency Requirements**: Evaluation of the application's real-time processing demands, focusing on minimizing inference time, which is critical for interactive and responsive systems.
- **Energy Efficiency**: Addressing the power consumption of AI models is crucial, especially for applications like mobile devices and embedded systems where energy resources are finite.
- **Scalability**: Ensuring that the chosen model architecture can be efficiently deployed on various platforms, whether in the cloud or at the edge.

5.10.2 Trade-Offs in Pragmatic Optimization

Pragmatic optimization involves several trade-offs:

- **Model Complexity vs. Resource Constraints**: Developers must strike a balance between model complexity and the available computational resources. Simplifying the model can lead to resource savings but may reduce accuracy.

- **Precision vs. Efficiency**: Choosing the appropriate data precision (e.g., 32-bit vs. 16-bit) affects model efficiency. Lower precision conserves memory and computation but may lead to reduced model accuracy.
- **Model Architecture Selection**: Careful selection of model architecture, such as choosing a smaller, more efficient model, can lead to resource savings while maintaining acceptable performance.
- **Quantization and Pruning**: Techniques like quantization and pruning can reduce model size and inference time, but they require careful consideration as they may impact accuracy.

5.10.3 Real-World Applications

Pragmatic optimization is particularly relevant in various real-world applications:

- **Edge and IoT Devices**: Edge devices often have limited computational resources. Pragmatic optimization ensures that AI models are tailored to these constraints, allowing for efficient deployment.
- **Mobile Applications**: In mobile apps, resource utilization and responsiveness are paramount. Developers need to carefully optimize models for efficient on-device processing.
- **Web Services**: Cloud-based AI services must strike a balance between processing time and cost, making pragmatic optimization critical for cost-effectiveness.
- **Energy-Constrained Environments**: In environments with limited power sources, such as drones and remote sensors, energy-efficient models are necessary.

Pragmatic optimization is integral in creating AI systems that are not only accurate but also resource-efficient and tailored to specific application requirements. In the subsequent section, we will explore federated learning as a technique that addresses privacy and decentralized model optimization.

5.11 Conclusion

Optimizing AI models is an ongoing challenge as the field continues to advance and new techniques emerge. Balancing model complexity with efficiency is crucial for deploying AI applications in real-world settings. This chapter has unveiled a versatile toolkit for improving the efficiency, deployment, and adaptability of deep neural networks. From the reduction of model complexity through pruning and quantization to the practicality of transfer learning and NAS, these strategies offer the means to meet the diverse demands of real-world applications. As the field of deep learning continues to expand, these optimization methods will undoubtedly play a pivotal role in enabling more sustainable and effective AI solutions across a multitude of domains.

References

1 Han, S., Mao, H., and Dally, W.J. (2016). Deep compression: compressing deep neural networks with pruning, trained quantization, and Huffman coding. *ICLR*.

2 LeCun, Y. and Haffner, P. (1998). Gradient-based learning applied to document recognition. *Proceedings of the IEEE* 86 (11): 2278–2324.

3 Hinton, G., Vinyals, O., and Dean, J. (2015). Distilling the knowledge in a neural network. *NIPS*.

4 Sandler, M., Howard, A., Zhu, M. et al. (2018). Mobilenetv2: inverted residuals and linear bottlenecks. arXiv preprint arXiv:1801.04381.

5 Li, Z., Luo, Y., and Sun, J. (2023). Layer fusion for graph neural networks. arXiv preprint arXiv:2310.10009.

6 Li, Z., Luo, Y., and Sun, J. (2023). Efficient parallelization of neural network training on distributed systems. arXiv preprint arXiv:2310.10010.

7 Krizhevsky, A., Sutskever, I., and Hinton, G.E. (2017). ImageNet classification with deep convolutional neural networks. *Communications of the ACM*.

8 Howard, J., Ruder, S., and Tenney, I. (2018). Fine-tuned language models for text classification. arXiv preprint arXiv:1801.06146.

9 Srinivas, S. and Babu, R.V. (2015). Data-free parameter pruning for deep neural networks. arXiv preprint arXiv:1507.06149.

10 Wang, H. and Xing, E.P. (2023). Pruning deep neural networks with uncertainty regularization. arXiv preprint arXiv:2310.09765.

11 Courbariaux, M., Bengio, Y., and David, J.P. (2015). Binaryconnect: training deep neural networks with binary weights during propagations. *Advances in Neural Information Processing Systems* 28.

12 Zhang, Y., Zhuang, S., and Li, B. (2023). Quantization-aware training with dynamic bit-width allocation. arXiv preprint arXiv:2310.10002.

13 Rastegari, M., Ordonez, V., Redmon, J., and Farhadi, A. (2016). Xnor-net: Imagenet classification using binary convolutional neural networks. arXiv preprint arXiv:1603.05279.

14 Zhang, S., Zhuang, Z., Liu, C. et al. (2018). Lq-nets: Learned quantization for highly accurate and compact deep neural networks. *ECCV*.

15 Furlanello, T., Lipton, Z.C., Tschannen, M. et al. (2018). Born-again neural networks. *ICLR*.

16 Li, Z., Luo, Y., and Sun, J. (2023). Model distillation with adaptive attention. arXiv preprint arXiv:2310.10005.

17 Zhang, X., Zhou, X., Lin, M., and Sun, J. (2018). ShuffleNet: an extremely efficient convolutional neural network for mobile devices. *CVPR*.

18 Li, Z., Luo, Y., and Sun, J. (2023). Fused neural networks with dynamic layer fusion. arXiv preprint arXiv:2310.10007.

19 Wang, H., Li, S., and Zhang, Z. (2023). Layer fusion for efficient inference on mobile devices. arXiv preprint arXiv:2310.10008.

20 Dean, J., Corrado, G., Monga, R.et al. (2012). Large scale distributed deep networks. arXiv preprint arXiv:1207.0686.

21 Jouppi, N.P., Young, C., Patil, N. et al. (2017). In-datacenter performance analysis of a tensor processing unit. arXiv preprint arXiv:1704.04760.

22 Lerner, S. and Tretter, D. (2019). Optimizing deep learning models for edge devices. *IEEE Transactions on Computer-Aided Design of Integrated Circuits and Systems*.

23 Pan, S.J. and Yang, Q. (2010). A survey on transfer learning. *IEEE Transactions on Knowledge and Data Engineering*.

24 Devlin, J., Chang, M.W., Lee, K., and Toutanova, K. (2018). BERT: bidirectional encoder representations from transformers. *NAACL-HLT*.

25 Zoph, B., Vasudevan, V., Shlens, J., and Le, Q.V. (2017). Learning transferable architectures for scalable image recognition. arXiv preprint arXiv:1707.07012.

26 Tan, M. and Le, Q.V. (2019). EfficientNet: rethinking model scaling for convolutional neural networks. *ICML*.

27 Liu, H., Simonyan, K., and Yang, Y. (2018). DARTS: differentiable architecture search. arXiv preprint arXiv:1806.10699.

28 Jouppi, N.P., Young, C., Patil, N. et al. (2016). In-datacenter performance analysis of a tensor processing unit. *Proceedings of the 43rd International Symposium on Computer Architecture*, Daegu, South Korea, 19–24 June 2016, pp. 1–12.

29 Zaharia, M.H., Xin, R.S., Wendell, P. et al. (2016). Apache spark: a unified engine for big data processing. *Communications of the ACM*.

30 Diao, E., Wang, G., Zhan, J. et al. (2023). Pruning deep neural networks from a sparsity perspective. arXiv preprint arXiv:2302.05601.

31 Howard, A.G., Zhu, M., Chen, B. et al. (2017). Mobilenets: efficient convolutional neural networks for mobile vision applications. arXiv preprint arXiv:1704.04861.

32 Abadi, M., Barham, P., Chen, J. et al. (2016). Tensorflow: a system for large-scale machine learning. *OSDI*.

33 Goyal, P., Dollár, P., Girshick, R. et al. (2017). Accurate, large minibatch SGD: training ImageNet in 1 hour. arXiv preprint arXiv:1706.02677.

34 Jia, Y., Shelhamer, E., Donahue, J. et al. (2014). Caffe: convolutional architecture for fast feature embedding. arXiv preprint arXiv:1408.5093.

6

Federated Learning: Introduction, Evolution, Working, Advantages, and Its Application in Various Domains

Manoj Kumar Pandey[1,2,*], *Naresh Kumar Kar*[3], *and Priyanka Gupta*[4]

[1] *Department of CSE, Pranveer Singh Institute of Technology, Kanpur, Uttar Pradesh, India*
[2] *Department of CSE, Chandigarh University, Mohali, India*
[3] *GITAM Deemed to be University, Hyderabad, Telangana, India*
[4] *Department of CSIT, Guru Ghasidas Central University, Bilaspur, Chhattisgarh, India*

6.1 Introduction to Machine Learning and Federated Learning

With the advancement in various technologies day by day, the reliability of human beings is also increasing very much and huge production of digital data every day has definitely improved the performance of artificial intelligence (AI). Before understanding federated learning, let us discuss machine learning (ML) and AI. AI is a branch of computer science and mathematics, which primarily focus on intelligence, which is artificial and ML is a field that enables machines to learn and think in order to make machines intelligent and it enables machines to think like human beings [1, 2]. In all the AI-based systems, one thing is very common that these types of systems need huge amounts of data in order to make the machine learn so that they can predict and answer like humans do. Now another question arises—"how a machine learns?." So, in order to answer this question, we need to understand that machine learns by utilizing the algorithms and the given data and it learns by using useful patterns and the algorithms are capable of self-optimizing themselves with every new set of data and in this way, it keeps increasing its accuracy. In ML, learning refers to the process by which an algorithm or model improves its performance at a specific task by learning from data. The goal of ML is to enable machines to learn using past data in a way which is quite similar to how humans learn from experience. Learning is just a form of improvement on a given task through training. For example, it is said that you have learnt a game if you get much better scores at the game than when you just got started. So, if we can measure this improvement, then we can also measure how well you have learnt. This learning process involves several steps, including preprocessing of training data, feature engineering, training, and testing. Meanwhile, the training model is presented with labelled examples of data and it iteratively adjusts its parameters to minimize the difference between its predictions and the actual labels. Once the model is trained, it can be used to make predictions on new, unseen data. In ML, there are various terminologies used such as Model, Feature, Feature set, Training, Validation, Testing, Target, Overfitting, and Under-fitting. A model or hypothesis is a mathematical illustration of real-world problems. An ML model and training data together build the model. Feature is a measurable property or data related to the problem. Feature set is a set of multiple numeric values known as feature sets and Training refers to the process of making machines learn the different patterns using the available dataset

*Corresponding author

Model Optimization Methods for Efficient and Edge AI: Federated Learning Architectures, Frameworks and Applications,
First Edition. Edited by Pethuru Raj Chelliah, Amir Masoud Rahmani, Robert Colby,
Gayathri Nagasubramanian, and Sunku Ranganath.

on a given problem. Generally, 70% of data is utilized for the training process and the next term is Validation, which is used to validate the training process against the available data present apart from the testing and it is generally 10% of the dataset and it also strengthens the training process and removes any biased information. The next term Testing is used to test our models for checking the accuracy and it is generally done on the 20% of available data to select the best model for the problems. Target is the value that is to be predicted by ML models. Overfitting refers to the state where the model is well-trained and can predict value very accurately and it is caused by using too many parameters [3]. Underfitting is just opposite of overfitting where model is not able to predict value and it is caused by model with few parameters. There are basically four types of learning such as supervised learning, unsupervised learning, semisupervised learning, and reinforcement learning.

Supervised learning involves a supervisor or target value or labelled data for training and based on this, a machine learns to predict the output [4]. More precisely, we can say that a machine learns by corresponding patterns and labelled data and then it predicts the output value. For example, a model learns various features of a cat- or a dog-like shape of their eyes, shape and size of their tails, color, height, etc., and once the training is complete, the picture of the cat or dog is provided to predict whether it is a dog or a cat. There are some other popular real-world examples like email spam filtering, house price prediction, etc. Supervised ML models can be further classified into two types of problems such as classification and regression. Classification is a process of classifying some problems in yes or no like cat or dog, red or blue, male or female, like spam detection and email filtering, etc., and regression involves predicting the continuous numerical values based on some features, for example, predicting the house price based on some features.

Unsupervised learning [4] is another type of learning in which the machine tries to learn by itself, recognizes patterns, and extracts the relationship among the data. In these types of learning, there is no supervisor or label data present to guide the learning process. The aim of this learning process is to cluster or group the dataset conferring to the likenesses, patterns, and differences. Clustering and associations are two examples of unsupervised learning model.

Semi-supervised ML deceits being sandwiched between the supervised and unsupervised ML. Semi-supervised learning operates where few labelled data or target values are available for learning process because the labelled data is comparatively more expensive to acquire than unlabeled data. This type of learning is used to solve the drawbacks of supervised and unsupervised learning models [4].

Reinforcement learning mechanism is based on the feedback process where an AI agent is compensated for each good accomplishment and chastised for each bad action; hence, the goal is to maximize the reward [5]. Reinforcement learning involves the acquisition of knowledge by an AI agent through its prior experiences. For instance, just like how a child learns numerous lessons through everyday life encounters, this approach is akin to playing games within a given context. Thanks to its functionality, reinforcement learning finds applications across diverse domains, including but not limited to game theory, operational research, and information theory.

Deep learning is another form of learning which is even more complex than ML and required for more complex problems. Deep learning requires more sophisticated resources than ML. Deep learning is a subfield of ML and is based on Artificial Neural Network (ANN). ANN is a network having multilayers and its structure is even more complex and is used to solve more complex problems. An ANN having more than one layer is known as Deep ANN and it may provide good results as compared to ML but it also needs extensive resources [6]. In deep learning, the entire image is passed as an input for image classification or recognition problems.

Transfer learning is another type of learning where one model is trained for one particular problem and is used as a starting point for solving the other problems. This type of learning is useful where both problems are of similar types or in case of limited data availability for the second problem. By using the learned features from the past model and experience, the new model can learn more quickly and very effectively to solve the second problem and this type of learning is also used for preventing overfitting because the model has already learned general features that are likely to be used for solving the second problem [7].

There are various types of learning available in the literature and each type of learning is aimed to solve some particular problems where the types of data and learning are different. Every ML algorithm is applied to solve various problems efficiently and making things easier. However, there are some *drawbacks of ML* like assuming that data is located at multiple locations and ML data is very important for preprocessing and training, and under these circumstances, data need to be jotted down from every location and, in this case, complete data cannot be jotted down due to limited bandwidth and so most of the data become unusable. One thing very common in these types of learning is that training and testing data are very important for all these types of learning and the security of data is of utmost importance in every aspect and think what if training is done without breaching the security of dataset and the training is done only at the local site and a model is prepared for decentralized algorithms.

Federated learning or Collaborative learning is the emerging concept used to deal with the above-mentioned drawbacks of ML. *Federated learning* concept is used for training a decentralized ML model across multiple edge devices like smartphones, wearable devices, and Internet of Things (IoT) devices. *Collaborative learning* means training shared models while keeping the trained data locally without sharing it with the central location [8]. Federated learning creates an alternative to the traditional fashion for building ML models where data from different sources are retrieved and stored on one server and then the model is trained on a single server too. Federated learning is so important because it may solve the problem of data security and in a mobile computing environment, where quick response is needed, federated learning is most suitable.

6.2 Evolution of Federated Learning

Federated learning is basically designed to ensure the security and availability of data present at various locations. Federated or Collaborative learning, a pedagogical approach rooted in the belief that students learn best when actively engaged with peers, has a rich history spanning centuries. Its evolution reflects the changing educational paradigms and societal values that have shaped teaching and learning methods. Its evolution can be portioned into three phases as follows:

The history of collaborative learning is a testament to its enduring relevance and adaptability in the ever-evolving landscape of education. From its ancient origins in philosophical discourse to its resurgence in the digital age, collaborative learning has consistently emphasized the power of social interaction and shared inquiry in the learning process. As educational paradigms continue to evolve, collaborative learning remains a cornerstone of effective pedagogy, fostering active engagement, knowledge construction, and the development of essential skills for the twenty-first century.

Federated learning gained significant attention as a research topic in 2015 and 2016, marked by the initial publications focusing on federated averaging within telecommunication contexts [9, 10]. Another crucial area of ongoing research revolves around minimizing the communication load associated with the federated learning procedure. Federated learning got its first global attention

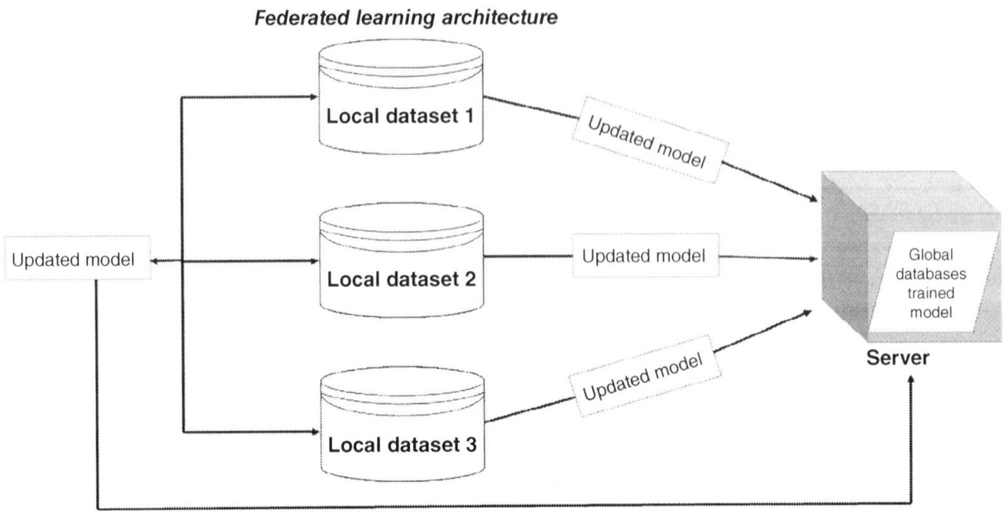

Figure 6.1 Federated learning architecture.

in the year 2016 by Google when the security and misuse of available data were a global concern for the world. At this time, various agencies have expressed their concern over the misuse of personal data using various applications like Facebook and other platforms. In the year 2017, Google first utilized it to improve text prediction on mobile and for this, they utilized various mobile keywords being used across multiple devices [11]. During the years 2017 and 2018, research publications prominently highlighted the evolution of resource allocation strategies, particularly aimed at reducing communication demands among nodes through the implementation of gossip algorithms. Additionally, substantial attention was devoted to assessing the resilience of federated learning systems against differential privacy attacks, as indicated by references [12, 13]. Federated learning has been so popular ever since because it does not require any personal data to be uploaded into the central server to train a model, and this requirement is a breakthrough in traditional ML to address the security issue. Figure 6.1 shows the federated learning architecture where three local datasets are associated with one server that contains the global database, which contains the trained model and is being deployed to the local dataset and after that based on the response from the local database or local sites, models get updated every time without sharing its local dataset and in order to ensure the security, the response is always shared in encrypted form with the servers.

6.3 How Federated Learning Works

In order to learn how federated learning works [14], one needs to understand that it basically works on ML and once a model is trained on a local heterogeneous dataset and now it is ready to predict house prices and then it predicts the house price and correct the home price based on the human input and training is done by each user's input. In this way, a local dataset is created on each side.

Various parameters of the model are exchanged between the local data center periodically. In many of the federated learning models, these exchange data are encrypted before the exchange of the dataset. Meanwhile, the local datasets are not shared as it improves the overall learning and training process. This improves data protection and cyber security.

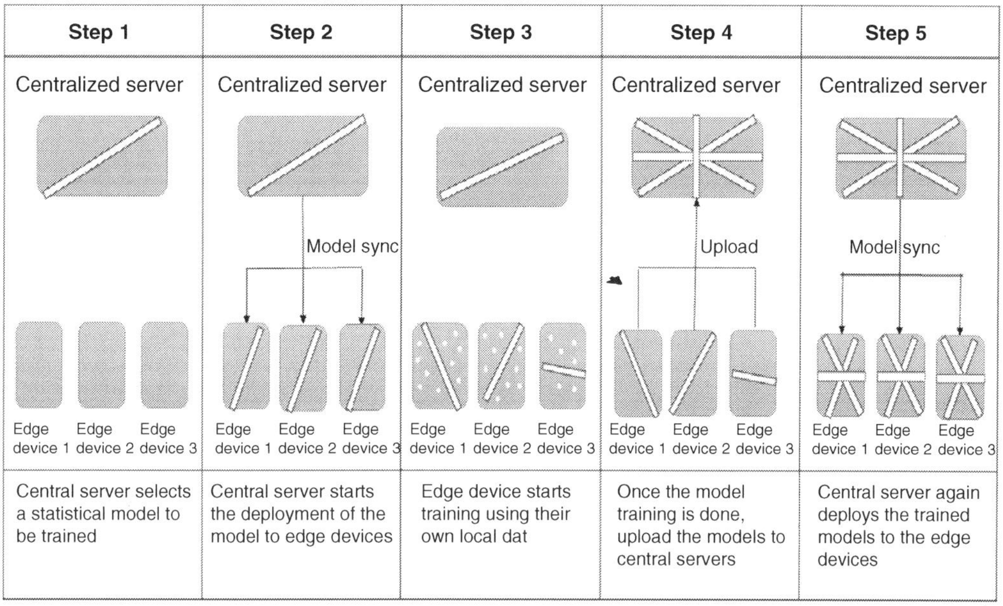

Figure 6.2 Working of federated learning in five steps.

In this way, a shared global model is built and is shared among all the sites and the characteristics of the global model are shared among all the local data models in order to integrate the global models into their ML local models.

Figure 6.2 shows the working of federated learning in detail and here in first step, central server selects the edge device for training the models; in the second step, central server starts the deployment of the model to the edge devices; in the third step, edge device starts training using their own local database and in the fourth step, once the model is trained at each edge device, it uploads the models to the central servers. In the fifth step, central server again deploys the trained models to the edge devices.

Federated learning encompasses two primary types: horizontal learning and vertical learning. In horizontal federated learning, the data across various clients exists within the same feature space, but the individual samples differ. For instance, this can be observed in scenarios where different hospitals possess standard tumor images, albeit from different sources. In contrast, vertical federated learning involves clients holding distinct features that pertain to the same sample, such as a patient's health insurance records and data from the hospital where they received treatment.

6.4 Some Scholarly Work Related to Federated Learning

To maintain overall control of gathered information, we can use Federated Learning. Ownership and management of the data will remain unchanged while utilizing it for ML purposes. In this case, transferring data to any central server is unnecessary. It will be decentralized in various nodes. Through this way, data breaches will be reduced. Collaborative data training can be achieved using parallel computing and coordination among the nodes, sensors, and data streams. Federated learning effectively distributes the model training and processing load across multiple computational nodes. This approach serves to minimize operational and computational overhead.

Simultaneously, it enhances data privacy by keeping sensitive data localized and reducing the need for centralized data processing.

Federated learning offers a solution to maintain complete control over data while harnessing its power for ML purposes. It allows organizations to retain ownership and oversight of their data without the need for centralizing it, thus reducing the risk of data breaches. One key aspect of federated learning is its ability to facilitate collaborative data training by coordinating and parallelizing computations across diverse computational nodes, sensors, and data streams. By distributing both models' training and processing responsibilities across multiple nodes, federated learning effectively minimizes operational and computational overhead while bolstering data privacy. A summary of some related recent works is as follows, which shows the advantage of utilizing federated learning in various domains:

In order to make it easier to train global AI (deep learning) models, Kavasidis et al. [15] established a platform that blends multiblockchain technology with federated learning capabilities. This platform is made to keep a transparent record of all training process-related data utilizing a number of blockchains, guaranteeing the training process's immutability. By using three cutting-edge federated learning algorithms on an industrial pharmaceutical dataset derived from two manufacturing lines, they were able to demonstrate the usefulness of this system. The findings showed good generalizability and quick convergence times, which were encouraging signs.

The objective of federated learning, a cooperative ML technique, is to train models without revealing sensitive user data. In a comprehensive analysis, Neto et al. [16] examined secure federated learning proposals designed to safeguard user privacy while improving model performance. For this investigation, data were screened and extracted from several electronic databases using a systematic review that adhered to predetermined criteria. The survey emphasized the critical role that security measures play in ensuring user privacy protection and the best model performance in the context of upcoming federated learning applications. Federated Learning has developed as a result of the drawbacks of the cloud-centric methodology. This collaborative learning approach tackles concerns like privacy protection and communication efficiency. Federated Learning accomplishes this by enabling the training of ML models utilizing real data from mobile devices while protecting privacy-sensitive data. It's significant because it enables collaborative ML algorithms without requiring the transfer of private data to a central cloud server. Consequently, Federated Learning plays a pivotal role in upholding data privacy in distributed environments. A developing area of research called Federated Learning makes use of the improved processing power of edge devices to maintain the confidentiality of participant data. These methods focus on building a collaborative learning model in which participants—who are usually mobile nodes—perform some of the learning process locally and contribute to a larger model. However, developing these collaborative learning models has its share of difficulties, including managing a range of participating devices and the statistical irregularities in the data they provide.

Tan et al. [17] focused on developing a federated learning system that stands apart from the traditional centralized learning approach. Here are the key innovations in their study: (a) Transfer Learning: To specifically extract data features from the region of interest (ROI) within photographs, they used transfer learning. This technique enhances data preprocessing and prepares it effectively for training purposes. (b) SMOTE Technique: They used the Synthetic Minority Oversampling Technique (SMOTE) to process the data. This technique's goal is to create a more balanced data classification, ultimately improving the accuracy of disease diagnosis predictions. (c) FeAvg-CNN + Mobile-Net: Within the federated learning framework, they utilized the FeAvg-CNN + MobileNet model. This choice ensures the privacy and security of customer data. (d) Experimental Results: The study presented comprehensive experimental results obtained using various deep learning, transfer learning, and federated learning models. These experiments encompassed both balanced and imbalanced mammography datasets. The findings clearly indicated that their approach consistently outperformed other methods, particularly in terms of classification accuracy. This suggests that their solution is highly suitable for applications in AI-driven healthcare.

In real-world scenarios, federated learning must not only ensure privacy but also fairness among participants. Sometimes, participants with ulterior motives might join the training process, contributing minimally while obtaining the global model. This can be unfair to those who actively contributed earlier. To address this issue, You et al. [18] introduced the Fed-ACC framework, incorporating a server-initiated global model accuracy control method. Fed-ACC works by measuring the cumulative contributions of new participants. It provides them with a model that matches their level of contribution. Importantly, it ensures the accuracy of participant gradients based on a model that gradually decays in accuracy. Under Fed-ACC, participants do not gain access to the full global model immediately upon joining. Instead, they must make a certain level of contributions before accessing the fully accurate model. To further protect privacy, an additional differential privacy mechanism is introduced. Experimental results demonstrate that Fed-ACC significantly enhances accuracy, achieving a 10–20% gain compared to state-of-the-art methods. Importantly, it achieves this while maintaining fairness, performance, and security in federated learning.

The healthcare data collaboration field offers enormous potential for improving patient care and medical research. However, the primary challenge lies in protecting the privacy and security of sensitive information, which poses significant hurdles. Abaoud et al. [19] has introduced an innovative solution to tackle these issues by implementing privacy-preserving federated learning models, opening up new possibilities in this complex research area. With this approach, healthcare organizations can jointly develop ML models using distributed data while maintaining the privacy of specific patient information. Modern privacy-preserving methods are used to protect sensitive data throughout the model's aggregation phase, including differential privacy and secure multi-party computation. To demonstrate the effectiveness of this solution, extensive simulations and evaluations were conducted, focusing on accuracy, computational efficiency, and privacy preservation. This method emphasizes the usefulness and efficacy of collaboration on healthcare data that is safely stored and privacy-preserving, and it offers a potential path further for the industry.

A decentralized ML strategy called federated learning enables devices to get model changes without relying on centralized cloud training. Asynchronous Federated Learning (AFL) is an extension of Federated Learning designed to reduce aggregation latency for improved efficiency. However, AFL can suffer from unstable learning performance due to poorly weighted local models. To tackle these challenges, Xu et al. [20] have presented a blockchain-based AFL system with a dynamic scalability factor. This novel method uses a committee-based consensus process built into the blockchain to increase reliability without requiring a lot of work. Furthermore, the dynamic scaling factor makes sure that AFL gives obsolete local models the proper weights. Through extensive tests conducted on a range of devices, this model has proven to have better learning performance, efficiency, and dependability when compared to existing methodologies.

In real-world scenarios, creating a robust network intrusion detection system (NIDS) to combat complex network security threats is challenging for individual enterprises. This challenge stems from the absence of high-quality attack instance data. To address this, dynamic weighted aggregation federated learning (DAFL), a successful intrusion detection system that is based on federated learning was introduced by Li et al. [21]. DAFL efficiently uses federated learning to safeguard data privacy. Additionally, it improves upon standard federated learning-based intrusion detection systems by including dynamic filtering and weighting techniques for local models. These strategies enhance DAFL's ability to detect network intrusions while reducing communication overhead. The proposed DAFL system is designed to overcome the limitations of single organizations with limited attack instances. It accomplishes this by expanding sample instances and constructing high-performance detection models suitable for complex network environments. This research presents a comprehensive system design and the key implementation aspects of DAFL. The effectiveness of DAFL is compared to that of existing methods through experimental evaluation, proving that it delivers superior detection performance with less communication overhead.

Efficient disaster classification and victim detection are crucial for facilitating effective rescue operations. Wong et al. [22] introduced a multitask learning (MTL) prototype that simultaneously addresses these two critical tasks. This MTL model offers several advantages, including reduced memory requirements (a savings of 12.8 MB) and improved disaster classification accuracy (a gain of 1–2%) while maintaining the same level of detection performance (average precision of 0.694) compared to traditional methods. The experimental findings demonstrated that Federated Learning-based approaches are on par with or even superior to centralized learning strategies. The research focused on optimizing Convolutional Neural Network (CNN) performance during training and inference stages. It began with designing an efficient MTL model capable of simultaneously handling disaster classification and victim detection. Because of AL, the training effort was reduced because the training algorithm could now actively retrieve and classify informative data from a collection of unlabelled datasets on each local IoT device.

Federated ML is a technology that allows resource-constrained appliances like IoT devices and smartphones to collaborate on creating a conveyed model while holding their raw data localized. This approach provides advantages like privacy preservation and economic benefits by avoiding centralizing data. However, the communication protocols used in federated ML can potentially be exploited by intruders to establish data poisoning attacks on other nodes, posing a significant threat to ML models. Liu et al. [23] set out to look into the security of federated ML models, particularly those employed in IoT systems. The federated MTL model is vulnerable to direct assaults that poison target nodes and indirect attacks that alter related nodes via the communication protocol, according to their experiments on diverse real-world datasets.

Device Selection and Resource Allocation (DSRA), also known as LAFLAS, is a unique technique that Lee [24] proposed for layer wise federated learning in wireless federated learning networks. This algorithm was developed with a focus on addressing the unique characteristics of both layer wise Federated Learning and wireless networks. The research delved into the crucial aspects of DSRA in the context of layerwise Federated Learning, aiming to optimize learning outcomes while considering the intricacies of wireless networks. Experimental findings from this study demonstrated that the model trained using LAFLAS outperformed models trained with DSRA algorithms that individually considered either the characteristics of layerwise Federated Learning or those of wireless networks.

A better way to manage the production, distribution, and consumption of electricity is made possible by the smart grid, which combines information and communication technology (ICT) with the traditional power infrastructure. However, it faces challenges in detecting abnormal activities within the grid. Detecting anomalies is vital for identifying issues like unusual power usage or equipment faults. Traditional systems call for exchanging data with a central server, which raises privacy issues and requires a lot of network and processing resources. To address this, Jithish et al. [25] introduced an approach where a global model is sent to smart meters for on-device training. Studies show that federated learning models provide equivalent anomaly detection performance compared to central models while respecting user privacy. These Federated Learning-based models work well and are useful in situations when there are limited resources, like with smart meters.

Federated learning techniques are utilized in IoT networks, notably in applications like intelligent healthcare systems, to provide safe monitoring services and privacy security. These methods are designed to protect sensitive data, but a centralized server typically manages their global aggregation. This centralized approach can be vulnerable to malicious attacks and privacy breaches, including inference attacks and free-riding, resulting in inefficient training models. Additionally, when patients upload analyzing parameters, participants may face the exposure of their personal data and the possibility of direct changes by the central server. To solve these limitations, Akter et al. [26] introduced a solution called the Federated Edge Aggregator, also known as Edge Intelligence. To balance model performance and privacy protection, this system uses an iteration-based CNN model and fake noise functions.

Federated learning seeks to make it easier to train an ML model collaboratively without requiring several clients to share raw data. Existing techniques for developing image segmentation models, however, rely on the assumption that each local client's training data is labeled in a consistent manner, i.e., with the same amount of image supervision. In practice, this assumption isn't always valid. To overcome this limitation, a label-neutral unified federated learning framework called FedMix was introduced by Wicaksana et al. [27], which is tailored for medical image segmentation. FedMix enables each client to update the federated model by effectively utilizing all available labeled data, spanning from strong pixel-level labels to weaker bounding box labels and even less informative image-level class labels. Studies using a variety of publicly accessible datasets show that FedMix greatly outperforms state-of-the-art techniques for segmenting medical images.

In the IoT context, data collected from various devices is conventionally aggregated and processed in centralized cloud environments to prepare ML models. However, this approach poses privacy risks for participants. To address this, Hassija et al. [28] proposed using federated learning, a privacy-preserving collective model training technique that utilizes data circulated across IoT users. To encourage participation in the federated learning process, participants need to be incentivized and rewarded for their contributions to the exercise of the federated model. This process of collective training occurs over an extended period with multiple iterations. Participants in the federated learning process may exhibit varying willingness to contribute, and there's the possibility of duplicate or low-quality data. Therefore, it's essential to assign rewards to participants based on their contributions within each process iteration. In this study, the researchers introduced a methodology for rewarding participants according to their contributions, and they employed an aggregation technique that uses Polyak-averaging to combine the weights of local models Based on each local model's precision on the test dataset, a weight is allocated to it. According to performance assessments, the federated learning model developed using this aggregation strategy obtains ML performance levels that increase with additional iterations and marginally surpass models developed using the Fed-Avg method. In addition, as compared to other profit-sharing plans, the incentivization methodology offers greater performance-based rewards.

6.5 Distinct Advantages of Federated Learning

In federated learning, a model is trained using the decentralized data and as a result, models have been trained using a variety of data with the minimum overhead and all these happened with the minimum latency and without compromising the privacy of the users. There are various examples of federated learning as follows [14].

6.5.1 Federated Learning is Collaborative in Nature

Federated learning is also known to be collaborative in nature and it allows several edge devices like mobile phones to allow models to learn and share predicted models together. Federated learning also helps keep the training data on the device rather than necessitating data to be uploaded to the central server. Collaborative learning works in a collaborative nature and is used for model training without taking any risk of sharing. For example, in the hospital sector, federated learning helps make models learn about the patient to predict various health updates of the patients.

6.5.2 Time-Saving

Federated learning saves lots of time as it can lead to the training of life-saving models in a collaborative nature while maintaining patient data privacy and developing results more quickly. There is no need to spend more time collecting and aggregating data from diverse sources each time. For example, in hospital sector, federated learning saves lots of time by keeping model learning in parallel fashion.

6.5.3 Security

Federated learning tends to secure the local data present in various edge devices like mobile phones of individual patients and in this way, servers cannot determine the address or source of the individual edge devices. For example, in hospital sector, federated learning provides security to the personal health data of the individual patients.

6.5.4 It Generates More Diverse Data

As federated learning involves training from different edge devices like mobile phones, it has diverse types of information related to the patients and, in this way, it can generate more diverse data. For example, in hospital sector, federated learning provides various diverse data for training the models so that models can perform well for different types of data. Figures 6.3 and 6.4 shows the federated learning in hospital sector and security of private data during federated learning in hospital respectively.

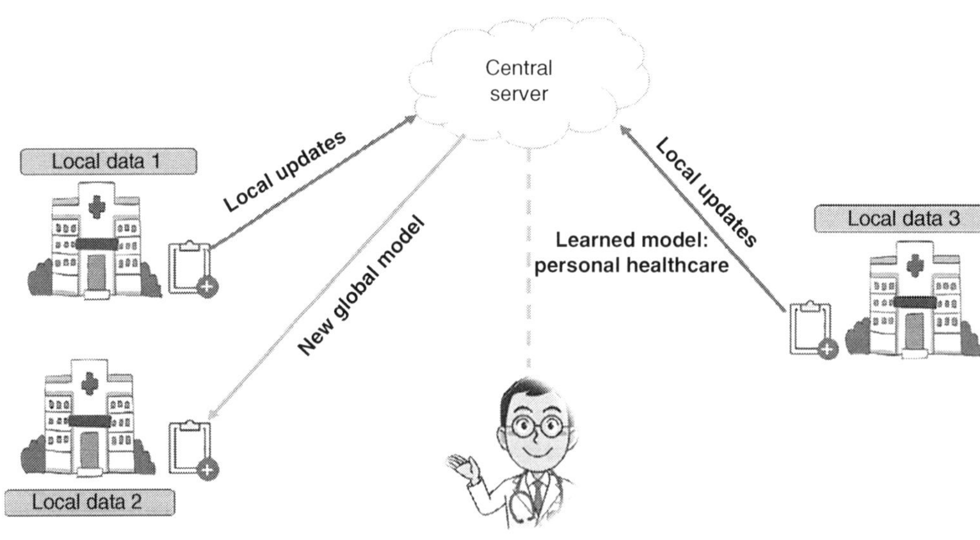

Figure 6.3 Federated learning in hospital sector.

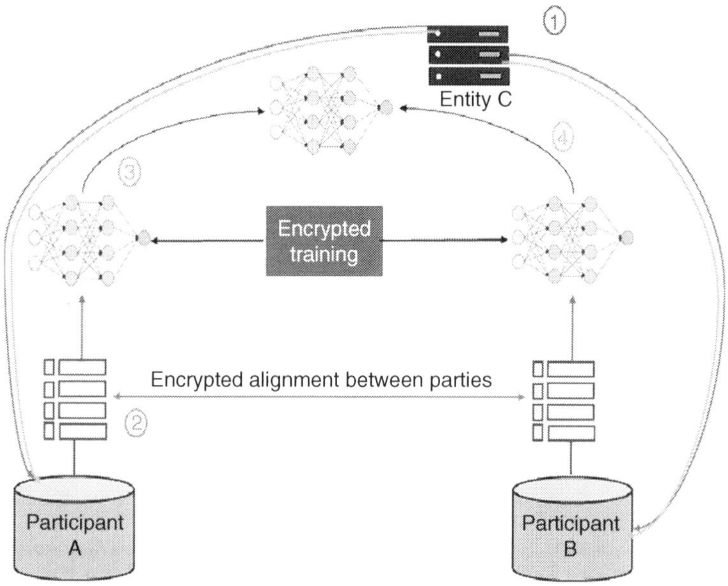

Figure 6.4 Security of private data during training in Federated Learning.

6.5.5 It Generates Real-Time Prediction

With the usages of federated learning, the prediction is done in real time using the edge device itself and even without the Internet connectivity, the edge devices are able to predict as the model is already downloaded in the edge devices. This also decreases the lag time that happens due to transferring the raw data from the edge device to the central server and then transferring the updated model back to the edge devices. For example, in hospital sector, it can generate real-time predictions for the patient using edge AI devices.

6.5.6 It Is Hands-Off and Noninvasive

One of the most important concerning things for any edge AI device is the draining of its battery due to its iterative training process for the models and, in fact, in these types of learning, the edge AI device is only involved in the training process while they are idle or free, or in the "no disturb" mode. For example, in hospital sector, a participating edge AI device like a mobile phone is only participating in the training process when it is idle or not being in use to save the battery drain.

6.5.7 Hardware Efficiency

One of the most important advantages is that this approach utilizes less complex hardware because federated learning models do not need one complex server to analyze the data.

6.5.8 Edge-Computing Capabilities

Federated Learning is well-suited for edge computing environments where data is processed near the data source (e.g., IoT devices). By enabling on-device model training, Federated Learning reduces latency and dependency on continuous Internet connectivity, making it ideal for real-time applications. Figure 6.5 shows the federated learning environment for providing more diverse data.

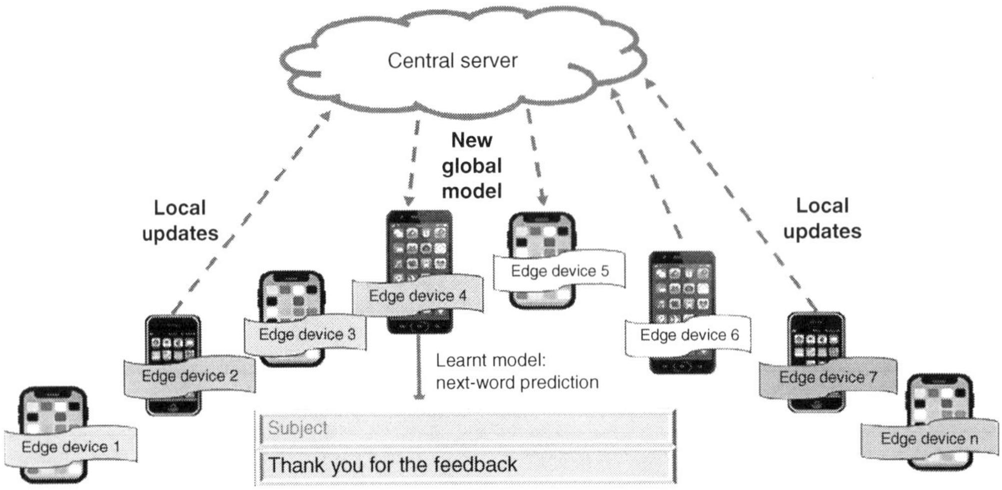

Figure 6.5 Federated learning for providing more diverse data.

6.5.9 Continuous Learning and Personalization

Federated Learning enables continuous model updates based on real-time user interactions. This allows applications to adapt and personalize user experiences dynamically without compromising privacy, making it invaluable in services like recommendation systems and personalized content delivery.

6.5.10 Scalability and Flexibility

Federated Learning is highly scalable, accommodating a growing number of devices and data sources. Its flexibility allows it to adapt to various domains, making it a versatile solution for a wide range of applications.

6.6 Applications of Federated Machine Learning

Federated learning, a decentralized approach to ML, has gained prominence in various domains due to its ability to address privacy concerns, reduce communication overhead, and enable collaborative model training. In this discussion, we will explore the applications of federated learning point by point, highlighting its significance in different sectors [29].

6.6.1 Healthcare

In the healthcare industry, privacy and security of patient data are paramount. Federated learning allows healthcare providers, research institutions, and pharmaceutical companies to collaborate on model training without sharing sensitive patient information. For instance, federated learning can be used to develop predictive models for disease diagnosis or treatment recommendations by aggregating insights from various hospitals and clinics while keeping patient data local.

6.6.2 Finance

Financial institutions deal with vast amounts of customer data, including transaction history and personal information. Federated learning can help banks and fintech companies develop fraud detection models collaboratively. Each institution trains its model locally using its data, and only model updates are shared, ensuring customer privacy. This approach improves the accuracy of fraud detection across the entire network.

6.6.3 Internet of Things

IoT devices generate enormous volumes of data, from smart home devices to industrial sensors. Federated learning enables these devices to collaboratively learn and adapt without sending data to a central server. For instance, smart thermostats in homes can learn temperature preferences without sharing specific temperature readings, enhancing energy efficiency.

6.6.4 Autonomous Vehicles

Federated learning can enhance the safety and intelligence of autonomous vehicles. Vehicles on the road can share insights into road conditions, traffic, and potential hazards without revealing sensitive information. This collective learning improves the decision-making capabilities of all vehicles on the network, making autonomous driving safer.

6.6.5 Natural Language Processing

In NLP applications, federated learning can be used to train models for sentiment analysis, language translation, or chatbots. Organizations with multilingual customer bases can collaborate to improve translation accuracy while respecting language data privacy.

6.6.6 Federated Healthcare Models

Federated learning is particularly relevant in healthcare for developing predictive models without compromising patient confidentiality. Institutions can collaborate to create models for predicting disease outbreaks, optimizing treatment plans, or monitoring public health trends. This approach has been crucial during the COVID-19 pandemic, where federated learning was used to analyze and predict the spread of the virus while preserving individual patient data.

6.6.7 Personalized Recommendations

Online platforms and streaming services often use federated learning to enhance user experience. By allowing devices or platforms to learn user preferences without sending detailed user data to central servers, personalized recommendations for products, movies, music, or content can be improved.

6.6.8 Defence and National Security

Federated learning is valuable for surveillance and reconnaissance missions. Unmanned aerial vehicles (UAVs), satellites, and ground sensors can collect vast amounts of data. Federated learning allows these devices to collaborate on object recognition, target tracking, and image analysis without transmitting raw data, ensuring data security and mission success.

6.6.9 Federated Learning in Education

In the field of education, federated learning can help customize learning experiences for students while preserving their privacy. Educators can collaborate to develop adaptive learning systems that recommend personalized resources and strategies for students, ensuring a more tailored and effective education.

6.6.10 Resource Allocation and Planning

Federated learning can assist in resource allocation and mission planning. Defence agencies can collaborate on models that optimize resource allocation during humanitarian missions, disaster response, or military operations. These models can adapt to changing conditions and requirements.

6.6.11 Edge Computing

Edge devices, such as smartphones and IoT devices, can benefit significantly from federated learning. These devices often have limited computational resources and connectivity. Federated learning enables on-device model training, making real-time, context-aware decisions possible without relying on a centralized server.

6.6.12 Research Collaborations

Academic and research institutions can leverage federated learning to collaborate on large-scale experiments. For example, scientists from different locations can collaboratively analyze data from particle accelerators, telescopes, or genomics research without sharing raw data, ensuring data sovereignty and security.

6.6.13 Personal Health Assistants

Virtual health assistants can provide personalized recommendations and health insights while preserving patient privacy. Federated learning allows these assistants to learn from a diverse range of healthcare data sources without centralizing sensitive health records.

6.6.14 Federated Learning in Energy Sector

Utilities and energy companies can use federated learning to optimize energy consumption, grid management, and predictive maintenance of equipment. This collaborative approach improves the efficiency and sustainability of energy systems.

6.6.15 Supply Chain and Manufacturing

Federated learning aids in optimizing supply chain operations, demand forecasting, and quality control without sharing proprietary production data among collaborating organizations. This is particularly relevant for global supply chains.

6.6.16 Environmental Monitoring

Federated learning can be applied to analyze environmental data collected from various sources, such as weather stations, satellites, and sensors. It aids in climate modelling, disaster prediction, and resource management.

6.6.17 Federated Learning in Law Enforcement

Law enforcement agencies can use federated learning for predictive policing while protecting sensitive information. Models can be trained collaboratively to identify crime patterns and allocate resources efficiently.

6.6.18 Secure Communication

In the field of secure communications, federated learning can help improve encryption and decryption algorithms. Different branches of the military or government agencies can collaborate to enhance cryptographic techniques while keeping the details of their encryption methods confidential. This leads to more robust and secure communication systems.

In summary, federated learning's applications span across various sectors, offering a privacy-preserving and efficient approach to collaborative ML. This decentralized paradigm ensures that data remains under the control of individual organizations or users while enabling them to benefit from collective intelligence and model improvements. Federated Learning, with its unique approach to decentralized ML, finds diverse applications across various domains, transforming how data is utilized and models are trained. In the healthcare sector, it enables collaborative research and the development of personalized treatment plans without compromising patient privacy, allowing hospitals and medical institutions to collectively advance medical science. In the finance industry, Federated Learning powers fraud detection systems by analyzing transaction patterns across multiple banks, enhancing security and protecting customer data. Smart devices in homes benefit from Federated Learning, as it enables them to learn and adapt to user preferences without transmitting sensitive information over networks, ensuring a seamless and secure user experience. In the realm of AI assistants, Federated Learning enhances speech recognition and language understanding by collaboratively training models on user interactions, all while preserving user privacy. Moreover, in sectors like manufacturing and supply chain management, Federated Learning optimizes processes, reduces inefficiencies, and ensures the security of proprietary information, enabling seamless collaboration between organizations. Additionally, in defence and intelligence applications, Federated Learning allows agencies to analyze vast amounts of data collectively, identifying patterns and threats to national security while safeguarding classified information. Its versatility and focus on data privacy position Federated Learning as a ground-breaking technology, fostering innovation and collaboration in a wide array of fields.

6.7 Conclusion

Federated Learning stands at the forefront of technological innovation, revolutionizing the way we approach ML in an era where data privacy and security are paramount. Figure 6.6 shows the various application of federated learning. Through this comprehensive exploration of its introduction, evolution, working principles, advantages, and diverse applications across various

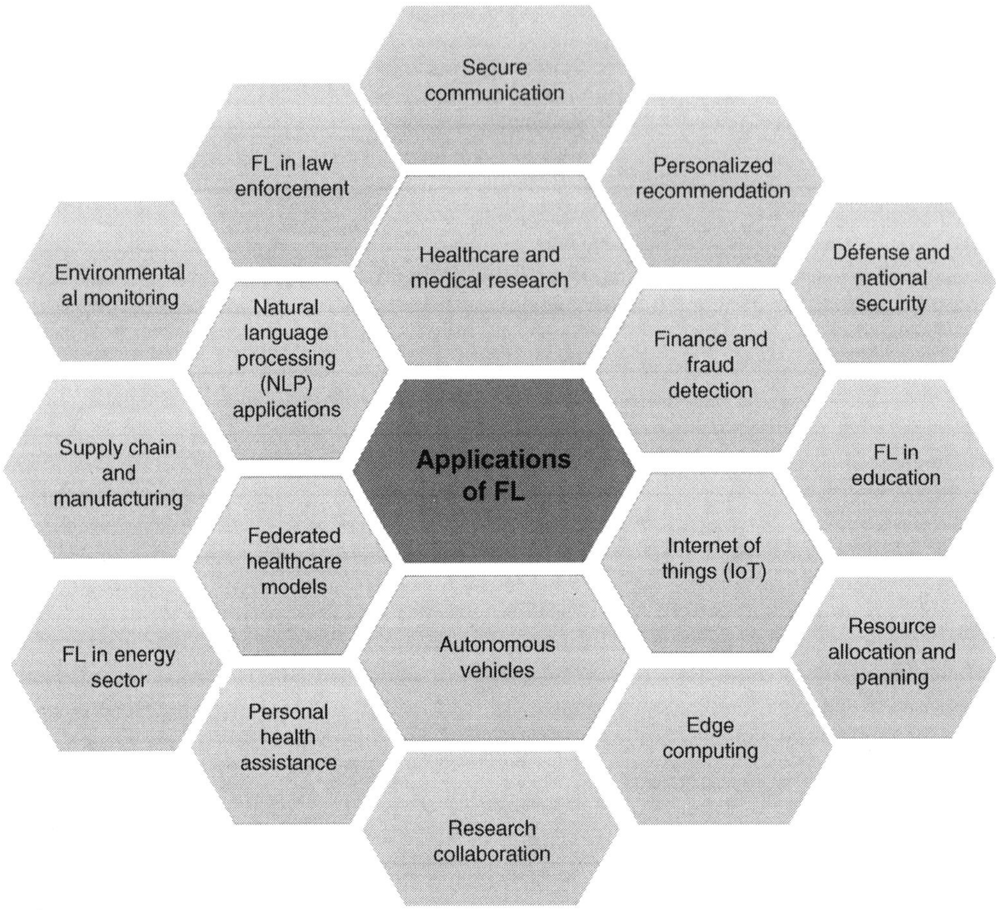

Figure 6.6 Applications of Federated Learning.

domains, it is evident that Federated Learning has transcended the conventional boundaries of centralized data processing. In essence, Federated Learning encapsulates the essence of a decentralized, privacy-preserving, and collaborative approach to ML. Its ability to harness the collective intelligence of decentralized devices, preserving individual privacy while enabling global model improvements, marks a significant milestone in the evolution of AI. The challenges associated with Federated Learning, including communication efficiency, model aggregation techniques, and privacy-preserving protocols, continue to be active areas of research, promising even more refined implementations in the future.

References

1 Beal, J. and Winston, P. (2009). The new frontier of human-level artificial intelligence. *IEEE Intelligent Systems* 24: 21–24. https://doi.org/10.1109/MIS.2009.75.

2 Lucci, S., Kopec, D., and Musa, S.M. (2022). *Artificial Intelligence in the 21st Century*. Germany: Mercury Learning and Information. ISBN: 9781683922247.

3 Yan, Y. (2022). Machine learning fundamentals. In: *Machine Learning in Chemical Safety and Health: Fundamentals with Applications*, 19–46. Wiley Online Library.
4 Jo, T. (2021). *Machine learning foundations. Supervised, Unsupervised, and Advanced Learning.* Cham: Springer International Publishing.
5 Ding, Z., Huang, Y., Yuan, H., and Dong, H. (2020). Introduction to reinforcement learning. In: *Deep reinforcement Learning: Fundamentals, Research and Applications*, 47–123. Springer.
6 Matsuo, Y., LeCun, Y., Sahani, M. et al. (2022). Deep learning, reinforcement learning, and world models. *Neural Networks* 152: 267–275.
7 Agarwal, N., Sondhi, A., Chopra, K., and Singh, G. (2021). Transfer learning: survey and classification. In: *Smart Innovations in Communication and Computational Sciences. Advances in Intelligent Systems and Computing*, vol. 1168 (ed. S. Tiwari, M. Trivedi, K. Mishra, et al.). Singapore: Springer https://doi.org/10.1007/978-981-15-5345-5_13.
8 Rahman, K.J., Ahmed, F., Akhter, N. et al. (2021). Challenges, applications and design aspects of federated learning: a survey. *IEEE Access* 9: 124682–124700.
9 Konečný, J., McMahan, B., and Ramage, D. (2015). Federated optimization: distributed optimization beyond the datacenter. arXiv:1511.03575.
10 Konečný, J., McMahan, H.B., Ramage, D., and Richtárik, P. (2016). Federated optimization: distributed machine learning for on-device intelligence.
11 Martineau, K. (2022). What-is-federated-learning. https://research.ibm.com/blog/what-is-federated-learning (accessed 21 September 2023).
12 McMahan, B., Moore, E., Ramage, D. et al. (2017). Communication-efficient learning of deep networks from decentralized data. *Proceedings of the 20th International Conference on Artificial Intelligence and Statistics*, 20–22 April 2017, Ft. Lauderdale, FL, USA. https://proceedings.mlr.press/v54/mcmahan17a.html.
13 Konečný, J., McMahan, H.B., Yu, F.X.(30 October 2017). Federated learning: strategies for improving communication efficiency. arXiv:1610.05492.
14 Dilmegani, C. (2023). What is federated learning? Use cases & benefits in 2024. https://research.aimultiple.com/federated-learning (accessed 21 September 2023).
15 Kavasidis, I., Lallas, E., Mountzouris, G. et al. (2023). A federated learning framework for enforcing traceability in manufacturing processes. *IEEE Access* 11: 57585–57597. https://doi.org/10.1109/ACCESS.2023.3282316.
16 Neto, H.N.C., Hribar, J., Dusparic, I. et al. (2023). A survey on securing federated learning: analysis of applications, attacks, challenges, and trends. *IEEE Access* 11: 41928–41953. https://doi.org/10.1109/ACCESS.2023.3269980.
17 Tan, Y.N., Tinh, V.P., Lam, P.D. et al. (2023). A transfer learning approach to breast cancer classification in a federated learning framework. *IEEE Access* 11: 27462–27476. https://doi.org/10.1109/ACCESS.2023.3257562.
18 You, X., Liu, X., Lin, X. et al. (2023). Accuracy degrading: toward participation-fair federated learning. *IEEE Internet of Things Journal* 10 (12): 10291–10306. https://doi.org/10.1109/JIOT.2023.3238038.
19 Abaoud, M., Almuqrin, M.A., and Khan, M.F. (2023). Advancing federated learning through novel mechanism for privacy preservation in healthcare applications. *IEEE Access* 11: 83562–83579. https://doi.org/10.1109/ACCESS.2023.3301162.
20 Xu, C., Qu, Y., Luan, T.H. et al. (2023). An efficient and reliable asynchronous federated learning scheme for smart public transportation. *IEEE Transactions on Vehicular Technology* 72 (5): 6584–6598. https://doi.org/10.1109/TVT.2022.3232603.

21 Li, J., Tong, X., Liu, J., and Cheng, L. (2023). An efficient federated learning system for network intrusion detection. *IEEE Systems Journal* 17 (2): 2455–2464. https://doi.org/10.1109/JSYST.2023.3236995.

22 Wong, Y.J., Tham, M.-L., Kwan, B.-H. et al. (2022). An optimized multi-task learning model for disaster classification and victim detection in federated learning environments. *IEEE Access* 10: 115930–115944. https://doi.org/10.1109/ACCESS.2022.3218655.

23 Liu, Z., Liu, Z., and Yang, X. (2023). Poisoning attack based on data feature selection in federated learning. *2023 13th International Conference on Cloud Computing, Data Science & Engineering (Confluence)*, Noida, India, pp. 106–110. 19–20 January 2023. https://doi.org/10.1109/Confluence56041.2023.10048854.

24 Lee, H.-S. (2022). Device selection and resource allocation for layer-wise federated learning in wireless networks. *IEEE Systems Journal* 16 (4): 6441–6444. https://doi.org/10.1109/JSYST.2022.3169461.

25 Jithish, J., Alangot, B., Mahalingam, N., and Yeo, K.S. (2023). Distributed anomaly detection in smart grids: a federated learning-based approach. *IEEE Access* 11: 7157–7179. https://doi.org/10.1109/ACCESS.2023.3237554.

26 Akter, M., Moustafa, N., Lynar, T., and Razzak, I. (2022). Edge intelligence: federated learning-based privacy protection framework for smart healthcare systems. *IEEE Journal of Biomedical and Health Informatics* 26 (12): 5805–5816. https://doi.org/10.1109/JBHI.2022.3192648.

27 Wicaksana, J., Yan, Z., Zhang, D. et al. (2023). FedMix: mixed supervised federated learning for medical image segmentation. *IEEE Transactions on Medical Imaging* 42 (7): 1955–1968. https://doi.org/10.1109/TMI.2022.3233405.

28 Hassija, V., Chawla, V., Chamola, V., and Sikdar, B. (2023). Incentivization and aggregation schemes for federated learning applications. *IEEE Transactions on Machine Learning in Communications and Networking* 1: 185–196. https://doi.org/10.1109/TMLCN.2023.3302811.

29 Aledhari, M., Razzak, R., Parizi, R.M., and Saeed, F. (2020). Federated learning: a survey on enabling technologies, protocols, and applications. *IEEE Access* https://doi.org/10.1109/access.2020.3013541.

7

Application Domains of Federated Learning

S. Annamalai[1], N. Sangeetha[2], M. Kumaresan[1], Dommaraju Tejavarma[1], Gandhodi Harsha Vardhan[1], and A. Suresh Kumar[1]

[1]*School of Computer Science and Engineering, Jain (Deemed-to-be University), Bangalore, Karnataka, India*
[2]*Department of Artificial Intelligence and Machine Learning, City Engineering College, Bangalore, Karnataka, India*

7.1 Introduction

7.1.1 Overview of Federated Learning

In 2016, Google used the phrase "Federated Learning" in a research paper that sought to address the issue of training a centralized machine learning model while the data is dispersed over millions of clients (in this context, mobile phones).

By training AI models without letting anyone else view or access your data, Federated Learning provides a mechanism for liberating data to enable new AI applications.

Horizontal, vertical, and federated transfer learning are the three types of data partitions that are supported by federated learning.

7.1.2 Importance and Advantages of Federated Learning

The amount of data needed to train typical machine learning models is considered to be an immense challenge, especially when that data is extremely sensitive to consumers. So, how can we quickly and accurately train ML models while also keeping consumers' privacy top of our minds? Here Federated Learning plays the role without transferring user data to cloud-based servers, federated learning includes building an ML model on user data [1].

Federated learning—also referred to as collaborative learning—trains an algorithm over a number of decentralized edge devices that store local data without transferring these datasets. As a result, it addresses the privacy issues raised previously and makes it possible to train models locally (on-device) considerably more quickly and effectively. In federated learning, only the Deep Neural Network (DNN) learning parameters are shared between the participants using different local datasets.

7.1.2.1 Why Federated Learning Is Crucial Right Now?

Accurate machine learning models are valuable to businesses, and the typical centralized machine learning systems have drawbacks such as combining isolated data on servers and a lack of continuous learning on edge devices. Federated learning helps eliminate these issues. In traditional ML, a

Model Optimization Methods for Efficient and Edge AI: Federated Learning Architectures, Frameworks and Applications, First Edition. Edited by Pethuru Raj Chelliah, Amir Masoud Rahmani, Robert Colby, Gayathri Nagasubramanian, and Sunku Ranganath.

center machine learning model is created with all training data that is accessible in a central space. This operates without any problems when predictions can be served by a central server.

Furthermore, consumers in mobile computing constantly want prompt responses, but the speed of communication between a user's device and a server may be insufficiently quick to provide the intended favorable user experience. The model may be put on the end-user device as a workaround, however, because the models are trained on entire datasets and the end-user device does not have access to them would make continued learning more difficult [2].

Another issue with classical machine learning is that user data is gathered in one place for training, which can violate privacy laws in some countries and increase the risk of data breaches. Federated learning eradicates these issues by permitting continuous learning on end-user devices and guaranteeing that end-user data stays on end-user devices.

7.1.2.2 Advantages of Federated Learning

By enabling companies to keep ownership and management of their data while also making it available for use with machine learning, federated learning can assist in maintaining total control over the data. Data breaches are less likely because data is kept decentralized and does not need to be moved to a centralized location. Federated learning is newer than centralized, conventional machine learning techniques, yet it already has a number of advantages. This includes:

- **Data Privacy and Security**: Preserving the training dataset locally on the devices will prevent the need for a data pool for the model [3].
- **Data Diversity**: Other problems than data security, such as network unavailability in edge devices, may obstruct businesses from merging datasets from different sources. Federated learning makes it simpler to access a variety of data, even when certain data sources can only communicate occasionally.
- **Increased Scalability and Cost Efficiency**: Federated learning speeds up the deployment of AI and IoT models and enhances the scalability of operations since it enables maintaining datasets in different locations and simultaneously training models on each piece of data.
- **Increased Adaptability**: As federated learning models can learn from past events and base predictions on them, they can typically be applied to new scenarios without the need for retraining. Additionally, this style of learning can be extended to different fields to enhance and gain knowledge in that topic.
- **Realtime Continuous Learning**: Models are continuously improved using client data instead of aggregating data as is required for continuous learning.
- **Hardware Effectiveness**: Because federated learning models don't require one, this method employs less sophisticated hardware.

7.1.3 Objectives of the Chapter

To provide a general outline of Federated Learning, needs and advantages of Federated Learning, and insights into its various application areas.

A few of the application areas are listed below and among them, some of the application areas will be discussed further.

- Healthcare
- Finance and Banking
- Internet of Things (IoT)

- E-commerce and Recommender Systems
- Telecommunications
- Automation Vehicle

7.2 Healthcare

7.2.1 Patient Privacy and Data Sharing Challenges

In healthcare system, to address concerns with data governance and privacy, federated learning is a technology created to support distributed machine learning. Through decentralized training of individual model copies using the proprietary data of participating healthcare facilities, federated learning fragments the growth of a machine learning model. Then, many models can be combined to produce the global model. Healthcare companies can use their shared, private data to safely train shared models using federated learning without having to exchange data [4].

Federated learning has significant consequences for patients because it is an effective technique for allowing data privacy protection. High-quality healthcare decisions would be provided regardless of location if a federated learning framework could be built on a large worldwide scale. The same high standard of healthcare decision-making would be available to patients in developing and undeveloped nations as well as in remotely located hospitals. Federated learning could aid medical professionals in treating unusual disorders and fending off new viruses before they spread globally.

7.2.2 Federated Learning for Clinical Decision Support Systems

Federated learning may be used to train a model in a clinical decision support system (CDSS) shown in Figure 7.1 that may recognize patients who are prone to a risk of readmission, recommend the optimal course of treatment, or anticipate the possibility that a patient will develop a given condition. Without needing the release of any personally identifiable information (PII), the model may be trained using data from many hospitals and clinics [5].

A few of the terminologies used in federated learning for CDSSS are mentioned as follows:

- **Client**: The hardware or software keeping the data and executing the local model training. Hospitals, clinics, or individual patients could all be considered consumers in the healthcare sector.

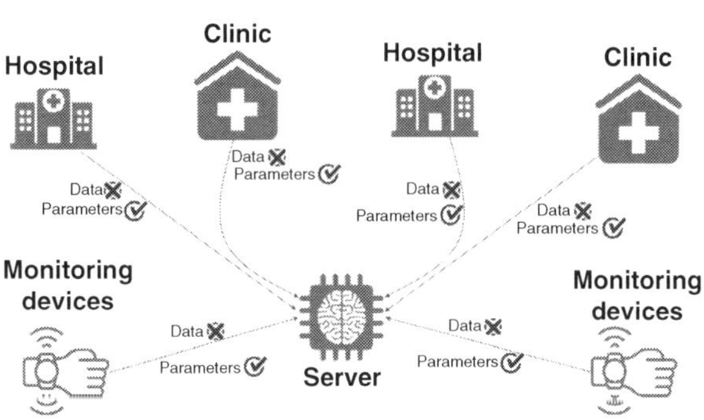

Figure 7.1 Clinical decision support systems.

- **Server**: The principal server in charge of managing the federated learning process. Although the server does not maintain any data, it does capture and aggregate client updates to update the global model.
- **Global Model**: The model that was trained utilized all of the client updates combined. The global model is available to all clients but is not retained on any one customer.
- **Local Model**: A model that was constructed using just one client's data. No other client or server may access the local model.

The approach that is applied to train the global model is termed the federated learning algorithm. The privacy of the client data is maintained by the federated learning algorithm [6].

Although the area of federated learning for CDSSs is still in its infancy, it has the ability to completely transform how healthcare is provided. Federated learning may aid in boosting the accuracy and effectiveness of CDSSs by allowing hospitals and clinics to work together on training ML models without having to share their data.

Following are some issues and constraints linked with federated learning for CDSSs:

Communication Overhead: If there are numerous customers, then controlling the volume of communication between clients and servers could be challenging.

Model Precision: The quality of the client-provided data may have an influence on the global model's accuracy.

7.2.3 Federated Learning for Disease Diagnosis and Prediction

Federated learning may be used to train a model that may identify people who are prone to a risk of contracting a given condition or anticipate the severity of a disease in the context of disease diagnosis and prediction. Without needing the release of any PII, the model may be trained using data from many hospitals and clinics.

The following points show how federated learning is used to detect and predict diseases:

1) The data is scattered across various devices, including mobile phones, tablets, and medical equipment.
2) Every device utilizes its own data to train a local model.
3) A global model is generated by integrating the local models.
4) Predictions are based on the global model.

The fundamental advantage of federated learning is that it preserves the confidentiality of patient data. Since the data never leaves the device, no one else can access it [7].

Although federated learning is still a nascent area of research, it has the potential to drastically revolutionize how illnesses are diagnosed and predicted. In addition to ensuring the security of patient data, it may aid in boosting the accuracy and effectiveness of sickness diagnosis and prediction models.

Following are some benefits of employing federated learning in the diagnosis and prognosis of diseases

- **Preserves Patient Privacy**: Since the data never leaves the device, no one else can access it.
- **Increases Accuracy**: The global model is more accurate since it was trained on more data than any one local model.
- **More Productive**: Because the local models are trained individually, they may be taught simultaneously. This may help you save time and money.

Federated learning is scalable and may be used to train models on big datasets, even when the data is scattered over numerous devices.

However, there are significant challenges with federated learning as well:

- **Communication Costs**: In order to aggregate their changes, the local models must interface with the server. The communication overhead may rise as a consequence.
- **Convergence**: Ensuring that the global model converges to a good solution could be problematic.
- **Security**: The server must prohibit unauthorized access to the aggregated updates.

7.2.4 Challenges and Opportunities in Healthcare Data Federations

A viable strategy for tackling the issues of data silos and privacy in healthcare is the use of healthcare data federations. Therefore, however, there are certain additional difficulties and chances that should be taken into account [8].

7.2.4.1 Challenges

- **Data Silos**: Healthcare data is often siloed, which means that it is kept in distinct systems that are not able to interact with one another. Due to this, it is challenging to communicate information and get a full picture of a patient's health.
  Patients are understandably worried about the privacy of their health information. By retaining the data on the user's device and only exchanging aggregated updates with the server, federated learning may assist in overcoming this problem. However, there are still certain privacy issues that must be resolved, such as how to guard against unwanted access to the data and how to guarantee that it isn't utilized for other reasons [9].
  Federated learning necessitates communication between the server and the devices, which adds to the overall communication overhead. This may increase communication overhead, particularly if there are many devices.
- **Convergence**: It might be challenging to make sure that the whole model converges on an appropriate response. This is due to the possibility that the local models weren't trained using identical data and may have distinct goals.
- **Security**: The server must prevent unauthorized access to the aggregated updates.

7.2.4.2 Opportunities

Federated learning may be used to enhance patient care by giving physicians and other healthcare professionals access to more data. They may use this to build more accurate diagnoses and treatment regimens [10].

- **New Opportunities for Research**: New healthcare research may be facilitated through federated learning. Researchers may acquire insights that would not be attainable otherwise by exchanging data without disclosing specific patient information.
- **Reduced Expenses**: Federated learning may save healthcare expenses by facilitating more effective data sharing. For hospitals, clinics, and insurance companies, this may mean cheaper expenses.
- **Enhanced Effectiveness**: By facilitating data sharing and research collaboration, federated learning may improve healthcare efficiency. This may hasten the creation of novel therapies and technologies.

In general, healthcare data federations provide a viable means of tackling the problems of data silos and privacy. Before this strategy can be extensively used, there are still a few issues that need to be resolved [11]. Here are a few strategies for overcoming the difficulties presented by healthcare data federations:

- **Data Standards**: Creating and implementing data standards may aid in facilitating data sharing across various systems.
- **Privacy-preserving Techniques**: A variety of privacy-preserving strategies may be employed to safeguard the confidentiality of health data while still enabling its sharing.
- **Communication Efficiency**: There are many methods that may be utilized to increase federated learning's communication effectiveness, including compression algorithms.
- There are many techniques that may be employed to guarantee that the whole model converges to a satisfactory solution.
- **Security**: Unauthorized access to the server must be prevented. Encryption and other security measures may be used for this.

Healthcare data federations may significantly influence both the quality of treatment and healthcare expenses by addressing these issues.

7.3 Finance and Banking

7.3.1 Privacy and Security Concerns in Financial Data

The banking sector is very concerned about privacy and security. Because it is often delicate and valuable, financial data is a target for hackers [12]. The model of Federated Learning in Financial Sector [12] is shown in Figure 7.2 and here are some issues with financial data privacy and security:

Data Breaches: In the banking sector, data breaches are a serious worry. Sensitive data is stolen or lost during a data breach. In addition to financial data like account numbers and credit card numbers, this might contain private information like names, residences, and social security numbers. The crime of obtaining another person's personal information and utilizing it to commit fraud is known as identity theft. This can include creating brand-new accounts, making erroneous transactions, or getting loans in the victim's name.

Fraud: In the financial sector, fraud is a significant threat as well. There are many different types of fraud, including phishing attempts, investment fraud, and credit card fraud.

Cyberattacks: The banking sector is increasingly at risk from cyberattacks. Cyberattacks are assaults on networks or computer systems. They may be used to extort money, disrupt business, or steal data.

The following terms pertain to issues with security and privacy in financial data:
Information that may be used to identify a specific person, such as their Name, Location, SSN, or date of birth, is known as PII.

Sensitive Data: Sensitive data is information that is regarded as private or confidential. This might contain personal information, financial information, or medical information.

Data Breach: A data breach occurs when private information is stolen or misplaced.

Fraud: In order to get anything, like money or property, by deceiving another person, it is a felony known as fraud.

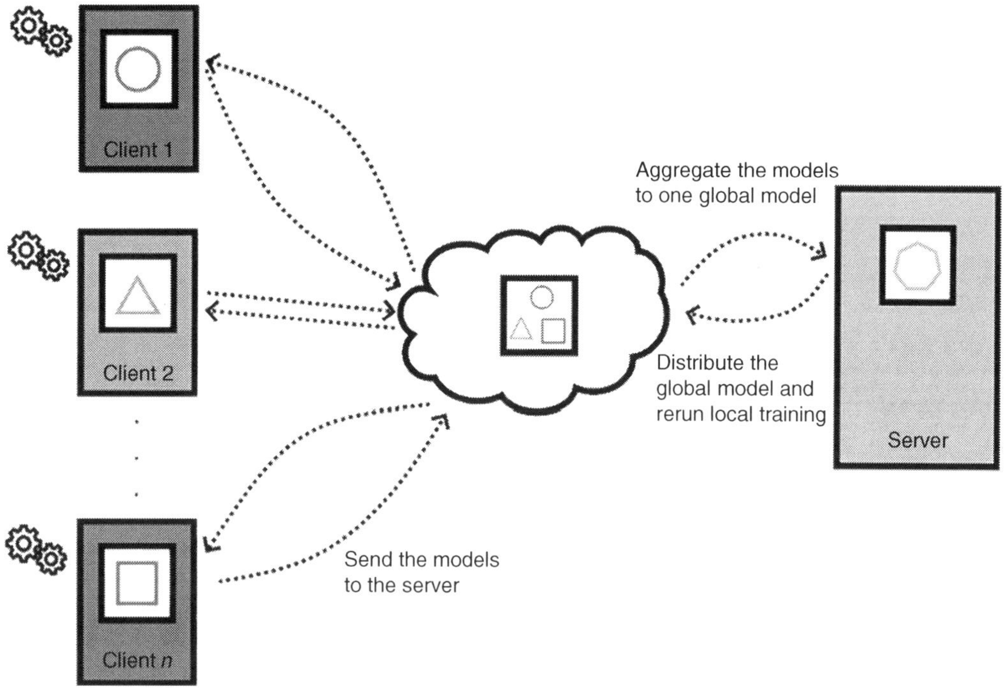

Figure 7.2 Federated learning in financial sector.

Following are a few instances of privacy and security issues with financial data:

- Customer Personal Information is taken if a bank's database is breached.
- The credit card information of a client is utilized for fraudulent transactions.
- To get a customer's personal information, a scammer calls them on the phone and claims to be from their bank.
- Money is taken from a customer's investment account once it has been breached.

7.3.2 Fraud Detection and Prevention with Federated Learning

Federated learning could be used to train a model that can recognize fraudulent transactions in the identification and prevention of fraud. Without sharing any specific data, the model may be trained using data from many banks and financial organizations [13]. In addition to helping to safeguard consumer data privacy, this enables banks to enhance their fraud detection skills [14] which is depicted in Figure 7.3.

Here is how federated learning works in terms of identifying and preventing fraud:

- There are several banks and financial institutions where the data is dispersed.
- Each bank uses its own data to develop a local model.
- A global model is then created by combining the local models.
- For making predictions regarding fraudulent transactions, the global model is employed.

The primary benefit of federated learning is that it safeguards the confidentiality of consumer information. Since the information never leaves the bank, no one else can access it. Although federated learning is still a young field of study, it has the potential to completely alter how fraud is

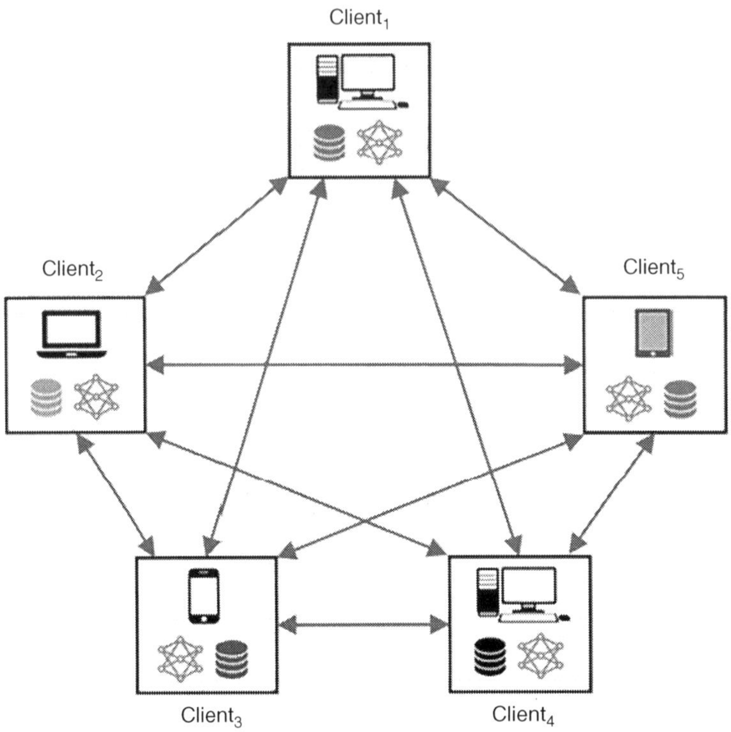

Figure 7.3 Fraud Detection and Prevention with Federated Learning.

found and avoided. It may aid in enhancing the efficacy and accuracy of fraud detection algorithms while safeguarding client data privacy [15–17].

Benefits of federated learning for preventing and detecting fraud include:

- **Protects Customer Privacy**: The data never leaves the bank, so it cannot be accessed by anyone else.
- **Improves Accuracy**: The global model is trained on more data than any single local model, so it can be more accurate.
- **More Productive**: Because the local models are trained separately, they may be trained concurrently. This may help you save time and money.
- **Scalable**: Federated learning allows for the training of models on big datasets, even when the data is dispersed over many banks.

Several difficulties with federated learning:

Cost of Communication: In order to aggregate their changes, the local models must connect with the server. The communication overhead may increase as a result.

Convergence: Ensuring that the global model converges to an appropriate response may be challenging.

The Server Needs to Protect the Aggregated Updates from Security: unauthorized access.

In general, federated learning is a promising strategy for identifying and preventing fraud. It offers the ability to safeguard client data privacy while enhancing the precision and efficacy of fraud detection methods [18, 19].

Examples of federated learning in the real world for detecting and preventing fraud in banking and finance:

Financial Services Education: A federated learning platform for financial services is being developed by Google. With the use of the framework, banks and other financial institutions may train machine learning models to spot fraud without disclosing their clients' personal information.

Citibank's Federated Learning for Fraud Detection: Citibank is using federated learning to train a model to detect fraudulent credit card transactions. Without sharing any specific data, the model is trained using data from a variety of Citibank clients.

Wells Fargo's Federated Learning for Identity Theft Prevention: Wells Fargo is training a model to prevent identity theft using federated learning. Without sharing any specific data, the model is trained using data from a variety of Wells Fargo clients.

7.3.3 Personalized Financial Services and Recommendations

Federated learning may be used to train a model that can offer goods and services to clients based on their unique requirements and preferences in the context of customized financial services and recommendations. Without sharing any specific data, the model may be trained using data from many banks and financial organizations [20, 21]. This enables the banks to provide more individualized services while also assisting in safeguarding the privacy of consumer data.

Here is how personalized financial services and recommendations use federated learning:

1) The information is spread out across several banks and financial entities.
2) Utilizing its own data, each bank builds a local model.
3) A global model is created by combining the local models.
4) The global model is used for customer suggestions.

The primary benefit of federated learning is that it safeguards the confidentiality of consumer information. Since the information never leaves the bank, no one else can access it. Although federated learning is still a young field of study, it has the potential to completely change how recommendations and tailored financial services are delivered. It may aid in enhancing the efficacy and accuracy of suggestions while safeguarding the confidentiality of consumer information [22].

Here are a few actual cases of federated learning for tailored financial services and suggestions in banking and finance:

Wells Fargo's Federated Learning for Personalized Banking: Wells Fargo is using federated learning to train a model to recommend products and services to customers based on their individual needs and preferences. Without sharing any specific data, the model is trained using data from a variety of Wells Fargo clients.

JPMorgan Chase's Federated Learning for Personalized Credit Card Offers: Federated learning is being used by JPMorgan Chase to build a model that will provide users with tailored credit card offers. The model is trained using data from several JPMorgan Chase clients; thus no specific customer data must be shared.

Bank of America's Federated Learning for Personalized Investing: Federated learning is being used by Bank of America to train a model that will suggest investments to consumers based on their unique risk tolerance and investing objectives. Without sharing any specific data, the model is trained using data from a variety of Bank of America clients.

Federated Learning have been used in many situations like customized financial services and recommendations in banking sector. As technology advances, it is expected to be employed in an increasing number of applications, assisting banks and other financial organizations in giving their consumers a better service.

7.3.4 Compliance and Risk Management in Federated Environments

Federated learning may be used to build a model that can recognize and stop fraudulent transactions in compliance and risk management. Without sharing any specific data, the model may be trained using data from many banks and financial organizations. This enables the banks to comply with rules and manage risks while also assisting in the privacy protection of consumer data [23, 24].

7.3.4.1 How Federated Learning Functions for Compliance and Risk Management Is Explained as Follows

1) The information is spread out across several banks and financial entities.
2) Utilizing its own data, each bank builds a local model.
3) A global model is created by combining the local models.
4) Fraudulent transaction detection and prevention are done using the global model.

The primary benefit of federated learning is that it safeguards the confidentiality of consumer information. Since the information never leaves the bank, no one else can access it.

Although federated learning is still a young field of study, it has the potential to completely alter how compliance and risk management are handled in the financial and banking industries. While preserving consumer data privacy, it may aid in enhancing compliance and risk management's accuracy and efficacy [25].

7.3.4.2 Some Advantages of Federated Learning for Compliance and Risk Management Are Listed as Follows

Protects Customer Privacy: The data never leaves the bank, so it cannot be accessed by anyone else.

Improves Accuracy: The global model is trained on more data than any single local model, so it can be more accurate.

More Productive: Because the local models are trained separately, they may be trained concurrently. This may help you save time and money.

Scalable: Federated learning allows for the training of models on big datasets, even when the data is dispersed over many banks.

Here are a few actual cases using federated learning in banking and finance for compliance and risk management:

- Google is creating a framework for federated learning specifically for the financial services industry. Without having to reveal their client's data, the framework is intended to assist banks and financial institutions in training machine learning models to adhere to rules and manage risks.
- Federated learning is being used by Citibank to build a model that will be used to identify fraudulent credit card transactions. Without sharing any specific data, the model is trained using data from a variety of Citibank clients [26].
- Wells Fargo's Federated Learning for Identity Theft Fire prevention involves Wells Fargo training a model to prevent identity theft using federated learning. Without sharing any specific data, the model is trained using data from a variety of Wells Fargo clients.

7.4 E-Commerce and Recommender Systems

7.4.1 Federated Learning for Personalized Recommendations

Federated Learning has gained attention in the realm of E-commerce and recommendation systems due to its potential to enhance personalized recommendations without compromising user privacy. Traditional recommender systems often depend on centralized data sources, which raises privacy issues and restricts the variety of available data. In order to train customized recommendation models, federated learning enables separate devices or platforms to work together while maintaining the decentralized nature of user data [27].

Federated learning for tailored suggestions looks like this:

1) The information is dispersed across many e-commerce sites.
2) Using its own data, each platform builds a local model.
3) A global model is created by combining the local models.
4) Recommendations to users are made using the global model.

The primary benefit of federated learning is that it safeguards user data privacy. Since the data never leaves the platform, no one else can access it.

Although federated learning is still a young field of study, it has the potential to completely change how tailored suggestions are offered on e-commerce and other online platforms. It may aid in enhancing the efficacy and accuracy of suggestions while safeguarding user data privacy.

Following are some advantages of federated learning for tailored recommendations:

- **Preserves User Privacy**: Since the information never leaves the platform, no one else can access it.
- **Increases Accuracy**: The global model is more accurate since it was trained on more data than any one local model.
- **More Productive**: Because the local models are trained separately, they may be trained concurrently. This may help you save time and money.
- **Scalable**: Even when data is dispersed across several platforms, federated learning may be utilized to train models on huge datasets.

However, there are certain difficulties with federated learning as well:

- **Communication Costs**: In order to aggregate their updates, the local models must interact with the server. The communication overhead may increase as a result.
- **Convergence**: Ensuring that the global model converges to a sound solution might be challenging.
- **Security**: The server must prevent unauthorized access to the aggregated updates.

7.4.2 Privacy-Preserving Customer Behavior Analysis

Privacy-preserving customer behavior analysis and federated learning can be used to train a model that can understand how customers interact with e-commerce platforms and recommender systems. The model shown in Figure 7.4 can be trained on data from multiple e-commerce platforms and recommender systems, without the need to share any individual data. This helps protect the privacy of customers' data, while also allowing the platforms and systems to improve their understanding of customer behavior [28].

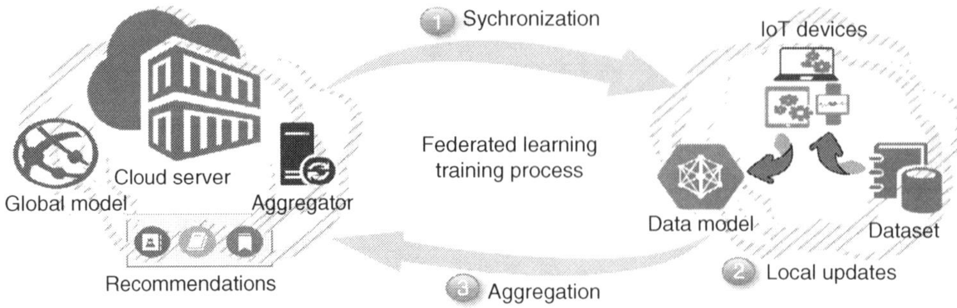

Figure 7.4 E-commerce and Recommender Systems.

7.4.2.1 Terminologies

1) **Client/Node/Device**: These words relate to the many companies, platforms, or organizations taking part in the Federated Learning process in the context of consumer behavior research. Each client has their own records of consumer activity.
2) **Server/Central Aggregator**: The Federated Learning procedure is coordinated by the central server. It gathers and combines client model updates, which helps advance the overall behavior analysis model.
3) **Model Aggregation**: To improve the overall behavior analysis model, model updates from diverse clients are combined. Aggregation techniques make a guarantee that the central model makes use of all participants' ideas.
4) **Local Training**: Using its own consumer behavior data, each client performs local model training. For analysis, one may use methods like clustering, sequence analysis, or neural networks.
5) **Global Behavior Analysis Model**: This is the main model that draws conclusions from the data of various clients without having direct access to the raw consumer behavior data. All participants gain from the pooled insights that are provided.

7.4.2.2 Customer Behavior Analysis Using Federated Learning While Protecting Privacy

For consumer behavior analysis that protects client privacy, federated learning provides a number of benefits:

1) **Data Privacy**: This addresses privacy issues connected with sharing raw data by keeping customer behavior data on the side of the respective clients. When handling private information like purchase history, browsing patterns, and personal preferences, this is especially crucial.
2) **Data Collaboration**: Federated Learning allows companies with comparable client bases to work together on model training while gaining access to confidential information.
3) **Regulatory Compliance**: Data protection laws are strictly enforced in several jurisdictions. By limiting data flow and maintaining data management, Federated Learning assists firms in adhering to requirements like GDPR.
4) **Personalized Insights**: The local model for each client is trained using the specific customer data from that client, resulting in individualized and nuanced insights for improved decision-making.
5) **Increased Accuracy**: Collaborative analysis across several platforms or enterprises results in more thorough and precise behavior analysis models.

7.4.2.3 Challenges and Considerations

1) **Communication Overhead**: Frequent contact between clients and the central server may result in delays and fees for communication.
2) **Model Heterogeneity**: Depending on the client, data quality and behavior may differ, necessitating cautious aggregation techniques.
3) **Bias Mitigation**: Since aggregated models might pick up biases from client data, it is important to keep an eye out for any possible biases.
4) **Security**: In order to prevent unwanted access, it is crucial to ensure the security of client devices and the central server.

Federated marketplaces and federated learning can be used to train a model that can recommend products to users based on their individual needs and preferences. Without sharing any specific data, the model may be trained using data from many federated markets. This enables the markets to provide more individualized suggestions while also assisting in preserving the privacy of user data [29, 30].

7.4.2.4 In Federated Markets, Federated Learning Operates as Follows

1) The information is dispersed across several federated markets.
2) Based on its own data, each market trains a local model.
3) A global model is created by combining the local models.
4) Recommendations to users are made using the global model.

The privacy of user data is protected, which is federated learning's main benefit in federated markets. Since the information never leaves the marketplace, no one else can access it.

Although federated learning is still a young field of study, it has the potential to completely alter how federated markets operate. It may aid in enhancing the efficacy and accuracy of suggestions while safeguarding user data privacy.

Following are some advantages of federated learning in federated marketplaces:

- **Preserves User Privacy**: Since the information never leaves the marketplace, no one else can access it.
- **Increases Accuracy**: The global model is more accurate since it was trained on more data than any one local model.
- **More Productive**: Because the local models are trained separately, they may be trained concurrently. This may help you save time and money.
- **Scalable**: Even when the data is dispersed over many markets, federated learning may be utilized to train models on big datasets.

But there are some difficulties with federated learning in federated markets as well:

- **Communication Costs**: In order to aggregate their updates, the local models must interact with the server. The communication overhead may increase as a result.
- **Convergence**: Ensuring that the global model converges to a sound solution might be challenging.
- **Security**: The server must prevent unauthorized access to the aggregated updates.

Federated learning is, all things considered, a viable strategy for federated markets. It has the ability to increase suggestion efficacy and accuracy while preserving user data privacy.

Here are some instances of federated learning in federated markets in the real world:

Amazon Marketplace, Walmart Marketplace, Alibaba Marketplace

7.4.3 Challenges and Performance Considerations in E-Commerce

Implementing Federated Learning in E-commerce comes with several challenges and performance considerations that need to be addressed to ensure its successful integration and effective operation. Following are some major issues to think about:

1) **Data Heterogeneity**: E-commerce systems often work with a variety of data types, such as information on products, consumer behavior, reviews, and more. The amount, quality, and distribution of data from various suppliers or product categories may differ. To ensure model convergence and efficacy, it is important to pay careful attention while aggregating such diverse data for federated learning.
2) **Communication Overhead**: In federated learning, the central server and participating clients must often communicate. This may result in higher network consumption and communication overhead, which might cause delays. To address this issue, effective communication tactics and optimization methods are required.
3) **Model Convergence**: It might be difficult to combine model updates from different clients while maintaining convergence. Variations in data distribution, quality, and training dynamics may affect the accuracy of the final model and the rate of convergence. This problem may be solved with the use of methods like weighted aggregation and sophisticated optimization techniques.
4) **Privacy and Security**: Sensitive information including customer profiles, transaction histories, and payment information is often included in e-commerce data. Federated Learning places a high priority on protecting data security and privacy. To avoid unwanted access or data breaches, protecting client data, safeguarding communication channels, and employing encryption measures are crucial.
5) **Fairness and Bias**: Federated Learning models may pick up on biases from client data. Biased recommendations in e-commerce may result in the unjust treatment of certain goods or consumers. Biases must be carefully monitored and reduced using fairness-aware aggregating approaches or pre-processing procedures.
6) **Cold Start Issue**: Limited data may be available for Federated Learning from recently onboarded suppliers or products. This may make it difficult to make significant contributions to the global model. This problem may be solved using techniques like transfer learning or information exchange among current suppliers.
7) **Scalability**: Because there are so many transactions and interactions on e-commerce platforms, there is often a lot of data involved. Federated Learning needs distributed computing infrastructure, effective parallelization, and load-balancing strategies to scale to manage this data.
8) **Vendor Engagement and Engagement**: In Federated Marketplaces, active vendor engagement is essential. It might be difficult to ensure that vendors share data and work together efficiently in the Federated Learning process; this requires reward systems and transparent communication routes.

9) **Hyperparameter Tuning**: For Federated Learning models to operate at their best, hyperparameters must be adjusted. Finding the ideal collection of hyperparameters, however, might be trickier than in conventional centralized training owing to the dispersed nature of the process.
10) **Model Deployment and Maintenance**: Careful orchestration is needed to deploy the global model back to the e-commerce platform while retaining real-time suggestions. Important factors to take into account include maintaining model consistency, version control, and timely updates.
11) **Resource Allocation**: Federated Learning needs computing power on the client devices as well as the central server. Good training and engagement need good resource allocation that takes into account elements like compute power, memory, and battery life.
12) **Evaluation and Monitoring**: It is crucial to continuously assess the performance of the global model and keep an eye out for drift or deterioration. The use of performance indicators and feedback loops aids in maintaining the caliber of suggestions.

7.5 Conclusion

7.5.1 Future Outlook for Application-Specific Federated Learning

The future outlook for application-specific Federated Learning is promising, as this approach continues to gain traction and evolve across various domains. Here are some key trends and expectations for the future of application-specific Federated Learning:

1) **Advancements in Privacy-Preserving AI**: Federated Learning will continue to play a vital part in the development of privacy-preserving AI technologies. As privacy concerns grow, there will be increased demand for models that can learn from decentralized data sources without compromising individual privacy.
2) **Healthcare and Medical Research**: In healthcare, Federated Learning will expand its applications, enabling medical institutions to collaborate on research and diagnostics while complying with strict data privacy regulations. Expect more breakthroughs in disease prediction, drug discovery, and personalized medicine.
3) **Financial Services and Fraud Detection**: The financial sector will increasingly adopt Federated Learning for fraud detection, credit scoring, and risk assessment. This will lead to more accurate models and reduced incidents of financial fraud.
4) **IoT and Edge Computing**: As IoT devices become more ubiquitous, Federated Learning will be instrumental in enabling edge devices to contribute to model training while maintaining low latency. Edge AI will become more powerful and efficient, benefiting industries like autonomous vehicles, smart cities, and industrial automation.
5) **Cross-Platform Personalization**: Federated Learning will continue to drive personalized experiences across various platforms, such as e-commerce, content recommendation, and digital advertising. Users will enjoy better-tailored content without compromising their data privacy.
6) **Natural Language Processing (NLP) and Chatbots**: The NLP field will see advancements in chatbots and virtual assistants that can leverage Federated Learning for improved

conversational abilities. These models will learn from interactions across different platforms and offer more context-aware responses.

7) **Education and Adaptive Learning**: Federated Learning will support adaptive learning platforms, allowing educational institutions to personalize education while respecting student data privacy. Expect better educational outcomes and improved learning experiences.
8) **Regulatory Frameworks**: Governments and regulatory bodies will develop and refine frameworks for the responsible use of Federated Learning, ensuring data security, privacy, and fairness. This will promote trust and wider adoption of the technology.
9) **Blockchain Integration**: Integration with blockchain technology may enhance the security and transparency of Federated Learning by providing immutable records of model updates and data contributions. This could be particularly important in highly regulated industries.
10) **Interoperability and Standardization**: Efforts to standardize Federated Learning protocols and interoperability between different platforms and devices will accelerate, making it easier for organizations to implement and collaborate on Federated Learning projects.
11) **Edge AI Hardware Advancements**: Hardware advancements, including more powerful and energy-efficient edge AI chips, will make Federated Learning more practical and scalable for edge devices, contributing to its widespread adoption.
12) **Research and Development**: Continuous research and development efforts in Federated Learning techniques, algorithms, and tools will lead to more efficient and effective models, further expanding their applicability.

In summary, the future of application-specific Federated Learning looks bright, driven by increasing concerns about data privacy, the need for collaboration in data-driven fields, and ongoing advancements in technology and regulation. Expect to see a growing number of industries leveraging Federated Learning to harness the power of decentralized data while respecting individual privacy and security concerns.

References

1 Huang, J., Cheng, X., Ji, Z. et al. (2022). AFLPC: an asynchronous federated learning privacy-preserving computing model applied to 5G-V2X. *Security and Communication Networks* 2022: 9334943.

2 Ma, J., Naas, S., Sigg, S., and Lyu, X. (2021). Privacy-preserving federated learning based on multi-key homomorphic encryption. arXiv preprint arXiv:2104.06824.

3 Park, J. and Lim, H. (2022). Privacy-preserving federated learning using homomorphic encryption. *Applied Sciences* 12 (2): 734.

4 Zhang, C., Li, S., Xia, J. et al. (2020). BatchCrypt: efficient homomorphic encryption for cross-silo federated learning. *Proceedings of the 2020 USENIX Conference on Usenix Annual Technical Conference*, USENIX Association, USA, 15–17 July 2020, pp. 493–506.

5 Tian, Y., Zhang, W., Simpson, A., and Jiang, Z. (2021). Defending against data poisoning attacks: from distributed learning to federated learning. *The Computer Journal* 66: 711.

6 Zhai, K., Ren, Q., Wang, L., and Yan, C. (2022). Byzantine-robust federated learning via credibility assessment on non- IID data. *Mathematical Biosciences and Engineering* 19 (2): 1659–1676.

7 Deng, Y., Lyu, F., Ren, J. et al. (2022). AUCTION: automated and quality-aware client selection framework for efficient federated learning. *IEEE Transactions on Parallel and Distributed Systems* 33 (8): 1996–2009.

8 Jin, H., Luo, Y., Li, P., and Mathew, J. (2019). A review of secure and privacy-preserving medical data sharing. *IEEE Access* 7 (99): 61656.

9 Shabisha, P., Braeken, A., Touhafi, A., and Steenhaut, K. (2019). Elliptic curve qu-vanstone based signcryption schemes with proxy re-encryption for secure cloud data storage. In: *Cloud Computing and Big Data: Technologies, Applications and Security*, vol. 21 (ed. M. Zbakh, M. Essaaidi, P. Manneback, and C. Rong), 1–19. Springer International Publishing https://doi.org/10.1007/978-3-319-97719-5_1.

10 Chen, Y., Lu, Z., Xiong, H., and Xu, W. (2018). Privacy-preserving data aggregation protocol for fog computing-assisted vehicle-to-infrastructure scenario. *Security and Communication Networks* 2018: 1–14.

11 Miao, Q., Lin, H., Hu, J., and Wang, X. (2022). An intelligent and privacy-enhanced data sharing strategy for blockchain-empowered Internet of Things. *Digital Communications and Networks* 88.

12 Stefanie, G., Stefanie, S., and Patrick, H. (2021). Federated Learning for Fraud detection in Accounting and Auditing. *ERCIM News* Special Theme Privacy Preserving Computation, 30–31.

13 Selvanambi, R., Natarajan, J., Karuppiah, M. et al. (2020). Lung cancer prediction using higher-order recurrent neural network based on glowworm swarm optimization. *Neural Computing and Applications* 32 (9): 4373–4386.

14 Mourad, B., Mohamed Ali, B., and Yacine, C. (2022). Security of Federated Learning: Attacks, Defensive Mechanisms, and Challenges. *Revue d'Intelligence Artificielle (RIA)*, 49–59.

15 Al-Fedaghi, S. (2011). A conceptual foundation for data loss prevention. *International Journal of Digital Content Technology and its Applications* 5 (3): 293–303.

16 Mamun, A.A., Hassan, M.K., and Van Lai, S. (2004). The impact of the Gramm-Leach-Bliley act on the financial services industry. *Journal of Economics and Finance* 28 (3): 333–347.

17 Buczak, A.L. and Guven, E. (2016). A survey of data mining and machine learning methods for cyber security intrusion detection. *IEEE Communications Surveys & Tutorials* 18 (2): 1153–1176.

18 Datardina, M. and Leung, K. (2009). Information leakage & data loss prevention. *IT Assurance Governance Inf*.

19 Janger, E.J. and Schwartz, P.M. (2001). The Gramm-Leach-Bliley act information privacy and the limits of default rules. *SSRN Electronic Journal* 86: 1219.

20 Dattolo, A., Ferrara, F., and Tasso, C. (2010). The role of tags for recommendation: a survey. *Proceedings of the 3rd International Conference on Human System Interaction (HIS'10)*, Rzeszow, Poland, pp. 548–555, 13–15 May 2010, https://doi.org/10.1109/HSI.2010.5514515.

21 Wang, Z., Wang, Y., and Wu, H. (2010). Tags meet ratings: improving collaborative filtering with tag-based neighborhood method. *Proceedings of the Workshop on Social Recommender Systems (SRS'10)*, January 2010, Hong Kong, China.

22 Dai, W., Yang, Q., Xue, G., Yu, Y. (2007). Boosting for transfer learning. *Proc. 24th Int'l Conf. Machine Learning*, pp. 193–200, 20 June 2007. Corvalis Oregon USA. https://doi.org/10.1145/1273496.1273521

23 Mooney, R.J. and Roy, L. (1999). Content-based book recommending using learning for text categorization. *Proceedings of DL00: Fifth ACM Conference on Digital Libraries*, 2–7 June 2000. San Antonio Texas USA. https://doi.org/10.1145/336597.33666, ISBN: 978-1-58113-231-1.

24 Breese, J.S., Heckerman, D., Kadie, C. (1988). Empirical analysis of predictive algorithms for collaborative filtering. *Proceedings of the 14th Annual Conference on Uncertainty in Artificial Intelligence (UAI98)*, 24–26 July 1998. Madison, WI, USA, 43–52, ISBN: 155860555X. https://dl.acm.org/doi/10.5555/2074094.2074100.

25 Wrethner, H., Hansen, R.H., and Ricci, F. (2007). Recommender systems. *IEEE the 40th Annual Hawaii International Conference on System Sciences 2007*, 03–06 January 2007. Waikoloa, HI, USA. https://doi.org/10.1109/HICSS.2007.459

26 Moshawrab, M.,Adda, M., Bouzouane, A. (2023). "Reviewing Federated Machine Learning and Its Use in Diseases Prediction" MDPI, Sensors 2023, Special Issue Smart Environments for Health and Well-Being.

27 Mothukuri, V., Parizi, R.M., Pouriyeh, S. et al. (2021, 2021). A survey on security and privacy of federated learning. *Future Generation Computer Systems* 115: 619.

28 Asad, M., Shaukat, S., Javanmardi, E. et al. (2023). A comprehensive survey on privacy-preserving techniques in federated recommendation systems. *MDPI Applied Science Journal* 13: 6201.

29 Shen, F. (2022). A study of federated learning in autonomous vehicle system. https://medium.com/@ys2223/a-study-of-federated-learning-in-autonomous-vehicle-system-ca9be70291fc (accessed 12 January 2024).

30 Asad, M., Moustafa, A., and Ito, T. (2020). FedOpt: towards communication efficiency and privacy preservation in federated learning. *MDPI Applied Science Journal* 10: 2864.

8

Advanced Architectures and Innovative Platforms for Federated Learning: A Comprehensive Exploration

Neha Bhati[1] *and Narayan Vyas*[2]

[1] *Department of Research & Development, AVN Innovations, Ajmer, Rajasthan, India*
[2] *Department of Computer Science and Application, Vivekananda Global University, Jaipur, Rajasthan, India*

8.1 Introduction

Federated Learning has carved a niche in the vast expanse of machine learning methodologies; unlike traditional approaches that centralized data for model training, Federated Learning advocates for a decentralized paradigm, allowing data to remain in its original locale, a device, or a server. This shift was not merely a technological whim but a response to growing concerns about data privacy, transmission costs, and the logistical challenges inherent in centralized systems. As Federated Learning began to gain traction, the spotlight inevitably turned to the architectures and platforms that underpin it [1]. These advanced architectures, meticulously crafted for Federated Learning, promise efficient training, minimal communication overhead, and stellar model performance. Simultaneously, the emergence of platforms tailored for Federated Learning has been instrumental. They offer the scaffolding, tools, and frameworks essential for the streamlined deployment and scalability of federated models, marking a significant stride in the journey of machine learning [2].

8.1.1 Background of Federated Learning

Federated Learning emerges as a transformative approach in machine learning, where model training occurs across multiple devices or servers while keeping data localized. Historically, traditional machine learning models required data centralized in one location, often leading to data privacy and transmission cost concerns. Federated Learning, however, offers a decentralized approach, allowing data to remain on its original device, thereby addressing many of these concerns [3].

8.1.2 Importance of Advanced Architectures and Platforms

The evolution and success of Federated Learning largely hinge on the development of advanced architectures and platforms. These architectures, tailored for Federated Learning, ensure efficient training, reduced communication overhead, and optimal model performance [4]. Concurrently, platforms designed for Federated Learning facilitate the seamless integration of these architectures, providing tools and frameworks that make the deployment of federated models more accessible and scalable [5].

Model Optimization Methods for Efficient and Edge AI: Federated Learning Architectures, Frameworks and Applications, First Edition. Edited by Pethuru Raj Chelliah, Amir Masoud Rahmani, Robert Colby, Gayathri Nagasubramanian, and Sunku Ranganath.

8.2 Basics of Federated Learning

By decentralizing model training, Federated Learning diverges from traditional machine learning. Instead of centralizing data, Federated Learning trains models across multiple devices without directly sharing raw data, ensuring enhanced data privacy. While this approach mitigates many concerns of centralized learning, it introduces challenges like device heterogeneity and communication overhead. Nonetheless, Federated Learning's emphasis on privacy and distributed learning positions it as a notable trend in contemporary machine learning [6].

8.2.1 Centralized vs. Federated Learning

Centralized learning, the traditional approach, involves collecting data from various sources and training a model on a centralized server. While being effective, this method raises concerns about data privacy, security, and the logistical challenges of data transmission [7]. On the other hand, Federated Learning stands as a decentralized alternative. In Federated Learning, the model is trained across multiple devices, with each device contributing to the model's learning without directly sharing raw data. This not only preserves data privacy but also leverages the computational capabilities of each participating device [8].

Table 8.1 contrasts centralized and federated learning approaches' key characteristics, advantages, and disadvantages.

8.2.2 Critical Challenges in Federated Learning

Despite its advantages, Federated Learning has challenges. The heterogeneity of devices, ranging from high-end servers to low-power IoT devices, can lead to discrepancies in computational capabilities and data distributions. Communication overhead, especially in bandwidth-limited scenarios, can hinder the efficiency of Federated Learning. Additionally, ensuring consistent model updates across devices and addressing potential security threats in a decentralized environment remain paramount concerns [9].

8.3 Advanced Architectures for Federated Learning

Federated Learning has spurred the development of specialized architectures, from neural network-based approaches that harness deep learning's prowess in decentralized settings to modular and layerwise techniques optimizing specific model components. Hybrid architectures blend

Table 8.1 Comparison of Centralized and Federated Learning.

Feature	Centralized Learning	Federated Learning
Data Location	Central server	Distributed across devices
Training	On central server	On individual devices
Data Privacy	Lower (data is moved)	Higher (data stays local)
Communication Overhead	Lower	Higher (due to aggregation)
Scalability	It depends on the server's capacity	Highly scalable

centralized efficiency with federated privacy, showcasing the dynamic evolution of Federated Learning's architectural landscape [10].

8.3.1 Evolution of Federated Learning Architectures

Over the years, the architectures underpinning Federated Learning have undergone significant evolution. Initially, simplistic models were adapted for federated settings, but as the complexities of decentralized data grew, so did the need for more sophisticated architectures tailored to address the unique challenges of Federated Learning [11].

8.3.2 Neural Network-Based Approaches

With their capacity to handle vast and complex data, neural networks have found a natural fit in the Federated Learning domain. These approaches leverage deep learning models, allowing for efficient data representation and learning decentralized, harnessing the power of devices in the federated network [12].

8.3.3 Modular and Layerwise Federated Learning

A more granular approach to Federated Learning involves modular and layerwise techniques. Here, different parts or layers of a model are trained on different devices, allowing for specialized learning and potentially reducing the communication overhead between devices [13].

8.3.4 Hybrid Architectures: Combining Centralized and Federated Approaches

The benefits and drawbacks of centralized and decentralized learning have led to the development of hybrid architectures. Centralized efficiency and the privacy advantages of federated learning are the goals of these designs [14].

8.4 Innovative Platforms for Federated Learning

With the growing popularity of Federated Learning, specialized systems have been created to facilitate its distributed methodology. The gap between Federated Learning's theoretical foundations and practical applications can be bridged with the help of several open-source and commercial platforms. By making Federated Learning accessible to a broader audience, open-source technologies like TensorFlow Federated and PySyft have facilitated the exploration, innovation, and refinement of Federated Learning by researchers and developers. On the other hand, commercial solutions provide the stability, scalability, and improved security required by large-scale, enterprise-level installations thanks to their purpose-built infrastructure [5].

8.4.1 Open-Source Platforms

- **TensorFlow Federated**: Building on the robust foundation of the TensorFlow framework, TensorFlow Federated has been designed with the nuances of Federated Learning in mind. It provides various tools and functionalities that empower developers to train models across many decentralized data sources, ensuring data privacy and efficient learning [15].

Table 8.2 Overview of Open-source Platforms for Federated Learning.

Platform	Main Features	Supported Architectures	Typical Use-Cases
TensorFlow Federated	Extension of TensorFlow, federated learning tools	Neural networks	General federated learning tasks
PySyft	Extends PyTorch and TensorFlow	Neural networks	Encrypted computations
FATE	Secure computing, collaborative ML	Custom architectures	Financial data, etc.

- **PySyft**: As a versatile library, PySyft augments popular frameworks like PyTorch and TensorFlow. Its capabilities extend beyond Federated Learning, encompassing encrypted computations, making it a go-to solution for scenarios demanding SMPCs.
- **FATE (Federated AI Technology Enabler)**: FATE is more than just a platform; it is a comprehensive ecosystem for Federated Learning. With a strong emphasis on secure computing, it facilitates collaborative machine-learning endeavors, ensuring data privacy and efficient model training [16].

Table 8.2 summarizes popular open-source platforms, highlighting their main features, supported architectures, and typical use cases.

8.4.2 Commercial Platforms and Solutions

The world of commercial Federated Learning platforms is vast and varied. These platforms, often backed by tech giants or specialized startups, bring to the table a suite of features tailored for businesses. Enhanced security protocols, scalability to handle vast networks of devices, and dedicated support are some of the hallmarks of these solutions [17]. Their turnkey nature ensures that organizations can seamlessly integrate Federated Learning into their operations without the steep learning curve.

8.4.3 Platform Selection Criteria: What to Look for?

The decision to adopt a particular Federated Learning platform is multifaceted. While scalability and security are paramount, other factors play a crucial role. The platform's ability to support diverse architectures, ease of integration into existing tech stacks, and the quality of community or vendor support can influence the decision-making process. Moreover, as industries have unique demands, the platform's flexibility and adaptability to cater to specific sectoral needs become essential. Whether healthcare's stringent data privacy requirements or retail's demand for real-time insights, the chosen platform should align with the industry's objectives and constraints.

8.5 Security and Privacy in Federated Learning

The allure of Federated Learning is deeply intertwined with its promise of data security and privacy. As data remains decentralized and resides on its original device, Federated Learning inherently protects against breaches [18]. However, the complexities of a federated system, coupled

Table 8.3 Security Mechanisms in Federated Learning.

Mechanism	Purpose	Strengths	Limitations
Homomorphic Encryption	Compute on encrypted data	High data privacy	Computational overhead
Differential Privacy	Add noise to data for privacy	Protects individual data points	It may reduce model accuracy
Secure Multiparty Computation (SMPC)	Multiple parties compute without revealing inputs	Collaborative computations	Requires trust among parties

with the ever-evolving landscape of cyber threats, necessitate the incorporation of additional security measures to fortify this framework [19].

Table 8.3 outlines various security mechanisms detailing their purpose, strengths, and potential limitations in the context of federated learning.

8.5.1 Homomorphic Encryption and Secure Aggregation

Homomorphic encryption stands as a revolutionary technique in the realm of data security. It permits computations on data while it is still encrypted, effectively ensuring that sensitive information remains shielded even during active processing. This is especially crucial in Federated Learning, where data from diverse sources converges for collective learning [20]. Complementing this is the concept of secure aggregation. Data from myriad devices must be aggregated in a federated network to refine the global model. Secure aggregation ensures this consolidation happens without divulging individual data points, thereby maintaining the sanctity of each device's data and further bolstering privacy.

8.5.2 Differential Privacy in Federated Learning

Differential privacy offers a nuanced approach to data protection. Introducing calibrated noise or randomness to data ensures that outputs (like query results) do not reveal specifics about individual entries. This obfuscation ensures that the specifics about an individual data point remain indiscernible even if data is accessed. Within the Federated Learning framework, differential privacy ensures that while models benefit from diverse data points, the individual privacy of each contributor remains uncompromised. It balances learning efficiency and data protection [21].

8.5.3 Secure Multiparty Computation

The essence of SMPC lies in its ability to facilitate computations involving multiple parties where each party's input remains concealed. Imagine a scenario where various devices, each with its data, contribute to a collective computation without revealing their data. SMPC makes this possible. In the context of Federated Learning, devices can collaboratively train a model, refining it with their data insights without the data ever leaving its native environment. This preserves data privacy and minimizes the risk of data breaches, making Federated Learning a more secure and robust framework [22].

8.6 Optimization Techniques for Federated Learning

The intricacies of Federated Learning present a unique set of challenges, distinct from traditional centralized learning. As the paradigm of Federated Learning gains momentum, the quest for refining its processes and enhancing its efficiency becomes paramount. This has led to the development and adoption of several optimization techniques explicitly tailored for the federated environment [23].

Table 8.4 enumerates different optimization techniques and quantifies their impact on various aspects of federated learning.

8.6.1 Communication-Efficient Algorithms

In the vast network of Federated Learning, where countless devices communicate and collaborate, the sheer volume of data exchanges can be overwhelming. While beneficial for data privacy, this decentralized structure can strain communication channels, leading to inefficiencies and delays. Communication-efficient algorithms come to the rescue by streamlining these data exchanges. They employ techniques like gradient sparsification and quantization to reduce the volume of data transmitted, ensuring that only the most pertinent information is communicated. This speeds up the learning process and conserves bandwidth, making Federated Learning more feasible even in resource-constrained settings.

8.6.2 Model Compression and Quantization

The diverse ecosystem of devices participating in Federated Learning, from powerful servers to modest IoT devices, means that storage and computational capabilities can vary widely. Model compression techniques, such as pruning and knowledge distillation, aim to reduce the complexity and size of models, ensuring they can be accommodated even on devices with limited resources. Quantization further complements this by representing model parameters with fewer bits, reducing the memory footprint without drastically affecting model accuracy. Together, these techniques ensure that Federated Learning remains inclusive, accommodating a wide range of devices without compromising learning efficacy.

8.6.3 Asynchronous Updates and Staleness Handling

The asynchronous nature of the natural world means that devices might only sometimes be in sync in a Federated Learning setting. Some might promptly provide model updates, while others

Table 8.4 Optimization Techniques and Their Impact.

Technique	Impact on Model Performance	Communication Overhead Reduction	Training Time Reduction
Communication-Efficient Algorithms	Varies based on implementation	Significant	Varies
Model Compression	Minimal loss if done correctly	Moderate	Faster deployments
Quantization	Minimal loss in accuracy	Significant	Faster computations

might lag due to network lags or operational downtimes. Asynchronous updates cater to this reality, allowing devices to contribute updates at their own pace. However, this can introduce 'staleness' to the model, where outdated or delayed updates might conflict with newer ones. Staleness handling techniques, such as adaptive learning rates and weighted averaging, ensure that these asynchronous updates harmoniously blend, ensuring the global model remains accurate and up-to-date.

8.7 Real-World Applications and Case Studies

The versatility of Federated Learning is evident in its widespread adoption across diverse sectors. Its unique blend of decentralized learning and data privacy makes it an attractive proposition for industries dealing with sensitive or voluminous data [24].

Table 8.5 showcases various applications of federated learning in different sectors, highlighting the specific challenges and solutions.

8.7.1 Healthcare: Patient Data Privacy and Collaborative Diagnostics

The healthcare sector is replete with sensitive patient data, where privacy is not just a preference but a mandate. Federated Learning emerges as a beacon in this context. By allowing data to remain on local devices, be it a hospital server or a patient's wearable device, Federated Learning ensures that patient confidentiality is never breached [25]. Beyond just data privacy, Federated Learning facilitates collaborative diagnostics. Different healthcare institutions can collaboratively refine diagnostic models without directly sharing patient data, leading to more accurate diagnoses and better patient outcomes.

8.7.2 Finance: Fraud Detection Across Institutions

With its intricate web of transactions, the financial sector is often a target for fraudulent activities. Traditional fraud detection methods often involve centralized data pools, which can be both a security risk and a logistical challenge [26].

Federated Learning offers a paradigm shift. Financial institutions, even competitors, can collaboratively train fraud detection models without exposing sensitive transaction details. This collective intelligence, powered by Federated Learning, ensures that fraud patterns are detected with higher accuracy, safeguarding both institutions and their customers.

Table 8.5 Real-World Federated Learning Applications Across Industries.

Industry	Challenges Faced	Federated Learning Solution
Healthcare	Data privacy, collaborative diagnostics	Local data processing, collaborative model training
Finance	Fraud detection, data privacy	Collaborative fraud pattern detection without data sharing
Smart Cities	Data aggregation, centralization	Decentralized data processing from multiple sensors
Retail	Personalization, data privacy	Local data processing for personalized recommendations

8.7.3 Smart Cities: Aggregating Data from Multiple Sensors Without Centralization

Modern urban landscapes, termed "smart cities," are dotted with sensors collecting data on everything from traffic patterns to air quality. Centralizing this data can be a logistical nightmare and a potential security risk. Federated Learning offers a solution. By processing data locally, at the sensor level, and then aggregating insights rather than raw data, innovative city management can gain a holistic view of the city's operations without the challenges of central data repositories [27].

8.7.4 Retail: Personalized Recommendations Without Sharing User Data

Personalization is the key to customer engagement in the age of e-commerce and digital marketplaces. Retailers strive to offer tailored recommendations, but this often comes at the cost of user data privacy [28]. With Federated Learning, retailers can process user data at the device level, whether a smartphone or a computer, to refine recommendation models. This ensures that users receive personalized shopping experiences without their data ever leaving their devices, striking a balance between personalization and privacy.

8.8 Challenges and Future Directions

Federated Learning stands at the intersection of innovation and practicality in machine learning. Its decentralized approach promises enhanced data privacy and localized processing. However, as with any pioneering technology, Federated Learning grapples with challenges even as it charts a course toward a promising future.

8.8.1 Scalability Concerns

The decentralized ethos of Federated Learning, while being its strength, also introduces scalability challenges. As the number of participating devices in the Federated Learning network grows, potentially reaching millions or even billions, ensuring that the system remains efficient becomes daunting. The sheer volume of data, the need for synchronized updates, and the variability in device capabilities necessitate the development of advanced algorithms and architectures. These must be adept at handling large-scale operations without compromising the speed or accuracy of the learning process.

8.8.2 Heterogeneity in Data and Devices

The Federated Learning landscape is characterized by various devices with capabilities and data profiles. The spectrum is vast, from robust data centers to everyday smartphones and even resource-constrained IoT devices. This diversity introduces challenges in ensuring consistent model training. The data distribution might vary across devices, leading to what is known as "non-IID" data, which can affect model accuracy. Solutions that can adapt to this diverse ecosystem, ensuring that each device, irrespective of its capabilities, contributes meaningfully to the global model, are essential.

8.8.3 Standardization and Interoperability

The nascent stage of Federated Learning means that many of its practices and protocols are still being defined. As more academic and commercial players enter the Federated Learning arena, the need for standardization becomes paramount. Interoperable solutions ensure that different

Federated Learning platforms, tools, and architectures can work in tandem, enhancing the collaborative spirit of Federated Learning. Standard protocols will streamline implementations and ensure that best practices are universally adopted, driving the success of Federated Learning initiatives.

8.8.4 The Future of Federated Learning: Predictions and Trends

The horizon for Federated Learning is expansive and bright. As global conversations around data privacy intensify, the decentralized model of Federated Learning offers a viable solution, promising both utility and privacy. The rise of edge computing, where computations move closer to the data source, further augments the relevance of Federated Learning. Innovations are on the horizon, with research focusing on enhancing security measures, refining optimization techniques, and developing tools that make Federated Learning more accessible. The confluence of these factors suggests that Federated Learning is not just a fleeting trend but a transformative force in machine learning's future.

8.9 Conclusion

8.9.1 Recap of Key Takeaways

The odyssey through the realm of Federated Learning has been a rich tapestry of insights, innovations, and implications. Federated Learning represents a paradigm shift at its core, challenging the traditional centralized machine learning model. This exploration has illuminated the multifaceted nature of Federated Learning, from its foundational tenets that prioritize data privacy and decentralized processing to the advanced architectures that underpin its operations. The real-world applications of Federated Learning, spanning sectors like healthcare, finance, and urban planning, underscore its practical relevance and transformative potential. However, like any pioneering technology, Federated Learning has its challenges. Scalability, heterogeneity, and the need for standardization emerge as critical focus areas. The comprehensive journey through these facets of Federated Learning offers a holistic understanding, highlighting its promises and challenges [29].

8.9.2 Encouraging Continued Exploration and Research

The narrative of Federated Learning is still being written. Its vast landscape, replete with opportunities and challenges, invites continued exploration. For researchers, it presents a fertile ground for innovation, pushing the boundaries of what is possible in decentralized machine learning. Practitioners, on the other hand, have the chance to implement and refine Federated Learning techniques, driving tangible impacts in various industries. Moreover, enthusiasts and students can contribute fresh perspectives, ensuring the Federated Learning community remains vibrant and diverse. As the digital world grapples with data privacy, security, and efficiency issues, Federated Learning emerges as a beacon, pointing toward a future where collaborative learning and data privacy coexist harmoniously. The journey ahead is filled with potential, and the call is clear: to delve, discover, and drive the next chapter in the Federated Learning saga.

References

1 Aledhari, M., Razzak, R., Parizi, R.M., and Saeed, F. (2020). Federated learning: a survey on enabling technologies, protocols, and applications. *IEEE Access* 8: 140699–140725.

2 Nguyen, D.C., Ding, M., Pathirana, P.N. et al. (2021). Federated learning for internet of things: a comprehensive survey. *IEEE Communications Surveys & Tutorials* 23 (3): 1622–1658.

3 Long, G., Tan, Y., Jiang, J., and Zhang, C. (2020). Federated learning for open banking. In: *Federated Learning*, vol. 12500 (ed. Q. Yang, L. Fan, and H. Yu), 240–254. Springer International Publishing. https://doi.org/10.1007/978-3-030-63076-8_17.

4 Kishor, K. (2022). Personalized federated learning. In: *Federated Learning for IoT Applications* (ed. S.P. Yadav, B.S. Bhati, D.P. Mahato, and S. Kumar), 31–52. Springer International Publishing. https://doi.org/10.1007/978-3-030-85559-8_3.

5 Issa, W., Moustafa, N., Turnbull, B. et al. (2023). Blockchain-based federated learning for securing internet of things: a comprehensive survey. *ACM Computing Surveys* 55 (9): 1–43. https://doi.org/10.1145/3560816.

6 Yang, Q., Liu, Y., Chen, T., and Tong, Y. (2019). Federated machine learning: concept and applications. *ACM Transactions on Intelligent Systems and Technology* 10 (2): 1–19. https://doi.org/10.1145/3298981.

7 Briggs, C., Fan, Z., and Andras, P. (2021). A review of privacy-preserving federated learning for the internet-of-things. In: *Federated Learning Systems*, vol. 965 (ed. M.H.U. Rehman and M.M. Gaber), 21–50. Springer International Publishing. https://doi.org/10.1007/978-3-030-70604-3_2.

8 Li, L., Fan, Y., Tse, M., and Lin, K.-Y. (2020). A review of applications in federated learning. *Computers & Industrial Engineering* 149: 106854.

9 Bonawitz, K., Eichner, H., Grieskamp, W. et al. (2019). Towards federated learning at scale: system design. *Proceedings of Machine Learning and Systems* 1: 374–388.

10 Burri, S.R., Agarwal, D.K., Vyas, N. and Duggar, R. A Machine Learning Framework for Accurate Prediction of Parkinson's Disease from Speech Data. *2023 3rd International Conference on Innovative Sustainable Computational Technologies (CISCT)*, Dehradun, India, 08 September 2023, pp. 1–6. https://doi.org/10.1109/CISCT57197.2023.10351422.

11 Zhang, C., Xie, Y., Bai, H. et al. (2021). A survey on federated learning. *Knowledge-Based Systems* 216: 106775.

12 Lo, S.K., Lu, Q., Zhu, L. et al. (2022). Architectural patterns for the design of federated learning systems. *Journal of Systems and Software* 191: 111357.

13 Zhu, S., Liu, T., and Sun, G. (2023). Layer-wise personalized federated learning with hypernetwork. *Neural Processing Letters* https://doi.org/10.1007/s11063-023-11420-z.

14 Chatterjee, S. and Hanawal, M.K. (2022). Federated learning for intrusion detection in IoT security: a hybrid ensemble approach. *International Journal of Internet of Things and Cyber-Assurance* 2 (1): 62. https://doi.org/10.1504/IJITCA.2022.124372.

15 Reina, G.A., Gruzdev, A., Foley, P. et al. (2022). OpenFL: an open-source framework for federated learning. *Physics in Medicine & Biology* 67 (21): 214001. https://doi.org/10.1088/1361-6560/ac97d9.

16 Moncada-Torres, A., Martin, F., Sieswerda, M. et al. (2020). VANTAGE6: an open source priVAcy preserviNg federaTed leArninG infrastructurE for secure insight eXchange. *AMIA Annual Symposium Proceedings, 870,* 25 January 2021. https://www.ncbi.nlm.nih.gov/pmc/articles/PMC8075508/

17 Khan, M., Glavin, F.G., and Nickles, M. (2023). Federated learning as a privacy solution overview. *Procedia Computer Science* 217: 316–325.

18 Mothukuri, V., Parizi, R.M., Pouriyeh, S. et al. (2021). A survey on security and privacy of federated learning. *Future Generation Computer Systems* 115: 619–640.

19 Ma, C., Li, J., Ding, M. et al. (2020). On safeguarding privacy and security in the framework of federated learning. *IEEE Network* 34 (4): 242–248.

20 Punugoti, R., Dutt, V., Kumar A., and Bhati, N. (2023). Boosting the accuracy of cardiovascular disease prediction through SMOTE. *2023 International Conference on IoT, Communication and Automation Technology (ICICAT)*, 23 June 2023, Gorakhpur, India, pp. 1–6, https://doi.org/10.1109/ICICAT57735.2023.10263703.

21 El Ouadrhiri, A. and Abdelhadi, A. (2022). Differential privacy for deep and federated learning: a survey. *IEEE Access* 10: 22359–22380.

22 Zhu, H. (2020). On the relationship between (secure) multi-party computation and (secure) federated learning. arXiv:2008.02609. arXiv. http://arxiv.org/abs/2008.02609.

23 Hanzely, F., Zhao, B., and Kolar, M. (2023). Personalized federated learning: a unified framework and universal optimization techniques. arXiv:2102.09743. arXiv. http://arxiv.org/abs/2102.09743.

24 Liu, Y., James, J.Q., Kang, J. et al. (2020). Privacy-preserving traffic flow prediction: a federated learning approach. *IEEE Internet of Things Journal* 7 (8): 7751–7763.

25 Qayyum, A., Ahmad, K., Ahsan, M.A. et al. (2022). Collaborative federated learning for healthcare: multi-modal COVID-19 diagnosis at the edge. *IEEE Open Journal of the Computer Society* 3: 172–184.

26 Kapadiya, K., Patel, U., Gupta, R. et al. (2022). Blockchain and AI-empowered healthcare insurance fraud detection: an analysis, architecture, and future prospects. *IEEE Access* 10: 79606–79627.

27 Shari, N.F.M. and Malip, A. (2022). State-of-the-art blockchain technology solutions for data dissemination in smart cities: a comprehensive review. *Computer Communications* 189: 120–147.

28 Sahoo, A.K., Mallik, S., Pradhan, C. et al. (2019). Intelligence-based health recommendation system using big data analytics. In: *Big Data Analytics for Intelligent Healthcare Management*, 227–246. Elsevier. https://www.sciencedirect.com/science/article/pii/B978012818146100009X.

29 Bashir, A. K., Victor, N., Bhattacharya, S. et al. (2023). A survey on federated learning for the healthcare metaverse: concepts, applications, challenges, and future directions. arXiv:2304.00524. arXiv. http://arxiv.org/abs/2304.00524.

9

Federated Learning: Bridging Data Privacy and AI Advancements

D. Sumathi, Likitha Chowdary Botta, Mure Sai Jaideep Reddy, and Avi Das

School of Computer Science and Engineering, VIT-AP University, Inavolu, Andhra Pradesh, India

9.1 Introduction

In an age characterized by the exponential growth of digital information and the ubiquity of artificial intelligence (AI) models, the paramount importance of data privacy has never been more pronounced. The 2023 Cisco Data Privacy Benchmark Study [1] reveals that a staggering 94% of organizations assert that their customers will refrain from transacting with them if data is inadequately protected. This statistic underscores the crucial role data privacy plays in fostering consumer trust and driving business sustainability. Furthermore, the emergence of machine learning and deep learning paradigms has revolutionized data-driven decision-making, promising enhanced efficiency and performance across various domains.

As organizations delve deeper into the world of ML and DL, safeguarding sensitive data becomes imperative. The multifaceted importance of data privacy during the training of ML and DL models is emphasized, underscoring the necessity of robust security measures. Moreover, AI models have proven instrumental in reshaping various fields. A study by [2] demonstrated the remarkable success of AI models in natural language processing, while DeepMind's AlphaGo [3] showcased the potential of AI in mastering complex games. Beyond these specific examples, AI models continue to drive innovation and efficiency in diverse applications, from healthcare [4] to autonomous vehicles [5].

The growing awareness of data privacy concerns has led to the development of innovative approaches to safeguarding sensitive information during model training. One such approach that has garnered considerable attention is federated learning. This study aims to deliver a holistic examination of the benefits associated with AI models, with a focus on Machine Learning and Deep Learning, and underscores the distinct advantages of federated learning over conventional models for preserving data privacy.

Federated learning, a groundbreaking paradigm in the field of AI, has emerged as a pioneering approach, enabling the training of machine learning models across decentralized and distributed environments [6]. This methodology departs from the conventional centralized training of AI models, offering unique advantages that extend beyond privacy considerations. In federated learning, model training occurs locally on individual devices or at the edge, with data distributed across these endpoints. The key distinction lies in the fact that data remains localized and is not centralized in a data center. This localized approach bears several distinct advantages. Unlike traditional AI models,

Model Optimization Methods for Efficient and Edge AI: Federated Learning Architectures, Frameworks and Applications, First Edition. Edited by Pethuru Raj Chelliah, Amir Masoud Rahmani, Robert Colby, Gayathri Nagasubramanian, and Sunku Ranganath.

which require extensive data aggregation in a centralized repository, federated learning minimizes data movement, reducing the need for large-scale data transfers [7]. This not only mitigates privacy concerns but also alleviates bandwidth and storage requirements, making it an efficient choice for resource-constrained environments.

Furthermore, federated learning fosters collaboration across devices and entities while preserving data sovereignty [8]. Each device independently updates the model with its local data, and the global model is improved through a federated averaging process. This decentralized approach ensures that data owners retain control over their information and participate in the model improvement process without sharing their data [9]. Such a cooperative, privacy-aware framework aligns with contemporary data protection requirements and regulatory standards. The distinctive characteristics of federated learning extend beyond data privacy and efficiency. It also offers the promise of continuous learning [10]. Models can be refined on an ongoing basis as more devices contribute their data, making it well-suited for dynamic and evolving environments. This stands in contrast to traditional AI models, which may require extensive retraining on centralized data, causing delays in adapting to changing conditions.

In the following subsections of this chapter, we shall explore in depth the subtle complexities of data privacy during machine learning and deep learning model training, explore the potential advantages of AI models in diverse applications, and highlight the manifold benefits of federated learning beyond data privacy. Federated learning has proven to be a promising paradigm in AI, offering not only robust data protection but also scalability, improved model generalization, and cost-effectiveness. This examination is rooted in a critical analysis of recent and pertinent literature, ensuring that the insights offered remain relevant in the evolving landscape of AI, where federated learning stands out as a pioneering approach with far-reaching implications.

9.2 Horizontal

Horizontal Federated Learning is an innovative architecture that has garnered significant attention in machine learning due to its privacy-preserving and decentralized nature. In Horizontal Federated Learning, participating entities collaborate to train machine learning models without sharing their raw data, thereby addressing privacy concerns. Horizontal Federated Learning's architecture is distinct from conventional centralized models, as it hinges on data localization and secure model update aggregation. It mitigates the need for data transmission, offering a robust solution to the privacy and security challenges associated with centralized data sharing.

Central to this architecture is the concept of decentralized model training, ensuring that data, typically originating from diverse and separate sources, remains localized and confidential. A foundational study by [11] elucidates the principal architectural underpinnings of Horizontal Federated Learning, emphasizing secure aggregation mechanisms that allow model updates to be shared and merged instead of raw data, thereby upholding individual privacy, an architectural premise crucial to Horizontal Federated Learning's viability.

Expanding upon this foundational work, McMahan et al. [8] investigate Horizontal Federated Learning's architectural intricacies, with a focus on data distribution and communication efficiency in a federated learning context. Furthermore, Yang et al. [12] offer an extensive exploration of Horizontal Federated Learning's architecture, encapsulating the unique challenges and considerations involved in data partitioning, model aggregation, and privacy preservation. Complementing these studies, Kairouz et al. [13] delve into the technical aspects of Horizontal Federated Learning, elucidating system design principles, including secure multiparty computation and encryption techniques. This corpus of research collectively contributes to a comprehensive understanding of Horizontal Federated Learning's architectural foundations, positioning it as a promising paradigm for privacy-preserving collaborative machine learning across multifarious data sources.

Horizontal Federated Learning serves is a valuable method for collaborative machine learning. It can be utilized in effectively handling issues related to data privacy, and is finding application in various domains, including healthcare, finance, telecommunications, and energy.

Application	Description	Citations
Healthcare	Improve predictive models for disease detection, prognosis, and treatment recommendations while preserving patient privacy.	[14]
Financial Services	Develop collaborative fraud detection algorithms without sharing individual transaction data, maintaining data privacy and security.	[11]
Telecommunications	Optimize network performance and quality of service by collaboratively optimizing network parameters, preserving user data privacy.	[15]
Energy Sector	Predict and prevent equipment failures in energy infrastructure to reduce maintenance costs and improve service reliability.	[16]

9.3 Vertical

Vertical Federated Learning is an innovative architectural paradigm designed to facilitate collaborative model training across decentralized data sources where the data is partitioned vertically, meaning different parties possess distinct subsets of features or variables. In the Vertical Federated Learning architecture, data privacy is paramount, with each participating entity retaining control over their local data while sharing model updates during training. These model updates are then collectively aggregated to improve the global model without ever revealing raw data.

Vertical Federated Learning is recognized as an innovative architectural framework for decentralized and privacy-focused collaborative machine learning, underpinned by a comprehensive body of research. Within the architectural framework of Vertical Federated Learning, data is vertically partitioned, with distinct subsets of features distributed across different parties, facilitating a powerful mechanism for data separation and privacy preservation. A seminal work by [17] underscores the critical components of Vertical Federated Learning architecture, emphasizing the essentiality of data partitioning, collaborative model training, and secure aggregation mechanisms in preserving the integrity and privacy of data while enabling decentralized model updates, forming a foundational cornerstone for the comprehension of Vertical Federated Learning's architectural design.

Complementing this foundational research, Diamos et al. [18] offer an expansive examination of Vertical Federated Learning's architectural intricacies, shedding light on cryptographic techniques, communication protocols, and robust security measures, and essential components that underpin Vertical Federated Learning's architecture, particularly in the context of privacy and data security. In parallel, Triastcyn et al. [19] expound on Vertical Federated Learning, encompassing architectural considerations, emphasizing practical implementations, and addressing the multifaceted challenges associated with data distribution, model training, and privacy preservation. Concurrently, Zhan et al. [20] contribute by exploring system design principles, providing an insightful analysis of secure aggregation methods, data partitioning strategies, and the collaborative model training process. These collective endeavors advance a comprehensive understanding of Vertical Federated Learning's architectural underpinnings, accentuating its role in maintaining data privacy and facilitating collaborative machine learning across diverse, privacy-sensitive data sources.

Vertical Federated Learning finds applications in various domains where data privacy and security are paramount.

Application	Description	Citations
Healthcare	Vertical federated learning enables collaborative predictive modeling for healthcare data, preserving data privacy and confidentiality.	[21]
Financial Services	Securely develop fraud detection models by collaborating on vertically partitioned financial data, ensuring data protection.	[22]
E-commerce	Collaboratively optimize product recommendations while maintaining the confidentiality of user data across e-commerce platforms.	[23]
Telecommunications	Enhance network performance and quality of service by aggregating vertically partitioned network data, respecting user data privacy.	[24]
Energy Sector	Develop predictive maintenance models for energy infrastructure by combining vertically partitioned sensor and maintenance data.	[25]

9.4 Federated Transfer Learning

Federated Transfer Learning is a novel approach and framework that addresses the challenge of knowledge sharing and model building in a data federation while preserving user privacy. Federated Transfer Learning allows for the transfer of knowledge across domains without compromising the privacy of individual use. This approach addresses the concerns of data owners by ensuring that sensitive information remains confidential while still allowing for effective use of deep learning algorithms in healthcare and other domains. Federated Transfer Learning can be modified to accommodate a range of secure multiparty machine learning undertakings and offers a resolution for institutions possessing restricted data access or inadequate managerial structures to develop precise AI models while adhering to the regulations associated with privacy and security of data.

Federated Transfer Learning employs transfer learning to optimize the advantages of federated learning through the dissemination of knowledge. It applies deep neural networks as feature transformation processes, projecting party-specific features into a shared feature subspace. The hidden representations produced by these networks serve as the basis for knowledge transfer. The Federated Transfer Learning algorithm integrates techniques that preserve privacy, including homomorphic encryption and secret sharing to ensure the security of data during the transfer process. Experimental evaluations on public datasets have demonstrated the effectiveness and scalability of Federated Transfer Learning, showing significant improvements in model performance compared to baseline self-learning models. Overall, Federated Transfer Learning provides a promising solution for federated machine learning, enabling organizations to collaborate and build accurate models while respecting privacy and security constraints.

The architecture of Federated Transfer Learning is designed to enable knowledge sharing and model building in a privacy-preserving manner within a data federation. Federated Transfer Learning addresses the challenge of combining data from multiple parties while respecting privacy and security constraints. The framework leverages transfer learning techniques to transfer knowledge from a source domain to a target domain. This permits a party within the target domain to develop efficient models, maximizing abundant labels from an originating domain while ensuring user privacy is not infringed. Federated Transfer Learning involves two parties, A and B, each holding their respective datasets. The architecture involves several key steps.

First, the parties compute the hidden representations of their data using deep neural networks. These hidden representations serve as the basis for knowledge transfer. Then, the parties securely exchange masked gradients and loss values, ensuring that no sensitive information is revealed. The model parameters are updated based on the received gradients, and the process iterates until convergence. Finally, during the prediction phase, the parties collaborate to compute prediction results for unlabeled data. Overall, the Federated Transfer Learning architecture ensures that the parties can collaborate and benefit from each other's knowledge while maintaining data privacy and security. It offers a viable and efficient solution for entities with restricted data or inadequate oversight to construct precise AI models in a decentralized environment [26].

9.5 Optimization Algorithms in Federated Learning

Algorithms for optimization are fundamentally important in federated learning, addressing critical problems presented by decentralized machine learning situations that prioritize privacy. Federated learning involves data dispersed over multiple devices or servers, rendering the centralization of such data for model training unfeasible or unwanted. These algorithms facilitate cooperative tutoring of machine learning models while ensuring protection of data confidentiality and security. They determine how model updates are aggregated, allowing for efficient communication and the coordination of learning across these decentralized nodes. The choice of the optimization algorithm significantly impacts convergence speed, communication overhead, and model performance, making it an essential consideration for the effectiveness and scalability of federated learning applications across diverse domains, from healthcare to edge computing. Furthermore, as privacy concerns become increasingly paramount in today's data-driven world, the development of optimization techniques that protect sensitive information during the training process remains a central challenge and underscores the ongoing importance of optimization algorithms in federated learning.

In federated learning, various optimization algorithms are used to improve the convergence and performance of the learning process without compromising data privacy. These algorithms can be broadly categorized into two types: standard optimization methods and adaptive optimization methods. Standard optimization methods, such as Federated Averaging (FEDAVG) [8], are commonly used in federated learning. FEDAVG involves clients performing multiple epochs of stochastic gradient descent (SGD) on their local datasets and then sending their model updates to the central server, which averages them to create a global model. While FEDAVG has been successful, it may exhibit convergence issues in certain scenarios and lacks adaptivity to handle heterogeneous data distributions. To address these limitations, adaptive optimization methods have been introduced in federated learning.

These methods incorporate adaptive learning rates that adjust based on past iterations to optimize the learning process. Some examples of adaptive optimizers used in federated learning include ADAGRAD, ADAM, and YOGI. These optimizers adaptively adjust the learning rates for each parameter, allowing for more efficient and effective optimization in the presence of heterogeneous data. The use of adaptive optimizers in federated learning has shown promising results in improving convergence and performance. They can better handle scenarios with heavy-tailed stochastic gradient noise distributions and exploit the sparsity in gradients that often arise in text data. By incorporating adaptivity, these optimizers can enhance the learning process and make it easier to tune the hyperparameters for optimal performance [27].

Optimization Algorithm	Characteristics
FedAvg	• Widely used baseline algorithm in federated learning. • The convergence speed is comparatively slower and the level of accuracy falls short in relation to other algorithms.
FedProx	• Adds a proximal term to improve stability. • Adaptively modifies the epochs of client training to address the issue of stragglers efficiently. • Increases computational overhead of clients.
FedAdp	• Addresses the client-drift issue in the FedAvg algorithm. • Maintains fast convergence. • Decreases convergence speed.
FedAwo	• Automatic weight optimization algorithm. • Considers quality of client data for weight allocation. • Reduces the cross-entropy loss to guarantee the most accurate prediction of local distribution in relation to global distribution. • Quick convergence and improved accuracy.
FedAwo	• Further reduces training cost by dynamically adjusting training epoch times. • Better convergence speed and model accuracy compared to baseline algorithms.
FedL	• Sampling technique based on the Graph Convolutional Network. • Enhances the precision of a universal model by studying the correlation between network characteristics and sampling nodes.
FedHQ	• Reduces the maximum rate of convergence as a consequence of heterogeneous quantization error. • Assigns different aggregation weights to different clients.
FedDyn	• Each client should be allocated a dynamic regularization optimizer. • Maintains approximately consistent distribution between local and global models.
FedNova	• Eliminates objective inconsistencies. • Ensure optimal convergence speed by applying normalization to local models.
FOLB	• Estimates gradient information of local models. • Performs weighted sampling based on client performance.

9.6 Federated Learning Applications

Federated learning has numerous applications in various fields, including healthcare, finance, and transportation. In healthcare, federated learning can be used to develop predictive models for disease diagnosis and treatment while preserving patient privacy. In finance, federated learning can be used to improve fraud detection and risk assessment. In transportation, federated learning can be used to improve traffic prediction and route optimization. Federated learning has also shown promise in the field of smart cities. By leveraging data from various sources, federated learning can be used to improve traffic flow prediction, energy consumption optimization, and waste management. Additionally, federated learning can be used to develop personalized services for citizens while preserving their privacy. The decentralized nature of federated learning makes it an ideal solution for smart cities, where data privacy and security are of utmost importance [28].

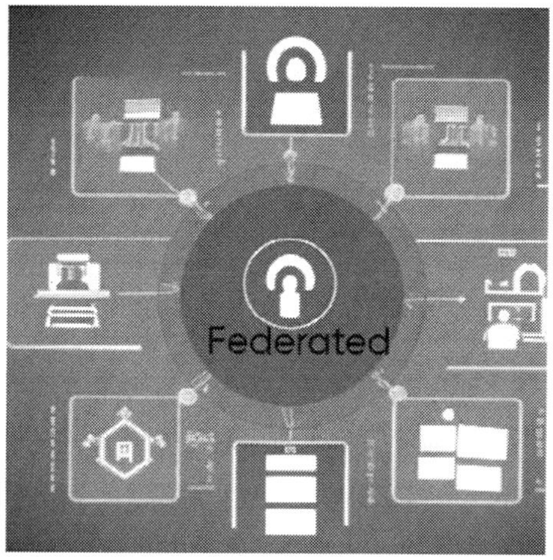

1) Federated learning is a technology that can enable a machine to analyze data both inside and outside the firewall. This machine can predict instances where sensitive data might be leaking from specific websites. Moreover, it can identify harmful websites that pose a threat to data security. Using the insights gained from these predictions, a model can be developed to proactively block such websites. This proactive approach helps protect the entire network of machines by preventing access to these harmful sites. By leveraging federated learning, organizations can collaborate across their networked devices to collectively enhance their security posture. This involves creating a shared model that learns from patterns and data without exposing sensitive information, ultimately bolstering cybersecurity defenses and safeguarding against data breaches.
2) Federated learning can be a powerful tool in preventing phishing attacks. The technology enables a model to scrutinize websites that request personal information, allowing it to learn how these websites operate and detect fraudulent messages. By collectively analyzing patterns and data from various sources, the model becomes adept at recognizing the tactics used by malicious websites to acquire personal data. If it detects a security breach, such as a leaked password, the system can automatically trigger notifications to users, prompt them to change their passwords, and proactively block access to harmful websites. This approach significantly enhances cybersecurity, protecting individuals and organizations from falling victim to phishing scams and mitigating the risks associated with data breaches. Leveraging federated learning in this manner empowers a collaborative defense against evolving threats while maintaining data privacy and security.
3) In the realm of Federated Learning, we find an application that holds immense potential for revolutionizing smart transportation systems. This technology allows for the seamless integration of various transportation modes, like buses and trains, into a unified predictive model. The primary objective is to enhance passenger experience by providing accurate and real-time information on the arrival of buses and trains. In today's world, rail systems often suffer from inefficiencies, particularly in terms of timing. Federated Learning comes to the

rescue by continuously monitoring and tracking the speed of each vehicle, enabling the system to automatically update estimated arrival times. This data-driven approach ensures that passengers can plan their journeys with confidence, knowing they can arrive at their destinations punctually. Moreover, Federated Learning goes a step further by predicting traffic conditions before the departure of buses or trains. This proactive traffic analysis allows for optimal route planning, reducing delays caused by congestion. By centralizing this predictive model, it ensures that all transportation facilities benefit from these insights, contributing to a more efficient and streamlined transportation network. Passengers are empowered to track the real-time status of their transportation, leading to a smoother and more predictable commuting experience. Overall, Federated Learning is a game-changer in the realm of smart transportation, reshaping the way people move, making it not just more efficient but also a more user-centric experience.

4) Federated learning can indeed be effectively implemented in a node topology, especially in the context of managing the massive influx of data generated by IoT (Internet of Things) devices. With the exponential growth of IoT appliances, the need for efficient data handling and analysis has become paramount. In a node-based approach, each IoT device serves as a node in the network, capable of running local machine learning models and algorithms. These devices process and analyze data at the edge, reducing the burden on centralized cloud servers and minimizing the need for continuous data transmission to the cloud. This approach not only alleviates the bandwidth constraints associated with sending massive volumes of data but also addresses latency issues, as real-time analysis can occur locally. Furthermore, it fosters better distribution and sharing of knowledge across the network and promotes optimized resource utilization. By enabling the deployment of federated learning models on these distributed nodes, updates and improvements can be made collaboratively while keeping sensitive or proprietary data on the local device. This paradigm shift in data processing and machine learning not only enhances the overall efficiency of IoT systems but also ensures data security and privacy, making it a fundamental framework for the future of IoT-driven applications and services. Local training is a fundamental component of federated learning, a cutting-edge approach with numerous applications that prioritize data privacy and distributed machine learning. In federated learning, local training occurs on individual devices, such as smartphones, IoT gadgets, or edge servers, without the need to transmit sensitive data to a central server. This decentralized approach allows multiple devices to collaboratively train machine learning models, ultimately enhancing the local model's accuracy and performance. During local training, the device leverages its own data to refine the machine learning model. This process can involve several iterations, with each iteration aiming to improve the model's understanding of the local data. As a result, the local model becomes more adept at making predictions, recommendations, or classifications based on the specific data available on the device. Local training has a range of applications across various domains. For instance, in healthcare, it enables wearable devices to continuously improve their health monitoring algorithms without compromising the privacy of individual health records. In predictive text input, your smartphone can learn your typing habits without transmitting your personal messages to a central server. In the realm of IoT, local training allows edge devices to adapt and respond to changing environmental conditions without sharing sensitive sensor data. The benefits of local training are two-fold: it enhances model accuracy by considering device-specific data and ensures data privacy by keeping sensitive information on the device. By enabling devices to learn and improve locally, federated learning strikes a delicate balance between innovation and safeguarding individual privacy, making it a powerful and ethical tool for the future of machine learning applications.

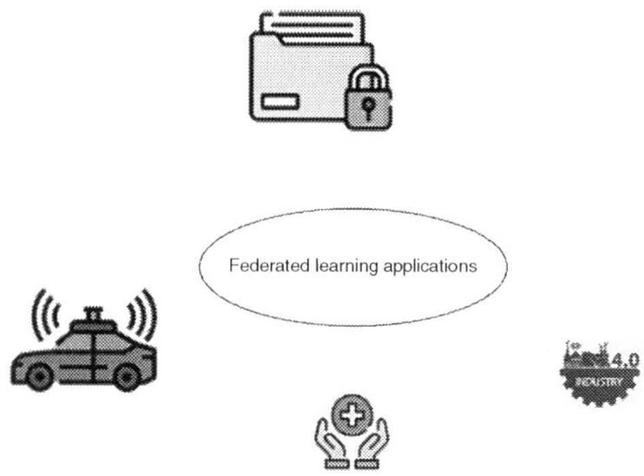

Federated learning presents a compelling application in the realm of route optimization, benefitting both logistics companies and public transportation services. This innovative approach leverages local data from various sources, such as GPS, weather conditions, and traffic, to create route optimization models that enhance the overall efficiency and safety of road networks. What makes federated learning particularly valuable in this context is its ability to provide significant improvements without centralizing sensitive location data. Logistics companies can harness federated learning to fine-tune their delivery and supply chain operations. By employing local data from GPS trackers on vehicles, these companies can collectively train models to develop optimal delivery routes. This helps reduce delivery times, fuel consumption, and environmental impact. Crucially, sensitive location data stays on the vehicles, ensuring data privacy and security. Public transportation services also stand to gain from federated learning's capabilities. With the collaboration of various transit vehicles, federated learning models can continuously analyze real-time traffic and weather conditions. This information feeds into route optimization models, allowing for more accurate predictions of arrival times and the ability to proactively adjust routes to avoid traffic congestion or adverse weather conditions. Passengers can enjoy a smoother and more predictable commute, and public transit systems can operate with greater efficiency. In essence, federated learning empowers a smarter and safer transportation network by capitalizing on decentralized data sources. It enables traffic updates, enhances road safety, and optimizes road network efficiency while respecting individual privacy and security concerns. By combining the power of localized data and collaborative machine learning, federated learning transforms the way we plan and experience trips, making travel more efficient, convenient, and safer for everyone.

Federated learning offers significant potential for predictive maintenance, particularly in the context of public transportation and infrastructure. By leveraging data from a multitude of sensors and maintenance logs from diverse sources, federated learning can develop predictive maintenance models that revolutionize the way we ensure the safety and efficiency of our critical infrastructure. In public transportation, such as bus fleets or train networks, federated learning enables the early detection of potential maintenance needs for vehicles. By analyzing data from sensors measuring engine performance, brake systems, and other components, these models can predict when maintenance is required, reducing the likelihood of breakdowns and service disruptions. This not only saves time and resources but also enhances the safety and reliability of public transportation systems. Moreover, federated learning can be applied to infrastructure components like bridges and railways. It enables the collection of data from various sources globally, encompassing details on the condition of bridges and their history. This data can be used to construct predictive models that assess the health of these structures and predict maintenance needs. By learning from past incidents, such as bridge collapses or disasters, federated learning contributes to designing more robust and disaster-resistant infrastructure. This approach ensures that new infrastructure is constructed with insights gained from past failures, making the world's critical infrastructure safer and more resilient. In essence, federated learning empowers a proactive approach to infrastructure maintenance and construction, making transportation systems and critical structures more reliable and secure. By aggregating data and knowledge from diverse sources, it not only predicts maintenance needs but also helps prevent catastrophic failures, ultimately leading to more sustainable and resilient infrastructure on a global scale.

Feedback mechanisms within federated learning are a crucial application that serves both user satisfaction and model improvement. These mechanisms enable users to report issues related to model performance and offer the option to opt out of federated learning if they have privacy concerns. By facilitating such feedback, federated learning models can consistently enhance their performance and ensure that every product or service aligns with customer expectations. This, in turn, contributes to better product creation and global market production.

Federated learning systems implement feedback loops that allow users to provide insights, comments, and critiques regarding their experiences. Users can report issues related to the model's predictions, suggestions, or behaviors. They can also voice privacy concerns and request more transparency about data usage. Allowing users to opt out ensures that individuals have control over their data and its utilization in the federated learning process.

The feedback collected is invaluable for model refinement and continuous improvement. It provides developers with real-world insights, helping them address shortcomings and refine algorithms. As a result, the federated learning models can adapt to changing user preferences and deliver more accurate, relevant, and reliable predictions. This user-centric approach not only fosters trust but also drives innovation in product development, making it more likely to meet customer expectations and succeed in the global market. By considering and acting upon user feedback, federated learning applications can fine-tune their offerings and continually enhance user satisfaction, product quality, and competitiveness on the global stage.

Federated learning, a revolutionary machine learning paradigm, has a broad spectrum of applications across various domains, enhancing data-driven processes while preserving data privacy and security. In the realm of cybersecurity, it equips machines to analyze both internal and external data, predicting data leaks and identifying malicious websites. This proactive defense helps safeguard networks from potential threats. Federated learning further proves its worth in preventing phishing attacks, allowing models to scrutinize websites for personal data requests, detect fraudulent messages, and automatically trigger responses to security breaches.

In the domain of transportation, federated learning optimizes routes, improves predictive maintenance, and enhances passenger experience by tracking vehicle speeds, predicting traffic conditions, and enabling real-time updates. The application extends to IoT devices, promoting efficient data processing while respecting data privacy, enabling smarter devices and better user experiences. Moreover, feedback mechanisms empower users to contribute to model performance and voice privacy concerns, ensuring continuous model refinement and the creation of products that align with customer expectations.

In essence, federated learning empowers a collaborative approach to machine learning, allowing data from diverse sources to collectively improve models without exposing sensitive information. This technology reshapes the way we optimize routes, protect against threats, and maintain critical infrastructure, making data-driven processes more efficient, secure, and user-centric across various industries.

9.7 Conclusion

In conclusion, this chapter has steered the complex situations in which there might be an immediate requirement for data privacy during the advancements of AI. It unveiled Federated Learning as a beacon of hope in this paradoxical realm, maintaining data privacy while harnessing the potential of AI across a spectrum of applications. This chapter has provided detailed knowledge about the types of Federated Learning in which each caters to the specific use cases that were handled and held along with optimization algorithms that drive this groundbreaking paradigm.

Through the examples and real-world case studies, various applications that emphasize its capacity to revolutionize collaborative, secure, and privacy-respecting model training could be observed. Finally, the remarkable strides in AI and the disruptive potential of Federated Learning have been elaborated.

References

1 Cisco Cisco (2024). Data Privacy Benchmark Study 2024. https://www.cisco.com/c/en/us/about/trust-center/data-privacy-benchmark-study.html#~about-the-study (accessed 23 May 2023).

2 Silver, D., Huang, A., Maddison, C.J. et al. (2016). Mastering chess and shogi by self-play with a general reinforcement learning algorithm. arXiv preprint arXiv:1712.01815.

3 Silver, D., Schrittwieser, J., Simonyan, K. et al. (2017). Mastering the game of go without human knowledge. *Nature* 550 (7676): 354–359.

4 Obermeyer, Z., Emanuel, E.J., and Cullen, M.R. (2016). Machine learning in medicine. *New England Journal of Medicine* 376 (26): 2506–2507.

5 Bojarski, M., Del Testa, D., Dworakowski, D. et al. (2016). End to end learning for self-driving cars. arXiv preprint arXiv:1604.07316.

6 Konečný, J., McMahan, H. B., Ramage, D., and Richtárik, P. (2015). Federated optimization: distributed machine learning for on-device intelligence. arXiv preprint arXiv:1511.03575.

7 Bonawitz, K., Ivanov, V., Kreuter, B. et al. (2016). Practical Secure Aggregation for Federated Learning on User-Held Data. *ArXiv*, abs/1611.04482.

8 McMahan, H.B., Moore, E., Ramage, D. et al. (2017). Communication-efficient learning of deep networks from decentralized data. Artificial intelligence and statistics.

9 Li, T., Sahu, A.K., Zaheer, M. et al. (2019). Federated optimization in heterogeneous networks. arXiv preprint arXiv:1912.04977.

10 Hoffer, E., Hubara, I., and Soudry, D. (2017). Train longer, generalize better: closing the generalization gap in large batch training of neural networks. In Proceedings of the 31st International Conference on Neural Information Processing Systems (NIPS'17). Curran Associates Inc., Red Hook, NY, USA, 1729–1739.

11 Bonawitz, K., Ivanov, V., Kreuter, B. et al. (2017). Practical Secure Aggregation for Privacy-Preserving Machine Learning. In Proceedings of the 2017 ACM SIGSAC Conference on Computer and Communications Security (CCS '17). Association for Computing Machinery, New York, NY, USA, 1175–1191. https://doi.org/10.1145/3133956.3133982.

12 Yang, Q., Liu, Y., Cheng, Y. et al. (2019). Federated learning. *Synthesis Lectures on Artificial Intelligence and Machine Learning* 13 (3): 1–102.

13 Kairouz, P., McMahan, B., Avent, B. et al. (2019). Advances and open problems in federated learning. arXiv preprint arXiv:1912.04977.

14 Nasajpour, M., Karakaya, M., Pouriyeh, S. et al. (2022). Federated Transfer Learning For Diabetic Retinopathy Detection Using CNN Architectures. In SoutheastCon 2022, Mobile, AL, USA, 2022, pp. 655–660, doi: 10.1109/SoutheastCon48659.2022.9764031.

15 Yang, Z., Chen, M., Wong, K. et al. (2021). Federated Learning for 6G: Applications, Challenges, and Opportunities. *ArXiv*, abs/2101.01338.

16 Shubyn, B., Kostrzewa, D., Grzesik, P. et al (2023). Federated Learning for improved prediction of failures in Autonomous Guided Vehicles. Journal of Computational Science. 68, 101956. https://doi.org/10.1016/j.jocs.2023.101956.

17 Sheller, M. J., Edwards, B., Reina, G. A. et al. (2020). Federated learning in medicine: facilitating multi-institutional collaborations without sharing patient data. *Scientific reports*, 10(1), 12598. https://doi.org/10.1038/s41598-020-69250-1.

18 Diamos, G.F., Vanhoucke, V., and Norouzi, M. (2020). Federated learning: strategies for improving communication efficiency. arXiv preprint arXiv:1610.05492.

19 Triastcyn, A., Troncoso-Pastoriza, J.R., Juels, A., and Reiter, M.K. (2021). Scalable privacy-preserving data sharing with tree-based models. *Proceedings on Privacy Enhancing Technologies* 2021 (4): 62–82.

20 Zhan, H., Li, L., An, Z. et al. (2021). Secureagg: secure and privacy-preserving federated learning with secure aggregation and benchmarking. arXiv preprint arXiv:2102.04361.

21 Lim, B., Yang, J., Park, D.H., and Kim, J. (2019). Vertical federated learning for predictive maintenance: a case study in industrial AI. *IEEE Access* 7: 178383–178391.

22 Qiu, X., Pan, H., Zhao, W. et al. (2023). Efficient Vertical Federated Learning with Secure Aggregation. 10.48550/arXiv.2305.11236.

23 Mammen, P. M. (2021). Federated learning: Opportunities and challenges. 10.48550/arXiv.2101.05428.

24 Khan, A., Thij, M.T., and Wilbik, A. (2022). Communication-Efficient Vertical federated learning. *Algorithms*, 15(8): 273. https://doi.org/10.3390/a15080273.

25 Ahn, J., Lee, Y., Kim, N. et al. (2023b). Federated Learning for Predictive Maintenance and Anomaly Detection Using Time Series Data Distribution Shifts in Manufacturing Processes. *Sensors*, 23(17): 7331. https://doi.org/10.3390/s23177331.

26 Liu, Y., Kang, Y., Xing, C. et al. (2020). A secure federated transfer learning framework. *IEEE Intelligent Systems* 35 (4): 70–82. https://doi.org/10.1109/MIS.2020.2988525.

27 Reddi, S.J., Charles, Z.B., Zaheer, M. et al. (2020). Adaptive Federated Optimization. *ArXiv*, abs/2003.00295.

28 Rauniyar, A., Hagos, D.H., Jha, D. et al. (2022). Federated learning for medical applications: a taxonomy, current trends, challenges, and future research directions. *IEEE Access* 10: 1–28.

10

Securing Edge Learning: The Convergence of Block Chain and Edge Intelligence

Rakhi Mutha

Department of IT, Amity University Rajasthan, Jaipur, Rajasthan, India

10.1 Introduction

The rise of edge computing and the advancements in machine learning have transformed the way data is processed, analyzed, and utilized. Edge devices, such as sensors, IoT devices, and smartphones, have become increasingly powerful, enabling data to be processed closer to the source, reducing latency, and enhancing real-time decision-making capabilities. This shift has given rise to the concept of edge intelligence, where machine learning and artificial intelligence algorithms are deployed directly on edge devices.

While edge intelligence offers numerous benefits, such as improved response time, reduced bandwidth usage, and enhanced privacy, it also introduces new security challenges. Edge devices are often resource-constrained and lack robust security mechanisms, making them vulnerable to various threats, including data breaches, tampering, and unauthorized access. Traditional security approaches, designed for centralized systems, are not well-suited for the distributed and heterogeneous nature of edge environments.

To address these security challenges, there is a growing interest in exploring emerging technologies that can enhance security in edge learning environments. One such technology is block chain, a decentralized and tamper-resistant distributed ledger technology (DLT). Block chain has gained significant attention in recent years for its potential to provide secure and transparent transactions in various domains, including finance and supply chain management. Its characteristics, such as immutability, transparency, and decentralization, make it an ideal candidate for securing edge learning environments.

10.2 Fundamentals of Block Chain

10.2.1 Block Chain Technology Overview

Block chain technology is a decentralized and distributed digital ledger that allows data to be recorded, shared, and maintained securely and transparently across a network of computers. The technology's main innovation is the creation of a chain of blocks as per Figure 10.1, where each block contains a batch of transactions and is linked to the previous block using cryptographic hashes, forming an immutable and tamper-proof record of data.

Model Optimization Methods for Efficient and Edge AI: Federated Learning Architectures, Frameworks and Applications, First Edition. Edited by Pethuru Raj Chelliah, Amir Masoud Rahmani, Robert Colby, Gayathri Nagasubramanian, and Sunku Ranganath.

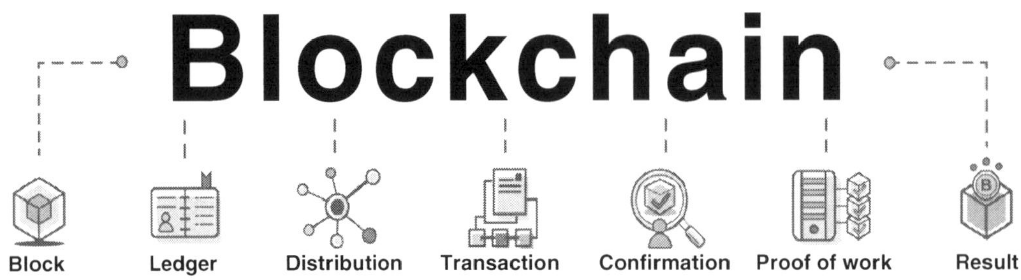

Figure 10.1 Block chain technology.

Key characteristics of block chain technology include:

Decentralization: Unlike traditional centralized systems, block chain operates on a decentralized network of nodes, where each node holds a copy of the entire block chain. This decentralization eliminates the need for a central authority, making the system more resilient and transparent.

Transparency: All transactions and data recorded on the block chain are visible to all participants in the network. Once a block is added to the chain, it cannot be altered or deleted, ensuring transparency and accountability.

Immutability: The data on a block chain is immutable, meaning it cannot be changed or tampered with after it has been added to the chain. This feature is achieved through cryptographic hashing and consensus mechanisms.

Security: Block chain's security is maintained through advanced cryptographic techniques. Each block contains a unique cryptographic hash that is generated based on the data in the block and the hash of the previous block. Any change to the data in a block would alter its hash, making it evident that tampering has occurred [1, 2].

Consensus Mechanisms: Block chain networks rely on consensus mechanisms to agree on the validity of transactions and the order in which they are added to the block chain. PoW and PoS are some common consensus mechanisms.

Smart Contracts: Block chain platforms like Ethereum introduced the concept of smart contracts. These are self-executing contracts with terms written in code. Smart contracts automatically execute when specific conditions are met, removing the need for intermediaries.

Cryptocurrencies: The most well-known application of block chain technology is in cryptocurrencies like Bitcoin, which operate on their respective block chains [2]. As per as Figure 10.2 cryptocurrencies enable secure, borderless, and peer-to-peer transactions without the need for traditional financial institutions.

Applications of block chain technology go beyond cryptocurrencies and include areas such as supply chain management, healthcare, identity verification, voting systems, real estate, and

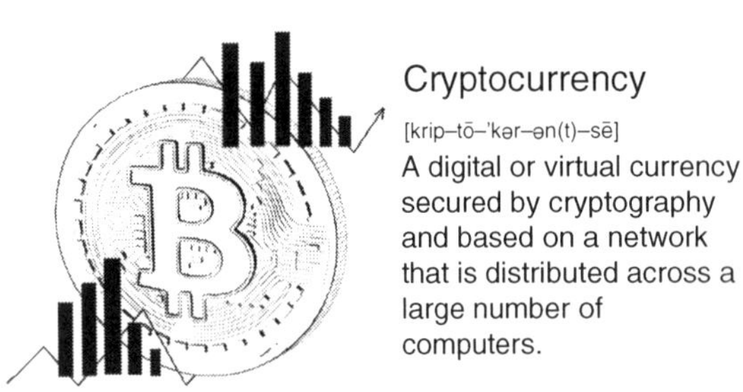

Figure 10.2 Cryptocurrency.

decentralized finance (DeFi). The technology's ability to provide trust, transparency, and security in a decentralized manner has garnered [3] significant interest across various industries and is continuously evolving as researchers and developers explore new use cases and improve the scalability and efficiency of block chain networks.

Block chain technology is the foundation of many decentralized systems and cryptocurrencies like Bitcoin and Ethereum. It is a DLT that enables secure and transparent recording of transactions across multiple participants in a network. The block chain consists of a chain of blocks, where each block contains a list of transactions or other data. Each block is linked to the previous block through cryptographic hashes, creating an immutable and tamper-resistant record.

10.2.1.1 Distributed Ledger Technology

The DLT forms the basis of block chain. It refers to the decentralized nature of the ledger, where multiple copies of the ledger are maintained and synchronized across a network of participants, known as nodes. This decentralized architecture eliminates the need for a central authority and allows for a trustless environment where participants can reach a consensus on the state of the ledger.

DLT is a broader term that encompasses block chain technology and other similar decentralized, distributed, and immutable data storage systems. DLT refers to a digital ledger or database that is maintained as per Figure 10.3 across multiple nodes or computers in a network. It allows for the recording, sharing, and synchronization of data among participants without the need for a central authority or intermediary.

Key characteristics of distributed ledger technology include:

Decentralization: DLT operates on a decentralized network of nodes, where each node holds a copy of the ledger. This eliminates the need for a central authority and allows for peer-to-peer communication and data exchange.

Consensus Mechanisms: DLT uses consensus mechanisms to achieve agreement among nodes on the validity of data and transactions. Various consensus mechanisms like PoW, PoS, PBFT, and others are used depending on the specific DLT implementation.

Immutability: Like block chain, DLT maintains data immutability, meaning once data is recorded on the ledger, it cannot be altered or deleted without consensus from the majority of nodes. This is achieved through cryptographic hashing and consensus algorithms.

Transparency: DLT provides transparency, as all participants in the network can access and view the entire ledger. This transparency enhances trust [4] and accountability in the system.

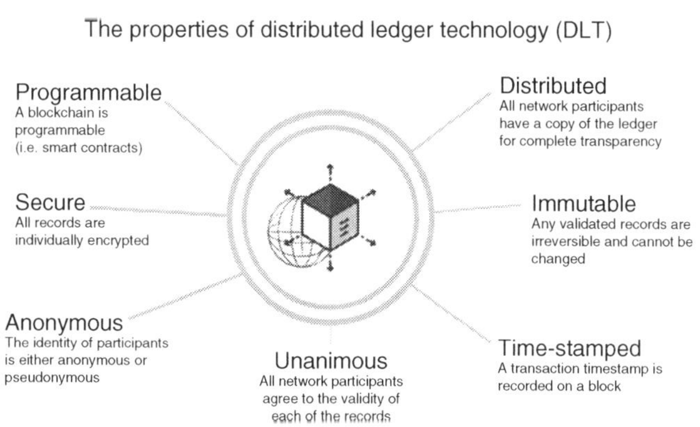

Figure 10.3 Distributed ledger technology.

Security: DLT employs cryptographic techniques to ensure the security of data and transactions. The distributed nature of the ledger makes it more resilient to attacks and single points of failure.

Smart Contracts: Many DLT platforms support smart contracts, self-executing code that automatically executes predefined actions when certain conditions are met. Smart contracts enhance automation and remove the need for intermediaries in various processes.

Permissioned and Permissionless: DLT can be either permissioned or permissionless. In permissioned DLT, access to the network and participation is restricted to authorized entities, while in permissionless DLT (like public block chains), anyone can join the network.

DLT finds applications in various industries, including finance, supply chain management, healthcare, voting systems, real estate, and more. It enables secure and transparent data sharing, reduces the reliance on intermediaries, and improves efficiency in processes that involve multiple parties [5].

While block chain is the most well-known implementation of DLT, other types of distributed ledgers, such as Directed Acyclic Graphs (DAGs) and Hash graph, have also gained attention for their unique characteristics and potential applications. As the technology continues to evolve, researchers and developers explore new DLT approaches to address scalability, privacy, and performance challenges in various use cases.

10.2.1.2 Consensus Mechanisms

Consensus mechanisms are protocols that enable participants in a block chain network to agree on the validity and order of transactions. Different consensus mechanisms exist, such as PoW, PoS, and PBFT. These mechanisms ensure that the majority of participants reach an agreement, preventing malicious actors from tampering with the ledger.

Consensus mechanisms are protocols or algorithms used in distributed systems, such as block chain networks, to achieve agreement among multiple nodes on the state of the system or the validity of transactions. These mechanisms play a crucial role in maintaining the integrity and security of decentralized networks. There are several consensus mechanisms as per Figure 10.4, and here are some of the most commonly known ones:

10.2.1.2.1 Proof of Work PoW is the consensus mechanism used in the Bitcoin block chain and many other cryptocurrencies. In PoW, miners compete to solve complex mathematical puzzles to add a new block to the block chain. The first miner to find the solution gets to add the block and is rewarded with newly minted coins and transaction fees. This process requires significant computational power, making it difficult for any single entity to control the network.

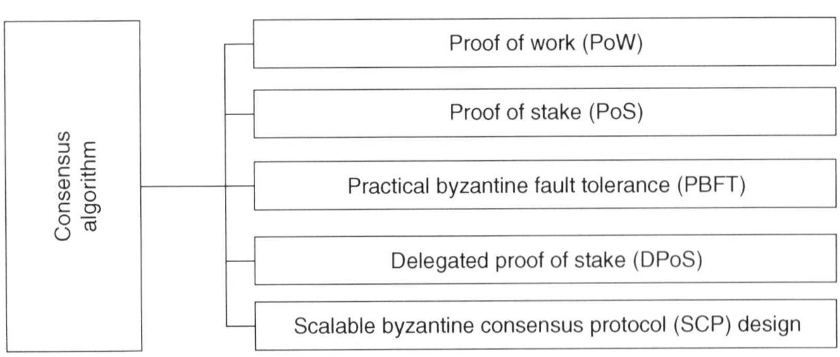

Figure 10.4 Consensus mechanisms.

10.2.1.2.2 Proof of Stake In the PoS mechanism, valuators are chosen to create new blocks based on the number of coins they "stake" or lock up as collateral. The selection of valuators is often determined by a combination of factors, such as the number of coins held and the length of time they have been held. PoS is considered more energy-efficient than PoW since it doesn't require intense computational work.

10.2.1.2.3 Delegated Proof of Stake DPoS is a variation of PoS where token holders vote for a set of delegates who are responsible for validating transactions and creating new blocks on behalf of the entire network. These delegates are typically selected based on the number of votes they receive. DPoS aims to increase scalability by having a smaller group of trusted valuators.

10.2.1.2.4 Practical Byzantine Fault Tolerance PBFT is used in some permissioned block chain networks and focuses on consensus in the presence of faulty or malicious nodes. It relies on a predetermined set of valuators who take turns proposing and voting on new blocks. To achieve consensus, a supermajority of valuators must agree on a proposed block.

10.2.1.2.5 Proof of Authority PoA is a consensus mechanism where block valuators are identified and authorized by a central entity or consortium of trusted parties. It is commonly used in private or consortium block chains, where the identity of valuators is known and reputation is important.

10.2.1.2.6 Proof of Burn PoB is a relatively less common consensus mechanism, where participants destroy or "burn" coins by sending them to an unspendable address. The act of burning coins provides evidence of the participant's commitment to the network, and they are rewarded with the right to mine blocks.

10.2.1.3 Smart Contracts

Smart contracts are self-executing contracts with predefined rules and conditions written as code on the block chain. They automatically enforce and execute the terms of an agreement between parties, eliminating the need for intermediaries. Smart contracts enable the automation of various processes and the creation of decentralized applications (DApps) that interact with the block chain.

Smart contracts are self-executing contracts with the terms of the agreement directly written into code. They run on block chain networks and automatically execute actions when specific conditions are met [6]. As per Figure 10.5, smart contracts eliminate the need for intermediaries, as the code itself enforces and executes the terms of the contract. This makes them tamper-proof, transparent, and secure.

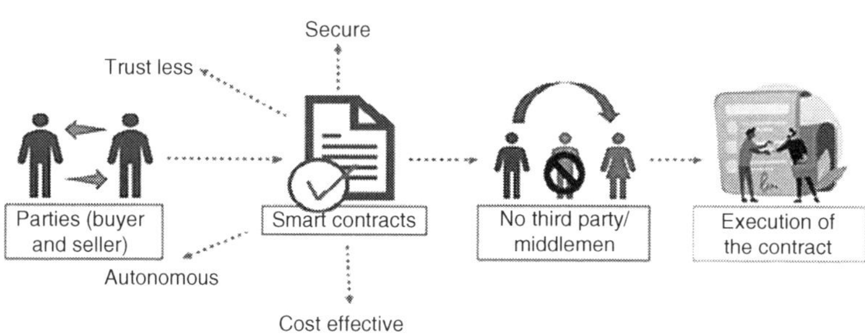

Figure 10.5 Smart contract.

Key characteristics of smart contracts include:

Autonomy: Once deployed, smart contracts operate autonomously, without any third-party involvement, and execute predefined actions when certain conditions are met.
Trust: Smart contracts rely on the security and transparency of block chain technology, providing all involved parties with confidence that the contract will be executed as intended.
Decentralization: Smart contracts run on a distributed network of computers (block chain nodes), ensuring that there is no single point of failure or control [8].
Immutability: Once a smart contract is deployed on the block chain, its code and terms cannot be changed or tampered with, ensuring the integrity of the contract.
Cost and Time Efficiency: By removing intermediaries and automating contract execution, smart contracts can reduce costs and speed up transaction times.

Smart contracts have various applications, including:

Decentralized Finance (DeFi): Smart contracts power various DeFi protocols, such as lending platforms, decentralized exchanges (DEXs), and yield farming, enabling permissionless and automated financial services.
Supply Chain Management: Smart contracts can track the movement of goods along the supply chain, automatically triggering actions like payments and inventory updates when certain milestones are reached.
Tokenization: Smart contracts are used to create and manage tokens, enabling the creation of cryptocurrencies, security tokens, and nonfungible tokens (NFTs).
Voting and Governance: Smart contracts can facilitate secure and transparent voting processes for decentralized organizations and governance systems.
Insurance: Smart contracts can automate insurance policies, enabling automatic claim processing and payouts based on predefined conditions.

One of the most well-known platforms for deploying smart contracts is Ethereum, which introduced the concept of "Turing-complete" smart contracts, allowing developers to create complex and versatile applications. However, other block chain platforms, such as Binance Smart Chain, Solana, and Cardano, also support smart contracts with varying degrees of capabilities and functionalities.

10.2.2 Block Chain Characteristics

Block chain possesses several key characteristics that make it suitable for securing edge learning environments.

10.2.2.1 Immutability

Once a block is added to the block chain, it becomes virtually impossible to alter or delete the information it contains [7]. The immutability of block chain ensures the integrity and tamper-resistance of data, making it an ideal solution for maintaining the integrity of edge learning models and data.

10.2.2.2 Transparency

Block chain provides transparency by allowing all participants to view and verify the transactions recorded on the ledger. This transparency enhances trust among participants and enables audit ability, as any changes or discrepancies can be easily detected.

10.2.2.3 Decentralization

Decentralization is a core principle of block chain technology. Instead of relying on a single central authority, block chain operates as a peer-to-peer network where multiple participants maintain and validate the ledger. Decentralization improves the resilience, fault tolerance, and security of the system, as there is no single point of failure.

10.2.3 Block Chain Security Considerations

Block chain technology offers several security advantages, but it also comes with its own set of considerations and challenges. Some of the key block chain security considerations include:

Immutability: While immutability is one of the strengths of block chain, it can also be a challenge if incorrect or malicious data is recorded on the block chain. Once data is added to a block and included in the chain, it becomes very difficult to alter or remove. Ensuring the accuracy and integrity of data before it is added to the block chain is crucial [9].

Consensus Mechanism Vulnerabilities: The consensus mechanism used in a block chain network is critical to its security. However, different consensus mechanisms have their vulnerabilities. For example, PoW block chains are susceptible to 51% of attacks, where a single entity gains control of the majority of the network's computational power. PoS block chains may be vulnerable to attacks if a large portion of the network's tokens is controlled by a single entity [11].

Smart Contract Security: Smart contracts are powerful tools, but they must be carefully designed and audited to avoid vulnerabilities. Bugs or errors in smart contract code can lead to significant financial losses, as demonstrated by various incidents in the history of block chain. Code audits, best practices, and secure development practices are essential to minimize risks.

Private Key Management: Block chain users are assigned private keys that grant access to their accounts and funds. Proper private key management is crucial, as the loss or theft of private keys can result in permanent loss of assets. Hardware wallets, secure key storage, and multifactor authentication (MFA) are some measures that can enhance private key security.

Distributed Denial of Service (DDoS) Attacks: Block chain networks, especially public ones, are susceptible to DDoS attacks, where malicious actors overwhelm the network with a flood of requests, leading to temporary disruption of service. Implementing DDoS mitigation strategies is important to maintain network availability [12].

Sybil Attacks: In public block chains, Sybil attacks occur when a single entity creates multiple identities or nodes to gain control or influence over the network. Sybil resistance mechanisms are necessary to prevent such attacks.

Governance and Consensus Protocol Upgrades: Decentralized block chains often require upgrades to improve functionality or address security issues. However, achieving consensus on such upgrades can be challenging, potentially leading to contentious forks and network splits.

Regulatory and Legal Compliance: Block chain networks must navigate regulatory requirements and compliance measures, especially in cases involving financial transactions, data privacy, and identity verification. Failure to comply with relevant regulations can lead to legal issues and reputational damage.

To enhance block chain security, ongoing research, third-party audits, community involvement, and industry collaboration are essential. Additionally, block chain developers and users should stay vigilant about emerging threats and security best practices to protect the integrity and reliability of block chain networks and applications.

While block chain offers robust security features, it is not immune to certain security risks and attacks. Understanding these considerations is vital when leveraging block chain for securing edge learning environments [13].

10.2.3.1 51% Attack

A 51% attack occurs when a single participant or a group of colluding participants control more than 50% of the total computing power in a block chain network. This control enables them to manipulate the consensus process and potentially modify transactions or double-spend digital assets. However, this attack becomes increasingly difficult as the network size grows.

A 51% attack, also known as a majority attack or double-spending attack, is a potential security issue that can occur in block chain-based cryptocurrencies as per Figure 10.6. It refers to a situation where a single entity or a group of colluding entities control more than 50% of the total hashing power (computing power) of a block chain network.

In a decentralized block chain, the consensus mechanism relies on the majority of participants agreeing on the validity of transactions and blocks. When an entity or group controls over 50% of the network's hashing power, they have the ability to manipulate the block chain in several ways, including:

Block Reorganization: The attackers can invalidate certain blocks and replace them with their own blocks, creating an alternative chain that becomes the longest chain in the network. This could cause a "reorganization" of the block chain, leading to previously confirmed transactions being reversed [13].

Preventing Confirmations: The attackers can prevent new transactions from being confirmed by consistently mining blocks faster than the rest of the network, resulting in a halt to transaction processing.

It is important to note that a 51% attack is more feasible on smaller, less secure, or less well-established block chains, as larger and more established networks typically have a more distributed and robust mining community, making it much more difficult for any single entity to accumulate such a large portion of the network's hashing power.

To prevent 51% attacks, block chain networks employ various consensus mechanisms, such as PoW and PoS, among others. These mechanisms aim to distribute decision-making power among a large number of participants, making it economically and computationally infeasible for any single entity to gain control of the majority of the network's resources. Additionally, block chain communities and developers continuously work to improve the network's security and detect any potential vulnerabilities [14].

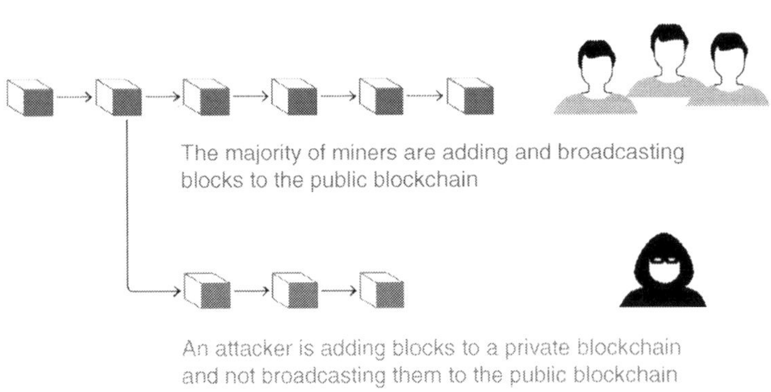

Figure 10.6 51% attack.

10.2.3.2 Sybil Attack

A Sybil attack involves an attacker creating multiple fake identities or nodes to gain control over a significant portion of the network. This attack can compromise the consensus mechanism and undermine the trust in the block chain network [15]. Mitigating Sybil attacks requires robust identity management mechanisms.

A Sybil attack is a type of security threat that occurs in decentralized networks, particularly in peer-to-peer systems and block chain networks as per Figure 10.7. The attack is named after the book "Sybil" by Flora Rheta Schreiber, which describes a case study of a woman diagnosed with dissociative identity disorder, having multiple personalities. In a Sybil attack, a single malicious entity creates multiple fake identities, known as "Sybils," and uses them to gain disproportionate influence or control over the network.

The primary goal of a Sybil attack is to undermine the integrity of the network by:

Gaining Unfair Influence: By controlling multiple fake identities, the attacker can increase their presence in the network and potentially gain more influence or voting power than honest participants. In decentralized systems that rely on voting or reputation mechanisms, this can lead to malicious actors having a disproportionate say in decision-making.

Manipulating Consensus: In block chain networks, where consensus mechanisms like PoW or PoS are used, the attacker can control a significant portion of the network's resources through their Sybil identities [16]. This could allow them to manipulate the consensus process, leading to potential double-spending, block withholding, or other malicious activities.

Degrading Network Performance: The presence of multiple fake identities can lead to increased overhead and reduce the overall efficiency of the network.

Preventing Sybil attacks is challenging but crucial for the security and fairness of decentralized systems. Some common defense mechanisms against Sybil attacks include:

Identity Verification: Requiring some form of real-world identity verification can limit the creation of multiple fake identities by a single entity. However, this may compromise the anonymity and privacy aspects of decentralized systems.

Proof of Work (PoW): In block chain systems that use PoW, an entity needs to invest significant computational power to create new identities, making it economically infeasible to perform a Sybil attack on a large scale [10].

Proof of Stake (PoS): PoS block chains require participants to stake cryptocurrency as collateral. An attacker would need to acquire a substantial amount of cryptocurrency to create multiple identities, which could be costly.

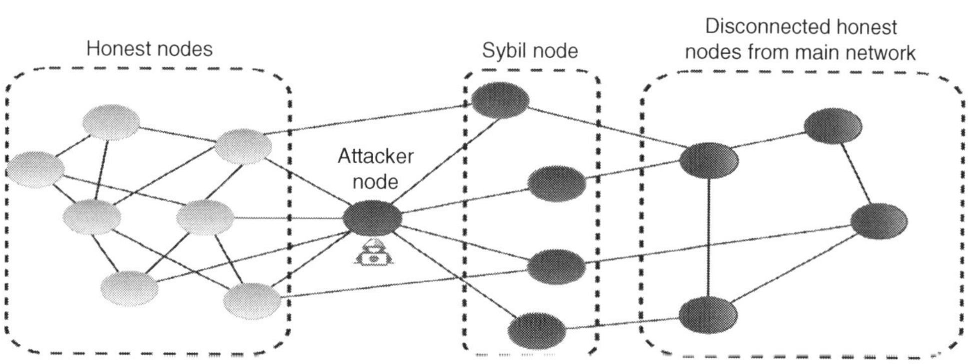

Figure 10.7 Sybil attack.

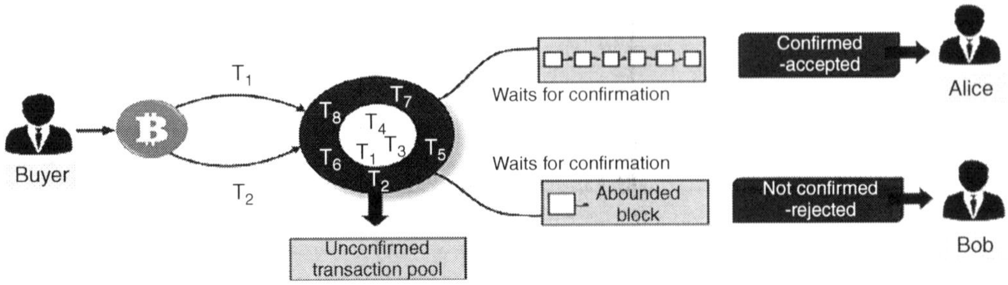

Figure 10.8 Double spending.

Reputation Systems: Implementing reputation-based mechanisms can help identify and ignore or penalize Sybil identities based on their behavior and interactions within the network.

While these defenses can mitigate the risk of Sybil attacks, no solution is entirely foolproof. Therefore, it's essential for decentralized networks to continuously improve their security measures and adapt to emerging threats to maintain the integrity of the system.

10.2.3.3 Double Spending

Double spending refers to the act of spending the same digital asset multiple times, exploiting the decentralized and distributed nature of the block chain. Block chain networks use consensus mechanisms to prevent double spending, ensuring that each transaction is valid and can only be recorded once.

Addressing these security considerations is essential when designing a block chain-based solution for securing edge learning, as they help ensure the integrity, authenticity, and trustworthiness of the system [16].

Double-Spending: The attackers can spend their cryptocurrency, create a separate chain with conflicting transactions, and then attempt to overtake the original chain, as per Figure 10.8 effectively invalidating the initial transactions and spending the same cryptocurrency again.

Double-spending is a term used in the context of digital currencies, especially cryptocurrencies, and refers to the act of spending the same unit of cryptocurrency more than once [17]. This is a significant concern in decentralized digital currency systems, where the validity and uniqueness of transactions must be ensured without relying on a central authority.

In traditional centralized payment systems (like credit cards or online banking), double-spending is prevented because all transactions are processed and verified by a central authority (e.g., a bank) that maintains a centralized ledger. This authority ensures that each unit of currency can only be spent once.

However, in decentralized cryptocurrencies like Bitcoin, there is no central authority. Instead, a distributed consensus mechanism (such as PoW) is used to validate transactions and create new blocks in the block chain. This raises the possibility of double-spending if someone can find a way to subvert the consensus mechanism.

The double-spending problem is effectively addressed in cryptocurrencies like Bitcoin by relying on the longest chain rule. When multiple conflicting transactions attempting to spend the same cryptocurrency appear in the network, the one that is included in the longest valid chain of blocks is considered the valid transaction [18]. Miners in the network compete to solve complex cryptographic puzzles and add new blocks to the block chain. The longest chain is typically the one that has the most accumulated computational work (PoW) put into it. Thus, it becomes infeasible for an attacker to rewrite the entire block chain with new double-spending transactions, as doing so would require an immense amount of computing power and would need to surpass the combined computational power of the honest nodes in the network.

As a result, the security of a block chain network against double-spending attacks is closely tied to the distribution of mining power and the overall network's hash rate. The more decentralized and robust the network is, the less susceptible it is to double-spending attacks.

10.3 Edge Intelligence

10.3.1 Edge Computing Overview

Edge computing is a distributed computing paradigm that brings computation and data storage closer to the edge devices, reducing the latency and bandwidth requirements associated with sending data to a centralized cloud infrastructure as per Figure 10.9. Edge computing utilizes the resources available at the network's edge, such as edge devices, gateways, and local servers, to process and analyze data in real-time.

10.3.1.1 Edge Devices and Infrastructure

Edge devices are the endpoint devices located at the edge of the network, such as sensors, cameras, smartphones, and IoT devices [19]. These devices collect and generate massive amounts of data. Edge infrastructure includes edge servers, gateways, and routers that provide processing power, storage, and connectivity to support edge computing capabilities.

10.3.1.2 Edge Computing Benefits

Edge computing offers several advantages for edge intelligence applications:

Reduced Latency: By processing data locally, near the source, edge computing minimizes the delay in transmitting data to a remote cloud server, enabling faster response times for real-time applications.

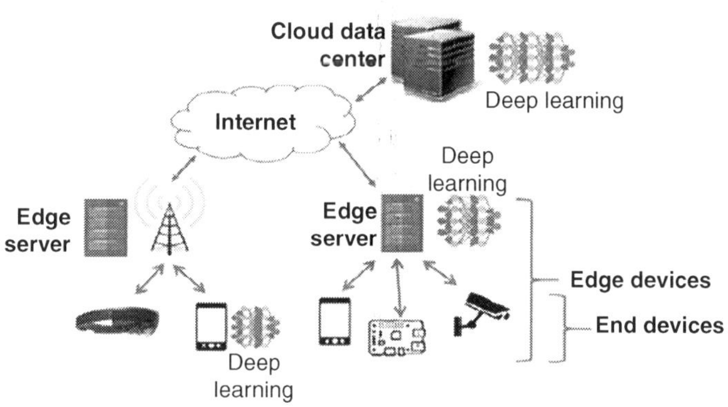

Figure 10.9 Edge intelligence.

Bandwidth Optimization: Edge computing reduces the amount of data transmitted to the cloud, optimizing bandwidth usage and reducing network congestion.

Offline Operation: Edge devices can continue to operate even in disconnected or low-connectivity environments, ensuring uninterrupted services.

Enhanced Privacy: Data can be processed and analyzed locally, minimizing the need to transmit sensitive or private information to a remote cloud, thereby enhancing data privacy and reducing security risks.

Scalability: Edge computing allows for scalable and distributed processing capabilities, leveraging the resources available at the edge, and enabling efficient scaling of applications.

10.3.2 Edge Learning

Edge learning involves deploying machine learning and artificial intelligence algorithms directly on edge devices to enable real-time analytics and decision-making at the edge. It leverages the computational power and storage capabilities of edge devices to perform tasks such as data preprocessing, model training, and inference as per Figure 10.10.

10.3.2.1 Edge Learning Architecture

The architecture of an edge learning system typically consists of three layers: the data acquisition layer, the edge processing layer, and the cloud layer. In this architecture, edge devices capture and preprocess the data, perform local analytics and model inference as per Figure 10.11, and can optionally communicate with the cloud for collaborative learning or long-term storage of aggregated insights.

10.3.2.2 Edge Learning Applications

Edge learning finds applications in various domains, including:

Autonomous Vehicles: Edge learning enables real-time perception, decision-making, and control in autonomous vehicles, allowing them to operate efficiently and safely.

Industrial IoT: Edge learning in industrial settings enhances predictive maintenance, anomaly detection, and optimization of manufacturing processes.

Healthcare: Edge learning can support remote patient monitoring, real-time diagnostics, and personalized healthcare services.

Smart Cities: Edge learning enables intelligent traffic management, environmental monitoring, and efficient energy consumption in smart city infrastructures.

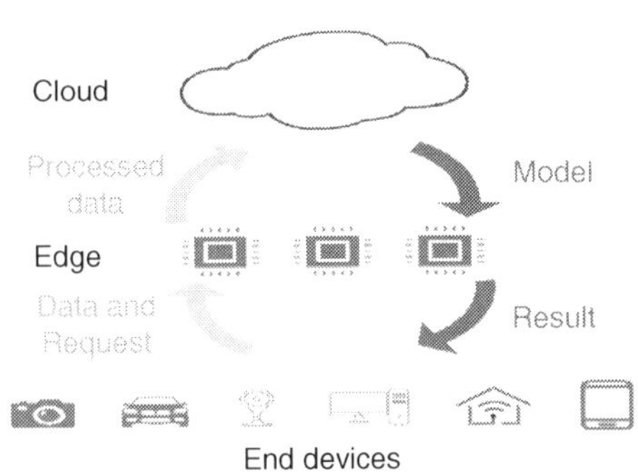

Figure 10.10 Edge learning.

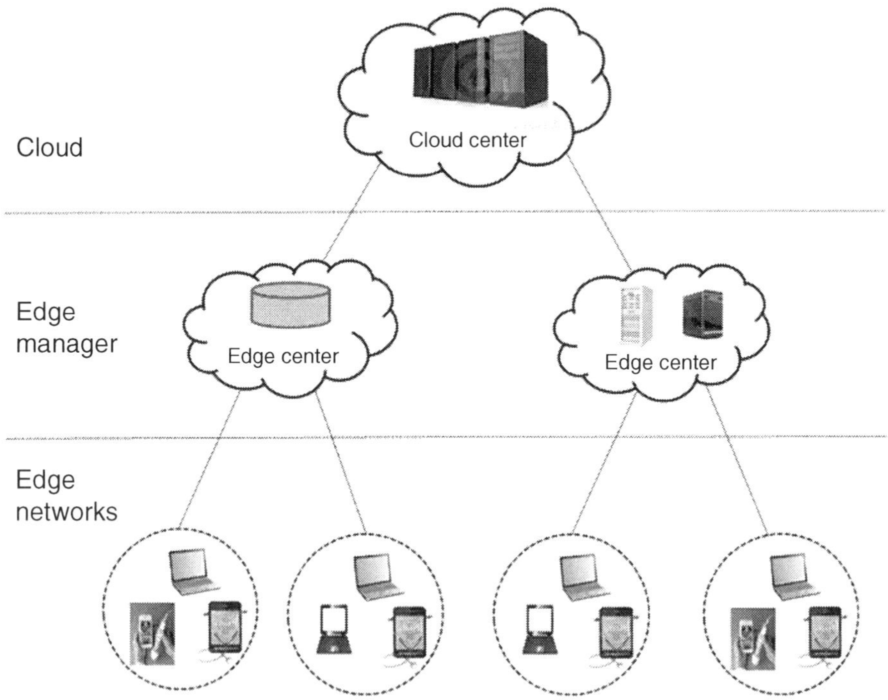

Figure 10.11 Edge learning architecture.

10.3.3 Security Challenges in Edge Learning

While edge intelligence provides numerous benefits, it also introduces unique security challenges that need to be addressed:

10.3.3.1 Data Privacy and Integrity

Edge devices often collect sensitive data, including personal information and critical industrial data. Ensuring data privacy and integrity is crucial to prevent unauthorized access, data breaches, and tampering during data transmission and storage.

10.3.3.2 Malicious Attacks

Edge devices are vulnerable to various malicious attacks, such as malware injection, denial-of-service attacks, and device impersonation. Securing edge learning systems requires protection mechanisms against these attacks to maintain the confidentiality, availability, and integrity of the system.

Edge learning refers to the practice of training and running machine learning models directly on edge devices, such as smartphones, IoT devices, and other edge computing devices, rather than sending data to a centralized server or cloud for processing. While edge learning offers several advantages, such as reduced latency, improved privacy, and bandwidth efficiency, it also introduces new security challenges. Some of the malicious attacks that can target edge learning systems include:

Model Poisoning: Model poisoning, also known as data poisoning, involves an attacker injecting malicious data into the training dataset used at the edge device. By poisoning the data, the attacker aims to manipulate the model's learning process and cause it to make incorrect predictions or behave in a harmful way.

Model Inversion Attacks: In model inversion attacks, an attacker attempts to infer sensitive information from the machine learning model deployed on the edge device. By analyzing the model's output, the attacker may be able to reconstruct private data that was used during the training process.

Adversarial Examples: Adversarial examples are inputs that are intentionally crafted to deceive a machine learning model. These inputs may look innocuous to humans but can cause the model to make incorrect predictions. Adversarial examples can be especially problematic in edge learning scenarios, where robustness and defense mechanisms may be limited due to resource constraints.

Evasion Attacks: Evasion attacks involve an adversary modifying input data to evade detection by the edge learning model. For example, in the context of intrusion detection or malware classification, attackers may try to craft inputs that are misclassified as benign.

Data Poisoning and Inference Attacks: In data poisoning attacks, an attacker manipulates the data sent from the edge devices to the central server during the model update process. Similarly, in inference attacks, an attacker may attempt to exploit information leaked during the inference process to infer sensitive data.

Model Extraction: Model extraction attacks involve an adversary trying to reverse-engineer or extract the machine learning model running on the edge device. The extracted model could be used for malicious purposes or to create a replica of the original model.

To mitigate these malicious attacks in edge learning, various techniques and security measures can be employed, such as:

- Using robust and adversarial trained machine learning models.
- Implementing encryption and secure communication protocols to protect data transmission.
- Employing anomaly detection techniques to identify and handle adversarial inputs.
- Monitoring and analyzing model performance for potential signs of attacks.
- Ensuring secure and trusted updates for edge devices and models.

It's important to recognize that security in edge learning is an ongoing research area, and as new attack vectors emerge, the defense mechanisms will continue to evolve to protect against them.

10.3.3.3 Trustworthiness and Accountability

With edge learning, decision-making occurs locally, making it essential to ensure the trustworthiness of the models and the accountability of the devices. Verifying the integrity and origin of the models and establishing trust among the participating devices are crucial for reliable and secure edge learning.

Addressing these security challenges is vital to unlock the full potential of edge learning. The convergence of block chain and edge intelligence offers promising solutions to enhance security in edge learning environments by providing secure data sharing, decentralized trust management, and improved accountability mechanisms.

10.4 Securing Edge Learning with Block Chain

Securing Edge Learning with block chain is a promising approach that combines the advantages of block chain technology with the decentralized nature of edge computing [21]. Edge learning refers to the process of running machine learning and artificial intelligence algorithms directly on edge devices, such as smartphones, IoT devices, and edge servers, instead of relying solely on centralized cloud servers. Integrating block chain into edge learning can offer several security benefits:

Data Privacy and Integrity: Edge devices often process sensitive data, and privacy is a major concern. Block chain's decentralized and immutable nature can help ensure that data remains private and tamper-proof. Instead of sending raw data to a centralized server for processing, edge devices can execute smart contracts on the block chain, allowing data owners to maintain control over their information.

Secure Data Sharing: Block chain can facilitate secure data sharing and access control in edge learning environments. Smart contracts can define rules for data sharing and enforce permissions, ensuring that only authorized entities can access specific data, while keeping the data encrypted and private.

Distributed Consensus: Block chain's consensus mechanism ensures that all nodes in the edge network agree on the validity of transactions and data. This consensus process adds a layer of trust to the edge learning environment, making it more resistant to attacks and manipulation [22].

Identity and Authentication: Block chain can provide a decentralized identity management system, allowing edge devices and users to have unique identities on the network. This helps prevent unauthorized access and ensures that only authenticated devices and users can participate in the edge learning process.

Smart Contract-Based Machine Learning Models: Block chain can host machine learning models as smart contracts, allowing edge devices to request and execute specific models directly on the block chain. This enables more efficient model updates and ensures that all devices are using the same verified model.

Resilience to Attacks: The decentralized nature of block chain makes the edge learning system more resilient to single points of failure and distributed denial-of-service (DDoS) attacks. Even if some edge nodes are compromised or go offline, the system can continue to function.

Transparency and Audit Ability: Block chain's transparency allows for the auditing of machine learning algorithms and their outcomes. This can be crucial in critical applications, such as healthcare or autonomous vehicles, where the performance and fairness of algorithms need to be verified.

Block chain technology offers several potential solutions for securing edge learning environments. In this section, we explore different aspects where block chain can be applied to enhance security in edge learning systems [23].

10.4.1 Block Chain for Data Integrity and Audit Ability

One of the fundamental challenges in edge learning is ensuring the integrity and audit ability of the data processed at the edge. Block chain can provide a robust solution in this regard.

10.4.1.1 Data Provenance and Traceability

Block chain enables the recording of data provenance and traceability by creating an immutable and transparent ledger of transactions. Each data transaction can be recorded as a transaction on the block chain, capturing information about the data source, processing steps, and any modifications made. This transparency allows for easy verification of the data's origin, ensuring its integrity and enabling effective audit ability.

10.4.1.2 Secure Data Sharing and Collaboration

Block chain can facilitate secure data sharing and collaboration in edge learning environments. By leveraging smart contracts, data owners can define access controls and permissions for their data. Block chain ensures that data sharing occurs based on predefined rules, preventing unauthorized access and ensuring data confidentiality. Moreover, block chain's decentralized nature eliminates the need for intermediaries, enhancing data privacy and reducing the risk of data breaches.

10.4.2 Block Chain for Decentralized Trust Management

Trust management is essential in edge learning environments to ensure the trustworthiness and accountability of the participating entities. Block chain can provide a decentralized trust management framework.

10.4.2.1 Identity and Access Management

Block chain can enable secure identity management for edge devices and users. Each device can have a unique digital identity stored on the block chain, allowing for secure and tamper-resistant authentication and access control. By leveraging cryptographic techniques, block chain can establish trust and enforce secure interactions among edge devices and users, mitigating the risk of unauthorized access.

Identity and Access Management (IAM) is a framework of policies, processes, and technologies used to manage and control access to resources within an organization's IT environment as per Figure 10.12. IAM is crucial for ensuring that the right individuals have the appropriate access to the right resources at the right time. It encompasses the management of user identities, authentication, authorization, and access permissions across various systems and applications.

Key components of identity and access management include:

Identification: This involves uniquely identifying users, devices, and entities within the system. Each user or entity is assigned a unique identity that serves as a digital representation within the IAM system.

Figure 10.12 Identity and access management.

Authentication: Authentication is the process of verifying the identity of a user or entity attempting to access a system. Common authentication methods include passwords, MFA, biometrics, smart cards, and more.

Authorization: After successful authentication, the IAM system determines the permissions and privileges that the authenticated user or entity has. Authorization ensures that users can only access the resources and perform actions that they are allowed to be based on their role and permissions.

Single Sign-On (SSO): SSO enables users to access multiple applications and services with a single set of credentials. Once authenticated, users can navigate between different systems without needing to re-enter their credentials for each application.

Role-Based Access Control (RBAC): RBAC is a common approach to managing access permissions based on the roles and responsibilities of users within an organization. Users are assigned specific roles, and each role has predefined access rights to certain resources.

Access Governance: Access governance involves regularly reviewing and managing user access rights to ensure they are appropriate and aligned with business needs. It helps identify and mitigate potential security risks caused by overprivileged or unnecessary access.

Privileged Access Management (PAM): PAM focuses on securing and managing privileged accounts, such as administrative accounts with elevated access rights. PAM solutions help control and monitor privileged access to critical systems and resources.

Federation: Federation enables secure sharing of identity and authentication information across different systems and organizations. It allows users from one organization to access resources in another organization without needing separate credentials.

Lifecycle Management: IAM includes user lifecycle management, which involves the automated provisioning and deprovisioning of user accounts based on changes in their roles or status (e.g., on boarding, transfers, and off-boarding).

IAM plays a vital role in enhancing security, productivity, and compliance within an organization's IT infrastructure. It helps prevent unauthorized access, identity theft, and data breaches while facilitating efficient access for authorized users. Implementing a robust IAM strategy is essential for modern organizations, particularly as they adopt cloud services, mobile devices, and other digital technologies that increase the complexity of managing identities and access across various systems.

10.4.2.2 Reputation Management

Block chain can facilitate reputation management for edge devices and participants. Reputation scores can be recorded on the block chain based on their past behavior and interactions. This reputation system enables participants to evaluate the trustworthiness and reliability of other devices and make informed decisions regarding data sharing and collaboration.

Reputation management is the practice of actively monitoring, influencing, and maintaining the perception and reputation of an individual, organization, brand, or entity in the eyes of the public. It involves strategies and actions taken to shape public opinion and manage the way others perceive and talk about the subject in question as per Figure 10.13. Reputation management is crucial in today's digital age, where information spreads quickly and widely through social media, online reviews, and news outlets.

Key aspects of reputation management include:

Monitoring: Reputation management starts with continuous monitoring of online mentions, social media conversations, customer reviews, and news articles. Monitoring helps identify both positive and negative sentiments surrounding the subject.

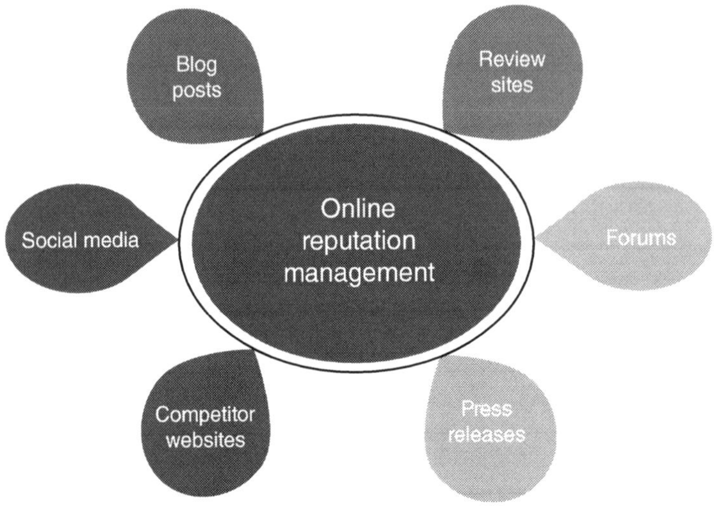

Figure 10.13 Reputation management.

Online Reviews and Testimonials: Managing online reviews and testimonials is essential, especially for businesses. Responding promptly and professionally to customer feedback, whether positive or negative, can significantly impact how the public perceives the brand.

Crisis Management: In times of crisis or negative publicity, reputation management focuses on addressing the issues, being transparent with stakeholders, and managing the narrative to mitigate damage to the reputation.

Social Media Management: Maintaining a strong presence on social media platforms is critical for reputation management. Engaging with followers, sharing valuable content, and responding to inquiries in a timely manner helps build a positive image.

Search Engine Optimization (SEO): Improving search engine results is part of reputation management. Positive and relevant content that showcases the subject in a favorable light can help push down negative search results.

Branding and Public Relations: A consistent and positive brand image, supported by effective public relations efforts, can contribute to a strong reputation.

Influencer Marketing: Collaborating with influencers and thought leaders can help increase brand credibility and reach a wider audience.

Customer Service: Providing excellent customer service is vital for reputation management. Addressing customer issues promptly and satisfactorily can turn dissatisfied customers into loyal advocates.

Employee Advocacy: Engaging employees to become brand advocates can positively influence the perception of the organization.

Reputation management is not just about hiding negative information but also about building a solid foundation of trust and credibility. It requires a proactive and ongoing effort to shape and maintain a positive image in the minds of the target audience. A good reputation can lead to increased customer loyalty, improved business opportunities, and a competitive advantage in the marketplace. On the other hand, neglecting reputation management or mishandling negative situations can result in significant damage to the subject's reputation, affecting their relationships with customers, partners, and stakeholders.

10.4.3 Block Chain for Consensus and Fault Tolerance

Consensus and fault tolerance mechanisms are crucial for ensuring the reliability and accuracy of edge learning systems. Block chain provides decentralized consensus mechanisms that can be leveraged in edge learning environments.

10.4.3.1 Consensus Mechanisms in Edge Learning

Block chain offers various consensus mechanisms, such as PoW or PoS, which can be adapted to secure the consensus process in edge learning networks. These mechanisms ensure agreement on the validity and order of transactions, preventing malicious actors from compromising the system.

Consensus mechanisms in edge learning play a crucial role in ensuring that edge devices can collectively and securely agree on the validity of transactions or decisions related to the training and inference of machine learning models. Since edge learning involves running machine learning algorithms directly on edge devices, the consensus mechanisms used in traditional block chain networks may not always be suitable due to the unique challenges of the edge environment, such as limited computational power, bandwidth, and intermittent connectivity.

Here are some consensus mechanisms that can be applied in edge learning scenarios:

Federated Learning: Federated Learning is a decentralized approach to model training where edge devices collaboratively participate in the training process. Instead of sending raw data to a central server, edge devices locally train the model and share only the model updates with the central server or other devices as per Figure 10.14. The central server aggregates the updates to create a global model, which is then distributed back to the edge devices. This consensus mechanism reduces the need for large data transfers and ensures data privacy.

PoW with Edge Hashing: In edge learning, PoW can be adapted by using lightweight cryptographic hashing algorithms that are feasible on resource-constrained edge devices. Edge Hashing allows devices to perform computational work to reach consensus on model updates or decisions while minimizing energy consumption.

Proof of Stake (PoS) with Trusted Nodes: In a PoS-based edge learning system, edge devices with a stake (e.g., based on computational resources or token holdings) are selected as trusted

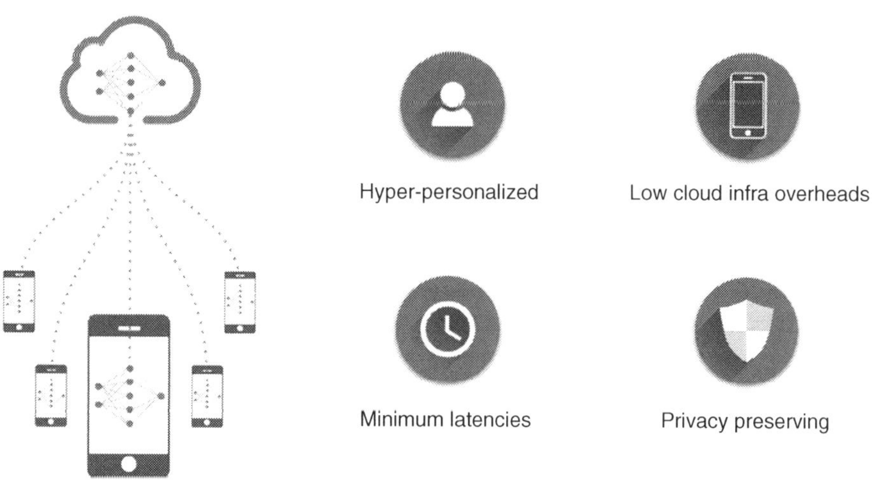

Figure 10.14 Federated learning.

nodes to collectively agree on model updates or decisions. PoS can reduce the energy requirements compared to PoW and increase the efficiency of the consensus process.

Delegated Proof of Stake (DPoS) for Edge Clusters: DPoS can be utilized in edge clusters, where a smaller set of nodes, called delegates, are responsible for validating model updates or decisions. Delegates are elected by the network, and their responsibility rotates over time, ensuring fairness and efficiency.

Practical Byzantine Fault Tolerance (PBFT) with Limited Nodes: PBFT can be adapted for edge learning scenarios by limiting the number of nodes involved in the consensus process. This ensures fast and deterministic agreement on model updates while improving the overall efficiency.

Directed Acyclic Graph (DAG) Consensus: DAG-based consensus mechanisms, like IOTA's Tangle, are designed for scalability and low-resource environments. They can be used in edge learning to reach consensus on model updates through a DAG structure.

Hybrid Consensus: A combination of different consensus mechanisms, such as a hybrid PoW/PoS approach, can be utilized to leverage the strengths of each mechanism while addressing the limitations of the edge environment.

The choice of consensus mechanism for edge learning depends on factors like network size, computational power, data privacy requirements, and the specific use case. The goal is to achieve secure and efficient consensus while minimizing resource consumption and latency on edge devices. As edge learning continues to evolve, new consensus mechanisms tailored to its unique challenges may also emerge.

10.4.3.2 Fault Tolerance and Byzantine Faults

Block chain's decentralized nature provides inherent fault tolerance capabilities, making it resilient to failures and attacks. Byzantine fault-tolerant consensus algorithms, such as PBFT, can be employed to ensure the correctness of the consensus even in the presence of malicious or faulty nodes in the network.

Fault tolerance and Byzantine faults are concepts related to the reliability and resilience of distributed systems, including block chain networks and other decentralized systems.

10.4.3.2.1 Fault Tolerance Fault tolerance is the ability of a system to continue operating and providing its intended functionality even in the presence of faults or failures. A fault can be any unexpected event that causes a component or node in the system to behave abnormally (Figure 10.15), leading to a failure. Faults can result from hardware malfunctions, software bugs, communication errors, or other unexpected events.

To achieve fault tolerance, distributed systems employ various techniques, such as redundancy, replication, and error detection and correction mechanisms. When a fault occurs, the system can continue to function properly by relying on backup resources or by reconfiguring the system to avoid faulty components. The goal of fault tolerance is to ensure that the system remains available and operational, providing a seamless experience to users even during challenging situations.

10.4.3.2.2 Byzantine Faults Byzantine faults refer to a type of fault in distributed systems where a node behaves in an arbitrary and malicious manner, deviating from the intended behavior of the system. These malicious nodes can send false or conflicting information to other nodes, leading to a breakdown in communication and coordination. Byzantine faults are particularly challenging because the malicious nodes can actively collude to deceive the rest of the network.

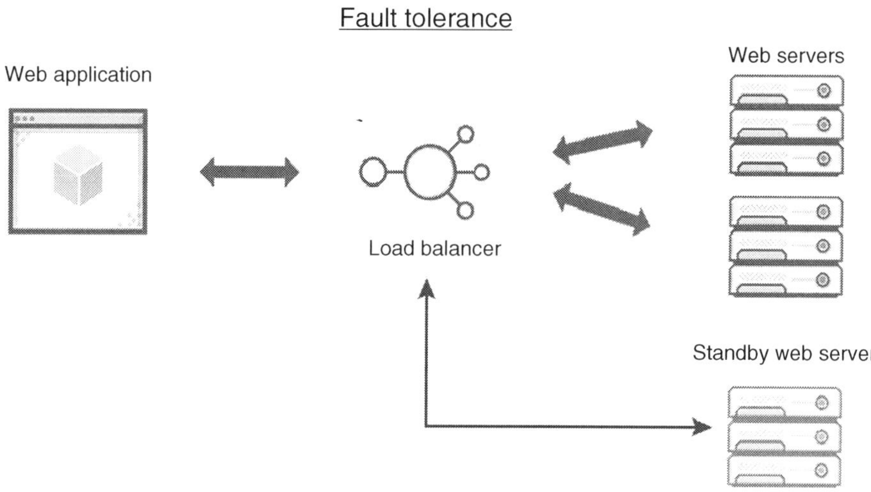

Figure 10.15 Fault tolerance.

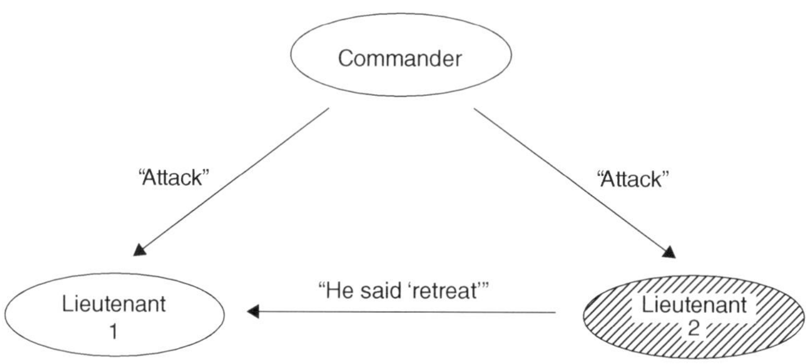

Figure 10.16 Byzantine faults.

The term "Byzantine faults" originates from the Byzantine Generals' Problem, a theoretical problem in computer science (Figure 10.16). In the problem, a group of Byzantine generals must agree on a coordinated attack or retreat, but some of the generals may be traitors who aim to undermine the consensus.

To handle Byzantine faults, consensus mechanisms like PBFT and Federated Byzantine Agreement (FBA) have been developed. These mechanisms are designed to allow the distributed system to reach a consensus even in the presence of malicious nodes. They achieve this by requiring a supermajority of nodes to agree on a decision, making it computationally infeasible for a group of malicious nodes to control the consensus.

Both fault tolerance and Byzantine faults are important considerations in the design and operation of distributed systems. Achieving fault tolerance and handling Byzantine faults are essential to ensuring the security, reliability, and continuity of block chain networks and other DApps.

10.4.4 Block Chain for Secure Edge Device Management

Managing and securing edge devices is critical to ensure the overall security of edge learning systems. Block chain can offer solutions for secure edge device management.

10.4.4.1 Secure Firmware Updates

Block chain can be utilized to securely distribute and verify firmware updates for edge devices. By recording firmware hashes and signatures on the block chain, device owners and users can verify the authenticity and integrity of firmware updates, preventing unauthorized modifications or malicious code injection.

10.4.4.2 Device Authentication and Authorization

Block chain-based identity management can be leveraged for secure device authentication and authorization. Edge devices can have their digital identities stored on the block chain, and smart contracts can enforce access control policies based on these identities. This ensures that only authorized devices can participate in edge learning activities, minimizing the risk of unauthorized access and malicious activities.

By integrating block chain technology into edge learning environments, these security mechanisms can enhance data integrity, privacy, trust management, consensus, and device management, contributing to the overall security and reliability of edge learning systems.

10.5 Existing Research and Case Studies

In this section, we provide an overview of existing research efforts and case studies that highlight the application of block chain in securing edge learning environments.

10.5.1 Block Chain-Based Secure Data Sharing in Edge Learning

Research has explored the use of block chain to ensure secure and trusted data sharing in edge learning scenarios. Block chain enables data owners to maintain control over their data while allowing secure and auditable sharing with authorized parties. Smart contracts can be utilized to define access controls, data usage policies, and payment mechanisms for data sharing in a decentralized and tamper-resistant manner. By leveraging block chain technology, data sharing in edge learning can be enhanced with improved security, privacy, and accountability [20].

10.5.2 Decentralized Trust Management in Edge Intelligence

Decentralized trust management is a crucial aspect of securing edge intelligence systems. Block chain offers a decentralized trust infrastructure where identity, reputation, and access management can be enforced. Research has explored the use of block chain-based identity management to establish trust among participating edge devices and users. By recording device identities and reputation scores on the block chain, trust can be established, and secure interactions can be facilitated. These approaches enhance the trustworthiness and accountability of edge learning systems.

10.5.3 Consensus Mechanisms for Edge Learning Networks

Consensus mechanisms play a vital role in ensuring the reliability and accuracy of edge learning networks. Research has investigated the adaptation of existing consensus mechanisms, such as PoW or PoS, for securing the consensus process in edge learning environments. These mechanisms enable participants to reach agreement on the validity and order of transactions or model updates, ensuring the integrity and consistency of the edge learning system. By leveraging block chain-based consensus mechanisms, edge learning networks can achieve secure and trustworthy operations.

10.5.4 Secure Device Management in Edge Computing

Managing the security of edge devices is crucial for the overall security of edge learning systems. Block chain can be leveraged for secure device management, including secure firmware updates, device authentication, and authorization. Research has explored the use of block chain to securely distribute and verify firmware updates for edge devices. By recording firmware hashes and signatures on the block chain, the authenticity and integrity of firmware updates can be ensured. Moreover, block chain-based identity management can be utilized for secure device authentication and authorization, preventing unauthorized access and ensuring the trustworthiness of participating devices.

These research efforts and case studies demonstrate the potential of block chain in enhancing security in edge learning environments. They highlight the benefits of block chain in secure data sharing, decentralized trust management, consensus mechanisms, and device management. By leveraging block chain technology, edge learning systems can address the unique security challenges posed by edge computing and enhance the overall security and trustworthiness of the systems.

10.6 Framework for Securing Edge Learning with Block Chain

In this section, we present a framework for securing edge learning environments using block chain technology. The framework encompasses the architecture, key design considerations, and implementation challenges.

10.6.1 Architecture and Components

The proposed framework consists of the following key components:

Edge Devices: These are the devices at the edge of the network where data is collected, processed, and analyzed. They can include sensors, IoT devices, and edge servers.

Block Chain Network: This is the distributed network of nodes that maintain the block chain ledger. It consists of participating edge devices and potentially cloud nodes acting as validators or miners.

Smart Contracts: These self-executing contracts define the rules and conditions for data sharing, access control, and trust management in the edge learning system.

Consensus Mechanism: The consensus mechanism ensures agreement on the validity and order of transactions within the block chain network. It can be adapted to suit the specific requirements of the edge learning environment.

Identity Management: This component manages the identities of edge devices and users participating in the edge learning system. It ensures secure authentication, authorization, and access control.

Data Management: This component handles data provenance, traceability, and secure data sharing among edge devices. It enables data integrity and privacy-preserving mechanisms.

Device Management: This component ensures secure device authentication, firmware updates, and secure device-to-device communication within the edge learning system.

10.6.2 Key Design Considerations

When designing a secure edge learning system using block chain, several key considerations should be taken into account:

Scalability: The framework should be designed to accommodate the large-scale deployment of edge devices and handle the increasing volume of data generated at the edge.
Performance: The design should optimize the performance of the system, considering the resource constraints of edge devices while ensuring real-time processing and decision-making capabilities.
Privacy and Data Confidentiality: The framework should incorporate mechanisms to protect sensitive data, including encryption, data obfuscation, and access control, to ensure privacy and confidentiality.
Interoperability: The system should facilitate interoperability between different edge devices, cloud nodes, and block chain networks to enable seamless data sharing and collaboration.
Trust and Security: The design should establish trust among participating devices and users through robust identity management, reputation systems, and secure communication protocols.
Governance and Compliance: The framework should adhere to regulatory requirements and incorporate governance mechanisms to ensure accountability, compliance, and ethical use of data.

10.6.3 Implementation Challenges and Trade-Offs

Implementing a block chain-based framework for securing edge learning systems comes with certain challenges and trade-offs:

Resource Constraints: Edge devices often have limited computational power, storage, and energy resources. The design should consider these constraints and optimize resource usage to ensure efficient operation.
Network Connectivity: Edge devices may operate in disconnected or low-connectivity environments. The framework should handle intermittent network connectivity and enable offline operation when necessary.
Latency: Block chain transactions and consensus mechanisms can introduce latency. Balancing the need for security with real-time processing requirements is crucial in designing the system.
Cost: Block chain technology may involve additional costs, such as storage, processing, and network bandwidth. The trade-offs between security, performance, and cost should be carefully considered.
Adoption and Integration: Integrating block chain into existing edge learning infrastructures may require careful planning, deployment strategies, and stakeholder engagement to ensure successful adoption.

Addressing these implementation challenges and trade-offs is vital for the successful implementation of a secure edge learning framework using block chain technology. By considering these considerations and carefully designing the architecture, organizations can leverage the benefits of block chain to enhance the security, trustworthiness, and privacy of edge learning systems.

10.7 Evaluation and Future Directions

In this section, we discuss the evaluation of the framework for securing edge learning with block chain, conduct a comparative analysis with traditional security mechanisms, and explore potential future research directions.

10.7.1 Performance Evaluation Metrics

To evaluate the performance of the framework, several metrics can be considered:

Throughput: This metric measures the number of transactions or data processing operations that can be handled per unit of time, indicating the system's capacity.

Latency: Latency measures the time taken for a transaction or data processing operation to complete, reflecting the system's responsiveness.

Scalability: Scalability refers to the system's ability to handle an increasing number of participants, data sources, and transactions while maintaining performance.

Resource Utilization: This metric assesses the efficient utilization of computational resources, storage, and network bandwidth, ensuring optimal use of available resources.

Security: The framework should be evaluated for its ability to protect against various security threats, including data breaches, unauthorized access, and malicious attacks.

Privacy: Privacy evaluation focuses on the effectiveness of mechanisms implemented to protect the confidentiality and integrity of sensitive data.

Fault Tolerance: The framework should be tested for its ability to withstand and recover from failures, ensuring the resilience and availability of the system.

10.7.2 Comparative Analysis with Traditional Security Mechanisms

A comparative analysis can be conducted to evaluate the effectiveness of the block chain-based framework against traditional security mechanisms commonly used in edge learning environments. key factors to consider include:

Security Strength: Assessing the robustness and effectiveness of security mechanisms in preventing and mitigating security threats.

Scalability: Comparing the scalability of block chain-based security mechanisms with traditional approaches, considering the increasing number of edge devices and the volume of data.

Privacy and Data Confidentiality: Evaluating the level of privacy and data confidentiality provided by block chain-based security mechanisms compared to traditional methods.

Trust and Accountability: Analyzing the ability of block chain-based mechanisms to establish trust, ensure accountability, and provide audit ability compared to traditional approaches.

Performance Overhead: Assessing the performance overhead introduced by block chain-based security mechanisms, such as transaction processing time and resource utilization, compared to traditional methods.

Implementation Complexity: Considering the complexity of implementing and maintaining the block chain-based framework compared to traditional security mechanisms.

10.7.3 Potential Future Research Directions

The framework opens up several potential future research directions for securing edge learning with block chain:

Optimization Techniques: Developing optimization techniques to enhance the performance and resource efficiency of block chain-based edge learning systems, considering the resource-constrained nature of edge devices.

Privacy-Preserving Mechanisms: Exploring advanced cryptographic techniques and privacy-preserving protocols to ensure the privacy and confidentiality of data in edge learning environments.

Hybrid Approaches: Investigating hybrid approaches that combine block chain with other emerging technologies, such as federated learning or secure multiparty computation, to achieve a balance between security, performance, and privacy.

Energy Efficiency: Developing energy-efficient mechanisms to reduce the energy consumption of block chain operations in edge learning systems, extending the battery life of resource-constrained edge devices.

Standardization and Governance: Contributing to the standardization efforts for block chain-based edge learning frameworks, addressing interoperability, governance, and regulatory compliance challenges.

Real-World Deployments: Conducting empirical studies and real-world deployments to validate the effectiveness and practicality of the proposed framework in diverse edge learning applications and use cases.

By pursuing these future research directions, the field can advance the security and trustworthiness of edge learning systems and address the unique challenges posed by edge computing environments, ultimately unlocking the full potential of edge intelligence.

10.8 Conclusion

This chapter explored the convergence of block chain and edge intelligence for securing edge learning environments. We discussed the fundamentals of block chain technology, including its distributed ledger nature, consensus mechanisms, and smart contracts. We also highlighted the characteristics of block chain, such as immutability, transparency, and decentralization, which make it suitable for securing edge learning. The chapter delved into the concept of edge intelligence, encompassing edge computing and edge learning. We discussed the benefits of edge computing, including reduced latency, improved privacy, and offline operation. We examined edge learning architecture and applications in various domains. Furthermore, we identified the security challenges in edge learning, including data privacy and integrity, malicious attacks, and trustworthiness. To address these challenges, we presented a framework for securing edge learning with block chain. The framework included components such as data integrity and audit ability, decentralized trust management, consensus and fault tolerance, and secure device management. We explored existing research and case studies related to block chain-based secure data sharing, decentralized trust management, consensus mechanisms, and secure device management in edge learning. These studies provided insights into the application of block chain in securing edge learning environments and showcased its potential benefits.

Implications and Contributions: The convergence of block chain and edge intelligence has significant implications for securing edge learning. By leveraging the characteristics of block chain, such as immutability, transparency, and decentralization, organizations can enhance the security, trust, and privacy of edge learning systems. The proposed framework contributes to the field by providing a comprehensive guide for implementing secure edge learning environments using block chain technology. It addresses key design considerations, identifies implementation challenges, and highlights potential trade-offs. The implications of this research are far-reaching. Secure edge learning can empower critical applications such as autonomous vehicles, industrial automation, and healthcare systems, ensuring the integrity of the learning models, protecting user privacy, and fostering trust among participants. The framework offers organizations a roadmap to navigate the complexities of securing edge learning environments and paves the way for the adoption of block chain in edge intelligence systems.

Securing edge learning is of paramount importance as organizations increasingly rely on edge computing and edge intelligence. Traditional security mechanisms are often insufficient to address the unique challenges of edge environments. The convergence of block chain and edge intelligence presents a promising solution, offering decentralized trust, data integrity, and privacy preservation.

References

1 Li, S., Xu, L.D., and Zhao, S. (2018). Secure and privacy-preserving edge computing in internet of things. *IEEE Network* 32 (1): 96–101.

2 Li, Y., Zhang, K., and Yang, Y. (2020). Block chain-enabled secure data sharing for industrial internet of things. *IEEE Transactions on Industrial Informatics* 16 (6): 4243–4251.

3 Xu, J., Zhu, L., Fu, X. et al. (2019). A survey of block chain technology applied to smart cities: research issues and challenges. *IEEE Access* 7: 22328–22348.

4 Yang, Y., Zhang, Y., Wang, X., and Jiang, L. (2021). A survey of edge intelligence in internet of things. *Journal of Network and Computer Applications* 179: 103127.

5 Yin, Y., Li, H., Guo, Z., and Zhang, J. (2020). Block chain technology in the internet of things: a systematic review. *Journal of Network and Computer Applications* 165: 113715.

6 Li, Q., Xu, Y., Xu, Z. et al. (2020). Block chain for edge intelligence: technologies, challenges, and open issues. *IEEE Network* 34 (5): 14–20.

7 Zhang, J., Li, C., and Zhou, L. (2021). Secure federated edge learning with block chain for internet of vehicles. *IEEE Transactions on Industrial Informatics* 17 (5): 3463–3472.

8 Zhang, S., Cui, P., and Chen, J. (2021). Privacy-preserving edge learning via block chain-based federated learning. *IEEE Transactions on Industrial Informatics* 17 (7): 4821–4830.

9 Li, X., Zhao, J., and Huang, K. (2020). Block chain-based secure and efficient edge learning in fog computing. *IEEE Internet of Things Journal* 8 (3): 1599–1610.

10 Zhang, W., Li, X., Zhang, L. et al. (2020). Securing federated edge learning using block chain for internet of vehicles. *IEEE Transactions on Industrial Informatics* 16 (11): 6982–6991.

11 Murtaza, M., Khan, R.A., Amin, R. et al. (2020). Securing edge computing: a survey, use cases, and future directions. *Future Generation Computer Systems* 107: 662–676.

12 Sato, H., Kobayashi, H., and Nakano, T. (2020). Edge computing with block chain-based trust management for internet of things. *IEEE Transactions on Network Science and Engineering* 7 (1): 698–710.

13 Li, X., Li, C., Liao, S., and Huang, X. (2020). Security and privacy in block chain-based edge computing: opportunities and challenges. *IEEE Network* 34 (5): 30–35.

14 Wang, S., Wang, X., and Shi, Y. (2020). An edge computing architecture for secure and privacy-preserving industrial internet of things. *IEEE Transactions on Industrial Informatics* 16 (6): 4029–4038.

15 Li, X., Cheng, X., Zhang, W. et al. (2020). Privacy-preserving deep learning via block chain for edge intelligence. *IEEE Internet of Things Journal* 8 (4): 2804–2814.

16 He, Y., Zhang, Z., Zhang, D., and Li, T. (2020). Block chain-assisted privacy-preserving edge computing in internet of things. *IEEE Transactions on Industrial Informatics* 17 (1): 647–656.

17 Zhang, Y., Xiao, X., Hu, W. et al. (2020). Privacy-preserving edge learning using block chain for industrial Internet of Things. *IEEE Transactions on Industrial Informatics* 16 (12): 7370–7379.

18 Yuan, Y., Xu, H., Li, L. et al. (2021). Federated block chain learning for secure edge intelligence in internet of things. *IEEE Transactions on Industrial Informatics* 17 (12): 8557–8566.

19 Li, X., Li, C., Gao, H., and Huang, X. (2021). A block chain-based framework for secure edge computing in smart cities. *IEEE Transactions on Industrial Informatics* 17 (7): 4992–5002.

20 Tian, Y., Hu, Y., Yang, K. et al. (2021). Secure and efficient edge computing enabled by block chain for smart grid. *IEEE Transactions on Industrial Informatics* 17 (9): 6253–6262.

21 Wang, J., Zhang, S., Wang, H. et al. (2021). A block chain-based framework for secure and efficient federated edge learning. *IEEE Internet of Things Journal* 8 (9): 7051–7063.

22 Chen, Y., Huang, D., Leng, M., and Wang, Y. (2021). Federated edge learning with block chain: a secure and privacy-preserving approach. *IEEE Transactions on Industrial Informatics* 17 (10): 6964–6973.

23 Yuan, Y., Xu, H., Hu, B., and Li, X. (2021). Privacy-preserving edge learning with block chain in distributed IoT systems. *Future Generation Computer Systems* 116: 672–681.

11

Training on Edge

Yamini Nimmagadda

Network and Edge, Intel, Portland, OR, USA

11.1 Introduction

In the dynamic field of machine learning (ML), training models on edge devices—like smartphones, IoT gadgets, and wearables—is becoming a game-changer. Unlike the traditional cloud-based approach, edge training offers distinct advantages such as lower latency, improved privacy, reduced bandwidth usage, and better energy efficiency. These benefits are particularly crucial for real-time applications in areas like healthcare, financial services, and autonomous vehicles.

This chapter explores the integration of edge training within the broader framework of Machine Learning Operations (MLOps). MLOps combine ML development with DevOps principles, streamlining the entire ML lifecycle from data preparation to model deployment and ongoing monitoring. We provide a detailed look at a typical MLOps pipeline, covering stages like data annotation, model training, deployment, post-deployment monitoring, and incremental training.

While edge training offers numerous benefits, it also comes with challenges. These include limited computational resources, constraints on model size, and increased security risks. This book addresses these challenges, offering insights into how edge training can revolutionize ML deployment as edge devices become more powerful and efficient. Through practical discussions and examples, you will learn the strategies and considerations necessary for successfully implementing training on edge devices.

11.2 The Field of Training on the Edge

Training on edge, also known as edge training, is the process of training an ML model on a device at the edge of the network, such as a smartphone, IoT device, or wearables. This is in contrast to training ML models in the cloud.

There are a number of benefits to training on edge:

- **Reduced Latency**: ML models trained on edge devices can provide responses to user queries in real time, without the need to send data to a cloud server. This is essential for applications where latency is a critical factor, such as self-driving cars and medical devices.

Model Optimization Methods for Efficient and Edge AI: Federated Learning Architectures, Frameworks and Applications, First Edition. Edited by Pethuru Raj Chelliah, Amir Masoud Rahmani, Robert Colby, Gayathri Nagasubramanian, and Sunku Ranganath.

- **Improved Privacy**: ML models trained on edge devices can be trained on sensitive data without the need to expose that data to the cloud. This is important for applications where data privacy is a concern, such as healthcare and financial services.
- **Reduced Bandwidth Usage**: ML models trained on edge devices can reduce the amount of data that needs to be sent to and from the cloud. This can save money and improve performance, especially for applications that are deployed in remote or underserved areas.
- **Energy Efficiency**: Edge devices are often more energy-efficient than cloud servers. This can make it more cost-effective to train and deploy ML models on edge devices, especially for applications that are used frequently.

Training on edge is an important part of MLOps used in production deployments. MLOps, or Machine Learning Operations [1], is a set of practices that combines ML development with DevOps principles to automate the deployment and management of ML models in production. The goal of MLOps is to streamline the ML lifecycle, from experimentation and development to deployment and monitoring, and to make it easier to put ML models into production and keep them running reliably.

MLOps [1] is important because it helps businesses deploy ML models faster and more reliably. By automating the deployment process, MLOps can help businesses get their ML models into production more quickly and with fewer errors. MLOps helps businesses monitor their ML models in production and identify and fix any problems that occur. This can help improve the quality and reliability of ML models over time. It also reduces the cost of developing and maintaining ML models. By automating the ML lifecycle, MLOps can help businesses reduce the cost of developing and maintaining ML models.

Figure 11.1 shows a high-level overview of MLOps. A generic MLOps pipeline consists of the following stages:

- **Data Preparation**: The first stage of the MLOps pipeline is data preparation. This involves gathering data from various sources, such as databases, sensors, and APIs. The data is then cleaned, transformed, and split into training and test sets.

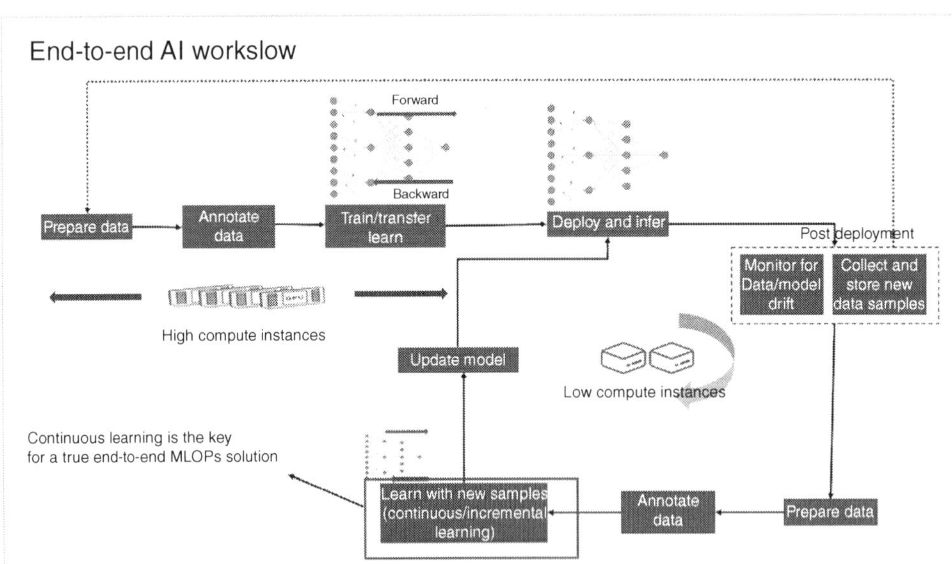

Figure 11.1 MLOps pipeline demonstrating various stages. After the new data samples are collected, the data can be sent back to high compute instances for full training or it can be incrementally trained on edge. Source: Kreuzberger et al. [1].

- **Data Annotation**: In some cases, the data may need to be annotated before it can be used to train the ML model. This involves labeling the data with the desired output. For example, if you are training a model to classify images of cats and dogs, you would need to label each image as either a cat or a dog.
- **Training**: Once the data is prepared and annotated, it can be used to train the ML model. This is done by feeding the data to the model and allowing it to learn the relationships between the input data and the output. If models are available for similar tasks, transfer learning can also be used to train the model for faster convergence.
- **Deployment/Inference**: Once the model is trained, it can be deployed to production. This means making the model available to users so that they can make predictions. The model can be deployed to a cloud server, an on-premises server, or an edge device.
- **Post-Deployment Monitoring**: Once the model is deployed, it is important to monitor its performance. This involves collecting metrics such as accuracy, precision, and recall. The model should also be monitored for bias and drift.
- **New Data Collection**: Over time, new data will become available. This data can be used to improve the model's performance by retraining it on the new data.
- **New Data Preparation and Annotation**: The new data will need to be prepared and annotated before it can be used to retrain the model. This follows the same steps as the initial data preparation and annotation stages.
- **Incremental Training**: Incremental training is a technique for retraining an ML model with new data without having to train the model from scratch. This can save time and resources, especially for large and complex models.
- **Model Updates**: Once the model is retrained, it needs to be updated in production. This can be done by replacing the old model with the new model or by using a technique called model blending to combine the old and new models.

As can be seen from Figure 11.1, there are two possibilities:

1) The newly collected data can be sent back to high compute instances, and training can be performed.
2) The newly collected data is used to incrementally train on edge [2].

The newly collected data can be added to previous data and the training can be performed on the entire data. Alternately, incremental training can be performed, where only the new data is used to train the model while preserving the information collected from the previous samples. If the size of the data is large, sending the data to high compute be more appropriate if training is performed on the entire data set. If the size of the data is small, performing the training incrementally may be more appropriate.

While there are numerous benefits with training on edge, there are also some unique challenges:

- **Limited Resources**: Edge devices typically have limited computing resources, such as CPU, memory, and storage. This can make it difficult to train large and complex ML models on edge devices.
- **Model Size**: ML models can be very large, and it can be difficult to deploy them on edge devices with limited storage capacity.
- **Security**: ML models deployed on edge devices are often more vulnerable to attacks than ML models deployed in the cloud. It is important to take steps to protect ML models from being tampered with or stolen.

Despite these challenges, training on edge is an emerging trend with the potential to revolutionize the way we develop and deploy ML applications. As edge devices become more powerful and efficient, we can expect to see more and more ML models being trained and deployed on edge devices.

In the following sections of this chapter, we will cover the benefits, challenges, and techniques for training on edge in more detail.

11.2.1 Benefits, Challenges, and Considerations

Edge training [2], also known as on-device training or federated learning, is an ML approach that occurs directly on edge devices, such as smartphones, IoT devices, and even on-premises servers. Unlike traditional cloud-based training, where data is sent to a centralized server for processing, edge training happens locally, making it a decentralized and privacy-centric technique. Below are the benefits of training on edge:

11.2.1.1 Privacy and Data Security

Edge training is a game-changer when it comes to privacy. Unlike cloud-based training, where sensitive data is transmitted and processed on remote servers, edge training keeps data localized. This means that personal and sensitive information remains on the user's device, reducing the risk of data breaches and privacy violations. Below are the detailed advantages:

- **Localized Data Processing**: Edge training ensures that data processing and model training occur locally on the user's device. This approach significantly reduces the risk associated with data transfer and storage on external servers. Sensitive information, such as personal identifiers or proprietary business data, stays within the confines of the user's device, reducing the exposure to potential security breaches.
- **Data Minimization**: Edge training encourages the principle of data minimization, where only the essential data required for model training remains on the edge device. This minimization reduces the potential attack surface, making it challenging for malicious actors to exploit the system. Users have greater control over their data, as they can choose which information is shared with the model.
- **End-to-End Encryption**: To further enhance privacy and security, end-to-end encryption can be implemented in edge training systems. This means that data is encrypted on the edge device and only decrypted for processing, ensuring that even if data were to be intercepted during transmission, it would remain unreadable to unauthorized parties.
- **Regulatory Compliance**: Edge training facilitates compliance with data protection regulations such as the General Data Protection Regulation (GDPR) [3] in Europe and the Health Insurance Portability and Accountability Act (HIPAA) [4] in the United States. These regulations require organizations to protect the privacy and security of sensitive data. By conducting model training on the edge, organizations can limit data exposure and maintain compliance with these stringent laws.
- **User Consent and Transparency**: Edge training systems often incorporate features that require explicit user consent for data processing. This transparency and consent-driven approach not only respects user privacy but also fosters trust in AI applications. Users are aware of how their data is used, and this empowers them to make informed decisions about data sharing.
- **Reduced Data Transfer**: With edge training, there is a reduced need for data to traverse the Internet to cloud servers. This means that the attack surface for potential data interception is minimized. The lower volume of data transferred also reduces the likelihood of data leaks or breaches during transmission.
- **Differential Privacy**: Some edge training techniques employ differential privacy, a method that adds random noise to the data during model training. This ensures that individual user data

remains anonymous while still contributing to model improvement. This adds an additional layer of privacy protection, making it extremely difficult to identify specific individuals in the training dataset.

- **Data Residency and Sovereignty**: Edge training allows organizations to adhere to data residency and sovereignty requirements, which are essential in certain sectors and geographic locations. For instance, healthcare institutions can keep patient data within a specific jurisdiction, complying with legal and regulatory mandates.

11.2.1.2 Low Latency

One of the most significant advantages of edge training is its low latency. Training models on the edge means that predictions can be made in real-time without the need for data to be sent to the cloud and back. This is crucial for applications such as autonomous vehicles, industrial automation, and IoT devices, where instantaneous decisions are critical. Below are the latency advantages provided by edge training:

- **Real-Time Decision Making**: One of the most significant advantages of low latency in edge training is real-time decision-making. For applications where timely decisions are critical, such as autonomous vehicles, medical devices, and industrial automation, low latency is indispensable. The ability to process data and make decisions in fractions of a second can be a matter of life and death, operational efficiency, or safety [5, 6].
- **Enhanced User Experience**: Low latency results in a smoother and more responsive user experience. In applications like virtual reality (VR), augmented reality (AR), and gaming, the reduction in latency ensures that user interactions and feedback happen with imperceptible delay. This translates to a more immersive and enjoyable user experience [5].
- **Improved Video and Audio Streaming**: Low latency in edge training greatly benefits video and audio streaming services. Video calls, live streaming, and online gaming platforms experience fewer interruptions and reduced buffering, leading to improved quality and user satisfaction. For telemedicine and remote collaboration tools, low latency is vital for natural and effective communication [5].
- **IoT and Sensor Networks**: In the realm of IoT and sensor networks, low latency enables near-instantaneous data analysis and response. This is crucial for applications like smart cities, where edge devices need to monitor and react to changing environmental conditions, traffic patterns, and security threats in real-time [6].
- **Reduced Bandwidth Usage**: Low latency also has the advantage of reducing bandwidth usage. Because data is processed locally on edge devices, there is no need for continuous data transmission to remote servers. This not only minimizes network congestion but also conserves data usage, which is particularly beneficial in regions with limited Internet access or for cost-conscious applications [6].
- **Edge-AI in Healthcare**: In healthcare, low latency is essential for telemedicine applications, remote monitoring of patients, and medical diagnostic devices. Low latency in edge training ensures that medical data can be analyzed promptly, allowing for faster diagnosis and treatment decisions [6].
- **Manufacturing and Automation**: In manufacturing and automation processes, low latency is critical for coordinating robots, machinery, and sensors. Quick decision-making at the edge improves production efficiency, reduces errors, and enhances worker safety [6].
- **Edge Computing for Edge Training**: Edge computing, which complements edge training, facilitates low latency by processing data closer to the source. Together, they provide a seamless

solution for real-time AI applications, as data is processed locally, minimizing the time it takes for data to travel to distant cloud servers and back [5, 6].

11.2.1.3 Bandwidth Efficiency and Low Cloud Costs

Edge training optimizes bandwidth usage. Since data remains on the device, there is less reliance on constant Internet connectivity. This can be especially important in remote or resource-constrained areas where network access is limited or expensive. Below are the detailed advantages of training on edge:

- **Localized Data Processing**: Edge training relies on processing data locally, eliminating the need to transfer large datasets to remote cloud servers. This localization significantly reduces the strain on network bandwidth, ensuring more efficient data handling [7, 8].
- **Reduced Data Transfer**: With data processing happening on edge devices, edge training minimizes the volume of data that needs to traverse the Internet. This not only reduces the potential for data leaks but also conserves network resources, making it invaluable in bandwidth-constrained scenarios [8, 9].
- **Real-Time Data Handling**: Edge training is particularly advantageous for real-time applications like autonomous vehicles and IoT devices. Local data processing minimizes the amount of data that must be transmitted to the cloud for analysis, optimizing bandwidth usage for instantaneous decision-making [9, 10].
- **Minimized Data Transfer Costs**: Cloud computing costs often stem from data transfer and storage. Edge training significantly reduces data transfer as data remains localized, resulting in substantial cost savings, especially for large-scale deployments [7, 8].
- **Lower Compute Costs**: Edge training shifts computation from resource-intensive cloud servers to edge devices. This means that cloud computing costs decrease as fewer resources are needed for model training and inference [8, 11].
- **Cost-Effective Scalability**: Edge training enables cost-effective scalability. Organizations can deploy AI models across numerous edge devices without exponential increases in cloud costs. This scalability facilitates business expansion without a corresponding spike in cloud computing expenses [7, 11].
- **Improved Resource Utilization**: Edge training optimizes resource utilization by performing tasks on edge devices rather than costly cloud servers. This efficient allocation of resources leads to lower overall cloud computing expenditure [10, 11].

11.2.1.4 Customization and Local Adaptation

Edge training allows models to be customized and adapted to local conditions. For instance, an autonomous vehicle can fine-tune its driving behavior based on local traffic patterns and road conditions, creating a more efficient and safe driving experience.

- **Personalization**: Edge training allows AI models to personalize their interactions with users. For instance, virtual assistants, like Siri or Google Assistant on smartphones, can learn and adapt to individual users' speech patterns, preferences, and requests. This customization enhances user satisfaction and provides a more tailored experience [12].
- **Enhanced Recommendations**: In applications like e-commerce, edge training enables AI systems to create more accurate product recommendations. By analyzing a user's behavior and purchase history locally, the system can offer relevant product suggestions in real-time, increasing the likelihood of successful sales and improving the user's shopping experience [13].

- **Local Language and Culture**: Edge training is particularly advantageous for applications that require understanding local languages and cultures. For instance, a translation app can adapt to regional dialects and idioms by learning from user interactions. This ensures that the translations are more accurate and culturally relevant [14].
- **IoT Devices**: In the Internet of Things (IoT) domain, local adaptation is crucial. IoT devices can learn and adapt to the specific environment in which they operate. For example, a smart thermostat can learn the user's temperature preferences and adjust settings to maintain a comfortable environment, reducing energy consumption [12].
- **Industrial Automation**: In manufacturing and industrial automation, edge-trained AI models can adapt to local conditions and variations. Robots and machinery can fine-tune their actions based on real-time data from sensors, ensuring optimal production processes and minimizing errors [15].
- **Dynamic Road Conditions**: In the context of autonomous vehicles, edge training allows the vehicle to adapt to changing and dynamic road conditions. The AI system can continuously learn from local traffic patterns, road quality, and even driver behavior, resulting in safer and more efficient autonomous driving [16].
- **Rapid Response to Environmental Changes**: For applications like environmental monitoring, edge-trained sensors and devices can adapt to rapidly changing conditions. They can detect local changes in temperature, humidity, and pollution levels, and respond accordingly in real-time, providing crucial data for environmental management [14].

Edge training, despite its many advantages, also presents a set of unique challenges that must be addressed to fully harness its potential. Below are the challenges:

- **Limited Computational Resources**: Edge devices, such as smartphones and IoT sensors, typically have limited computational power and memory compared to powerful cloud servers. Training complex ML models with these constraints can be challenging. Model architectures need to be lightweight, and training algorithms must be optimized for resource-efficient learning.
- **Data Heterogeneity**: Edge environments often feature a wide variety of devices and data sources. This data heterogeneity poses challenges for training models that can effectively generalize across diverse data types and formats. Ensuring compatibility and consistency in data preprocessing is a significant challenge [15].
- **Data Privacy and Security**: Edge training involves processing data locally, which raises concerns about data privacy and security. Ensuring that sensitive information is adequately protected on edge devices is a challenge. Federated learning and differential privacy are techniques used to address these issues but require careful implementation [12].
- **Lack of Continuous Connectivity**: Edge devices may not have continuous Internet connectivity. Handling intermittent or low-bandwidth connections while still enabling efficient training is a challenge. Techniques like data buffering and asynchronous training are employed to mitigate these issues.
- **Model Updates and Version Control**: Managing model updates and versions across a large number of edge devices can be complex. Keeping models up to date, ensuring compatibility, and deploying new features without disrupting existing applications requires careful planning and robust version control mechanisms [15].
- **Scalability and Maintenance**: Deploying edge-trained models at scale can pose challenges in terms of device management and maintenance. Ensuring that models function correctly,

monitoring performance, and handling software updates across a vast network of devices can be resource-intensive [13].

- **Local Anomalies and Bias**: Training models on the edge means they may learn from local anomalies and biases present in the data from a specific environment. Addressing these challenges and ensuring that models generalize well to different conditions is a significant concern.
- **Edge Device Diversity**: Edge devices come in various forms, from smartphones to sensors to edge servers, each with different hardware capabilities and operating systems. Developing models that can run efficiently on this diverse range of devices can be challenging [12].
- **Power Consumption**: Many edge devices are constrained by limited battery life. Running resource-intensive ML models can quickly deplete the device's power. Balancing model accuracy with power efficiency is a crucial challenge in edge training.
- **Latency and Real-Time Processing**: While edge training offers low latency for inference, training models in real-time on edge devices can be challenging. This is especially important for applications like autonomous vehicles, where rapid decision-making is critical.
- **Quality of Data Labeling**: Supervised learning often requires labeled data for training, and the quality of these labels can vary. Ensuring the accuracy of labels is essential for model training, and handling noisy or erroneous data is a challenge [14].
- **Regulatory Compliance**: Edge training must comply with data protection regulations such as GDPR [3] and HIPAA [4]. Navigating the legal and regulatory landscape to ensure that data privacy is upheld is a complex challenge, particularly in cross-border deployments.
- **Network Synchronization**: In federated learning scenarios, where models are trained on distributed edge devices, ensuring network synchronization and coordination without compromising data privacy can be a complex challenge.

While edge training offers remarkable benefits, it comes with a set of intricate challenges that need to be addressed for successful implementation. These challenges range from resource constraints on edge devices to data privacy, model updates, and device diversity. Overcoming these challenges is essential to fully realize the potential of edge training in various applications, and it requires a combination of innovative algorithms, robust infrastructure, and careful consideration of security and privacy concerns.

11.3 Incremental Training on Edge

Incremental learning on edge devices is an ML approach that enables models to continuously learn and adapt to new data while residing on resource-constrained edge devices. This process is crucial in applications where data continuously evolves, and it's not practical to retrain models from scratch due to limitations in processing power, memory, or bandwidth [16, 17].

Incremental learning can be categorized into task, class, and domain incremental learning (DIL) based on how the data is presented to the model during training and inference as shown in Figure 11.2.

- **Task Incremental Learning (TIL)**: In TIL, the task identity is known during training and inference. This means that the model knows which task it is performing at any given time. TIL techniques are typically used in applications where the tasks are clearly distinguishable, such as classifying images of different object categories or translating text from one language to another [16, 17].

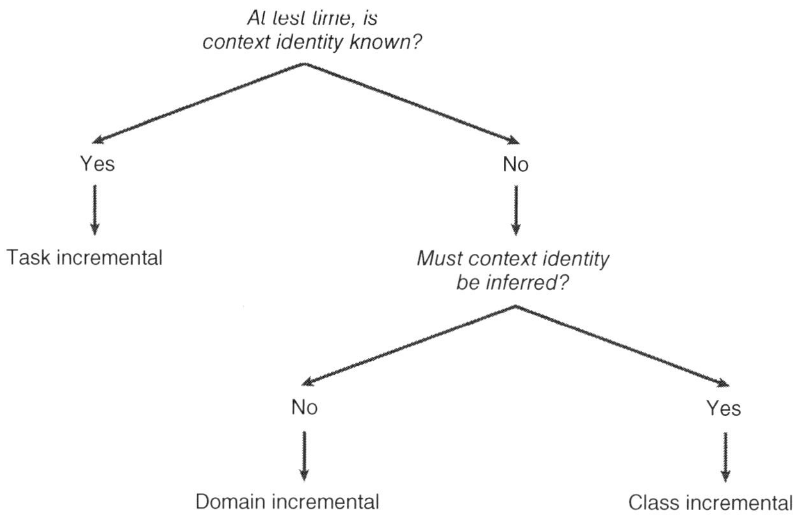

Figure 11.2 Different types of incremental learning.

- **Class Incremental Learning (CIL)**: In CIL, the task identity is known during training, but it is not known during inference. This means that the model does not know which task it is performing when it makes a prediction. CIL techniques are typically used in applications where the tasks are similar, such as classifying images of different breeds of dogs or translating text from different languages to English [16, 17].
- **Domain Incremental Learning (DIL)**: In DIL, the task identity is not known during training or inference. This means that the model does not know which task it is performing or which data distribution it is operating on. DIL techniques are typically used in applications where the tasks are similar and the data distributions are different, such as classifying images of the same object category taken in different lighting conditions or classifying images of the same object category taken with different cameras [16, 17].

11.3.1 Task Incremental Learning

In TIL, the task identity is known during training and inference. This means that the model knows which task it is performing at any given time. TIL techniques are typically used in applications where the tasks are clearly distinguishable, such as classifying images of different object categories or translating text from one language to another.

Some common TIL techniques include:

- **Learning Without Forgetting (LwF)**: Learning without forgetting (LwF) [18] is a regularization-based TIL technique that penalizes the model for changing the parameters that have been learned from previous tasks. This helps to prevent the model from forgetting the knowledge it has learned from previous tasks when it is trained on new tasks. LwF is typically implemented by adding a regularization term to the model's loss function. This regularization term penalizes the model for the difference between its predictions on the new task data and its predictions on the data from previous tasks. The strength of the regularization term can be controlled using a hyperparameter, which is typically tuned using a validation set. LwF has been shown to be effective at preventing forgetting in a variety of TIL tasks, including image

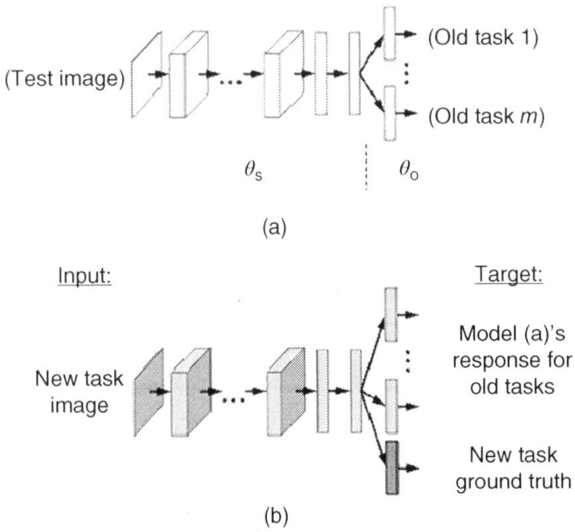

Figure 11.3 Illustration for learning without forgetting. Source: Zhizhong Li, Derek Hoiem Learning without Forgetting. © [2017] IEEE. (a) Original model, (b) learning without forgetting.

classification, natural language processing, and machine translation. However, it is important to note that LwF can also lead to a decrease in the model's performance on the new task. This is because the regularization term can prevent the model from learning the new task as effectively as possible. Figure 11.3 illustrates the original model and LwF.

- **Pseudo-Rehearsal**: Pseudo-rehearsal [19] is a TIL technique that trains a model on new tasks by generating pseudo-labels for the new data using the model's predictions on the old data. This allows the model to learn from the new data without forgetting the knowledge it has learned from the old data. To implement pseudo-rehearsal, the model is first trained on the old data. Then, the model is used to generate pseudo-labels for the new data. The pseudo-labels are the model's predictions on the new data, but with a high confidence threshold. This means that the model only generates pseudo-labels for the data that it is confident in. The model is then trained on the new data using the pseudo-labels. This process is repeated for each new task. Pseudo-rehearsal has been shown to be effective at preventing catastrophic forgetting in a variety of TIL tasks, including image classification, natural language processing, and machine translation. However, it is important to note that pseudo-rehearsal can also lead to a decrease in the model's performance on the new task. This is because the pseudo-labels may not be accurate, and because the model may start to learn the new task at the expense of the old tasks.
- **Knowledge Distillation**: Knowledge distillation (KD) [20, 21] is a powerful technique that can be used to prevent catastrophic forgetting in incremental learning. In incremental learning, the model is trained on a new task without forgetting its existing knowledge. This can be a challenging task, as the new data can overwrite the old data in the model's memory. KD can help to prevent catastrophic forgetting by transferring the knowledge of a pretrained model (teacher model) to a new model (student model). The student model is trained to mimic the outputs of the teacher model on the new data. This helps the student model to learn the new task without forgetting its existing knowledge. Figure 11.3 in Chapter 5 illustrates knowledge distillation technique. There are a number of different ways to use KD in incremental learning. One common approach is to use a distillation loss in addition to the regular cross-entropy loss. The distillation loss penalizes the student model for deviating from the predictions of the teacher model. Another approach is to use a gradient episodic memory (GEM) [22]. A GEM is a memory buffer that stores a subset of the training data from previous tasks. The student model is trained to minimize the

difference between its predictions on the new data and its predictions on the data in the GEM. KD has been shown to be effective at preventing catastrophic forgetting in a variety of incremental learning tasks, including image classification, natural language processing, and machine translation.

11.3.2 Class Incremental Learning

In CIL, the task identity is known during training, but it is not known during inference. This means that the model does not know which task it is performing when it makes a prediction. CIL techniques are typically used in applications where the tasks are similar, such as classifying images of different breeds of dogs or translating text from different languages to English. Some common CIL techniques include:

- **Class Isolation**: Class isolation incremental learning (CIL) is a type of incremental learning where the task identity is known during training, but it is not known during inference. This means that the model does not know which task it is performing when it makes a prediction. CIL techniques are typically used in applications where the tasks are similar, such as classifying images of different breeds of dogs or translating text from different languages to English. One common CIL technique is to train a separate model for each class. This allows the model to learn the specific features of each class without having to worry about interference from other classes. However, this approach can be computationally expensive and memory-intensive, especially for large numbers of classes. Another CIL technique is to use a shared model architecture for all classes. This can be more efficient in terms of computation and memory usage, but it can be more difficult to prevent the model from forgetting the knowledge it has learned from previous classes [23].
- **Feature Augmentation**: Feature augmentation class incremental learning (FA-CIL) is a type of CIL that uses feature augmentation to improve the performance of the model. Feature augmentation is a technique that generates new training data by transforming existing training data. This can be done by applying a variety of different transformations to the data, such as cropping, flipping, rotating, and adding noise. FA-CIL works by first generating augmented training data for the new class. This augmented data is then used to train the model in the new class without forgetting the knowledge it has learned from previous classes. There are a number of different ways to generate augmented data for FA-CIL. One common approach is to use a pretrained feature extractor. The feature extractor is used to extract features from the existing training data. The extracted features are then transformed using a variety of different techniques to generate new training data. Another approach is to use a generative adversarial network (GAN). A GAN is a type of neural network that can be used to generate realistic synthetic data. GANs can be used to generate augmented training data for FA-CIL by training them to generate data that is similar to the existing training data [23].
- **Few-Shot Learning**: Few-shot learning incremental learning (FS-IL) is a type of incremental learning where the model is trained on a few examples of the new task without forgetting its existing knowledge (Figure 11.4). FS-IL is a challenging task, as the model needs to be able to learn the new task from a very limited amount of data. There are a number of different FS-IL techniques. One common approach is to use a pretrained model and fine-tune it on the new task data. This approach can be effective, but it can also lead to catastrophic forgetting, where the model forgets its existing knowledge when it is trained on the new task data. Another approach to FS-IL is to use a meta-learning algorithm. Meta-learning algorithms learn to learn quickly

Figure 11.4 Incremental Few Shot Learning (FS-IL). Source: mtruchon/Adobe Stock.

by adapting to new tasks and data distributions. This can be effective for FS-IL, as it allows the model to learn the new task from a very limited amount of data [24].

11.3.3 Domain Incremental Learning

In DIL, the task identity is not known during training or inference. This means that the model does not know which task it is performing or which data distribution it is operating on. DIL techniques are typically used in applications where the tasks are similar and the data distributions are different, such as classifying images of the same object category taken in different lighting conditions or classifying images of the same object category taken with different cameras. Some common DIL techniques include:

- **Domain Adaptation**: Domain adaptation incremental learning (DA-IL) [26–28] is a type of incremental learning where the model is trained on a new domain (e.g., a new dataset with different data distribution) without forgetting its existing knowledge from previous domains. DA-IL is a challenging task, as the model needs to be able to adapt to the new domain without forgetting its existing knowledge. There are a number of different DA-IL techniques. One common approach is to use a domain adaptation algorithm to adapt the model to the new domain. Domain adaptation algorithms work by finding a common representation between the source domain (i.e., the domain where the model was trained) and the target domain (i.e., the new domain). Once a common representation has been found, the model can be updated to work on the new domain. Another approach to DA-IL is to use a meta-learning algorithm. Meta-learning algorithms learn to learn quickly by adapting to new tasks and data distributions. This can be effective for DA-IL, as it allows the model to adapt to the new domain very quickly.
- **Active Learning**: Active learning incremental learning (AL-IL) [27, 29] is a type of incremental learning where the model actively selects the most informative data from the new task to learn from, in order to minimize the need for labeled data and improve its performance. AL-IL is a challenging task, as the model needs to be able to identify the most informative data from the new task without forgetting its existing knowledge. There are a number of different AL-IL techniques. One common approach is to use a query-by-committee (QBC) algorithm. QBC algorithms work by building a committee of models and then querying the model with the highest uncertainty about the next data point to label. This approach can be effective, as it allows the model to focus

on the most informative data. Another approach to AL-IL is to use a Bayesian active learning (BALD) [30] algorithm. BALD algorithms work by estimating the uncertainty of each data point and then querying the data point with the highest uncertainty. This approach can be effective, as it allows the model to focus on the data points that are most likely to help it learn.

- **Transfer Learning**: Transfer learning incremental learning (TL-IL) [31] is a type of incremental learning where the model leverages the knowledge it has learned from previous tasks to learn new tasks more quickly and efficiently. TL-IL is a powerful technique that can be used to train ML models on new tasks with very little data, making it a promising approach for real-world applications. There are a number of different TL-IL techniques. One common approach is to use a pretrained model as a starting point for the new task. The pretrained model can be fine-tuned on the new task data to learn the specific features and patterns that are important for the new task. Another approach to TL-IL is to use a knowledge distillation technique. Knowledge distillation allows the model to transfer the knowledge it has learned from previous tasks to the new task without having to fine-tune the pretrained model. This can be useful for tasks where the new task data is limited or where the pretrained model is too large or complex to fine-tune.

11.3.4 Challenges of Incremental Learning

Incremental learning, which involves updating and refining ML models with new data while retaining knowledge from previous data, presents several significant challenges. These challenges can affect the model's performance, scalability, and efficiency. Here are the key challenges associated with incremental learning [21, 32]:

- **Catastrophic Forgetting**: Catastrophic forgetting [21] can occur when a model is trained on new data and unintentionally overwrites previously learned knowledge. Techniques such as regularization, weight consolidation, and progressive neural networks help mitigate this challenge. Continual learning models need to be designed to balance new learning without compromising existing knowledge, ensuring that both old and new tasks can be performed accurately [32].
- **Concept Drift**: Concept drift can be gradual or sudden, and it poses a significant challenge in domains with evolving data patterns. Adaptive learning algorithms, ensemble methods, and concept drift detection mechanisms are used to recognize and respond to these shifts effectively. The ability to continuously monitor and adjust to concept drift is essential for maintaining model accuracy [32, 33].
- **Limited Memory**: Edge devices and resource-constrained systems may have very limited memory, making it critical to implement efficient data storage and management. Techniques like data compression, feature selection, and memory-efficient data representations are used to maximize the utilization of available memory resources. Data retention policies need to be established to ensure valuable historical data is preserved while minimizing memory consumption.
- **Model Complexity**: As models accumulate knowledge over time, their complexity can increase, leading to higher computational and memory requirements. Model compression techniques, such as quantization and knowledge distillation, are used to reduce model size. Additionally, dynamic model architectures can help manage complexity by adjusting model structure based on the specific learning task.
- **Data Labeling**: Obtaining high-quality labels for new data is a crucial component of incremental learning. Active learning strategies, which intelligently select the most informative instances for labeling, can reduce the labeling burden. Semi-supervised learning approaches leverage both labeled and unlabeled data to improve model performance, especially when labeled data is scarce.

- **Hyperparameter Tuning**: Fine-tuning hyperparameters for different tasks and data distributions requires automated techniques such as Bayesian optimization or grid search. Adaptive learning rate schedules can be employed to allow models to adapt their learning rates during incremental learning, facilitating better convergence and performance.
- **Model Overfitting**: Overfitting to new data is a concern in incremental learning, especially when data is limited. Regularization techniques, early stopping, and the use of transfer learning from pretrained models are applied to mitigate overfitting. Strategies that balance the model's adaptation to new data while preserving old knowledge are essential [32].
- **Task Selection and Scheduling**: Task selection and scheduling strategies depend on the specific application. Factors like the importance of tasks, the availability of new data, or user-defined criteria guide task prioritization. Continuous monitoring for task relevance and impact on existing knowledge is crucial when making decisions about when to update the model for each task.
- **Memory Replay**: Memory replay mechanisms store and sample past experiences to train on. Careful selection of replayed data, along with policies for memory management, is essential for balancing the preservation of old knowledge and effective learning from new data. Balancing memory usage with performance improvements is a continuous concern.
- **Data Skew**: Imbalanced data distributions can lead to models being biased toward more frequently occurring classes. Techniques for handling data skew include re-sampling strategies, adjusting class weights, and the use of cost-sensitive learning algorithms. Ensuring that the model does not disproportionately focus on newer, more prevalent data is crucial.
- **Privacy and Security**: Protecting data privacy and security is of paramount importance, particularly in applications with sensitive information. Techniques like federated learning, homomorphic encryption, and secure multi-party computation allow models to learn from decentralized, privacy-protected data sources without exposing individual data points or compromising privacy [32].
- **Model Version Control**: Managing model versions is essential to ensure the correct deployment of model updates. Version control systems and continuous integration and deployment (CI/CD) practices are employed to keep track of model versions, enabling smooth transitions between different iterations of the model.
- **Resource Management**: In distributed systems, managing resource allocation, load balancing, and task scheduling is essential to optimize the use of computational resources. Effective resource management ensures that incremental learning can be performed efficiently across multiple devices or nodes, taking into account the varying computational capabilities of each component [32].

11.4 Data at the Edge

At the heart of the MLOps cycle is data, which is often generated at the edge, where devices and sensors produce a continuous stream of information. Handling new data at the edge is crucial for training and deploying ML models for real-time decision-making.

11.4.1 Data Acquisition at the Edge

11.4.1.1 Data Sources

The process begins with data ingestion and collection at the edge, where data is generated by devices or sensors. This data can be in the form of images, text, numerical values, sensor readings, or any other data type. Effective data collection mechanisms are crucial, ensuring that data is continuously and reliably transmitted to the pipeline for processing. Edge devices, such as IoT sensors, cameras, industrial machines, and vehicles, generate a variety of data types, including:

- **Numerical Sensor Data**: Readings from environmental sensors, accelerometers, gyroscopes, and more.
- **Image and Video Data**: Captured by cameras for applications like surveillance, computer vision, and autonomous vehicles.
- **Text Data**: Logs, messages, and textual information generated at the edge.
- **Audio Data**: Captured by microphones for tasks like voice recognition and sound analysis.

11.4.1.2 Data Ingestion

The process begins with data ingestion, where data generated at the edge is collected and transmitted to the MLOps pipeline. Data ingestion mechanisms need to be robust and efficient, ensuring that data is transmitted securely and reliably. Several methods can be employed, including:

- **Real-time Data Streaming**: Data is transmitted as it is generated, often via protocols like MQTT or Apache Kafka.
- **Batch Data Ingestion**: Data is periodically collected and transmitted in batches, suitable for scenarios with less stringent real-time requirements.

11.4.1.3 Data Preprocessing

Once data is ingested, preprocessing steps are applied to prepare it for further analysis. These steps can include:

- **Data Cleansing**: Handling missing or erroneous data points.
- **Data Normalization**: Scaling data to a consistent range.
- **Feature Extraction**: Deriving relevant features from raw data.
- **Data Transformation**: Converting data into a format suitable for analysis.

11.4.2 Data Storage at the Edge

11.4.2.1 Storage Infrastructure

Storing data at the edge requires infrastructure capable of handling the data generated by edge devices [34, 35]. This infrastructure can include:

- **Local Storage**: Edge devices may have local storage, such as onboard memory or attached storage devices like SD cards. This is suitable for short-term storage but may have limited capacity.
- **Edge Servers**: In more complex edge environments, edge servers with greater storage capacity may be deployed. These servers can store and process data locally.
- **Distributed Edge Storage**: In larger edge deployments, data may be distributed across multiple edge devices or servers to ensure redundancy and scalability.

11.4.2.2 Data Retention Policies

Data storage at the edge must consider data retention policies. These policies determine how long data is stored, when it is archived, and when it is deleted. Data retention is influenced by factors such as:

- **Regulatory Compliance**: Some industries have regulations that dictate data retention periods [36].

- **Use Case Requirements**: The specific use case may require data to be retained for different durations.
- **Storage Costs**: Storing data, especially in large volumes, can incur costs, so retention policies must balance cost and data availability [35].

11.4.2.3 Data Backup and Redundancy

Data stored at the edge should be backed up to prevent data loss due to device failure or other unforeseen events. Redundancy strategies, such as mirroring data across multiple devices or employing RAID (Redundant Array of Independent Disks) configurations, can be crucial for ensuring data integrity and availability [34].

11.4.2.4 Data Security

Data security is paramount when storing data at the edge, as edge devices are often more vulnerable to physical and network-based attacks. Security measures may include:

- **Data Encryption**: Encrypting data at rest and in transit to protect against unauthorized access.
- **Access Control**: Implementing role-based access control to restrict data access to authorized personnel.
- **Security Updates**: Regularly updating edge device firmware and software to patch known vulnerabilities [35].

11.4.2.5 Data Lifecycle Management

Effectively managing data at the edge involves considering its lifecycle. The data lifecycle consists of several stages:

- **Data Collection**: Data is generated and ingested at the edge.
- **Data Storage**: Data is stored in local storage, edge servers, or distributed storage systems.
- **Data Processing**: Data may be processed locally for real-time decision-making or transmitted to a central server or cloud for more complex analysis.
- **Data Retention**: Data is retained for a specified period based on retention policies.
- **Data Archiving**: Data that needs to be preserved for historical or compliance reasons is archived.
- **Data Deletion**: Data that is no longer needed or has exceeded its retention period is securely deleted [36].

11.4.3 Edge Data Synchronization

In cases where data is processed and analyzed both at the edge and centrally (in the cloud or on-premises), data synchronization mechanisms must be in place. These mechanisms ensure that data generated at the edge is consistent with the central data repository, enabling seamless analysis and decision-making [37].

11.4.3.1 Edge Computing and Data Processing

Edge computing involves processing data locally at the edge, closer to the data source, rather than transmitting it to a central server or the cloud for processing. This approach minimizes latency, conserves bandwidth, and enables real-time decision-making. When adopting edge computing, it is crucial to have adequate computing resources and processing capabilities at the edge [34].

11.4.3.2 Challenges in Edge Data Management

Managing data at the edge presents several challenges:

- **Resource Constraints**: Edge devices often have limited storage and computational resources, necessitating careful resource allocation [38].
- **Security Risks**: Edge devices are more exposed to physical and network-based security threats, requiring robust security measures [37].
- **Data Consistency**: Ensuring that data is consistent across edge devices and centralized repositories can be challenging [34].
- **Data Backup**: Implementing effective data backup and redundancy strategies can be complex in edge environments [38].
- **Data Privacy**: Complying with data privacy regulations, especially when data is generated at the edge, is crucial [36].

11.4.4 Data Annotation

Data annotation is a fundamental step in ML, allowing models to learn and make informed decisions. However, acquiring labeled data is often labor-intensive and costly. To address this challenge, unsupervised learning techniques are deployed in MLOps pipelines to annotate unlabeled data at the edge. In this detailed exploration, we will discuss the processes and techniques involved in this crucial phase.

The core of handling unlabeled data involves annotation or labeling. This process can take several forms, including manual annotation, active learning, or unsupervised learning techniques. The choice of method depends on factors such as resource availability, data volume, and the desired level of accuracy.

11.4.4.1 Unsupervised Learning Techniques for Annotation

Unsupervised learning techniques are central to the annotation process, as they enable data categorization without the need for labeled examples. The following techniques are used:

- **K-Means Clustering**: K-Means groups data into clusters based on similarity or distance, making it valuable for image segmentation, text clustering, and customer segmentation.
- **Principal Component Analysis (PCA)**: PCA reduces data dimensionality while preserving essential information. It simplifies annotation and is applied in feature reduction, image compression, and anomaly detection.
- **Autoencoders**: Autoencoders are neural network models used for feature learning, data reconstruction, and image generation. They assist in feature extraction, denoising, and content generation.
- **Isolation Forests**: This algorithm excels at anomaly detection, making it essential for spotting outliers in datasets, such as fraud detection, quality control, and anomaly annotation.
- **DBSCAN (Density-Based Spatial Clustering of Applications with Noise)**: DBSCAN clusters data based on density, useful in spatial data clustering, customer behavior analysis, and image segmentation [39].
- **Latent Dirichlet Allocation (LDA)**: LDA is applied to text data for topic modeling, simplifying content categorization, and annotation in content recommendation systems [40].

11.4.4.2 Model Training and Deployment

Once the data is annotated, ML models are trained to perform specific tasks, such as classification, clustering, or anomaly detection. These models are then deployed at the edge, allowing for real-time predictions or decision-making.

11.4.4.3 Monitoring and Feedback Loop

Continuous monitoring is a core aspect of MLOps. It involves evaluating model performance, detecting data distribution drift, and assessing model response to evolving unlabeled data. Monitoring ensures the model's effectiveness over time and enables feedback-driven improvements.

11.4.4.4 Human-in-the-Loop

In scenarios where manual annotation is involved, a human-in-the-loop approach can be implemented. Automation assists human annotators, reducing their workload while maintaining high-quality annotations. Active learning techniques may also be integrated to prioritize data samples for annotation, optimizing the process.

In conclusion, managing unlabeled data at the edge within an MLOps framework is a complex but critical process. It involves data collection, preprocessing, annotation, model training, deployment, and monitoring, while considering resource constraints, data privacy, scalability, and a feedback loop for adaptation. MLOps practices ensure that the pipeline is efficient, secure, and effective in making real-time decisions at the edge.

11.4.4.5 Conclusion

The acquisition and storage of new data at the edge within an MLOps pipeline are foundational processes for data-driven decision-making. Handling data generated at the edge, including data ingestion, preprocessing, storage, security, and synchronization, is a complex but essential task. Effective edge data management ensures that valuable insights can be derived from data generated at the edge, facilitating real-time decision-making and enhancing operational efficiency.

11.5 Distributed Edge Training

Distributed edge training is an ML training paradigm that involves training a model on multiple edge devices simultaneously. This can be used to train large models more quickly and efficiently, and to deploy models to edge devices that do not have the resources to train a model from scratch.

Data and model parallelism are two common techniques used for distributed edge training [41]. Data parallelism involves dividing the training data into shards and assigning each shard to a different edge device. Model parallelism involves dividing the model into sub-models and assigning each sub-model to a different edge device.

11.5.1 Data Parallelism

Data parallelism is a simple and efficient technique for distributed edge training. However, it can be challenging to implement data parallelism on edge devices, as they typically have limited resources. One common approach is to use a centralized server to coordinate the training process. The server distributes the training data to the edge devices and collects the updated model parameters from the edge devices.

Another approach to data parallelism on edge devices is to use a peer-to-peer network. In a peer-to-peer network, the edge devices communicate directly with each other to share the training data and the updated model parameters. This approach can be more efficient than using a centralized server, as it does not require all of the data and model parameters to be sent to and from the server.

11.5.2 Model Parallelism

Model parallelism is a more complex technique than data parallelism, but it can be more efficient for training large models. Model parallelism can be implemented on edge devices using a variety of techniques, such as pipeline parallelism, block parallelism, and parameter parallelism.

- Pipeline parallelism involves dividing the model into a sequence of layers and assigning each layer to a different edge device. The edge devices process the training data in a pipeline fashion, with each edge device passing the output of its layer to the next edge device in the pipeline.
- Block parallelism involves dividing the model into a set of blocks and assigning each block to a different edge device. The edge devices process the training data in parallel, with each edge device updating the model parameters for its assigned block.
- Parameter parallelism involves dividing the model parameters into shards and assigning each shard to a different edge device. The edge devices update the model parameters in parallel, with each edge device updating the shard that it is assigned.

11.5.3 Challenges of Distributed Edge Training

Distributed edge training has a number of challenges, including:

- **Limited Resources**: Edge devices typically have limited resources, such as computing power and memory. This can make it difficult to train large models on edge devices.
- **Network Heterogeneity**: Edge devices can vary widely in terms of their network connectivity and bandwidth. This can make it difficult to efficiently communicate between edge devices.
- **System Complexity**: Distributed edge training systems are complex and can be difficult to implement and debug.

There have been a number of recent advances in distributed edge training. One advance is the development of new algorithms for data and model parallelism that are more efficient and scalable. Another advance is the development of new hardware platforms for edge devices that are designed for ML training.

Distributed edge training is a promising new paradigm for ML training. It has the potential to revolutionize the way that models are developed and deployed in a wide range of applications.

11.6 Use Cases

Incremental learning, also known as lifelong learning or continual learning, is an ML paradigm that focuses on adapting and extending models over time as new data becomes available. This approach is vital in scenarios where data is continually evolving, and models need to stay up-to-date and adaptive. Below, we explore several detailed use cases of incremental learning, highlighting its importance and practical applications across various domains.

11.6.1 Use Case 1: Autonomous Vehicles

Autonomous vehicles rely on a diverse range of sensors to navigate and make real-time decisions. Incremental learning is crucial in this context because the environment is dynamic, and the vehicle encounters a multitude of situations. For example, a self-driving car can continually update its models to recognize new traffic signs, adapt to changing road conditions, and improve its decision-making based on data from other connected vehicles.

In this use case, the vehicle's perception, object recognition, and path planning models are incrementally updated. As the vehicle collects data from sensors and cameras, it refines its understanding of the environment, enabling it to make safer and more reliable driving decisions.

11.6.2 Use Case 2: Healthcare Predictive Analytics

In the healthcare sector, incremental learning plays a critical role in predictive analytics. For example, a predictive model used to detect early signs of diseases, such as diabetes or heart conditions, should continually adapt and improve its accuracy as new patient data becomes available.

As more patient data is collected over time, the model can refine its predictions, identify evolving risk factors, and adjust treatment recommendations. This dynamic approach can lead to better patient outcomes, early intervention, and personalized healthcare recommendations.

11.6.3 Use Case 3: Fraud Detection in Finance

Financial institutions continually face new and evolving fraud techniques. Incremental learning is invaluable for fraud detection systems that need to adapt to changing attack methods. The system can learn from historical data and adjust its fraud-detection algorithms to identify new patterns and trends in fraudulent activities.

For instance, a credit card company can use incremental learning to detect and prevent fraudulent transactions in real-time. The system adapts to new tactics used by fraudsters, enhancing its ability to protect customers and reduce financial losses.

11.6.4 Use Case 4: Natural Language Processing (NLP) Applications

In natural language processing applications, incremental learning allows models to keep pace with evolving language and user behavior. Consider a virtual assistant like Siri or Alexa, which relies on NLP to understand and respond to user queries.

Over time, these assistants need to adapt to new words, phrases, and language trends. Incremental learning enables them to continuously update their language models, improving their conversational abilities and comprehension of user requests.

11.6.5 Use Case 5: Manufacturing and Predictive Maintenance

In manufacturing, incremental learning is applied to predictive maintenance. Machines and equipment generate vast amounts of sensor data, which can be used to predict when maintenance is required to avoid costly breakdowns [42].

The model learns from historical data and sensor readings, allowing it to adapt to the unique behavior of each machine. Over time, the model becomes more accurate in predicting maintenance needs, optimizing production and reducing downtime.

11.6.6 Use Case 6: Customer Behavior Analysis in E-Commerce

E-commerce platforms benefit from incremental learning to enhance customer behavior analysis. They continuously gather data on user interactions, purchases, and preferences. With incremental learning, these platforms can refine their recommendation systems, offering users more relevant product suggestions.

As users' preferences change and new products are introduced, the recommendation system adapts to provide a personalized shopping experience, ultimately increasing customer satisfaction and sales.

11.6.7 Use Case 7: Agriculture and Crop Yield Prediction

In the agricultural sector, incremental learning is used for crop yield prediction. Sensors and IoT devices collect data on soil conditions, weather, and crop health. The model continually updates itself with new data, enabling it to make more accurate predictions about crop yields.

Farmers can use this information for better resource allocation, optimizing crop management practices, and maximizing their harvests.

11.6.8 Use Case 8: Social Media Content Moderation

Social media platforms employ incremental learning to improve content moderation. With user-generated content constantly evolving and new forms of objectionable content emerging, moderation models must adapt.

These models learn to recognize new types of harmful content and adapt their filtering algorithms, ensuring a safer and more compliant online environment.

11.6.9 Use Case 9: Energy Management in Smart Grids

Incremental learning plays a critical role in energy management within smart grids. These grids use sensors and data analytics to optimize electricity distribution. Over time, the grid learns to anticipate power demand, dynamically allocate resources, and adapt to fluctuations in energy generation.

This approach enhances energy efficiency and grid stability while accommodating the growth of renewable energy sources.

11.6.10 Use Case 10: Quality Control in Manufacturing

In manufacturing, incremental learning can be applied to quality control. Machines on the production line continuously capture data on product quality. The model adapts to changing manufacturing processes and product variations to improve its ability to detect defects and maintain product quality.

This use case ensures that the production process remains efficient and cost-effective while upholding high-quality standards.

11.6.11 Use Case 11: Content Recommendation in Streaming Services

Streaming platforms leverage incremental learning for content recommendation. They continually analyze user viewing behavior and preferences, updating their recommendation models to provide users with personalized content.

As viewership patterns change and new content is added, the system adapts, enhancing user engagement and content discovery.

11.6.12 Use Case 12: Supply Chain Optimization

Supply chain management relies on incremental learning to adapt to changing market conditions, demand fluctuations, and logistical challenges. Models analyze historical data, adapt to evolving trends, and improve inventory management, delivery routes, and demand forecasting.

This use case leads to more efficient and cost-effective supply chain operations.

11.6.13 Use Case 13: Personalized Education

In the field of education, incremental learning is used to personalize the learning experience. Learning platforms continuously monitor student progress and adapt the curriculum to meet individual needs.

The system identifies areas where students need more support and adjusts the content accordingly, ensuring a tailored and effective learning experience.

11.6.14 Use Case 14: Customer Churn Prediction

Businesses that rely on subscription models use incremental learning for customer churn prediction. As they collect customer data and transaction history, the model updates itself to recognize changing patterns and behaviors.

This enables businesses to take proactive measures to retain customers and reduce churn rates.

11.6.15 Use Case 15: Predictive Maintenance in Healthcare Equipment

Hospitals and healthcare facilities use incremental learning for predictive maintenance of critical medical equipment. This ensures that devices such as MRI machines and ventilators are well-maintained and minimizes downtime.

The models adapt to the specific usage patterns of each device, optimizing maintenance schedules.

In each of these use cases, incremental learning demonstrates its utility by allowing ML models to adapt and extend their capabilities over time. This adaptability is essential in dynamic environments where data and patterns are continually evolving, enabling more accurate predictions, improved decision-making, and enhanced user experiences across various domains.

11.7 Conclusion

Training ML models on edge devices is emerging as a powerful alternative to the traditional cloud-based approach. It offers significant advantages, including faster response times, better privacy, reduced data transfer needs, and greater energy efficiency. These benefits are especially important for applications requiring real-time data processing and those dealing with sensitive information. By integrating edge training into the Machine Learning Operations (MLOps) framework, organizations can streamline the entire ML workflow, from data preparation to deployment and monitoring, making it easier and more reliable to deploy models. However, it's crucial to navigate the challenges of limited computational resources, model size constraints, and security risks. With continuous advancements in technology, edge training is set to play a pivotal role in the future of ML deployment, enhancing responsiveness and security across various industries. This chapter has laid a solid foundation by exploring the benefits, challenges, and techniques of edge training, preparing readers for more in-depth discussions and practical implementations in the upcoming sections.

References

1 Kreuzberger, D., Kühl, N., and Hirschl, S. (2023). Machine learning operations (MLOps): overview, definition, and architecture. *IEEE Access* 1–1. https://doi.org/10.1109/ACCESS.2023.3262138.

2 Khouas, A.R., Bouadjenek, M.R., Hacid, H., and Aryal, S. (2024). Training machine learning models at the edge: a survey. ArXiv, abs/2403.02619.

3 GDPR (2023). General Data Protection Regulation, https://gdpr-info.eu/.

4 US Department of Health and Human Services (2023). https://www.hhs.gov/hipaa/index.html.

5 Domeke, A., Cimoli, B., and Monroy, I.T. (2022). Integration of network slicing and machine learning into edge networks for low-latency services in 5G and beyond systems. *Applied Sciences* 12 (13): 6617. https://doi.org/10.3390/app12136617.

6 Yu, Y. and Sun, S. (2022). Scalable and low-latency federated learning with cooperative mobile edge networking. arXiv preprint arXiv:2205.13054. https://arxiv.org/abs/2205.13054.

7 Amanda Saunders (2022). Building an edge strategy: cost factors. Retrieved from NVIDIA Developer.

8 Digi International (2023). Edge computing: saving bandwidth, time and money. Retrieved from Digi.

9 EdgilityOS (2023). Reducing latency and bandwidth costs with edge computing. Retrieved from EdgilityOS.

10 Valero, A. (2023). Edge computing: empowering real-time data processing and analysis. Retrieved from SUSE.

11 Xailient (2022). A comprehensive guide to edge AI. Retrieved from Xailient.

12 Jang, S. (2023). Empower intelligence at the edge with AI deployment. Retrieved from TechTarget.

13 Microsoft Learn (2023). Deploy AI and machine learning at the edge—Azure Architecture Center. Retrieved from Microsoft Learn.

14 Zewe, A. (2022). Learning on the edge. Retrieved from MIT News.

15 Shah, A. and Patel, C. (2023) Mastering LLM techniques: customization. Retrieved from NVIDIA Developer.

16 MIT News (2023). Learning on the edge. Massachusetts Institute of Technology. Retrieved from https://news.mit.edu/2023/learning-on-edge-0314.

17 Qin, Z., Yu, F., and Chen, X. (2019). TeAM—task-adaptive incremental learning for intelligent edge devices. ArXiv. Retrieved from https://arxiv.org/abs/1910.03122.

18 Oren, G. and Wolf, L. (2021). In defense of the learning without forgetting for task incremental learning. *2021 IEEE/CVF International Conference on Computer Vision Workshops (ICCVW)*, pp. 2209–2218.

19 Mellado, D., Saavedra, C., Chabert, S., and Salas, R. (2017). Pseudorehearsal approach for incremental learning of deep convolutional neural networks. In: *Computational Neuroscience.*

LAWCN 2017. Communications in Computer and Information Science, vol. 720 (ed. D. Barone, E. Teles, and C. Brackmann). Cham: Springer. Porto Alegre, Brazil , November 22-24, 2017. https://doi.org/10.1007/978-3-319-71011-2_10.

20 Hinton, G.E., Vinyals, O., and Dean, J. (2015). Distilling the knowledge in a neural network. arXiv preprint arXiv:1503.02531.

21 Feng, T., Wang , M., and Yuan, H. (2022). Overcoming catastrophic forgetting in incremental object detection via elastic response distillation. *2022 IEEE/CVF Conference on Computer Vision and Pattern Recognition (CVPR)*, 19–24 June 2022, pp. 9417–9426, New Orleans, LA.

22 Lopez-Paz, D. and Ranzato, M. (2017). Gradient episodic memory for continual learning. *Proceedings of the Neural Information Processing Systems*, Long Beach, CA, USA (4–9 December 2017).

23 Tao, X., Hong, X., Chang, X. et al. (2020). Few-shot class-incremental learning. *2020 IEEE/CVF Conference on Computer Vision and Pattern Recognition (CVPR)*, Seattle, WA, USA, pp. 12180–12189, doi: 10.1109/CVPR42600.2020.01220. keywords: {Power capacitors;Training;Task analysis;Topology;Adaptation models;Neural networks;Network topology}.

24 Zhang, J., Liu, L., Silvén, O. et al. (2023). Few-shot class-incremental learning: a survey. ArXiv, abs/2308.06764.

25 Karthik Bhaskar, K. (2021). Incremental Few Shot Learning. https://www.kbhaskar.com/post/fsil/ (accessed 09 February 2023).

26 Kamath, U., Liu, J., and Whitaker, J. (2019). Transfer learning: domain adaptation. In: *Deep Learning for NLP and Speech Recognition*, 495–535. Cham. https://doi.org/10.1007/978-3-030-14596-5_11: Springer.

27 Huang, Y., Bert, C., Gomaa, A. et al. (2023). A survey of incremental transfer learning: combining peer-to-peer federated learning and domain incremental learning for multicenter collaboration. ArXiv, abs/2309.17192.

28 Moradi, M., Rahmanimanesh, M., and Shahzadi, A. (2024). Unsupervised domain adaptation by incremental learning for concept drifting data streams. *International Journal of Machine Learning and Cybernetics*. https://doi.org/10.1007/s13042-024-02135-1.

29 van de Ven, G.M., Tuytelaars, T., and Tolias, A.S. (2022). Three types of incremental learning. *Nature Machine Intelligence* 4: 1185–1197. https://doi.org/10.1038/s42256-022-00568-3.

30 Houlsby, N., Huszár, F., Ghahramani, Z., and Lengyel, M. (2011). Bayesian active learning for classification and preference learning. ArXiv, abs/1112.5745.

31 Boukli Hacene, G., Gripon, V., Farrugia, N. et al. (2018). Transfer incremental learning using data augmentation. *Applied Sciences* 8 (12): 2512. https://doi.org/10.3390/app8122512.

32 Avcontentteam (2024). Incremental learning: benefits, implementation and challenges. https://www.analyticsvidhya.com/blog/2023/08/incremental-learning/\ignorespaces(accessed\ignorespaces25\ignorespacesMay\ignorespaces2024).

33 Ali Awan, A.A. (2023). What is incremental learning? https://www.datacamp.com/blog/what-is-incremental-learning\ignorespaces(accessed\ignorespaces10\ignorespacesMay\ignorespaces2023).

34 Velimirovic, A. (2022). What is an edge server and how does it work? https://phoenixnap.com/blog/edge-server (accessed 25 September 2023).

35 Symonds, J. (2021). An architect's guide to edge storage. https://thenewstack.io/storage/an-architects-guide-to-edge-storage/ (accessed 25 October 2023).

36 Iyengar, A. and Rao, A. (2022). Distributed cloud: empowerment at the edge. https://www.ibm.com/blog/distributed-cloud-empowerment-at-the-edge/ (accessed 02 September 2023).

37 Gabriel, D.J. (2024). Edge Computing and the Power of Data Centers for Real-Time Decision-Making. https://www.sify.com/data-centers/edge-computing-and-the-power-of-data-centers-for-real-time-decision-making/ (accessed 25 May 2024).

38 Williams, D. (2019). Edge Computing: Inherent Challenges and How to Address Them https://www.rfcode.com/blog/edge-computing-inherent-challenges-and-how-to-address-them (accessed 20 August 2023).

39 Matei, D. (2023). Clustering: How does it work? https://www.baeldung.com/cs/dbscan-algorithm (accessed 09 January 2023).

40 Kapadia, S. (2019). Topic modeling in python: latent Dirichlet allocation (LDA). https://towardsdatascience.com/end-to-end-topic-modeling-in-python-latent-dirichlet-allocation-lda-35ce4ed6b3e0 (accessed 25 August 2023).

41 Sen, T. and Shen, H. (2022). Distributed training for deep learning models on an edge computing network using shielded reinforcement learning. *2022 IEEE 42nd International Conference on Distributed Computing Systems (ICDCS)*, Bologna, Italy, pp. 581–591, 10–13 July 2022. https://doi.org/10.1109/ICDCS54860.2022.00062.

42 Theissler, A., Pérez-Velázquez, J., Kettelgerdes, M., and Elger, G. (2021). Predictive maintenance enabled by machine learning: use cases and challenges in the automotive industry. *Reliability Engineering & System Safety* 215: 107864.

12

Architectural Patterns for the Design of Federated Learning Systems

Vijay Anand Rajasekaran[1]*, Jayalakshmi Periyasamy*[1]*, Madala Guru Brahmam*[1]*, and Balamurugan Baluswamy*[2]

[1]*School of Computer Science Engineering & Information Systems, Vellore Institute of Technology, Vellore, Tamil Nadu, India*
[2]*Shiv Nadar University, Delhi-National Capital Region (NCR), India*

12.1 Introduction

The introduction to this book chapter embarks on an illuminating journey into the intricate realm of federated learning, illuminating its pivotal role in the landscape of modern machine learning paradigms. Federated learning, a cutting-edge approach, empowers machine learning models to glean knowledge from diverse and decentralized data sources, all without the need for centralizing sensitive information. This innovative technique stands as a bulwark against privacy breaches while simultaneously mitigating the risks associated with the concentration of training data samples [1].

In the following pages, this chapter meticulously navigates through the architectural intricacies and design patterns that underpin federated learning systems, offering readers a comprehensive and in-depth understanding of this transformative technology [2]. The architectural landscape is dissected into three primary domains: Client-based, Server-based, and Hybrid architectures, each with its own distinctive features and advantages. Client-based architectures prioritize data privacy and locality, placing model management predominantly at the client level. On the other hand, Server-based architectures focus on centralized model learning, emphasizing robustness and scalability [3]. Meanwhile, Hybrid architectures harmoniously blend both client and server-based learning to harness the strengths of each paradigm, thereby optimizing flexibility, efficiency, and overall performance.

12.1.1 Background and Motivation

In recent years, the field of machine learning has witnessed a transformative paradigm shift—one that places data privacy and decentralization at the forefront of model development. The advent of federated learning has emerged as a groundbreaking response to the growing need to train machine learning models across a multitude of decentralized devices or servers while safeguarding the integrity of sensitive data samples.

Second, this chapter aims to foster a deeper understanding of the theoretical underpinnings and practical nuances inherent to federated learning. By exploring various architectural

Model Optimization Methods for Efficient and Edge AI: Federated Learning Architectures, Frameworks and Applications, First Edition. Edited by Pethuru Raj Chelliah, Amir Masoud Rahmani, Robert Colby, Gayathri Nagasubramanian, and Sunku Ranganath.

frameworks and design patterns, readers will be equipped with the knowledge required to make informed decisions when considering the adoption of federated learning in their own projects and applications.

Furthermore, this chapter addresses a critical issue in the realm of machine learning—the challenge of data isolation. In many real-world scenarios, data is distributed across a multitude of sources, each governed by distinct ownership and privacy considerations [4]. Federated learning offers a compelling solution to this challenge by enabling model training on the edge, where data resides, thus circumventing the need for centralized data storage and transmission.

12.1.2 Scope and Objectives

The scope of this book chapter is to provide an extensive and in-depth exploration of architectural patterns for the design of federated learning systems. It encompasses a wide spectrum of topics, ranging from the fundamental concepts and motivations behind federated learning to the practical application of various architectural frameworks and design patterns. The overarching objectives of this chapter can be summarized as follows:

1) **Comprehensive Coverage**: This chapter aims to comprehensively cover the landscape of federated learning. It will explore the foundational principles and theoretical underpinnings, ensuring that readers gain a thorough understanding of the core concepts that govern federated learning systems.
2) **Architectural Patterns**: The chapter meticulously categorizes and elucidates the architectural patterns essential for federated learning systems. These patterns, including Client-based, Server-based, and Hybrid architectures, are examined in detail, providing readers with insights into their strengths, weaknesses, and practical applications.

12.1.3 Importance of Federated Learning

Federated learning emerges as a revolutionary paradigm in the realm of machine learning, underscored by its profound importance in addressing some of the most pressing challenges and opportunities of our data-driven era [5]. This section explores the pivotal role and significance of federated learning across several critical dimensions.

1) **Preservation of Data Privacy**: One of the foremost reasons for the rising importance of federated learning is its ability to preserve data privacy. In an age where data breaches and privacy infringements are all too common, federated learning stands as a beacon of hope. It enables machine learning models to be trained on decentralized data sources without the need for data to leave its original location.
2) **Decentralization of Learning**: Federated learning transforms the landscape by shifting the paradigm from centralized learning to decentralized learning. In traditional machine learning approaches, data from various sources is aggregated into a central repository. Federated learning, however, empowers data sources (clients or servers) to train models locally, reducing the need for centralization and data movement [6].

12.2 Federated Learning Fundamentals

To delve into the intricate world of federated learning, it's essential to establish a solid foundation by exploring the fundamental principles that underpin this transformative approach. This section serves as a primer on federated learning fundamentals, elucidating the core concepts that define its architecture and operation.

1) **Decentralized Data Sources**: At the heart of federated learning lies the concept of decentralized data sources. Unlike traditional machine learning, where data is typically aggregated into a central repository, federated learning operates on data that remains distributed across multiple devices or servers.
2) **Model Collaboration**: Federated learning facilitates model training across these decentralized data sources while preserving the privacy of the data itself. Instead of sharing raw data, federated learning involves sharing model updates or gradients.

12.2.1 Definition and Concept

Federated learning is a groundbreaking paradigm in the field of machine learning, revolutionizing the way models are trained and knowledge is aggregated. This section delves into the core definition and underlying concepts of federated learning, providing a clear understanding of its essence [7].

12.2.1.1 Definition

Federated learning is an advanced machine learning technique that enables the collaborative training of a global machine learning model across a network of decentralized and often geographically distributed data sources, such as client devices or servers, without the need to transfer or centralize raw data samples.

12.2.1.2 Conceptual Framework

The concept of federated learning is founded on several key principles:

1) **Decentralization**: At its core, federated learning embraces the idea of decentralized data sources. In contrast to traditional centralized machine learning, where data is typically aggregated into a central repository, federated learning leverages data that remains distributed across multiple devices or servers.
2) **Privacy Preservation**: A central tenet of federated learning is the preservation of data privacy. By design, federated learning does not require the sharing of raw data samples. Instead, data sources collaborate by sharing model updates or gradients, which are aggregated to update the global model. Privacy-enhancing techniques, such as differential privacy, are often employed to further protect individual data sources' contributions.

12.2.2 Key Advantages

Federated learning introduces a transformative paradigm in machine learning that offers a multitude of advantages over traditional centralized approaches. In this section, we explore the key benefits that make federated learning a compelling choice for a wide range of applications and industries [8].

1) **Preservation of Data Privacy**: Perhaps the most significant advantage of federated learning is its unparalleled ability to preserve data privacy. In a world where data breaches and privacy infringements are rampant, federated learning stands as a bastion of security. It ensures that sensitive and confidential data remains localized, never leaving the data source.
2) **Decentralization**: Federated learning promotes the decentralization of machine learning, which is particularly relevant in today's world of distributed data sources. Instead of

aggregating data into a central repository, federated learning allows models to be trained at the edge, where data resides.

12.2.3 Challenges and Limitations

While federated learning offers numerous advantages, it is not without its challenges and limitations. In this section, we delve into the complexities and constraints that practitioners and researchers must contend with when implementing federated learning systems [9].

1) **Communication Overhead**: Federated learning relies on communication between data sources and a central server or among peers. This communication overhead can be substantial, particularly in scenarios with a large number of participating devices or when model updates are frequent. High communication costs can lead to delays, increased network utilization, and scalability issues.
2) **Privacy-Preserving Techniques**: Ensuring data privacy is a core objective of federated learning. However, privacy-preserving techniques, such as differential privacy or secure aggregation, introduce computational overhead and may require additional processing power. Striking the right balance between privacy and performance can be a nontrivial task [10].

12.3 Architectural Frameworks

12.3.1 Overview of Architectural Patterns

Architectural patterns form the backbone of federated learning systems, defining the structures and processes that enable decentralized, privacy-preserving machine learning. This section provides an overview of the architectural patterns that play a pivotal role in federated learning, categorizing them into three main segments: Client-based, Server-based, and Hybrid architectures [11].

12.3.1.1 Client-Based Architectures

In Client-based architectures, the majority of the model management and execution occurs at the client level, typically on user devices or local servers. This approach prioritizes data privacy and locality, making it particularly well-suited for applications where sensitive data must remain on the device, such as personal devices, edge devices, and mobile applications [12]

- **On-Device Learning**: Client-based architectures enable on-device machine learning, allowing models to be trained directly on user devices without transmitting data to external servers. This approach is ideal for scenarios like predictive text input, voice recognition, and personalization.
- **Enhanced Privacy**: By keeping data on the client side, Client-based architectures offer enhanced privacy. Model updates are computed locally, and only aggregated updates are shared with a central server or peers, minimizing the risk of data exposure.

12.3.1.2 Server-Based Architectures

Server-based architectures centralize the model learning process predominantly on a central server. This approach is well-suited for applications requiring robustness, scalability, and the ability to handle large-scale data sources, such as cloud-based machine learning platforms and data centers [13].

- **Centralized Aggregation**: In Server-based architectures, the central server plays a pivotal role in aggregating model updates from data sources. This centralized aggregation simplifies the coordination of updates and ensures a comprehensive view of the learning process.
- **Scalability**: Server-based architectures excel in scalability, making them suitable for applications with extensive data sources or when real-time model updates are required. They can efficiently handle large-scale deployments.

12.3.1.3 Hybrid Architectures

Hybrid architectures seek to strike a balance between the advantages of Client-based and Server-based approaches. They amalgamate both client and server-based learning to harness the strengths of each paradigm, optimizing flexibility, efficiency, and overall performance [14].

- **Flexibility and Efficiency**: Hybrid architectures offer flexibility by allowing data sources to train models locally while also centralizing some aspects of the learning process. This flexibility ensures adaptability to diverse scenarios.
- **Privacy and Scalability**: By combining local and centralized learning, Hybrid architectures aim to address the privacy concerns of Client-based architectures while leveraging the scalability benefits of Server-based architectures.

12.3.2 Client-Based Architectures

Client-based architectures represent a fundamental approach in federated learning where most of the model management and execution occurs at the client level, typically on user devices or local servers. This section provides an in-depth exploration of Client-based architectures, highlighting their characteristics, advantages, challenges, and practical applications [15].

Characteristics of Client-Based Architectures:

1) **Local Model Training**: In Client-based architectures, model training occurs locally on user devices or edge servers. Data sources independently compute model updates based on their local data samples.
2) **Privacy Emphasis**: Privacy is a central tenet of Client-based architectures. Data remains on the client side, and only model updates or gradients are shared with a central server or peers, ensuring the confidentiality of sensitive data samples.

Advantages of Client-Based Architectures:

1) **Privacy Preservation**: Client-based architectures offer enhanced data privacy. User data remains on the device, reducing the risk of data exposure or unauthorized access during model training.
2) **Low Data Transfer**: Since raw data does not need to be transmitted to a central server, Client-based architectures minimize data transfer requirements. This is particularly advantageous for scenarios with limited network bandwidth.

Challenges of Client-Based Architectures:

1) **Limited Computational Resources**: User devices may have limited computational resources, which can pose challenges for model training. Strategies for efficient local model updates and training are essential.
2) **Heterogeneous Devices**: Client-based architectures must accommodate heterogeneous device capabilities and operating systems. Ensuring compatibility and optimizing model performance on a wide range of devices is a complex task.

Practical Applications of Client-Based Architectures:

1) **Predictive Text Input**: Mobile keyboards use federated learning to improve predictive text input. Models are trained locally on the device, learning from the user's typing habits and language preferences without exposing sensitive text data samples.
2) **Voice Assistants**: Voice assistants like Siri and Google Assistant employ federated learning for speech recognition and natural language understanding. Model updates occur on user devices, allowing voice assistants to adapt to individual speech patterns.

12.3.3 Server-Based Architectures

Server-based architectures represent a core approach in federated learning where the primary model management and execution are centralized on a central server. This section provides an in-depth exploration of Server-based architectures, highlighting their characteristics, advantages, challenges, and practical applications [16].

Characteristics of Server-Based Architectures:

1) **Centralized Model Management**: In Server-based architectures, model management and aggregation predominantly occur on a central server. Data sources send their model updates to this central entity for aggregation and model improvement.
2) **Global Model View**: The central server maintains a comprehensive view of the learning process and model updates from all participating data sources. This centralized oversight simplifies the coordination of model updates.

Advantages of Server-Based Architectures:

1) **Centralized Aggregation**: Server-based architectures centralize the aggregation of model updates. This simplifies coordination and ensures a comprehensive view of the learning process, which can be valuable for achieving convergence.
2) **Scalability**: The centralized nature of Server-based architectures makes them highly scalable. They can efficiently handle large-scale deployments with numerous data sources.

Challenges of Server-Based Architectures:

1) **Data Privacy**: Data sources must transmit their data or model updates to a central server, which raises privacy concerns. Ensuring secure data transfer and compliance with data protection regulations is essential.
2) **Communication Overhead**: Server-based architectures entail communication overhead, especially when a large number of data sources are involved. This can lead to increased network utilization and delays.

Practical Applications of Server-Based Architectures:

1) **Healthcare Data Analysis**: Healthcare organizations use federated learning with Server-based architectures to analyze patient data securely. Hospitals and clinics send encrypted model updates to a central server for the development of predictive models without sharing individual patient records.
2) **Financial Fraud Detection**: Financial institutions employ federated learning to detect fraudulent activities across multiple banks. Server-based architectures enable the central aggregation of transaction data for improved fraud detection models.

12.3.4 Hybrid Architectures

Hybrid architectures represent a versatile and adaptive approach in federated learning, combining elements from both Client-based and Server-based architectures. This section provides an in-depth exploration of Hybrid architectures, elucidating their characteristics, advantages, challenges, and practical applications [17].

Characteristics of Hybrid Architectures:

1) **Balanced Model Management**: In Hybrid architectures, model management is distributed between client devices or local servers and a central server. This balance enables a flexible approach to federated learning, with certain aspects of the learning process decentralized and others centralized.
2) **Adaptive Learning**: Hybrid architectures excel in adaptability. They dynamically distribute computation between clients and servers based on factors such as device capabilities, network conditions, and data sensitivity.

Advantages of Hybrid Architectures:

1) **Flexibility and Efficiency**: Hybrid architectures offer flexibility by allowing data sources to train models locally while also centralizing some aspects of the learning process. This adaptability ensures adaptability to diverse scenarios.
2) **Privacy Enhancement**: Hybrid architectures provide an effective means to enhance privacy. Data remains primarily on client devices, reducing data transfer requirements and privacy risks associated with centralized data storage.

Challenges of Hybrid Architectures:

1) **Complexity**: Designing and managing Hybrid architectures can be complex, requiring careful consideration of data flow, synchronization, and aggregation strategies to optimize the learning process effectively.
2) **Communication Overhead**: Like Server-based architectures, Hybrid architectures may involve communication overhead, especially when a substantial number of data sources are involved. Strategies for efficient communication are crucial.

Practical Applications of Hybrid Architectures:

1) **Telecommunications Network Optimization**: Telecommunications companies employ federated learning with Hybrid architectures to optimize network performance. Local servers on cell towers perform initial model training, while a central server aggregates updates for global model improvements.
2) **Retail Customer Personalization**: Retailers use federated learning with Hybrid architectures to personalize customer experiences. Local devices in stores gather data and perform initial model updates, while a central server aggregates insights to enhance product recommendations.

12.4 Design Patterns in Federated Learning

Design patterns play a crucial role in shaping the architecture and implementation of federated learning systems. They offer reusable solutions to common challenges and provide a structured approach to designing and deploying federated learning solutions. In this section, we explore several design patterns that are instrumental in the successful development of federated learning systems [18].

1) **Horizontal Federated Learning**:
   - **Pattern Description**: Horizontal federated learning is used when data sources have the same feature space but different data samples. This pattern is particularly common in applications where user data is distributed across multiple devices or servers.
   - **Use Cases**: Predictive text input, collaborative filtering, and sentiment analysis are examples of use cases where horizontal federated learning is valuable.
2) **Vertical Federated Learning**:
   - **Pattern Description**: Vertical federated learning is applied when data sources have the same data samples but different feature spaces. This pattern is useful when combining data sources with complementary information.
   - **Use Cases**: Medical diagnosis, fraud detection, and recommendation systems often benefit from vertical federated learning.
3) **Transfer Learning**:
   - **Pattern Description**: Transfer learning leverages pretrained models on one dataset to accelerate training on another related dataset. In federated learning, this can involve transferring knowledge from a global model to individual data sources.
   - **Use Cases**: Image classification, natural language processing, and speech recognition can benefit from transfer learning in federated settings.
4) **Ensemble Learning**:
   - **Pattern Description**: Ensemble learning combines multiple models to improve prediction accuracy and robustness. In federated learning, ensemble methods can be applied to aggregate model updates from multiple data sources.
   - **Use Cases**: Financial forecasting, disease diagnosis, and anomaly detection often utilize ensemble learning in federated settings.
5) **Adaptive Learning Algorithms**:
   - **Pattern Description**: Adaptive learning algorithms dynamically adjust model hyperparameters or learning rates based on the convergence status of each data source. This ensures that the federated learning process adapts to varying data quality and convergence rates [19].
   - **Use Cases**: Adaptive learning is valuable in applications where data sources have different levels of data quality or may experience concept drift.

12.4.1 Horizontal Federated Learning

Horizontal federated learning patterns are instrumental when data sources have the same feature space but different data samples [20]. This pattern is well-suited for scenarios where user data is distributed across multiple devices or servers, and the goal is to collaboratively train a model without sharing individual data samples.

Characteristics of Horizontal Federated Learning Patterns:

1) **Common Feature Space**: In horizontal federated learning, all participating data sources share a common feature space. This implies that the same set of features or attributes is used to represent the data across all sources.
2) **Different Data Samples**: While the feature space is consistent, each data source possesses a unique set of data samples. These data samples are typically drawn from different subsets of the overall dataset, reflecting the decentralized nature of federated learning.

Advantages of Horizontal Federated Learning Patterns:

1) **Privacy Preservation**: Horizontal patterns inherently prioritize privacy since individual data samples are not shared. Instead, model updates are exchanged, enabling data source collaboration while safeguarding sensitive information.
2) **Decentralized Training**: Data remains at its respective source, reducing the need for central data storage or transmission. This decentralization aligns with privacy and security requirements while minimizing data transfer overhead.

Practical Applications of Horizontal Federated Learning Patterns:

1) **Predictive Text Input**: Mobile keyboards employ horizontal federated learning to enhance predictive text input. Each device collects typing data, such as keystrokes and text inputs, and collaboratively trains a model to improve text predictions without sharing personal text data samples [21].
2) **Spam and Fraud Detection**: Email service providers use horizontal federated learning to detect spam and fraudulent emails. By training models locally on users' email behaviors and content, the detection accuracy improves without exposing individual email contents.

12.4.2 Vertical Federated Learning

Vertical federated learning patterns are essential when data sources have the same data samples but different feature spaces. This pattern is particularly valuable when combining data sources with complementary information. In this section, we delve into the intricacies of vertical patterns in federated learning, exploring their characteristics, advantages, and practical applications [22].

Characteristics of Vertical Federated Learning Patterns:

1) **Shared Data Samples**: In vertical federated learning, all participating data sources share the same data samples or examples. This implies that the data sources are dealing with identical instances or records.
2) **Different Feature Spaces**: Despite sharing the same data samples, each data source uses a distinct set of features or attributes to represent those data samples. These feature spaces are typically tailored to the specific characteristics or requirements of each source.

Advantages of Vertical Federated Learning Patterns:

1) **Complementary Information**: Vertical federated learning patterns enable the combination of data sources with complementary information. By having different feature spaces, data sources can provide unique perspectives on the same data samples, leading to improved model accuracy.
2) **Privacy Preservation**: Just like horizontal patterns, vertical patterns prioritize privacy since individual data samples are not shared among sources. Model updates are exchanged instead, allowing for collaborative learning while protecting sensitive information [23].

Practical Applications of Vertical Federated Learning Patterns:

1) **Healthcare Data Integration**: In healthcare, vertical federated learning is used to integrate data from various sources such as electronic health records, medical imaging, and genomics.
2) **Financial Risk Assessment**: Financial institutions employ vertical federated learning to assess credit risk. Various data sources, including transaction histories, credit scores, and

customer behavior, contribute to assessing an individual's creditworthiness without sharing sensitive financial details.

12.4.3 Transfer Learning

Transfer learning is a powerful technique that leverages knowledge gained from one domain or task to accelerate learning in another related domain or task [22]. In the context of federated learning, transfer learning-based patterns can be invaluable for improving model performance, especially when dealing with limited data resources or when trying to adapt models to varying local conditions. This section explores the principles, advantages, and practical applications of transfer learning-based federated learning patterns.

Principles of Transfer Learning in Federated Learning:

1) **Pretrained Model**: In transfer learning-based federated learning, a pretrained model is used as the starting point. This model has already learned valuable features or representations from a large dataset in a related domain.
2) **Fine-tuning**: The pretrained model is fine-tuned using data from the federated learning participants. Fine-tuning allows the model to adapt its learned features to the specific characteristics of the local data sources.

Advantages of Transfer Learning-Based Federated Learning Patterns:

1) **Data Efficiency**: Transfer learning enables models to learn from relatively small amounts of local data. Instead of training from scratch, models can benefit from pre-existing knowledge, making them data-efficient [24].
2) **Improved Generalization**: Pretrained models often possess robust feature representations that can lead to improved generalization. This is especially valuable when dealing with limited data resources or highly specialized tasks.

Practical Applications of Transfer Learning-Based Federated Learning Patterns:

1) **Image Classification in Edge Devices**: Transfer learning-based federated learning is used in edge devices such as smartphones and surveillance cameras for image classification tasks. Pretrained models for object recognition are fine-tuned with local data to adapt to specific environments, such as monitoring wildlife or detecting anomalies.
2) **Natural Language Processing**: In NLP applications, pretrained language models (e.g., BERT, GPT) are fine-tuned using text data from various sources. In federated learning settings, these models can be adapted to different dialects or regional languages while retaining their language understanding capabilities [25].

12.4.4 Other Relevant Patterns

In addition to the primary architectural and design patterns discussed earlier, federated learning also benefits from several other relevant learning patterns and techniques that can enhance its effectiveness and adaptability [26]. These patterns address various challenges and requirements, ensuring the successful deployment of federated learning systems in diverse domains. Here, we explore some of these additional learning patterns:

1) **Self-Supervised Learning**:
   - **Pattern Description**: Self-supervised learning is a technique where a model learns to predict parts of its own input data. In federated learning, it can be used to improve model robustness and adaptability by training models on local data for self-supervised tasks.

- **Use Cases**: Self-supervised learning can be valuable in applications where labeled data is scarce or expensive to obtain, such as in healthcare or manufacturing.

2) **Reinforcement Learning**:
   - **Pattern Description**: Reinforcement learning enables models to learn from interactions with an environment. In federated settings, reinforcement learning can be used to train agents that interact with local data sources and adapt their behavior based on feedback.
   - **Use Cases**: Reinforcement learning is useful in scenarios where models need to make sequential decisions, such as in autonomous systems or recommendation engines.
3) **Semi supervised Learning**:
   - **Pattern Description**: Semi supervised learning combines labeled and unlabeled data to improve model performance. In federated learning, semi supervised techniques can make use of limited labeled data from clients to enhance model accuracy.
   - **Use Cases**: Semi supervised learning is applicable in situations where acquiring labeled data is costly, such as in fraud detection or content classification.
4) **Meta-Learning**:
   - **Pattern Description**: Meta-learning focuses on training models to learn how to learn. In federated settings, meta-learning can help models quickly adapt to new data sources by leveraging past learning experiences.
   - **Use Cases**: Meta-learning is beneficial in scenarios where data sources frequently change or where models need to adapt rapidly to evolving conditions.

12.4.5 Practical Examples for Different Patterns

Federated learning patterns have found practical applications across various domains, and their social impacts are increasingly profound. Here, we provide practical examples that highlight different federated learning patterns and their significant societal implications [27].

1) **Horizontal Federated Learning—Predictive Text Input**:
   - **Pattern**: Horizontal federated learning
   - **Example**: Mobile keyboard apps employ horizontal federated learning to improve predictive text input. Each user's typing data, including keystrokes and text inputs, is used locally to enhance the keyboard's predictions. Collaborative model training happens without sharing personal text data. The social impact is substantial, as improved typing suggestions lead to increased productivity and user satisfaction. Moreover, user privacy is preserved.
2) **Vertical Federated Learning—Healthcare Data Integration**:
   - **Pattern**: Vertical federated learning
   - **Example**: In healthcare, vertical federated learning is applied to integrate data from different sources, such as electronic health records, medical imaging, and genomics. Each source contributes data with different feature representations. The social impact is profound as it leads to more accurate disease diagnosis and treatment recommendations, ultimately improving patient care and healthcare outcomes while preserving patient privacy.
3) **Transfer Learning—Autonomous Vehicles**:
   - **Pattern**: Transfer learning-based federated learning
   - **Example**: In autonomous vehicles, transfer learning is employed to adapt models for different driving conditions and environments. Pretrained models for object detection and path planning are fine-tuned using data from various vehicles and locations. The social impact is significant, as it enhances road safety and facilitates the widespread adoption of autonomous driving, reducing accidents and congestion.

12.5 Applications of Federated Learning

Federated learning has gained traction across a wide range of industries and applications, offering a privacy-preserving and efficient approach to machine learning in decentralized environments. This section explores diverse applications of federated learning, showcasing its versatility and impact on various domains [28].

1) **Healthcare**:
   - **Medical Image Analysis**: Federated learning is applied to analyze medical images, such as X-rays, MRIs, and CT scans, across multiple hospitals and clinics. Models are trained collaboratively to improve diagnostic accuracy while ensuring data privacy.
   - **Disease Prediction**: Healthcare institutions use federated learning to predict diseases based on patient data from different sources. Federated models enhance early diagnosis and personalized treatment recommendations.
2) **Finance**:
   - **Credit Scoring**: Financial institutions utilize federated learning to assess credit risk by aggregating credit histories and transaction data from various sources. This approach enhances credit scoring accuracy while maintaining data privacy.
   - **Fraud Detection**: Federated learning helps detect financial fraud by analyzing transaction data from multiple banks without centralizing sensitive financial information. Fraud detection models benefit from collective intelligence.

12.6 Technologies and Methodologies

Federated learning relies on a combination of technologies and methodologies to ensure secure and efficient model training in decentralized environments. These components play a critical role in the success of federated learning systems [29].

1) **Encrypted Computation**:
   - **Pattern Description**: Encrypted computation techniques, such as secure multiparty computation (SMPC) and homomorphic encryption, enable data sources to perform computations on encrypted data without revealing sensitive information. In federated learning, this technology protects individual data samples while allowing for collaborative model updates.
   - **Advantages**: Encrypted computation ensures data privacy, making it suitable for applications in healthcare, finance, and personalization where sensitive information must be safeguarded. It enables secure aggregation of model updates without exposing raw data.
2) **Decentralized Algorithms**:
   - **Pattern Description**: Decentralized algorithms govern the coordination of model updates across distributed data sources. These algorithms ensure that models converge effectively in federated learning settings, even when data sources have varying data quality or distributions.
   - **Advantages**: Decentralized algorithms enable federated learning systems to operate efficiently across a diverse range of data sources. They adapt to local conditions and facilitate robust model training, making federated learning suitable for real-world scenarios.
3) **Communication Protocols**:
   - **Pattern Description**: Communication protocols define the rules and procedures for data sources to exchange model updates securely. Custom protocols or industry standards, such as federated learning communication (FLC), are employed to ensure efficient and secure communication.

- **Advantages**: Communication protocols are essential for coordinating model updates and aggregating them securely. They minimize communication overhead, enable fault tolerance, and ensure the confidentiality and integrity of model updates.

12.7 Implementing Federated Learning Systems

The successful implementation of federated learning systems involves a structured approach that encompasses various components and stages. This section provides an overview of the key steps and considerations in implementing federated learning systems, including system architecture and components, data preprocessing, data distribution, model training, evaluation, inference, and deployment [30].

12.7.1 System Architecture and Components

- **Pattern Description**: A federated learning system consists of multiple components, including clients (data sources), a central server, and associated infrastructure. The server coordinates model training, while clients perform local computations and contribute model updates.
- **Considerations**: Designing an effective system architecture involves defining the roles and responsibilities of clients and the central server. It includes selecting suitable hardware and software infrastructure, ensuring scalability, and establishing secure communication channels.

12.7.2 Data Preprocessing and Data Distribution

- **Pattern Description**: Data preprocessing is a crucial step in federated learning, as data sources may have varying data quality and characteristics. Preprocessing involves data cleaning, feature engineering, and data transformation. Data distribution determines how data is partitioned and distributed to clients.
- **Considerations**: Data preprocessing should address data heterogeneity and ensure that local data sources are prepared for model training. Data distribution strategies must consider fairness, data imbalance, and data privacy concerns.

12.7.3 Model Training and Evaluation

- **Pattern Description**: Model training occurs iteratively in federated learning. Clients train local models using their data and send model updates to the central server, where they are aggregated to update the global model. Model evaluation involves assessing the performance of the global model and the convergence of the training process.
- **Considerations**: Model training strategies include selecting optimization algorithms, setting hyperparameters, and managing model complexity. Model evaluation metrics should align with the application's objectives and may include accuracy, loss, or domain-specific metrics [31].

12.8 Emerging Trends and Future Scopes

Federated learning continues to evolve, driven by emerging trends and expanding possibilities. This section explores key trends and future scopes in federated learning, including adaptive learning algorithms, interoperable standards, novel applications, ethical and regulatory implications, and challenges in adoption and scalability [32].

12.8.1 Adaptive Learning Algorithms

- **Trend Description**: Adaptive learning algorithms are gaining prominence in federated learning. These algorithms enable models to dynamically adjust their behavior based on changing data distributions, client capabilities, or privacy constraints. This adaptability enhances model performance and robustness.
- **Future Impact**: Adaptive learning algorithms will enable federated learning systems to handle dynamic environments, accommodate varying data sources, and maintain model accuracy over time. They are essential for applications with evolving data characteristics [33].

12.8.2 Interoperable Standards

- **Trend Description**: The development of interoperable standards for federated learning is becoming a priority. Standardization efforts aim to establish common protocols, APIs, and data formats that facilitate interoperability between federated learning systems and enable seamless collaboration.
- **Future Impact**: Interoperable standards will promote the widespread adoption of federated learning by simplifying system integration and fostering collaboration across industries. Standardized interfaces will enable federated learning to become a core component of AI ecosystems [34].

12.8.3 Novel Applications

- **Trend Description**: Federated learning is finding applications in novel domains and use cases. These applications include fields such as climate modeling, wildlife conservation, and environmental monitoring, where privacy, data decentralization, and collaboration are crucial [35].
- **Future Impact**: Federated learning's expansion into novel applications will contribute to addressing global challenges while preserving privacy and data sovereignty. It will enable the development of AI solutions for domains previously constrained by data sharing limitations [36].

12.9 Conclusion

In conclusion, federated learning represents a transformative paradigm shift in the field of machine learning, offering the promise of collaborative model training across decentralized data sources while preserving data privacy and security. This chapter has provided a comprehensive exploration of the intricate world of federated learning, spanning its architectural patterns, design principles, underlying technologies, applications, and ethical considerations.

The chapter then elucidated the fundamentals of federated learning, defining the concept and highlighting its key advantages, such as privacy preservation, data locality, and scalability. We also delved into the challenges and limitations that federated learning encounters, emphasizing the need for robust solutions and ongoing research in addressing these challenges.

In essence, this chapter serves as a comprehensive guide for researchers, practitioners, and enthusiasts in the field of federated learning, offering a holistic overview of the architectural patterns, design principles, underlying technologies, applications, and ethical considerations pivotal to the design and implementation of federated learning systems. The amalgamation of theoretical insights, practical examples, and future prospects underscores its significance as an invaluable resource in the evolving landscape of federated learning.

References

1 Singh, A., Chiu, W.-Y., Manoharan, S.H., and Romanov, A.M. (2022). Energy-efficient gait optimization of snake-like modular robots by using multiobjective reinforcement learning and a fuzzy inference system. *IEEE Access* 10: 86624–86635. https://doi.org/10.1109/ACCESS.2022.3195928.

2 Borrageiro, G., Firoozye, N., and Barucca, P. (2022). The recurrent reinforcement learning crypto agent. *IEEE Access* 10: 38590–38599. https://doi.org/10.1109/ACCESS.2022.3166599.

3 Kazemi, E., Taherkhani, F., and Wang, L. (2023). Semisupervised learning for noise suppression using deep reinforcement learning of contrastive features. *IEEE Sensors Letters* 7 (4): 1–4. https://doi.org/10.1109/LSENS.2023.3264998.

4 Chen, Z., Nakamura, Y., and Ishiguro, H. (2023). Outperformance of mall-receptionist android as inverse reinforcement learning is transitioned to reinforcement learning. *IEEE Robotics and Automation Letters* 8 (6): 3350–3357. https://doi.org/10.1109/LRA.2023.3267385.

5 Samarasinghe, D., Barlow, M., and Lakshika, E. (2022). Flow-based reinforcement learning. *IEEE Access* 10: 102247–102265. https://doi.org/10.1109/ACCESS.2022.3209260.

6 Amin, S., Uddin, M.I., Alarood, A.A. et al. (2023). Smart E-learning framework for personalized adaptive learning and sequential path recommendations using reinforcement learning. *IEEE Access* 11: 89769–89790. https://doi.org/10.1109/ACCESS.2023.3305584.

7 Pateria, S., Subagdja, B., Tan, A.-H., and Quek, C. (2022). End-to-end hierarchical reinforcement learning with integrated subgoal discovery. *IEEE Transactions on Neural Networks and Learning Systems* 33 (12): 7778–7790. https://doi.org/10.1109/TNNLS.2021.3087733.

8 Borrageiro, G., Firoozye, N., and Barucca, P. (2022). Reinforcement learning for systematic FX trading. *IEEE Access* 10: 5024–5036. https://doi.org/10.1109/ACCESS.2021.3139510.

9 Fu, K., Yu, Y., and Li, B. (2023). Multi-feature supervised reinforcement learning for stock trading. *IEEE Access* 11: 77840–77855. https://doi.org/10.1109/ACCESS.2023.3298821.

10 Liu, J., Wang, G., Guo, X. et al. (2022). Deep reinforcement learning task assignment based on domain knowledge. *IEEE Access* 10: 114402–114413. https://doi.org/10.1109/ACCESS.2022.3217654.

11 Abdulazeez, D.H. and Askar, S.K. (2023). Offloading mechanisms based on reinforcement learning and deep learning algorithms in the fog computing environment. *IEEE Access* 11: 12555–12586. https://doi.org/10.1109/ACCESS.2023.3241881.

12 Rapetswa, K. and Cheng, L. (2023). Towards a multi-agent reinforcement learning approach for joint sensing and sharing in cognitive radio networks. *Intelligent and Converged Networks* 4 (1): 50–75. https://doi.org/10.23919/ICN.2023.0005.

13 Zhu, W., Guo, X., Owaki, D. et al. (2023). A survey of sim-to-real transfer techniques applied to reinforcement learning for bioinspired robots. *IEEE Transactions on Neural Networks and Learning Systems* 34 (7): 3444–3459. https://doi.org/10.1109/TNNLS.2021.3112718.

14 Kochdumper, N., Krasowski, H., Wang, X. et al. (2023). Provably safe reinforcement learning via action projection using reachability analysis and polynomial zonotopes. *IEEE Open Journal of Control Systems* 2: 79–92. https://doi.org/10.1109/OJCSYS.2023.3256305.

15 Koch, R., Kühl, S., Schairer, W. et al. (2022). High-generalizability reinforcement learning based capacity optimization in WDM long-haul networks. *IEEE Photonics Technology Letters* 34 (17): 891–894. https://doi.org/10.1109/LPT.2022.3191171.

16 Fei, H., Zhang, Y., Ren, Y., and Ji, D. (2022). Optimizing attention for sequence modeling via reinforcement learning. *IEEE Transactions on Neural Networks and Learning Systems* 33 (8): 3612–3621. https://doi.org/10.1109/TNNLS.2021.3053633.

17 Liu, H. and Wu, W. (2022). Federated reinforcement learning for decentralized voltage control in distribution networks. *IEEE Transactions on Smart Grid* 13 (5): 3840–3843. https://doi.org/10.1109/TSG.2022.3169361.

18 Jiang, J., Wang, J., Chu, H. et al. (2023). Whale swarm reinforcement learning based dynamic cooperation clustering method for cell-free massive MIMO systems. *IEEE Transactions on Vehicular Technology* 72 (3): 4114–4118. https://doi.org/10.1109/TVT.2022.3222756.

19 Yuan, Z., Hall, A.W., Zhou, S. et al. (2022). Safe-control-gym: a unified benchmark suite for safe learning-based control and reinforcement learning in robotics. *IEEE Robotics and Automation Letters* 7 (4): 11142–11149. https://doi.org/10.1109/LRA.2022.3196132.

20 Bokade, R., Jin, X., and Amato, C. (2023). Multi-agent reinforcement learning based on representational communication for large-scale traffic signal control. *IEEE Access* 11: 47646–47658. https://doi.org/10.1109/ACCESS.2023.3275883.

21 Hong, S.-H. and Lee, H.-S. (2023). Robust energy management system with safe reinforcement learning using short-horizon forecasts. *IEEE Transactions on Smart Grid* 14 (3): 2485–2488. https://doi.org/10.1109/TSG.2023.3240588.

22 Rothmann, M. and Porrmann, M. (2022). A survey of domain-specific architectures for reinforcement learning. *IEEE Access* 10: 13753–13767. https://doi.org/10.1109/ACCESS.2022.3146518.

23 Wang, D. and Hu, M. (2023). Contrastive learning methods for deep reinforcement learning. *IEEE Access* 11: 97107–97117. https://doi.org/10.1109/ACCESS.2023.3312383.

24 Prabhakar, S.K. and Lee, S.-W. (2022). SASDL and RBATQ: sparse autoencoder with swarm based deep learning and reinforcement based Q-learning for EEG classification. *IEEE Open Journal of Engineering in Medicine and Biology* 3: 58–68. https://doi.org/10.1109/OJEMB.2022.3161837.

25 Martinez, A.D., Del Ser, J., Osaba, E., and Herrera, F. (2022). Adaptive multifactorial evolutionary optimization for multitask reinforcement learning. *IEEE Transactions on Evolutionary Computation* 26 (2): 233–247. https://doi.org/10.1109/TEVC.2021.3083362.

26 Qu, X., Ong, Y.-S., and Gupta, A. (2022). Frame-correlation transfers trigger economical attacks on deep reinforcement learning policies. *IEEE Transactions on Cybernetics* 52 (8): 7577–7590. https://doi.org/10.1109/TCYB.2020.3041265.

27 Yuan, Z., Li, G., Wang, Z. et al. (2023). RL-CSL: a combinatorial optimization method using reinforcement learning and contrastive self-supervised learning. *IEEE Transactions on Emerging Topics in Computational Intelligence* 7 (4): 1010–1024. https://doi.org/10.1109/TETCI.2021.3139802.

28 Leiva, L., Torrents-Barrena, J., and Vázquez, M. (2023). FPGA-based accelerator for AI-toolbox reinforcement learning library. *IEEE Embedded Systems Letters* 15 (2): 113–116. https://doi.org/10.1109/LES.2022.3218168.

29 Gong, C., Sou, W.-K., and Lam, C.-S. (2023). Reinforcement learning based sliding mode control for a hybrid-STATCOM. *IEEE Transactions on Power Electronics* 38 (6): 6795–6800. https://doi.org/10.1109/TPEL.2023.3247835.

30 Yau, K.-L.A., Chong, Y.-W., Fan, X. et al. (2023). Reinforcement learning models and algorithms for diabetes management. *IEEE Access* 11: 28391–28415. https://doi.org/10.1109/ACCESS.2023.3259425.

31 Liu, Y. and Wang, Z. (2022). Reinforcement learning-based tracking control for a class of discrete-time systems with actuator fault. *IEEE Transactions on Circuits and Systems II: Express Briefs* 69 (6): 2827–2831. https://doi.org/10.1109/TCSII.2021.3131360.

32 Huang, C., Zhang, H., Wang, L. et al. (2022). Mixed deep reinforcement learning considering discrete-continuous hybrid action space for smart home energy management. *Journal of Modern Power Systems and Clean Energy* 10 (3): 743–754. https://doi.org/10.35833/MPCE.2021.000394.

33 Wang, Z., Wang, X., and Zhao, C. (2023). Event-triggered containment control for nonlinear multiagent systems via reinforcement learning. *IEEE Transactions on Circuits and Systems II: Express Briefs* 70 (8): 2904–2908. https://doi.org/10.1109/TCSII.2023.3250250.

34 Kim, K. (2022). Enhancing reinforcement learning performance in delayed reward system using DQN and heuristics. *IEEE Access* 10: 50641–50650. https://doi.org/10.1109/ACCESS.2022.3174361.

35 Huang, F., He, Y., Deng, X., and Jiang, W. (2023). A novel discount-weighted average fusion method based on reinforcement learning for conflicting data. *IEEE Systems Journal* 17 (3): 4748–4751. https://doi.org/10.1109/JSYST.2022.3228015.

36 Yanhao, Z., Abhishek, N.V., and Gurusamy, M. (2022). RAVEN: resource allocation using reinforcement learning for vehicular edge computing networks. *IEEE Communications Letters* 26 (11): 2636–2640. https://doi.org/10.1109/LCOMM.2022.3196711.

13

Federated Learning for Intelligent IoT Systems: Background, Frameworks, and Optimization Techniques

Partha Pratim Ray

Department of Computer Applications, Sikkim University, Gangtok, Sikkim, India

13.1 Introduction

The dawn of the digital age has continually reshaped the contours of our world, propelling us into an era where technology doesn't merely support our lives but intricately intertwines with it. At the forefront of this transformative journey stands the IoT, an ecosystem where devices seamlessly communicate, creating a symphony of interconnected operations [1, 2]. From humble beginnings with rudimentary interdevice communications to today's expansive network, the IoT paradigm is rapidly evolving, fuelled by the twin engines of intelligence and decentralized learning [3, 4]. As we stand at this pivotal juncture, this chapter endeavors to navigate the intricate nexus of the ITIoT and federated learning, offering insights into their confluence, challenges, and the monumental potential they collectively harbor.

13.1.1 Background of IoT and the Need for Intelligence

The IoT has dramatically transformed the landscape of digital connectivity. It symbolizes a transformative shift from the earlier era where the Internet was predominantly a network of interlinked computers [5–7]. Today, the paradigm has expanded to a sprawling web of devices, from the innocuous sensors on our refrigerators to the complex systems controlling entire smart cities. This monumental progression is evident in forecasts like the one from Cisco, which anticipates over 500 billion devices to be interconnected by 2030. Such predictions invite us to envision a world where nearly every tangible entity, be it a car, toaster, or traffic light, plays its part in the vast digital symphony of the Internet. With this immense proliferation of devices, we find ourselves awash in an unprecedented deluge of data. Each of these devices continuously emits streams of information, painting a picture of its environment, user interactions, and internal processes. Deciphering this ocean of data is not just about extracting useful insights; it's about enabling these devices to be more than just passive data loggers. This brings us to the concept of intelligence in IoT [8, 9]. When we talk about intelligence in IoT devices, we're referring to their capability to be proactive rather than reactive. An intelligent device doesn't just sense; it understands, learns, and makes decisions. For instance, consider a factory's robotic arm. A basic IoT-enabled arm can perform tasks and report any mechanical faults. But introduce intelligence, and the arm becomes capable of recognizing patterns, foreseeing potential wear and tear, adjusting its movements based on the items it's handling, and even collaborating with other robotic entities in its vicinity.

Model Optimization Methods for Efficient and Edge AI: Federated Learning Architectures, Frameworks and Applications, First Edition. Edited by Pethuru Raj Chelliah, Amir Masoud Rahmani, Robert Colby, Gayathri Nagasubramanian, and Sunku Ranganath.

This nuance can be understood through three primary capacities of intelligent IoT devices as follows [10–12].

- **Adaptability**: At its core, an intelligent device possesses the dexterity to modify its responses or actions rooted in the data it processes. Picture an agricultural drone that, upon identifying a parched section of a farm, modulates its water dispensing rate to adequately cater to that area.
- **Predictive Prowess**: One of the hallmarks of intelligence is the ability to anticipate. A smart intelligent conveyor belt, for example, might predict an impending breakdown based on the vibrational patterns it records, thereby notifying supervisors in advance.
- **Enhanced Interaction**: The age of impersonal gadgets is giving way to an era of devices that understand us. Personalization is becoming the norm. A smart speaker that not only plays music but also discerns the mood of the room, or a thermostat that adjusts not just based on the room's temperature but also its occupants' preferences, exemplifies this trend.

The push toward such intelligence is driven by advances in Machine Learning (ML) and Artificial Intelligence (AI). These technologies, with their potential to sift through data and discern patterns, are being integrated more than ever into IoT systems. However, this union is not without its share of hurdles [13, 14]. Centralized processing of data brings forth pressing concerns about user privacy. The computational resources required to handle such vast quantities of data are substantial. Not to mention, real-time processing demands can introduce significant latency, which, in scenarios like autonomous driving, can be critical.

13.1.2 Emergence of Federated Learning

The ML has witnessed rapid evolution, but its reliance on data to function optimally has raised intricate issues. Historically, ML's modus operandi involved centralized systems, where data from a multitude of sources would be amassed at a central repository [15, 16]. Once aggregated, intensive computations ensued, leading to the development of a global model. This traditional method, often seen as a foundational brick in ML's establishment, is however riddled with significant obstacles [17, 18].

- **Data Privacy Concerns**: The privacy implications of this centralized approach cannot be overstated. Every fragment of data, from mundane to the most intimate, travels from its source (like a user's device) to a central repository. This transfer reveals potentially sensitive insights. A case in point is health data from wearable devices. Information like heart rate, sleep patterns, and other biometrics are invaluable for medical research and health applications [19]. However, by relaying this raw data to a centralized system, we risk exposing deeply personal aspects of a user's life, making them vulnerable to potential data breaches or misuse.
- **Latency Issues**: As the IoT ecosystem has burgeoned, so has the volume of data generated. Transmitting this vast amount of data from countless devices to a single point invariably leads to latency [20]. In many applications, especially those requiring real-time responses like autonomous vehicles or medical devices, even the slightest delay can have dire consequences.
- **Bandwidth and Resource Strain**: Beyond latency, there's the sheer logistic challenge of transmitting vast volumes of data. This process consumes extensive bandwidth, posing both an infrastructural challenge and leading to increased operational costs [21]. Furthermore, centralized servers must be equipped with substantial computational power to process this influx, leading to resource inefficiency.

- Recognizing these challenges, there was a pressing need for a new approach, one that could harness the benefits of ML while circumventing its inherent challenges. The solution materialized in the form of federated learning [22].
- **Conceptualizing Federated Learning**: Federated Learning isn't just a technical solution; it's a paradigm shift. At its heart, Federated Learning reimagines how ML models are trained. Instead of the data traveling to the model, the model essentially travels to the data. Devices, or nodes, undertake local computations and generate model updates based on their individual datasets. These updates, typically much smaller in size than the raw data, are then transmitted to the central server [23, 24]. The central system aggregates these diverse updates, refines the global model, and redistributes it back to the devices.
- **Decentralization and Empowerment**: What Federated Learning offers is a sense of decentralization. Devices no longer play a passive role, merely contributing data. They are now active participants in the learning process, processing information locally and contributing to the global model's intelligence [25, 26]. This decentralization not only addresses the challenges of privacy, latency, and bandwidth but also brings a sense of empowerment to the edge devices, making them stakeholders in the broader learning process.
- **Evolutionary Context**: It's also essential to understand the evolutionary context of Federated Learning. As industries and academia realized the limitations of centralized learning, there was a concerted effort toward developing decentralized techniques [27]. Federated Learning emerged from this crucible of research, a culmination of efforts to make ML more sustainable, efficient, and respectful of user privacy.

13.1.3 Significance of Federated Learning in IoT

The IoT encapsulates a broad spectrum of devices, each churning out data incessantly. As we seek to derive value from this data, conventional centralized learning methods reveal their limitations, particularly in the IoT context [28, 29]. Federated Learning emerges as a transformative solution in this landscape, tailored to address the intricacies of IoT ecosystems. The significance of Federated Learning for IoT is multifaceted.

- **Data Privacy**: A cornerstone of Federated Learning's appeal is its staunch commitment to data privacy. IoT devices often handle sensitive data—from security cameras monitoring our homes to wearable devices tracking our biometrics. By keeping the raw data localized to the device, Federated Learning ensures that this sensitive information isn't exposed to potential external vulnerabilities. Instead, only model updates, which are abstract representations that do not divulge specifics about the underlying data, are shared [30]. This architectural decision inherently shields user data from potential breaches and misuse.
- **Reduced Latency**: The very nature of many IoT applications demands rapid responses. Whether it's a medical device adjusting drug dosages or a smart traffic light optimizing flow in real-time, latency can be a critical impediment. By emphasizing local data processing, federated learning ensures that the majority of analytical tasks happen on the device itself, allowing instantaneous decision-making [31]. The need to transmit data, await processing, and then receive feedback is eliminated, paving the way for truly real-time IoT applications.
- **Bandwidth Efficiency**: The transmission of vast volumes of raw data over networks is both resource-intensive and costly. In an IoT landscape with billions of devices, this could lead to significant network congestion. Federated Learning elegantly sidesteps this by only transmitting compact model updates, substantially reducing the data load on networks [32]. This not only

translates to monetary savings but also ensures a smoother, more efficient network operation, crucial for the seamless functioning of IoT ecosystems.

- **Scalability**: Traditional centralized systems encounter scalability challenges, especially as the number of devices and the volume of data surge. Central servers can become bottlenecks, hampering system performance [33]. Federated Learning, with its distributed nature, is inherently scalable. Each device operates semi-independently, and the system is not bogged down by central data storage or processing constraints. This means as the IoT landscape expands, Federated Learning systems can gracefully scale to accommodate even billions of devices without significant overhauls.
- **Enhanced Personalization**: A unique strength of Federated Learning in the IoT space is its ability to cater to specific nuances of individual devices or users. Since data processing occurs locally, each device has the capacity to tailor its operations based on the data it collects [34]. A smart refrigerator, for instance, can adapt to the dietary habits of its users. A home automation system can tweak its settings based on the unique routines of its residents. This localized learning ensures that IoT devices are not just smart but also personalized, offering bespoke experiences to users.
- **Redefining the Role of IoT Devices**: At a more philosophical level, Federated Learning reshapes our perception of IoT devices [35, 36]. No longer are they merely sensors or data loggers; they are transformed into intelligent entities capable of local computation and decision-making. They shift from being passive data repositories to active contributors in a broader ML ecosystem. This not only amplifies their utility but also sets the stage for a more decentralized, democratic digital future.

13.1.3.1 Key Contributions

In this chapter, we thoroughly navigate the interwoven paths of the ITIoT and federated learning. The narrative underscores four primary contributions:

- The chapter commences by painting the backdrop of IoT's emergence and how intelligence has become a cornerstone for modern devices [37]. It narrates the transformation from mere interconnected devices to entities that exhibit adaptability, predictive prowess, and enhanced interactivity. This evolution is spurred by advancements in ML and AI, setting the context for the subsequent emergence of federated learning.
- An in-depth exploration of federated learning is a pivotal segment of this discourse. The chapter dives into the limitations of traditional ML frameworks, emphasizing concerns like data privacy, latency, and resource constraints [38, 39]. In response, federated learning emerges as a paradigm shift, decentralizing the learning process and enabling devices to participate actively in model training. This is not just a technical adjustment but a reimagining of the broader ML landscape.
- This chapter accentuates federated learning's pivotal role in revolutionizing the IoT domain. It emphasizes the profound benefits, from staunch data privacy and reduced latency to enhanced personalization and scalability [40]. By localizing data processing and reducing the need for raw data transmission, federated learning addresses both the technical and philosophical challenges in IoT, paving the way for a future where devices are both intelligent and autonomous. Figure 13.1 presents the organization of the paper.

13.2 Basics of Federated Learning

As we venture deeper into the realms of data-driven decision-making and ubiquitous connectivity, understanding the foundational principles of emerging technologies becomes paramount.

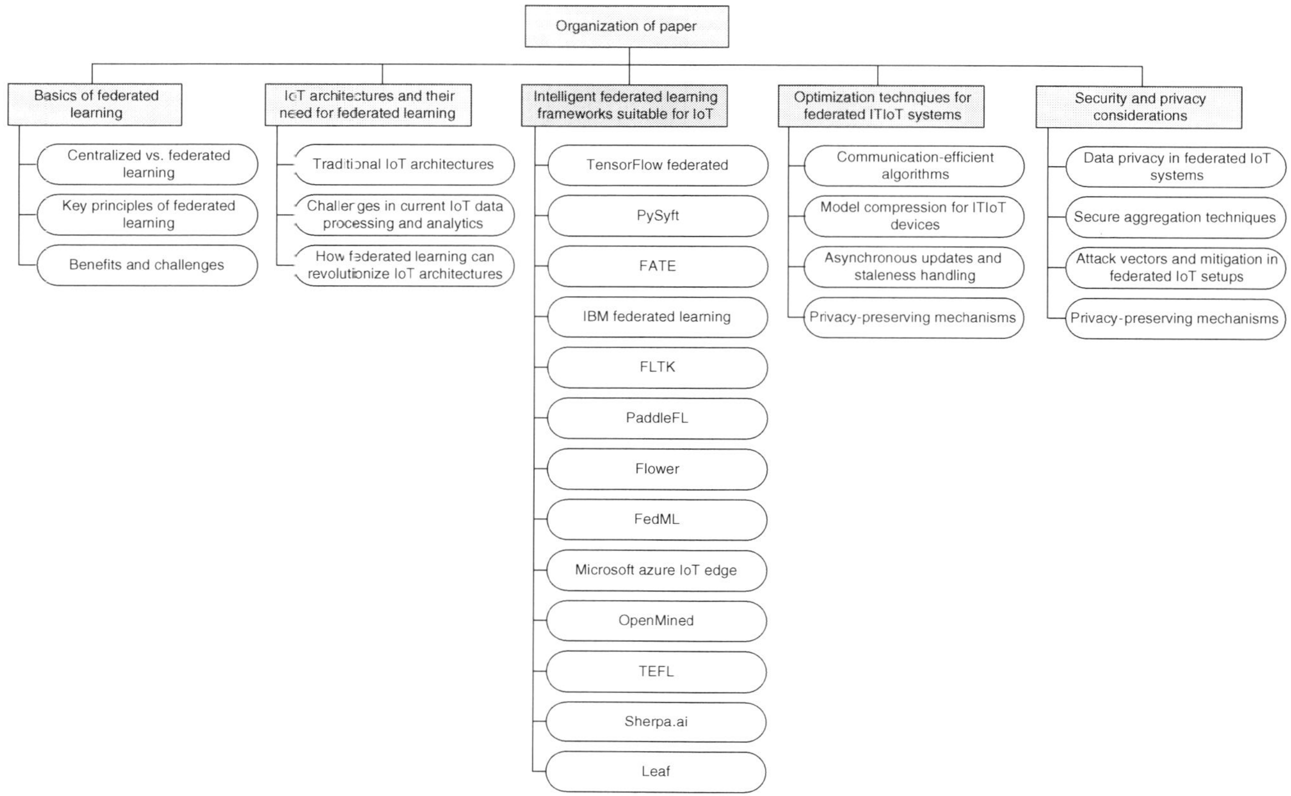

Figure 13.1 Organization of paper.

Among these, Federated Learning stands out as a key player in redefining how we approach ML, particularly in distributed environments [41–43].

13.2.1 Centralized vs. Federated Learning

- Centralized Learning
  At its core, centralized learning represents the more traditional model of ML. Here, data from various sources—be they sensors, users, or databases—is collected and sent to a central location or server [44]. This server undertakes the responsibility of parsing, processing, and training the ML model on this aggregated data. It's a monolithic approach, where the central system holds the computational and decision-making power.
  - **Data Flow**: The flow of information in centralized learning is linear. Data generated at the source is transmitted across networks to reach the central repository. After processing and model training, the intelligence (the trained model or insights derived from the data) is dispatched back to the source or to other relevant endpoints [45].
  - **Infrastructure**: Such a system is heavily reliant on powerful central servers. These servers must be equipped with substantial computational resources, both in terms of processing power and storage [46]. They serve as the brain of the operation, handling the major chunk of analytics and ML tasks.
  - **Privacy Concerns**: However, the centralized approach isn't without its pitfalls. Transmitting raw data across networks exposes sensitive and private information. As the data is stored and processed centrally, it becomes a potential target for breaches, unauthorized access, or cyber-attacks [47]. Such a concentrated pool of data is, by nature, a lucrative target for malicious entities.
  - **Scalability Issues**: Centralized systems also grapple with scalability. As data generation rates skyrocket, there's mounting pressure on the central servers to process this data in a timely fashion [48]. This can lead to performance bottlenecks, delays, and increased latency, especially as the system scales up.
- Federated Learning
  Federated Learning emerges as a counterpoint to centralized learning. In this approach, the emphasis shifts from the center to the edges [49]. Instead of sending raw data for centralized processing, devices or nodes process data locally and train models on the device itself. The central server's role changes from being a processor to an aggregator.
  - **Data Flow**: The data trajectory in Federated Learning is more decentralized [50]. Raw data stays on the device, ensuring immediate processing. Instead of data, devices send model updates or gradients to a central server. These updates, devoid of raw data, are then aggregated to refine a global model, which may be sent back to the devices for further enhancements.
  - **Infrastructure**: The infrastructure in Federated Learning is spread out. While there's still a central server, its primary role shifts to aggregating model updates. The heavy lifting, in terms of computation, is distributed among the edge devices [50]. These devices, be they smartphones, IoT devices, or edge servers, become mini computation hubs, collaboratively contributing to the learning process.
  - **Enhanced Privacy**: One of the standout features of Federated Learning is its intrinsic privacy preservation [51]. Since raw data never leaves the source device, user data's exposure risk is substantially minimized. What travels across the network are model updates, which are typically nonreversible and don't carry the explicit details of the raw data.

- **Natural Scalability**: Federated Learning's distributed design is inherently scalable. As more devices join the network, each brings its computational resources [52]. The system is adaptive, accommodating a growing number of devices without the need for massive central infrastructure upgrades. This decentralized approach ensures a more resilient, flexible, and adaptive learning system.

13.2.2 Key Principles of Federated Learning

- **Local Computation**: At the heart of Federated Learning is the principle of local computation. Devices, whether they are smartphones, sensors, or other IoT devices, utilize their computational power to process data. This is crucial not just for speed and responsiveness, but also because it ensures data does not unnecessarily travel across networks, thereby enhancing data privacy [53]. Leveraging local device capabilities means that data can be processed in near real-time, even in environments where latency is critical.
- **Collaborative Learning**: While devices work on their localized data, they're not working in silos. Each device contributes to a more substantial, global understanding [54]. This amalgamation ensures that while a device learns and adapts to its immediate environment or user, it also benefits from broader patterns identified across the network of devices. It's a harmonious blend of local insights and collective wisdom.
- **Data Stays Local**: One of Federated Learning's defining principles is the retention of raw data at its source [55]. This approach stands in stark contrast to traditional models where data is consistently sent to central servers. By keeping data local, Federated Learning directly addresses numerous concerns, from privacy issues to the actual logistical and bandwidth challenges associated with large-scale data transfer.
- **Weight Aggregation**: Once the local data is processed, and learning occurs on the device, what travels back to the central server isn't the data but the learnings from it [56]. In technical terms, these are often the model weights or updates. The central server's task is to aggregate these updates from all devices, synthesizing them into a consensus model that encapsulates the collective learning of all devices.
- **Continuous Iteration and Update**: The process of learning and updating in a Federated Learning system is continuous. As devices gather more data, they process and learn from it, sending their updates to the central system [57]. In return, they may receive refinements based on the broader network's learnings. This iterative process ensures that the AI models in use are dynamic, adaptable, and reflective of the latest data.
- **Secure Communication**: With numerous devices communicating with central servers and potentially with each other, secure communication is non-negotiable [58]. Federated Learning systems often integrate advanced encryption techniques, ensuring that the data in transit (model updates) is secure. Techniques like differential privacy can also be applied, adding noise to the shared updates, further ensuring that sensitive information cannot be reverse-engineered.

13.2.3 Benefits and Challenges

- Benefits
  - **Data Privacy**: Federated Learning's architecture is a boon for data privacy. By ensuring that raw data remains on the source device, the paradigm minimizes risks associated with breaches during data transfer or unauthorized access in centralized storage systems [59]. This decentralized approach aligns with strict data protection regulations worldwide, offering a solution for companies to employ ML without compromising user privacy.

  - **Efficiency**: The traditional approach of constantly transmitting voluminous data to central servers puts immense strain on network bandwidth and resources [60]. Federated learning, by focusing on transmitting lightweight model updates or parameters, significantly reduces this burden. The result is not only monetary savings due to reduced data transfer costs but also a smoother and more efficient network operation.
  - **Scalability**: Federated learning is inherently designed for scale. Its decentralized nature means that as the number of devices grows, the system can adapt without massive overhauls [61]. This scalability ensures that as IoT ecosystems expand or as more users adopt digital devices, Federated Learning systems can handle the growth seamlessly.
  - **Real-Time Responsiveness**: Decision-making at the edge, or on the device, ensures that responses are almost instantaneous [62]. This speed is crucial in situations where delays could be detrimental. For example, in healthcare, wearable devices employing Federated Learning can quickly detect anomalies in a patient's vitals and alert them or their caregivers, potentially saving lives.
  - **Personalization**: Each device in Federated Learning can tailor its model based on the unique interactions and behaviors of its user [63]. This ability means that over time, devices become highly attuned to individual user preferences, needs, and behaviors, allowing for an unparalleled personalized experience.
- Challenges
  - **Heterogeneity**: The diversity of devices in a Federated Learning system presents a significant challenge. Different devices have varying computational powers, battery lives, and data generation patterns. Ensuring a balanced learning process across this diverse ecosystem requires sophisticated algorithms and protocols [64, 65].
  - **Communication Overhead**: While the data transfer is reduced in Federated Learning compared to centralized systems, the frequent exchange of model updates introduces its communication challenges [66]. Ensuring timely and efficient communication, especially in large-scale systems, can be complex.
  - **Model Aggregation Complexity**: One of the central tasks in Federated Learning is the aggregation of model updates from different devices into a cohesive global model. This process can be computationally intensive, especially when dealing with diverse or conflicting updates [67].
  - **Security Concerns**: A distributed system increases the surface area for potential attacks. Beyond just ensuring secure communication channels, there's a need to safeguard against malicious actors who might introduce skewed updates to sabotage the global model [68]. Ensuring the integrity of updates and protecting against potential adversarial attacks is crucial.
  - **Technical Maturity**: Federated Learning, as a concept, is still maturing. Many of its processes and techniques are subjects of active research [69]. Adopters need to stay abreast of the latest advancements and be prepared for iterative implementations as the field evolves and new best practices emerge.

13.3 IoT Architectures and Their Need for Federated Learning

13.3.1 Traditional IoT architectures

IoT architectures have traditionally been established on a centralized paradigm where most of the heavy computational tasks are offloaded to centralized data centers or clouds. This design has its roots in the early days of computing when resources were scarce, and centralization was the most efficient way to manage and process data [70, 71]. However, as the Internet of Things grew in size and complexity, this architecture began to show its limitations. Let's delve deeper into the layers of this traditional structure.

- **Device Layer**: The Device Layer stands out as a marvel of diversity, a testament to the wide array of gadgets that are part of our day-to-day existence. This includes everything from the smart appliances in our homes—like refrigerators, wearables, and light fixtures—to the advanced sensors utilized in various industries [72]. As the bedrock of the IoT landscape, these devices primarily function as the conduits for real-world data collection. Consider, for example, the IoT setups in agriculture: they may utilize soil moisture sensors, vigilantly monitoring the moisture to assist in streamlined irrigation practices. However, it's crucial to understand that these devices aren't without their restrictions. Many face challenges linked to energy consumption, computational capacity, and storage space [73]. This means a battery-operated soil sensor can't endlessly transmit data or undertake intricate computational tasks. But advancement doesn't halt; the dawn of localized intelligence in some contemporary devices bestows them with the prowess to carry out basic data processing. Today's cutting-edge smart cameras can discern motion, activating recording only during periods of movement. When discussing these devices, energy management remains paramount, especially given that many are reliant on batteries [74]. Adaptive energy-saving mechanisms, which prompt devices to awaken only during specific scenarios or intervals, ensure that battery longevity is maintained. This layer has several characteristics as below.
  - **Nature and Variety**: The Device Layer is the most diverse in terms of the types of devices it incorporates. From smart refrigerators, wearables, and light bulbs in a household setting to large-scale sensors in industries, these devices serve as the primary data sources.
  - **Functionality**: The primary function is to capture real-world data. For instance, agricultural IoT setups might involve soil moisture sensors, which constantly monitor moisture levels to aid in efficient irrigation.
  - **Constraints and Limitations**: Devices at this layer often operate under energy, computation, and storage constraints. A battery-powered soil sensor, for instance, can't afford to be constantly transmitting data or running complex algorithms.
  - **Localized Intelligence**: Some devices, especially the newer ones, come equipped with a degree of computational capability allowing them to process data at a rudimentary level. For instance, modern smart cameras can detect motion and only record when there's activity.
  - **Energy Management**: Given that many IoT devices are battery-operated, energy management is a pivotal concern. Some devices employ energy-saving modes, waking up only at intervals or under specific conditions, to extend battery life.
- **Gateway and Network Layer**: Acting as the linchpin of the IoT system, the Gateway and Network Layer serves a critical purpose, creating a nexus between the device layer and the expansive realm of the cloud. By aggregating data from myriad sources, it takes on the pivotal role of determining the nuances of data transmission to the cloud, be it the timing or method. This isn't just a mindless pass-through, though [75]. There's room for elementary data processing here; for instance, rather than incessantly sending every minuscule temperature change recorded by a sensor, the gateway could strategically compile hourly readings, sending a mean value or only signaling when a predetermined temperature boundary is breached. The significance of gateways also extends into the realms of security [76]. By adopting encryption methodologies and adhering to specialized communication protocols like Message Queuing Telemetry Transport (MQTT)

or Constrained Application Protocol (CoAP), they fortify IoT communications. The evolution of edge computing is pushing gateways to not merely transmit but also analyze data, paving the way for on-the-spot decisions without a perpetual reliance on a central cloud. This has life-saving implications [77, 78]: an astute IoT gateway, for instance, could identify an irregularity in a machine and promptly deactivate it even before communicating with the primary system. Ensuring seamless, uninterrupted operations is paramount, so redundancy measures are in place. If a gateway encounters failure, backup systems can seamlessly take the reins. Moreover, the strategic choice of network topologies—be it star, mesh, or even a hybrid approach—is driven by specific deployment scenarios. For large-scale operations like monitoring expansive farmlands, mesh networks, where devices collaborate to relay information, might be the most apt. This layer possesses key characteristics as below.

- **Bridging the Gap**: This layer often acts as a bridge between the device layer and the cloud. It aggregates data from multiple sources and acts as a relay point, determining when and how data should be transmitted to the cloud.
- **Processing and Filtering**: Some preliminary data processing can happen at this level. For example, instead of sending every temperature reading from a sensor, the gateway might aggregate hourly readings and send an average or only transmit when the temperature crosses a specific threshold.
- **Security and Protocols**: Gateways play a crucial role in ensuring secure data transfer. They often employ encryption techniques and use specific communication protocols, like MQTT or CoAP, optimized for IoT communications.
- **Role of Edge Computing**: Recently, there's been a push toward edge computing, where the gateway doesn't just transmit but also processes data. This means decisions can sometimes be made without always consulting the central cloud, thus offering faster responses. An intelligent IoT gateway might detect an anomaly in machinery and shut it down immediately before consulting the central system.
- **Redundancy and Reliability**: This layer often employs redundant systems to ensure data integrity and continuous operation. If one gateway fails, another can take over to prevent data loss or downtime.
- **Network Topologies**: Depending on the deployment scenario, different network topologies like star, mesh, or hybrid might be adopted. For instance, in a vast agricultural farm with sensors spread out, a mesh network could be more efficient as devices relay data through each other.

• **Cloud/Data Center Layer**: When it comes to the Cloud/Data Center Layer, it's all about monumental computational prowess. Nestled within the servers of these centers lies an unparalleled capacity for intricate tasks, be it intricately training ML models or delving deep into data analytics [79, 80]. Storage isn't a challenge here; these hubs can effortlessly accommodate petabytes of data. Visualize an IoT mechanism diligently overseeing an entire metropolis's vehicular movements—the quantum of data generated daily would be colossal. Safely and efficiently storing this data is this layer's forte. But storage is just the beginning; this data, once processed, morphs into actionable insights [81]. Taking the instance of health wearables, not only can the cloud layer trace health trajectories of users, but it can also juxtapose them with larger data sets, proffering tailored recommendations or even forecasting potential health challenges. This isn't a one-way street. Processed insights can journey back to the devices, enhancing their utility [82]. The health wearable, upon discerning a user's sedentary lifestyle, might nudge them toward more frequent physical activity. Leveraging virtualization, these centers can astutely use resources, allowing a plethora of virtual systems to cohabit on a singular physical server. Tools of orchestration ensure this coexistence remains harmonious. Given the sacrosanct nature of data, redundancy isn't just a luxury but a necessity. Ensuring data's sanctity means creating replicas across diverse locations, acting as a bulwark against unforeseen adversities. But the cloud's role isn't insular; it often dovetails with other systems—whether they're industrial Enterprise Resource Planning (ERP) setups or hospital-based health management systems—ensuring a holistic utilization of IoT data [83]. Finally, the culmination of these processes is often relayed to end-users via intuitive dashboards or applications. Envision an energy oversight system offering real-time energy consumption visualizations, empowering administrators to make judicious decisions. It comprises of several key aspects as mentioned below.

- **Massive Computational Power**: This is where the most robust computational tasks occur. Cloud servers have the luxury of vast computational resources, making them ideal for tasks like training ML models or running sophisticated data analytics.
- **Storage**: These centers house vast storage systems capable of storing petabytes of data. An IoT system monitoring an entire city's traffic, for instance, would generate massive amounts of data daily. Storing such data efficiently and reliably is a primary task of this layer.
- **Analytics and Insights Generation**: Once data is stored and processed, it's turned into actionable insights. For the smart health wearable example, the cloud might not only track a user's health trends but also compare them with other users, provide recommendations, or even predict potential health issues.
- **Feedback Loop**: The processed data or insights can be sent back to the devices. For instance, if a user's health trends indicate a need for more exercise, their smartwatch might start offering more frequent exercise reminders or tailored fitness challenges.
- **Virtualization and Orchestration**: To efficiently utilize resources, these centers often employ virtualization techniques, allowing multiple virtual systems to run on a single physical server. Orchestration tools ensure that these virtual entities interact seamlessly.
- **Data Redundancy and Backup**: Given the critical nature of data, cloud centers often have backup mechanisms. Data might be replicated across multiple physical locations to safeguard against failures or disasters.
- **Integration with Other Systems**: These centers aren't standalone entities. They often integrate with other IT systems, be it ERP systems in industries or health management systems in hospitals. This integration ensures that IoT data can be holistically used for varied purposes.
- **User Interface and Interactivity**: The insights and data visualizations generated here are often presented to end-users through dashboards or apps. For instance, an energy management system might provide building managers with a real-time view of energy consumption patterns, helping them make informed decisions.

13.3.2 Challenges in Current IoT Data Processing and Analytics

As the digital landscape evolves, IoT ecosystems' growing complexity highlights the limitations of traditional data processing and analytics mechanisms [84–86]. Here's a comprehensive dive into these challenges [87, 88].

- Latency
  - **Real-Time Decision Making**: In many IoT scenarios, the ability to process data and make decisions in real-time is crucial. Consider healthcare wearables that detect irregular heart rhythms; any latency in processing and responding could be detrimental.

- **Geographical Challenges**: Sometimes, the distance between the IoT devices and the centralized servers can be vast. If sensors in a remote agricultural field in Asia are communicating with servers in North America, the physical distance can introduce noticeable delays.

- Bandwidth Overload
  - **Data Granularity**: Not all data generated by IoT devices is equally relevant. For example, a surveillance camera might record hours of footage where nothing significant happens. Transmitting all this to the cloud is not just unnecessary but can also strain bandwidth.
  - **Peak Times**: There can be periods when multiple devices try to communicate simultaneously, leading to traffic spikes. Such peaks can overwhelm network resources unless properly managed.
- Security and Privacy
  - **Data In-Transit**: While data at rest (stored data) can be encrypted and protected, data in transit (during transmission) can be intercepted if not adequately secured. Man-in-the-middle attacks, where attackers secretly relay or possibly alter the communication between two parties, are concerns here.
  - **Device Vulnerabilities**: Many IoT devices might not have robust built-in security features. These can become weak links in the chain, where attackers can compromise a device to gain access to the broader network.
- Scalability
  - **Heterogeneity**: The IoT world is incredibly diverse. Devices from different manufacturers, running different software versions, and having varied computational capabilities all need to coexist and communicate seamlessly. Scaling such a diverse ecosystem can be daunting.
  - **Infrastructure Limitations**: There are physical limits to how much data a centralized server or data center can handle. As more devices come online, there's a risk of these centers becoming overwhelmed unless they're continuously upgraded.
- Cost
  - **Operational Expenses**: Running vast data centers 24/7 involves significant costs, not just in terms of electricity but also cooling, maintenance, and staffing. As data volumes grow, these operational expenses can skyrocket.
  - **Data Redundancy**: To safeguard against data loss, many systems store multiple copies of data. While this is good for reliability, it increases storage costs and demands more extensive data management strategies.
  - **Network Costs**: Continuously transmitting vast volumes of data across networks isn't free. Especially in scenarios where cellular or satellite networks are involved, the costs can accumulate quickly.

13.3.3 How Federated Learning Can Revolutionize IoT Architectures

The vast IoT ecosystem, characterized by billions of devices and sensors constantly communicating and generating data, demands an adaptive approach to data processing and analytics [89, 90]. Federated learning, with its emphasis on local computation and collaborative learning, can pave the way for a more resilient and efficient IoT landscape [91–93].

- Edge Computing and Decentralization
  - **Reduced Central Dependency**: Federated Learning decentralizes the learning process, reducing dependency on central servers. This becomes especially crucial in areas with limited or unreliable Internet connectivity. For example, agricultural IoT devices in remote farms

could benefit immensely from local data processing without constantly relying on distant data centers.
- **Leveraging Local Insights**: Some data patterns might be extremely localized and might get diluted when looked at at a broader level. By processing data at the edge, specific local patterns can be identified and acted upon immediately.

- Enhanced Data Privacy
  - **User Trust**: As more consumers become wary of how their data is used, devices that guarantee on-device data processing can become more appealing. A fitness tracker that processes health data locally, for instance, might be preferred over one that constantly sends data to unknown servers.
  - **Regulatory Compliance**: Various regions have introduced stringent data protection regulations. Devices that minimize data transfer can more easily comply with such rules, reducing legal and operational challenges for manufacturers.
- Bandwidth Efficiency
  - **Smart Data Transfer**: Instead of flooding the network with voluminous raw data, devices can send compact, meaningful model updates. For instance, a network of environmental sensors spread across a forest might only send alerts or significant changes, rather than continuously streaming all data.
  - **Network Longevity**: By conserving bandwidth, the longevity and efficiency of network resources are enhanced, ensuring smoother communication even as the number of devices multiplies.
- Real-Time Decision Making
  - **Proactive Responses**: In situations where timely responses are crucial, local data processing can be a lifesaver. Imagine smart sensors in a factory detecting machine anomalies. With Federated Learning, the sensors could shut down a malfunctioning machine immediately, preventing potential accidents or further damage.
  - **Reduced Feedback Loop**: Without the need to communicate with a central server, the feedback loop for many applications gets significantly shortened. An IoT-enabled climate control system in a building could adjust temperatures in real-time based on occupant preferences and external weather conditions, ensuring optimal comfort levels.
- Cost-Efficiency
  - **Infrastructure Savings**: By reducing the dependency on massive centralized servers, organizations can make significant infrastructure savings. It would also cut down on expenses related to data storage, backup, and retrieval.
  - **Adaptive Models**: Federated Learning not only reduces the amount of data being sent but also ensures that the models are continuously updated and refined based on real-world insights from each device. This dynamic learning can lead to more efficient operations and less wastage.

13.4 Intelligent Federated Learning Frameworks Suitable for IoT

IoT, with its diverse range of devices and vast data generation, necessitates frameworks that can handle the intricacies of federated learning at scale. Several federated learning frameworks have been developed to meet the unique challenges posed by IoT environments [94, 95]. These frameworks help in decentralizing the training process, optimizing model updates, and ensuring secure and efficient communication among devices [96, 97].

13.4.1 TensorFlow Federated

TensorFlow Federated (TFF) is a cutting-edge open-source framework spearheaded by Google, constructed as a specialized extension to the TensorFlow platform [98]. This enhancement was done specifically to cater to the intricacies and demands of federated environments.

- Features for ITIoT
  - **Scalability**: At its core, TFF is powered by TensorFlow, a highly scalable ML framework. When applied to the Intelligent Internet of Things (ITIoT) sector, TFF's scalability ensures it can effortlessly handle vast networks of interconnected devices, making it an invaluable asset in large-scale intelligent operations.
  - **Edge Compatibility**: One of the significant breakthroughs of TFF is its compatibility with TensorFlow Lite. TensorFlow Lite is a lightweight solution, designed explicitly for on-device ML. This means that TFF, in conjunction with TensorFlow Lite, can facilitate training and inference right on the edge devices. For ITIoT, this feature is indispensable. Given the remote and often decentralized nature of many intelligent devices, being able to process and infer data on the spot can make operations smoother and more efficient.
  - **Flexible API**: The dynamic nature of industries requires solutions that are not one-size-fits-all. Recognizing this, TFF offers a flexible API that grants developers the liberty to architect custom federated algorithms. This flexibility goes beyond just the standard training routines, ensuring that diverse intelligent challenges can be addressed with tailored solutions.
- Use Cases
  - **Real-Time Quality Checks**: In modern manufacturing lines, the ability to identify defects or issues in real-time can save significant costs and time. Using TFF, devices on the manufacturing line can be trained to instantly detect any anomalies, ensuring that only top-quality products make it to the end of the line.
  - **Predictive Maintenance**: Heavy machinery is a staple in many industries, and their maintenance can be both time-consuming and expensive. TFF can be used to train devices to monitor the machinery's operational metrics continuously. By doing so, these devices can predict when a machine might fail or require maintenance, allowing for timely interventions.
  - **Supply Chain Optimization**: The supply chain is a complex network of interconnected stages. By deploying TFF in the ITIoT devices scattered across the supply chain, industries can gain real-time insights into the flow of goods, potential bottlenecks, and areas of inefficiency. With these insights, the entire supply chain can be optimized for maximum productivity and profitability.

13.4.2 PySyft

PySyft is an advanced library that augments the PyTorch framework to facilitate encrypted, privacy-preserving ML [99]. It was developed by the OpenMined community, which is dedicated to democratizing access to private data for ML purposes without violating the data's privacy. PySyft represents a paradigm shift in how data access is approached, emphasizing data privacy and security while still harnessing the value locked in the data.

- Features for ITIoT
  - **Encrypted Computation**: PySyft enables data operations and computations to be carried out on encrypted data. This means models can be trained and inferences made without ever having to access the raw, unencrypted data. This approach, often termed "encrypted deep learning," is critical in ITIoT where data security and privacy are paramount.

  - **Support for Diverse Privacy Techniques**: Beyond encrypted computations, PySyft also supports a plethora of other privacy-enhancing techniques. This includes differential privacy, which ensures individual data points cannot be singled out, and federated learning, where model training happens across multiple devices without centralizing data. Such techniques make PySyft a versatile tool in privacy-preserving ML.
  - **Modular Design**: The architecture of PySyft is modular, allowing developers to harness its capabilities for various use cases. It provides tools and hooks to extend the functionalities of deep learning frameworks, such as PyTorch, and ensures that developers can tailor solutions according to specific ITIoT needs.
  - **Active Community Support**: Given its backing by the OpenMined community, PySyft benefits from regular updates, a vast array of documentation, tutorials, and an active community ready to assist with queries. This ensures the framework remains up-to-date with the latest in privacy-preserving techniques.
- Use Cases for ITIoT
  - **Encrypted Analytics in Smart Grids**: Smart grids generate vast amounts of data. With PySyft, energy companies can perform analytics on this data without compromising user privacy, allowing for optimized energy distribution and forecasting.
  - **Private Collaborative Learning Among Different Manufacturing Units**: In a scenario where multiple manufacturing units of a large conglomerate want to collaboratively train an ML model without sharing proprietary data, PySyft can facilitate such collaborations.
  - **Secure Data Pooling in Supply Chains**: Companies in a supply chain can leverage PySyft to pool data for analytics without revealing sensitive information to other parties, leading to optimized logistics and operations.

13.4.3 Federated AI Technology Enabler

FATE is an open-source federated learning platform, created to offer secure and privacy-preserving distributed computing capabilities [100]. Initiated by Webank's AI Department, FATE aims to provide a robust foundation to support collaborative computations in heterogeneous environments, especially where data privacy and security are crucial.

- Features for ITIoT
  - **Robust Security Protocols**: FATE heavily emphasizes secure multiparty computation (SMPC). SMPC allows multiple parties to compute functions over their inputs while keeping those inputs private. This is a monumental shift from traditional methods where data needs to be centralized or openly shared to derive insights.
  - **Cross-Platform Compatibility**: FATE boasts a design which allows it to operate seamlessly across various hardware setups and infrastructures. Whether it's edge devices, on-premise servers, or cloud platforms, FATE ensures federated learning operates efficiently. This is particularly beneficial in ITIoT scenarios where device heterogeneity is the norm.
  - **High-Performance Computing**: To cater to the high data-generation rates and processing needs of ITIoT environments, FATE is optimized for speed and efficiency. Its distributed architecture ensures computational workloads are spread effectively, making the most of available resources.
  - **Interoperability**: FATE can be interfaced with popular ML frameworks, ensuring a smoother integration with existing ML pipelines and applications.

  - **Granular Control Over Data**: FATE provides data owners with the ability to control how their data is used in federated computations, down to the finest granularities. This ensures compliance with regulatory requirements and ethical considerations.
- Use Cases for ITIoT
  - **Interfactory Analytics Without Sharing Raw Data**: Consider multiple factories wanting to optimize their operations based on collective insights. With FATE, they can collaboratively train ML models without exposing their proprietary or sensitive data to each other.
  - **Joint Venture Collaborations**: In industries where joint ventures or collaborations between competitors are common for specific projects, FATE allows these entities to benefit from combined data insights without revealing trade secrets.
  - **Research in Intelligent Contexts**: For R&D departments aiming to improve processes or develop new products, FATE provides a platform where data from various sources can be utilized without direct access, ensuring data privacy while fostering innovation.

13.4.4 IBM Federated Learning

IBM Federated Learning is an innovative approach to ML model training developed by IBM [101]. Unlike traditional centralized training methodologies, IBM's federated learning framework focuses on training models where the data resides. This decentralized approach allows organizations to harness insights from vast and diverse datasets spread across multiple nodes without having to centralize the data, ensuring privacy and security.

- Features for ITIoT
  - **Integration with IBM Cloud**: IBM Federated Learning is deeply integrated with IBM Cloud, offering a robust platform that benefits from the security, flexibility, and scale of IBM's cloud infrastructure. This tight integration is especially beneficial for ITIoT scenarios that rely on hybrid cloud architectures for operational efficiency.
  - **Model Lifecycle Management**: IBM FL provides a comprehensive set of tools and workflows to manage the entire lifecycle of federated learning models—from model development and training to deployment and monitoring. This holistic approach ensures that ITIoT applications are always powered by the most up-to-date and optimized models.
  - **Differential Privacy**: A standout feature of IBM's federated learning approach is its commitment to ensuring data privacy. By supporting differential privacy techniques, IBM ensures that data contributing to the learning process remains private, abstracted, and nonidentifiable.
  - **Scalability**: Given the diverse and large-scale nature of ITIoT deployments, IBM's federated learning framework is designed to efficiently scale and handle vast networks of devices, sensors, and nodes.
  - **Robust Security Mechanisms**: In addition to differential privacy, IBM FL employs encrypted communications, robust authentication and authorization procedures, and other state-of-the-art security mechanisms to protect data and model integrity.
- Use Cases for ITIoT
  - **IoT-Driven Asset Management**: Industries can use IBM FL to derive insights from sensors placed on various assets across different locations without centralizing the data, optimizing asset utilization, and predictive maintenance schedules.
  - **Real-Time Analytics for Shipping and Logistics**: By analyzing data from sensors on vehicles, containers, and infrastructure across various nodes, logistics companies can optimize routes, reduce operational costs, and increase efficiency.

- **Cloud-to-Edge AI Deployments**: Factories and other intelligent settings can use IBM Federated Learning to train models on the cloud using data from multiple edge devices, and then deploy the trained models back to the edge for real-time decision-making.
- **Collaborative Research and Development**: Companies can collaborate on R&D projects without sharing raw data, instead, they can share insights derived from federated learning processes, ensuring proprietary information remains confidential.

13.4.5 Federated Learning Toolkit

The FLTK emerges as a comprehensive solution designed to expedite the development, training, and deployment of federated learning models [102]. With the increasing demand for decentralized models due to privacy and data decentralization, FLTK fills the gap by providing a structured framework tailored to these challenges.

- Features for ITIoT
  - **Rapid Prototyping**: In the fast-paced realm of ITIoT, speed is paramount. FLTK stands out by offering predefined templates and modules that enable developers to quickly design, train, and deploy federated learning models tailored to specific ITIoT scenarios. This accelerates time-to-value for ITIoT projects.
  - **Extensive Documentation**: A significant challenge with the adoption of new technologies in an intelligent setup is the learning curve associated with them. FLTK addresses this by providing thorough documentation, tutorials, and best practice guides, ensuring that ITIoT professionals can swiftly and confidently adopt the toolkit for their use cases.
  - **Optimized for Performance**: Recognizing the computational demands of ITIoT, where data is often vast and time-sensitive, FLTK places a high emphasis on performance. It ensures efficient computations, streamlined communication, and minimal latency, critical for real-time ITIoT applications.
  - **Modularity and Customization**: Every intelligent setup has its unique challenges and requirements. FLTK's modular design allows businesses to tailor the toolkit to their specific needs, whether it's integrating with existing systems or accommodating unique data structures.
  - **Security Enhancements**: With data security being paramount in ITIoT, FLTK incorporates advanced security protocols to ensure data integrity and privacy during the federated learning process.
- Use Cases for ITIoT
  - **Rapid Deployment in Manufacturing**: Factories can swiftly implement FLTK to deploy federated learning models that optimize production lines, detect anomalies, and predict maintenance needs without centralizing sensitive operational data.
  - **AI-driven Quality Control**: In industries where product quality is paramount, FLTK can power real-time quality checks. By processing data at the source (e.g., cameras or sensors on the production line), decisions about product quality can be made instantaneously, reducing waste and ensuring consistent product quality.
  - **Supply Chain Optimization**: With sensors deployed throughout the supply chain, from raw materials to finished product delivery, FLTK can help businesses derive insights from each node independently, optimizing inventory management, reducing costs, and improving delivery times.

- **Energy Management**: For industries with significant energy consumption, FLTK can analyze data from various sensors to optimize energy use, reduce costs, and decrease carbon footprints.

13.4.6 PaddleFL

PaddleFL is an extension of the PaddlePaddle deep learning platform, specifically designed to cater to federated learning applications [103]. PaddlePaddle itself is a comprehensive platform developed by Baidu, and with PaddleFL, the focus shifts toward decentralized and privacy-preserving ML, making it a suitable contender in the ITIoT landscape.

- Features for ITIoT
  - **Comprehensive Federated Learning Algorithms**: PaddleFL offers a rich set of federated learning algorithms out-of-the-box. This diverse toolkit allows ITIoT applications to choose the best-fit approach for their specific needs.
  - **Vertical and Horizontal Federated Learning Support**: Depending on the data distribution and collaboration needs, ITIoT setups might require vertical or horizontal federated learning. PaddleFL supports both, ensuring adaptability to various intelligent scenarios.
  - **Privacy-Preservation**: Recognizing the importance of data privacy in intelligent applications, PaddleFL integrates cryptographic techniques and differential privacy, ensuring that data remains secure during the federated learning process.
  - **Flexible Scheduling**: PaddleFL provides a scheduler for federated learning tasks, ensuring efficient coordination among various devices in ITIoT systems. This scheduler guarantees that learning tasks are executed in an orderly and optimized manner.
  - **High Scalability**: PaddleFL can efficiently handle the vast amount of data generated in ITIoT environments, thanks to its optimized algorithms and efficient communication protocols.
- Use Cases for ITIoT
  - **Predictive Maintenance**: Using PaddleFL, industries can deploy federated models that predict when machinery might fail or require maintenance. By analyzing data at the source, businesses can minimize downtime and improve operational efficiency.
  - **Supply Chain Collaboration**: Companies across a supply chain can collaboratively optimize operations without sharing sensitive data. PaddleFL allows for joint model training where each entity retains its data, ensuring both collaboration and data privacy.
  - **Energy Consumption Analysis**: For sectors like utilities or manufacturing, PaddleFL can process data from various locations to optimize energy consumption without centralizing sensitive operational data.
  - **Smart Factory Optimization**: In factories with a myriad of sensors and devices, PaddleFL can assist in real-time decision-making, optimizing production lines and ensuring product quality, all while preserving data at the edge.
  - **Safety and Compliance Monitoring**: By analyzing data locally in factories or intelligent setups, PaddleFL-powered models can ensure that operations adhere to safety and regulatory standards.

13.4.7 Flower

Flower is a versatile open-source federated learning framework designed to facilitate efficient and scalable ML in decentralized environments [104]. Built with an aim to make federated learning more accessible and standardized, Flower encourages developers and researchers to implement and experiment with federated learning strategies across diverse scenarios, including ITIoT.

- Features for ITIoT
  - **Framework Agnostic**: One of Flower's standout features is its ability to be agnostic to the underlying ML framework. This means that industries can leverage Flower irrespective of whether they are using TensorFlow, PyTorch, or any other platform.
  - **Highly Scalable**: Designed with scalability in mind, Flower can effortlessly accommodate a large number of devices, making it apt for the expansive device landscape of ITIoT.
  - **Efficient Communication Protocols**: Flower emphasizes efficient data communication to minimize bandwidth consumption. This is particularly beneficial for ITIoT, where excessive data transfers can be costly and resource-intensive.
  - **Customizable Federated Learning Strategies**: Flower allows developers to tailor federated learning processes to fit specific requirements, granting flexibility for diverse ITIoT applications.
  - **Robust Security Features**: While it emphasizes streamlined communications, Flower does not compromise on security. It offers mechanisms to ensure data remains private and secure during the federated learning process.
- Use Cases for ITIoT
  - **Asset Performance Monitoring**: Industries can use Flower to create federated models that continuously monitor the performance of assets spread across different locations, training locally and aggregating insights without sharing raw data.
  - **Real-Time Process Optimization**: Flower can help industries in real-time optimization by training models on the edge, analyzing data from various sensors, and making immediate adjustments to processes.
  - **Supply Chain Efficiency**: Through federated analytics, industries within a supply chain can collaboratively optimize operations using Flower without exposing sensitive data to each other.
  - **Safety Protocols Adherence**: In intelligent setups, ensuring safety can be paramount. Flower can be leveraged to develop models that monitor and ensure adherence to safety protocols across decentralized units.
  - **Energy Management**: For industries with significant energy consumption, Flower-powered models can analyze data from multiple facilities to strategize and optimize energy use without central data pooling.

13.4.8 FedML

FedML is an open-source research library and benchmarking platform that's dedicated to federated learning [105]. Designed with researchers and developers in mind, it promotes advancements and innovations in the domain of federated learning, distributed optimization, and ML in general. While it's versatile across various domains, its features can be particularly beneficial for the ITIoT landscape.

- Features for ITIoT
  - **Multiplatform Support**: FedML can operate on a wide range of platforms, from local standalone machines to distributed GPU clusters to mobile devices. This versatility ensures its readiness for varied ITIoT environments.
  - **Extensibility**: FedML is designed to be modular and extensible, making it easier for developers to integrate new algorithms, models, or datasets tailored to specific ITIoT needs.

 - **Optimized Algorithms**: The framework brings optimized federated learning algorithms out of the box, ensuring efficient training and performance, crucial for real-time ITIoT applications.
 - **Diverse Communication Backends**: With support for different communication backends like MPI, WebSocket, and more, FedML can adapt to various networking conditions, ensuring robustness in diverse intelligent scenarios.
 - **Rich Dataset and Model Zoo**: FedML provides a collection of datasets and pretrained models, allowing for quick experimentation and prototyping, thereby accelerating ITIoT application development.
- Use Cases for ITIoT
 - **Decentralized Intelligent Quality Control**: By deploying federated models across manufacturing units, industries can ensure quality standards are maintained without the need for centralized data collection.
 - **Predictive Maintenance**: FedML can facilitate the development of models that predict equipment failures or maintenance needs based on data from sensors spread across different locations.
 - **Optimized Resource Allocation**: In industries where resource allocation is crucial, federated models can analyze data from various units to allocate resources efficiently.
 - **Energy Consumption Analysis**: Industries can employ FedML to develop federated models that provide insights into energy consumption patterns across different facilities, promoting energy-saving measures.
 - **Supply Chain Collaboration**: Different entities within a supply chain can utilize FedML to collaboratively optimize logistics and inventory management without sharing sensitive data.

13.4.9 Microsoft Azure IoT Edge

Microsoft Azure IoT Edge is a part of Microsoft's Azure cloud service, focusing on pushing intelligence and cloud analytics capabilities to the edge, close to the source of data [106]. While not exclusively a federated learning platform, Azure IoT Edge provides tools and services that facilitate decentralized computations and analyses, making it relevant in the context of federated models and ITIoT.

- Features for ITIoT
 - **Modular Architecture**: Azure IoT Edge operates on a module-based system where each module is a discrete unit of computation. This design is particularly beneficial for ITIoT as industries can deploy specific modules tailored to individual tasks or machines.
 - **Seamless Integration with Azure Services**: It allows direct integration with other Azure services like Azure Stream Analytics, Azure Functions, or ML services. This means industries can utilize cloud-based tools when necessary while processing critical data locally.
 - **Containerization**: Azure IoT Edge uses Docker-compatible containers for its modules, ensuring consistency across the development, testing, and deployment phases. This feature is crucial for ITIoT environments to maintain uniformity across diverse devices and systems.
 - **End-to-End Security**: The platform provides layered security from edge to cloud. It offers features like secure boot, signed containers, and full device authentication, ensuring data privacy and integrity in ITIoT scenarios.
 - **Offline and Intermittent Connectivity**: Recognizing the varied networking conditions of ITIoT environments, Azure IoT Edge is designed to operate even when intermittent connectivity issues arise, ensuring no disruption in data processing and analytics.

- Use Cases for ITIoT
  - **Streamlined Operations in Remote Locations**: Industries located in remote areas, where consistent Internet connectivity is a challenge, can leverage Azure IoT Edge to process data on-site and sync with the cloud when connections are available.
  - **Real-time Data Analysis**: With the capability to process data at the source, industries can get real-time insights into operations, aiding in immediate decision-making. This is crucial for sectors like manufacturing where timely decisions can prevent mishaps.
  - **Reduced Operational Costs**: By processing data locally, industries can save on bandwidth costs and reduce the load on central servers, optimizing overall operational costs.
  - **Custom Solutions with Edge Modules**: Industries can develop bespoke solutions tailored to their needs using the modular architecture of Azure IoT Edge, ensuring flexibility and adaptability.
  - **Enhanced Security in Sensitive Environments**: In industries where data security is paramount, such as defense or critical infrastructure, Azure IoT Edge's security features can ensure data remains protected and uncompromised.

13.4.10 OpenMined

OpenMined is an open-source community whose primary goal is to ensure that privacy-preserving AI tools are accessible to a broader audience [107]. This initiative works toward advancing decentralized and secure ML techniques, which are of great significance in ITIoT contexts where data security and privacy are paramount.

- Features for ITIoT
  - **Privacy-Preserving Toolset**: OpenMined offers a suite of tools, such as PySyft, to enable encrypted, private ML. This ensures that data from intelligent devices remains secure during the entire computational process.
  - **Interoperability**: The tools under the OpenMined umbrella are designed to integrate seamlessly with popular deep learning frameworks like PyTorch and TensorFlow, ensuring that developers and industries can transition without steep learning curves.
  - **Federated Learning Support**: OpenMined champions federated learning, which aligns well with ITIoT, allowing devices to compute models locally without sending raw data.
  - **Differential Privacy**: Techniques to add noise to data or queries, ensuring that individual data points cannot be reverse-engineered, are fundamental to OpenMined's ethos. This is particularly useful in industries where data confidentiality is crucial.
  - **Secure Multiparty Computation (SMPC)**: OpenMined tools allow multiple parties to collaboratively compute functions over their datasets without revealing their raw data. This is pivotal in collaborative intelligent settings where data sharing might be a concern.
- Use Cases for ITIoT
  - **Secure Data Aggregation**: Industries with multiple units can securely aggregate operational data to gain insights without revealing sensitive information from individual units.
  - **Private Predictive Analytics**: ITIoT devices can utilize OpenMined tools to predict equipment failures or maintenance needs without sending sensitive data to a central server.
  - **Collaborative R&D**: Companies that collaborate on research and development projects can utilize OpenMined to share insights without exposing proprietary datasets.
  - **Regulatory Compliance**: In sectors where data handling and processing are under strict regulations (e.g., the energy sector), OpenMined can ensure that data analytics comply with privacy standards.

  - **Supply Chain Privacy**: Companies can analyze supply chain operations, demand forecasts, and inventory statuses across multiple partners without revealing individual operational data.

13.4.11 TinyML Federated Learning

TinyML, or TinyML, refers to the deployment of ML capabilities on ultra-low power, small footprint devices [108]. TEFL focuses on the combination of federated learning techniques with the principles of TinyML. This enables ML models to be trained across a myriad of small devices without centralized data aggregation, catering particularly to the constraints of tiny devices.

- Features for ITIoT
  - **Low-Power Consumption**: TEFL is optimized for devices with minimal computational capabilities and power budgets, making it ideal for ITIoT environments where devices often run on batteries or energy-harvesting sources.
  - **Lightweight Models**: The models designed under TEFL are compact and efficient, ensuring that they can run on devices with limited memory and storage.
  - **Decentralized Training**: Just as in traditional federated learning, TEFL allows for model training across a network of devices without the need for raw data centralization.
  - **On-Device Inference**: After training, devices can perform inferences locally, ensuring real-time decision-making without the need for cloud connectivity.
  - **Adaptive Learning**: Given the dynamic nature of ITIoT environments, TEFL models can adapt and learn from new data streams continuously.
- Use Cases for ITIoT
  - **Sensor Networks**: In industries with vast sensor networks monitoring parameters like temperature, humidity, or pressure, TEFL can enable these sensors to collaboratively learn and adapt to environmental changes without constantly sending data to a central hub.
  - **Wearable Intelligent Devices**: For safety and health monitoring wearables in industries, TEFL can provide real-time feedback to the wearer, such as alerting about hazardous conditions or health anomalies.
  - **Asset Tracking**: In scenarios where assets are tagged with low-power devices to monitor their location or condition, TEFL can enable these trackers to make local decisions, like sending alerts when an asset leaves a designated area.
  - **Energy Management**: In energy-conscious intelligent settings, TEFL can be deployed on devices monitoring energy consumption, allowing them to make local decisions to optimize energy usage.
  - **Predictive Maintenance**: Devices can monitor the health and operational status of machinery, learn patterns over time, and predict potential failures or maintenance needs using TEFL.

13.4.12 Sherpa.ai

Sherpa.ai is known for its efforts in advancing the frontier of AI by integrating state-of-the-art techniques like federated learning and differential privacy [109]. Their Federated Learning and Differential Privacy Framework is designed to provide robust, scalable, and secure ML solutions, especially in scenarios where data privacy and decentralized learning are paramount.

- Features for ITIoT
  - **Data Privacy First**: Sherpa.ai's framework is built around the principles of differential privacy, ensuring that individual data records are obfuscated in a way that the overall model still benefits from the data, but individual entries can't be reverse-engineered.

  - **Decentralized Learning**: Adhering to the principles of federated learning, the framework enables devices to train local models without sharing raw data, with only model parameters or updates communicated back to a central server.
  - **Scalable Architecture**: Designed to handle the complexities and scale of ITIoT, the framework can be deployed across a wide variety of devices, from powerful intelligent computers to resource-constrained sensors.
  - **Flexible Deployment**: It offers versatility in deployment scenarios, whether it's cloud, on-premises, or edge environments, ensuring that industries can align it with their existing infrastructural setups.
  - **Real-Time Monitoring and Updates**: The framework provides tools for real-time monitoring of the learning process and can accommodate dynamic updates, essential for the ever-evolving landscape of ITIoT.
- Use Cases for ITIoT
  - **Sensitive Data Handling**: In industries where data might be of a sensitive nature (e.g., nuclear power plants or chemical factories), the differential privacy feature ensures that data remains private while still contributing to global model improvements.
  - **Decentralized Predictive Maintenance**: Facilities with multiple machinery units can use the framework to predict maintenance needs locally at each unit, without pooling raw data, thus ensuring faster and localized decision-making.
  - **Supply Chain Optimization**: For industries with complex supply chains, Sherpa.ai's framework can be used to gain insights and optimize operations without revealing individual data points, thereby protecting business strategies.
  - **Energy Grid Management**: In scenarios like smart grids where there's a need to optimize energy distribution based on usage patterns, the framework can ensure that individual consumption data remains private while still enabling global optimization.
  - **Safety and Compliance Monitoring**: In industries like mining or construction, where safety is a top priority, the framework can enable wearable devices to monitor and ensure compliance with safety protocols without violating worker privacy.

13.4.13 Leaf

Leaf is an open-source ML framework for enabling federated learning in a decentralized manner [110]. While it was not primarily designed with ITIoT in mind, its characteristics make it a suitable choice for intelligent contexts that demand privacy and distributed data processing.

- Features for ITIoT
  - **Interoperability**: Leaf is designed to be platform-agnostic, making it suitable for diverse ITIoT environments with various devices and systems. This allows for smooth integration without the need to overhaul existing systems.
  - **Decentralized Framework**: Leaf champions the decentralized training of models, allowing ITIoT devices to process and learn from data locally, which aligns with the principles of edge computing prevalent in ITIoT.
  - **Flexible Client-Server Communication**: The framework enables asynchronous communication between client devices and the central server, allowing for more flexible and efficient model updates.
  - **Lightweight**: Considering that ITIoT devices can range from powerful machines to resource-constrained sensors, Leaf's lightweight design ensures it can be deployed across a wide range of devices.

  - **Data Efficiency**: In scenarios where transmitting data can be costly or slow, Leaf's ability to process data on-device and only transmit model updates can lead to bandwidth and cost savings.
- Use Cases for ITIoT
  - **Distributed Sensor Networks**: In environments like smart factories where numerous sensors collect data, Leaf can be employed to locally process and analyze this data, reducing the need to transmit vast amounts of raw sensor data.
  - **Predictive Maintenance**: By processing data on machinery in real-time, Leaf can help in predicting when a machine is likely to fail or requires maintenance, ensuring uninterrupted intelligent processes.
  - **Quality Assurance**: In manufacturing setups, Leaf can be used to train models that assess the quality of products on the production line in real-time without compromising on proprietary production data.
  - **Energy Management**: In sectors like renewable energy where data from various sources like wind turbines or solar panels is crucial, Leaf can aid in optimizing energy output based on local conditions.
  - **Supply Chain Optimization**: By processing data at different points in a supply chain, Leaf can offer insights into optimization without revealing sensitive supply chain data.

Table 13.1 offers a detailed comparison across various frameworks in terms of their features and capabilities. Starting with TensorFlow Federated (TFF), this open-source platform seamlessly integrates with TensorFlow and supports on-device training, distributed training, and aggregation.

Table 13.1 Comparison of federated learning platforms for ITIoT.

Framework/ Feature	Open Source	Support for On-device Training	Distributed Training and Aggregation	Integration with Popular ML Libraries	Support for Heterogeneous Devices	Ease of Deployment in ITIoT Environment	Active Community and Documentation
TensorFlow Federated	Yes	Yes	Yes	TensorFlow	Yes	High	High
PySyft	Yes	Yes	Yes	PyTorch	Yes	Medium	High
FATE	Yes	Yes	Yes	Yes	Yes	High	High
IBM Federated Learning	No	Yes	Yes	Various	Yes	High	Medium
FLTK	Yes	Yes	Yes	Yes	Yes	High	Medium
PaddleFL	Yes	Yes	Yes	PaddlePaddle	Yes	High	High
Flower	Yes	Yes	Yes	Yes	Yes	Medium	Medium
FedML	Yes	Yes	Yes	Various	Yes	High	High
Microsoft Azure IoT Edge	Yes	Yes	Yes	Azure ML libraries	Yes	Very high	High
OpenMined	Yes	Yes	Yes	PyTorch	Yes	Medium	High
TFFL	Yes	Yes	Yes	TinyML	Yes	High	Medium
Sherpa.ai	Yes	Yes	Yes	Yes	Yes	High	Medium
Leaf	Yes	Yes	Yes	Yes	Yes	High	Medium

Impressively, it caters to heterogeneous devices and boasts a high ease of deployment in the ITIoT environment, coupled with an active community and comprehensive documentation. Similarly, PySyft stands out as another open-source platform, aligning well with PyTorch and offering comparable capabilities to TensorFlow Federated (TFF), though its deployment in ITIoT settings is considered to be of medium ease. FATE is also an open-source option that doesn't limit itself to a specific ML library but offers wide-ranging compatibility. It mirrors the high ease of deployment seen in TensorFlow Federated (TFF) and boasts an active community. A notable exception in the open-source domain is IBM Federated Learning, which, despite its proprietary nature, supports training on various devices, distributed training, and aggregates while integrating with a myriad of ML libraries. It also ensures high ease of deployment in the ITIoT landscape, but its community and documentation support is rated medium. Other frameworks like FLTK, PaddleFL, FedML, and Sherpa.ai continue the trend of being open-source platforms that support a broad range of ML libraries and maintain high ease of deployment in ITIoT contexts. However, Flower diverges slightly, offering medium ease in both deployment and community/documentation support. Microsoft Azure IoT Edge presents an interesting case. Although primarily aligned with Azure ML Libraries, it offers the highest ease of deployment among the platforms. Its open-source status is conditional, indicated by an asterisk. Frameworks like OpenMined, TFFL, and Leaf all offer open-source capabilities, align with varying ML libraries, and maintain a high to medium ease of deployment in ITIoT scenarios. In summation, while most federated learning platforms for ITIoT offer open-source capabilities, they differ in their integration with ML libraries, deployment ease, and the robustness of their community and documentation. These differences are crucial for industries to consider based on their specific requirements and priorities.

13.5 Optimization Techniques for Federated ITIoT Systems

The rapid expansion of ITIoT devices combined with the rising implementation of federated learning in these ecosystems underscores the importance of optimizing federated ITIoT structures [111, 112]. These systems are distinctively challenged by the sheer volume of devices they encompass, the imperative of seamless communication, the necessity to shrink models to accommodate devices with limited resources, the task of synchronizing potentially inconsistent updates, and the duty to protect data privacy [113–115].

13.5.1 Communication-Efficient Algorithms

In the sprawling networks of federated ITIoT systems, the need for effective communication is paramount [116, 117]. The exchange of data, especially when large in size, can exacerbate latency issues and inflate energy consumption. This challenge becomes even more pronounced in intelligent contexts, where devices could span across diverse geographies and might communicate over unstable network links [118–120]. Overcoming such challenges requires devising algorithms that enhance communication efficiency [121].

- **Local Update Aggregation**: The essence of federated learning lies in training models on local devices and sending the updates to a central server for global aggregation. However, when dealing with countless ITIoT devices, sending updates from each device independently can congest the network. To alleviate this, edge computing can be instrumental [122]. By employing edge servers—intermediate servers situated closer to the IoT devices—local updates can be first collected and aggregated at these edge servers. This means that instead of a barrage of individual updates flooding the central server, a consolidated update from each edge server is sent. Not only does this reduce the network load, but it also leads to a faster global model aggregation due to the reduced number of incoming transmissions.

- **Sparse Updates**: Every update sent from a local device to the central server carries information about the changes made to the model during local training. However, not all of these changes are significant [123]. Some might have a negligible effect on the model's performance. Sparse updates focus on transmitting only the crucial parts of these changes. By defining certain thresholds or criteria, insignificant updates (like those causing very minor weight adjustments) can be filtered out. This ensures that only substantial, meaningful updates are sent over the network, reducing the data volume and consequently, the transmission time.
- **Quantization**: In the realm of ML, model weights are often represented using floating-point numbers to maintain precision. However, these precise representations can be bulky [124, 125]. Quantization provides a solution to this by approximating these floating-point numbers into simpler formats, such as lower-bit integers. The primary advantage is a drastic reduction in the size of the data being communicated. For instance, transitioning from 32-bit floating-point representation to 8-bit integer values can lead to a 4× reduction in data size. Such compression not only speeds up the communication but also economizes bandwidth usage. Furthermore, many ITIoT devices support hardware-level acceleration for integer arithmetic, making quantized models run faster during local training and inference. However, it's crucial to balance the level of quantization with the permissible loss in model accuracy. Too much compression can degrade the model's performance, so it's essential to strike the right balance for optimal results.

13.5.2 Model Compression for ITIoT Devices

ITIoT devices often differ from traditional computing environments in terms of their computing power, memory, and storage capabilities. Especially in edge environments, ITIoT devices may be resource-constrained, making it challenging to run large and complex ML models directly. This limitation underscores the importance of model compression for ITIoT devices, as it facilitates the deployment of powerful ML capabilities on devices without overwhelming their limited resources.

- **Pruning**: Pruning is a technique that focuses on removing certain parts of a neural network, like weights or even entire neurons, that have minimal contribution to the model's performance [126, 127]. By doing so, the model becomes sparser, leading to reduced memory footprint and faster inference times.
- **Weight Pruning**: Weight pruning zeroes out certain individual weights in a neural network based on pre-defined criteria. After pruning, the model is usually retrained for a few epochs to regain any lost accuracy. By reducing the number of active weights, computational demands decrease. This aids in faster processing and less memory consumption during inference. Weight pruning is commonly used in scenarios where model efficiency is more critical than achieving the highest possible accuracy.
- **Neuron/Filter Pruning**: This method eliminates entire neurons or filters, reducing the overall dimensionality of the corresponding layer in the network.

  Drastic reductions in model size can be achieved, often without a significant drop in performance. It can lead to decreased computation during forward and backward passes. It's particularly valuable for devices with stringent memory and computational power constraints, like edge devices in ITIoT.

- **Knowledge Distillation**: A larger, trained model (teacher) is used to guide the training of a smaller model (student). The student learns from the teacher's predictions (usually soft probabilities) rather than the ground truth [128]. The compact student model can achieve performance levels close to the more substantial teacher model but with reduced computational overhead. It is widely applied when deploying models on mobile or edge devices where computation and storage are limited.
- **Quantization**: Quantization reduces the numerical precision of the neural network's parameters and intermediate activations, thereby saving memory and computation.
  - **Weight Quantization**: The continuous set of weight values is mapped to a smaller set. For example, instead of 32-bit float values, weights might be represented using 8-bit integers. Reduces model size and accelerates model inference, especially on hardware that has optimized support for low-precision computations.
  - **Activation Quantization**: Neural network activations during inference are constrained to a reduced precision format, like 8-bit integers. Speeds up inference, especially during operations like matrix multiplications.
- **Weight Sharing**: Weight sharing is an optimization technique tailored for neural networks, primarily aimed at reducing their memory footprint [129]. The essence of this method lies in the identification of weights that have closely-matched values. Instead of maintaining these as distinct entities, they are grouped and replaced by a singular representative value. This consolidation drastically reduces the number of unique weights that need to be stored, effectively streamlining the model's size. Furthermore, this method employs indexing to store these representative values, ensuring efficient memory usage. Such an approach is particularly beneficial for deployments on edge devices, which often grapple with stringent memory constraints. By utilizing weight sharing, these devices can run complex models more efficiently without compromising on performance.
- **Compact Architectures**: Compact architectures represent a breed of neural network designs meticulously crafted to be diminutive in size but robust in function. The cornerstone of these designs is the emphasis on minimizing computational overhead without egregiously eroding model performance. MobileNets, for instance, incorporate depthwise separable convolutions, substantially curtailing computational demands. Meanwhile, SqueezeNets integrates "fire modules" that judiciously economize on parameters. These architectures strike a judicious balance between performance efficacy and computational thriftiness, carving a niche for themselves in real-time applications, especially on devices where computational resources are at a premium.
- **Binary and Ternary Networks**: In the realm of neural networks, Binary and Ternary Networks epitomize the quest for minimalism. They operate under a regime where weights and activations in the network are confined to a scant two or three distinct values. For instance, a ternary network might restrict its palette to −1, 0, and 1. This austerity translates to a significant reduction in model size. Moreover, the computational advantage is palpable: the often-expensive matrix multiplication operations are frequently supplanted by nimble accumulation tasks. However, this simplicity comes with trade-offs. Particularly for intricate tasks, these networks may grapple with a diminution in accuracy.
- **Huffman Coding**: Huffman coding is an algorithmic marvel in the world of lossless data compression. It operates on a deceptively simple premise: encode recurring values using shorter codes while assigning protracted codes to rarer values. This asymmetry in code lengths ensures that the overall representation is succinct. In the context of neural networks, when Huffman coding is harmoniously combined with other optimization techniques like pruning and quantization, the results can be transformative. The ensuing model compression can be substantial, making

it an ideal final touch before deploying models on ITIoT devices. The overarching aim is clear: ensure that these models, while powerful, have a minimal storage footprint.

13.5.3 Asynchronous Updates and Staleness Handling

In the context of Federated ITIoT systems, devices or nodes are often geographically distributed and operate in diverse network conditions. As such, the synchronized update of global models can be a challenging endeavor.

- **Asynchronous Updates**: In federated learning, particularly in ITIoT systems with varied device capabilities and network conditions, asynchronous updates offer a flexible approach to model aggregation [130]. The central server doesn't mandatorily await all local model updates to arrive simultaneously. Instead, as soon as a local update is received, it's integrated into the global model. This "on-the-go" approach ensures that devices with faster computation or better network connectivity aren't held back by slower devices, thus enhancing overall system efficiency. However, this efficiency comes at a cost. The risk is that the global model may diverge or fail to converge to an optimal solution, especially if updates from certain devices are consistently delayed or if those updates are based on stale versions of the model.
- **Staleness Handling**: As devices send their local updates at different times in an asynchronous setting, some updates might be based on older versions of the global model, leading them to be termed as "stale" updates [131, 132]. These stale updates, if not managed properly, can negatively impact the convergence and performance of the global model. There are several strategies to address this staleness.
  - **Version Tracking**: With version tracking, each local update is tagged with a version number, signifying which iteration or state of the global model the update was based upon. When the central server receives the update, it can reference the version number. This reference informs the server about the "freshness" of the update. Fresh, recent updates might be given more priority, while stale updates could be weighed down or treated differently during aggregation.
  - **Delayed Averaging**: Rather than rushing to average and incorporate every received update, the central server can employ patience. Updates based on much older versions of the global model can be set aside temporarily. Once the server collects a sufficient number of such "older" updates, they can be averaged among themselves first before being integrated into the global model. This strategy ensures that older updates do not unduly influence the global model with potentially outdated information.
  - **Adaptive Learning Rates**: The learning rate, a critical hyperparameter in model training, determines the step size in updating model parameters. In the context of staleness handling, the central server can adjust the learning rate based on the age or version of the received update. More recent updates might be integrated with a standard or larger learning rate, ensuring their influence is adequately captured. Conversely, stale updates could be integrated with a reduced learning rate, limiting their potential negative impact on the global model's convergence and performance.

13.5.4 Privacy-Preserving Mechanisms

The federated nature of ITIoT systems, where data remains localized, inherently offers better privacy compared to centralized systems. However, to ensure that the shared model updates do not leak individual data information, additional privacy-preserving mechanisms become essential [133].

- **Differential Privacy**: Differential privacy introduces controlled randomness to the data or model updates to protect individual data points. By adding carefully calibrated noise to data or the results of computations, it guarantees that the outputs remain nearly identical, whether or not a single data point is included. In federated learning, applying differential privacy means that the updates shared by individual nodes have a degree of uncertainty [134]. Consequently, malicious actors can't infer specifics about any individual data point based solely on the shared updates. This method is a cornerstone for maintaining data anonymity and ensuring that sensitive information remains obfuscated even when used for collaborative computations.
- **Homomorphic Encryption**: Homomorphic encryption is a form of encryption allowing computation on ciphertexts, generating an encrypted result which, when decrypted, matches the result of the operations performed on the plaintext. In federated learning, this is incredibly valuable because nodes can encrypt their local model updates before transmitting them [135]. The central server, or another entity, can then perform computations (like aggregating the updates) directly on this encrypted data. The advantage is twofold: data remains private throughout the process, and there's no need for frequent decryption and re-encryption, which would be computationally expensive and could introduce potential security vulnerabilities.
- **Secure Multiparty Computation (SMPC)**: SMPC offers a method for several parties to jointly compute a function over their individual inputs, all while ensuring that those individual inputs remain private [136]. Each participant doesn't learn anything about the others' inputs, apart from what can be inferred from the output. Within federated learning, SMPC can be a game-changer, especially in environments where entities might not fully trust each other. Nodes can collaborate to compute the aggregated model update without any single node revealing its specific update to the others. This preserves the privacy of each participant's data and the corresponding model update.
- **Federated Learning with Trusted Aggregator**: While federated learning inherently promotes decentralized computations, there are scenarios where mutual distrust exists among nodes or between nodes and the central server [137]. To navigate this, a trusted third-party aggregator can be incorporated into the workflow. In this setup, nodes encrypt their local model updates and send them to the trusted aggregator. The aggregator, acting as a neutral entity, decrypts these updates, aggregates them to produce the global model update, and then re-encrypts this for transmission to the central server. This architecture ensures that the central server never directly accesses any individual node's update, bolstering privacy and trust in the system.
- **Model Masking**: Model masking is a privacy-enhancing technique where model updates from individual nodes are obscured by adding random values, termed "masks." When a node is ready to share its update, it first combines (or masks) this update with a random value. After receiving masked updates from all nodes, these random values are mathematically cancelled out during aggregation, resulting in a correct global update [138]. Throughout this process, the actual updates from individual nodes remain hidden from both the central server and other nodes, thereby ensuring that the specifics of any individual's update cannot be discerned or reverse-engineered.

13.6 Security and Privacy Considerations

The integration of federated learning with the IoT ecosystem brings forward a unique set of challenges. These systems are often inherently distributed, with nodes ranging from high-power servers to low-power sensors. Ensuring security and privacy across such diverse setups becomes critical.

13.6.1 Data Privacy in Federated IoT Systems

Traditional ML paradigms predominantly focus on centralizing data, making it seemingly simpler to implement a uniform layer of security measures. This approach, however, can lead to single points of failure where a breach could expose vast amounts of sensitive information [139–141]. Federated IoT systems radically deviate from this model by distributing data across various devices. The primary advantage of such decentralization is that raw data stays localized, reducing the risks associated with transmitting and centralizing data. However, this landscape is not without its own set of unique challenges.

- Localized Data Exposure
  - **Nature of Devices**: Federated IoT encompasses a wide range of devices from intelligent sensors, and smart home appliances to wearables. The security features available on a high-end server are often not feasible on, say, a temperature sensor due to size, power, or cost constraints.
  - **Physical Access**: Edge devices, by virtue of their placement, might be more accessible, making them vulnerable to physical tampering or unauthorized access.
  - **Outdated Firmware**: Many IoT devices run on outdated firmware or software, which might have known vulnerabilities that haven't been patched. The onus is often on manufacturers and users to ensure regular updates, which, unfortunately, doesn't always occur.
- Model Inferences
  - **Model Reconstruction**: By accessing multiple model updates over time, a malicious entity might attempt to reconstruct parts of the local dataset, gaining insights into sensitive information.
  - **Differential Attacks**: By observing changes in the model updates before and after a specific input is introduced or modified, attackers might infer the nature of that input.
  - **Noise Insufficiency**: While adding noise to data is a common technique to prevent direct inference, inadequately calibrated noise could either fail to protect privacy or render the shared updates useless.
- Data Leakage through Side-Channels
  - **Timing Attacks**: Observing how long a device takes to compute and send its updates might reveal information about the amount or nature of the data it processes.
  - **Power Analysis**: Variations in a device's power consumption during computation can, at times, be used to deduce information about the data or operations it's performing.
  - **Electromagnetic Eavesdropping**: Sophisticated adversaries might use equipment to capture electromagnetic emissions from a device, potentially decoding valuable information. For instance, different data patterns might lead to discernible emission patterns.

13.6.2 Secure Aggregation Techniques

Secure aggregation in federated learning environments, especially within the domain of the IoT, ensures that individual device updates are not directly exposed to either the central server or other devices, thereby maintaining data privacy. Here's an in-depth look at secure aggregation techniques tailored for federated IoT scenarios [142–144].

- **Threshold Cryptography**: This involves splitting cryptographic keys into multiple parts and setting a threshold number of parts required to perform decryption [137]. Model updates can be encrypted such that decryption (and thus aggregation) is only possible when a threshold number of updates are present. This prevents the server from viewing individual updates but allows it to

decrypt the aggregated result once enough updates have been collected. It introduces a layer of protection against both external attacks and potential misbehavior by the central server.
- **Noise Addition and Differential Privacy**: This involves adding calibrated noise to data or updates to mask exact values [145]. Before transmitting their updates, devices can add noise, ensuring that the transmitted data doesn't precisely represent the raw local updates. When these noisy updates are aggregated at scale, the noise tends to average out, leaving a meaningful signal behind. It provides a balance between privacy and utility, ensuring individual data remains obscured while the aggregate data is useful.
- **Commitment Schemes**: A commitment scheme allows one to commit to a chosen value while keeping it hidden, with the ability to reveal it later. Devices can "commit" to their updates without revealing them. Once all commitments are collected, devices can then reveal their updates, ensuring honest behavior. It provides a method to detect and prevent dishonest behavior in the network, particularly beneficial for ensuring trustworthiness in decentralized environments.

13.6.3 Attack Vectors and Mitigation in Federated IoT Setups

The decentralized nature of federated IoT systems brings forth a unique set of challenges concerning attack vectors. Below are some common attack vectors specific to federated IoT setups and their respective mitigation techniques.

- **Model Poisoning Attacks**: In the federated learning paradigm, devices or nodes independently train local models and then share their model updates with a central server. Model poisoning attacks occur when malicious nodes introduce tainted or manipulated updates. These poisoned updates can have several motives: to degrade the overall performance of the aggregated global model, to introduce specific vulnerabilities, or to embed backdoors. This type of attack is especially insidious because it exploits the collaborative nature of federated learning, turning the system's decentralized advantage into a potential vulnerability.
- **Eavesdropping or Interception Attacks**: As devices in federated IoT systems communicate their model updates or other sensitive information to the central server or other nodes, there's an opportunity for malicious actors to eavesdrop or intercept these communications. Eavesdropping involves passively listening to these communications, aiming to extract or learn sensitive information. The decentralized nature of federated IoT setups, combined with potentially unsecured communication channels, can make these systems particularly susceptible to such attacks.
- **Sybil Attacks**: In a Sybil attack, a single adversary takes on multiple identities or creates several fake nodes within the network. By flooding the system with these counterfeit identities, the attacker aims to disproportionately influence the aggregated model or gain undue insights into the data of honest nodes. Given the distributed nature of federated IoT systems, detecting and preventing Sybil attacks can be challenging, as the central authority might not have a clear view of all the nodes participating at any given moment.
- **Replay Attacks**: Replay attacks in federated IoT systems involve capturing valid transmissions between nodes and the central server and then resending them at a later time. The intention might be to confuse the system, invalidate genuine updates, or even to manipulate the global model in certain scenarios. Because the data being sent is authentic, these attacks can be particularly tricky to detect.
- **Inference Attacks**: Inference attacks are particularly concerning from a privacy perspective. Even though federated learning is designed to keep raw data localized on each node, the model

updates that are shared can sometimes provide clues about the underlying data. In an inference attack, adversaries analyze these model updates to deduce information about the original datasets on local devices. This could lead to unintended disclosures of private or sensitive information.

- **Denial-of-Service (DoS) Attacks**: A DoS attack on federated IoT systems aims to disrupt the normal functioning of the network, often by overwhelming the central server or nodes with an excessive amount of data or requests. The objective is not to gain unauthorized access or steal information but to render the system unusable. In the context of federated learning, this could halt the process of model aggregation, preventing the system from benefiting from collaborative learning.
- **Physical Attacks on Devices**: While many of the attacks mentioned are software-based or target the digital components of a system, physical attacks involve direct tampering with the IoT devices themselves. Malicious actors might physically access devices to extract data, introduce corrupted updates, or disrupt their operation. Given the vast and distributed nature of IoT networks, ensuring the physical security of every device can be a formidable challenge.

13.7 Conclusion

The future of ITIoT is intrinsically linked with federated learning, promising enhanced performance, scalability, and data privacy. However, with great potential also come great challenges. As IoT devices continue to proliferate, the need to optimize these systems while ensuring data privacy and system security becomes more critical. From communication-efficient algorithms to handling asynchronous updates, we need intricate strategies that respect device constraints. Privacy-preserving mechanisms, such as differential privacy and homomorphic encryption, have paved the way for secure federated learning. Yet, being vigilant of various attack vectors and proactive in implementing robust security measures remains crucial. As we advance further into the era of decentralized learning and processing, a cohesive approach that intertwines optimization with robust security and privacy protocols will be the cornerstone of successful and safe IoT implementations.

Acknowledgment

Author used Claude3 for initial drafting and design of the chapter.

References

1 Nguyen, D.C., Ding, M., Pathirana, P.N. et al. (2021). Federated learning for internet of things: a comprehensive survey. *IEEE Communications Surveys & Tutorials* 23 (3): 1622–1658.

2 Imteaj, A., Thakker, U., Wang, S. et al. (2021). A survey on federated learning for resource-constrained IoT devices. *IEEE Internet of Things Journal* 9 (1): 1–24.

3 Khan, L.U., Saad, W., Han, Z., Hossain, E., and Hong, C.S. (2021). Federated learning for internet of things: recent advances, taxonomy, and open challenges. *IEEE Communications Surveys & Tutorials* 23 (3): 1759–1799.

4 Savazzi, S., Nicoli, M., and Rampa, V. (2020). Federated learning with cooperating devices: a consensus approach for massive IoT networks. *IEEE Internet of Things Journal* 7 (5): 4641–4654.

5 Mills, J., Jia, H., and Min, G. (2019). Communication-efficient federated learning for wireless edge intelligence in IoT. *IEEE Internet of Things Journal* 7 (7): 5986–5994.

6 Wu, Q., He, K., and Chen, X. (2020). Personalized federated learning for intelligent IoT applications: a cloud-edge based framework. *IEEE Open Journal of the Computer Society* 1: 35–44.

7 Pang, J., Huang, Y., Xie, Z. et al. (2020). Realizing the heterogeneity: a self-organized federated learning framework for IoT. *IEEE Internet of Things Journal* 8 (5): 3088–3098.

8 Zhan, Y., Li, P., Qu, Z. et al. (2020). A learning-based incentive mechanism for federated learning. *IEEE Internet of Things Journal* 7 (7): 6360–6368.

9 Lu, Y., Huang, X., Dai, Y. et al. (2019). Blockchain and federated learning for privacy-preserved data sharing in industrial IoT. *IEEE Transactions on Industrial Informatics* 16 (6): 4177–4186.

10 Mothukuri, V., Khare, P., Parizi, R.M. et al. (2021). Federated-learning-based anomaly detection for IoT security attacks. *IEEE Internet of Things Journal* 9 (4): 2545–2554.

11 Kholod, I., Yanaki, E., Fomichev, D. et al. (2020). Open-source federated learning frameworks for IoT: a comparative review and analysis. *Sensors* 21 (1): 167.

12 Zhang, T., He, C., Ma, T. et al. (2021). Federated learning for internet of things. *Proceedings of the 19th ACM Conference on Embedded Networked Sensor Systems*, Coimbra, Portugal.

13 Ghimire, B. and Rawat, D.B. (2022). Recent advances on federated learning for cybersecurity and cybersecurity for federated learning for internet of things. *IEEE Internet of Things Journal* 9 (11): 8229–8249.

14 Rey, V., Sánchez, P.M.S., Celdrán, A.H., and Bovet, G. (2022). Federated learning for malware detection in IoT devices. *Computer Networks* 204: 108693.

15 AbdulRahman, S., Tout, H., Mourad, A., and Talhi, C. (2020). FedMCCS: multicriteria client selection model for optimal IoT federated learning. *IEEE Internet of Things Journal* 8 (6): 4723–4735.

16 Du, Z., Wu, C., Yoshinaga, T. et al. (2020). Federated learning for vehicular internet of things: recent advances and open issues. *IEEE Open Journal of the Computer Society* 1: 45–61.

17 Zhang, W., Yang, D., Wu, W. et al. (2021). Optimizing federated learning in distributed industrial IoT: a multi-agent approach. *IEEE Journal on Selected Areas in Communications* 39 (12): 3688–3703.

18 Yuan, B., Ge, S., and Xing, W. (2020). A federated learning framework for healthcare iot devices. arXiv preprint arXiv:2005.05083.

19 Banabilah, S., Aloqaily, M., Alsayed, E. et al. (2022). Federated learning review: fundamentals, enabling technologies, and future applications. *Information Processing & Management* 59 (6): 103061.

20 Rahman, S.A., Tout, H., Talhi, C., and Mourad, A. (2020). Internet of things intrusion detection: centralized, on-device, or federated learning? *IEEE Network* 34 (6): 310–317.

21 Gao, Y., Kim, M., Abuadbba, S., Kim, Y. et al. (2020). End-to-end evaluation of federated learning and split learning for internet of things. arXiv preprint arXiv:2003.13376.

22 Kwon, D., Jeon, J., Park, S. et al. (2020). Multiagent DDPG-based deep learning for smart ocean federated learning IoT networks. *IEEE Internet of Things Journal* 7 (10): 9895–9903.

23 Cui, L., Qu, Y., Xie, G. et al. (2021). Security and privacy-enhanced federated learning for anomaly detection in IoT infrastructures. *IEEE Transactions on Industrial Informatics* 18 (5): 3492–3500.

24 Zhao, Y., Zhao, J., Jiang, L. et al. (2020). Privacy-preserving blockchain-based federated learning for IoT devices. *IEEE Internet of Things Journal* 8 (3): 1817–1829.

25 Singh, S., Rathore, S., Alfarraj, O. et al. (2022). A framework for privacy-preservation of IoT healthcare data using federated learning and blockchain technology. *Future Generation Computer Systems* 129: 380–388.

26 Zhang, W., Lu, Q., Yu, Q. et al. (2020). Blockchain-based federated learning for device failure detection in industrial IoT. *IEEE Internet of Things Journal* 8 (7): 5926–5937.

27 Ferrag, M.A., Friha, O., Hamouda, D. et al. (2022). Edge-IIoTset: a new comprehensive realistic cyber security dataset of IoT and IIoT applications for centralized and federated learning. *IEEE Access* 10: 40281–40306.

28 Alam, T. and Gupta, R. (2022). Federated learning and its role in the privacy preservation of IoT devices. *Future Internet* 14 (9): 246.

29 Chen, H., Huang, S., Deyou, Z. et al. (2022). Federated learning over wireless IoT networks with optimized communication and resources. *IEEE Internet of Things Journal* 9 (17): 16592–16605.

30 Nguyen, V.-D., Sharma, S.K., Vu, T.X. et al. (2020). Efficient federated learning algorithm for resource allocation in wireless IoT networks. *IEEE Internet of Things Journal* 8 (5): 3394–3409.

31 Campos, E.M., Saura, P.F., González-Vidal, A. et al. (2022). Evaluating federated learning for intrusion detection in internet of things: review and challenges. *Computer Networks* 203: 108661.

32 Weinger, B., Kim, J., Sim, A. et al. (2022). Enhancing IoT anomaly detection performance for federated learning. *Digital Communications and Networks* 8 (3): 314–323.

33 Ali, M., Karimipour, H., and Tariq, M. (2021). Integration of blockchain and federated learning for internet of things: recent advances and future challenges. *Computers & Security* 108: 102355.

34 Elayan, H., Aloqaily, M., and Guizani, M. (2021). Sustainability of healthcare data analysis IoT-based systems using deep federated learning. *IEEE Internet of Things Journal* 9 (10): 7338–7346.

35 Zhao, Y., Barnaghi, P., and Haddadi, H. (2022). Multimodal federated learning on iot data. *2022 IEEE/ACM Seventh International Conference on Internet-of-Things Design and Implementation (IoTDI)*, Milan, Italy. IEEE.

36 Javed, A.R., Hassan, M.A., Shahzad, F. et al. (2022). Integration of blockchain technology and federated learning in vehicular (iot) networks: a comprehensive survey. *Sensors* 22 (12): 4394.

37 Zhao, Y., Zhao, J., Yang, M. et al. (2020). Local differential privacy-based federated learning for internet of things. *IEEE Internet of Things Journal* 8 (11): 8836–8853.

38 Li, Z., He, Y., Yu, H. et al. (2022). Data heterogeneity-robust federated learning via group client selection in industrial iot. *IEEE Internet of Things Journal* 9 (18): 17844–17857.

39 Liu, Y., Garg, S., Nie, J. et al. (2020). Deep anomaly detection for time-series data in industrial IoT: a communication-efficient on-device federated learning approach. *IEEE Internet of Things Journal* 8 (8): 6348–6358.

40 Fu, A., Zhang, X., Xiong, N. et al. (2020). VFL: a verifiable federated learning with privacy-preserving for big data in industrial IoT. *IEEE Transactions on Industrial Informatics* 18 (5): 3316–3326.

41 Attota, D.C., Mothukuri, V., Parizi, R.M., and Pouriyeh, S. (2021). An ensemble multi-view federated learning intrusion detection for IoT. *IEEE Access* 9: 117734–117745.

42 Rjoub, G., Wahab, O.A., Bentahar, J., and Bataineh, A. (2022). Trust-driven reinforcement selection strategy for federated learning on IoT devices. *Computing* 106: 1–23.

43 Nguyen, T.D., Rieger, P., Miettinen, M., and Sadeghi, A.-R. (2020). Poisoning attacks on federated learning-based IoT intrusion detection system. *Proc. Workshop Decentralized IoT Syst. Secur. (DISS)*, San Diego, CA, USA.

44 Boobalan, P., Ramu, S.P., Pham, Q.-V. et al. (2022). Fusion of federated learning and industrial internet of things: a survey. *Computer Networks* 212: 109048.

45 Yang, H., Yuan, J., Li, C. et al. (2021). BrainIoT: brain-like productive services provisioning with federated learning in industrial IoT. *IEEE Internet of Things Journal* 9 (3): 2014–2024.

46 Friha, O., Ferrag, M.A., Shu, L. et al. (2022). FELIDS: federated learning-based intrusion detection system for agricultural internet of things. *Journal of Parallel and Distributed Computing* 165: 17–31.

47 Chhikara, P., Singh, P., Tekchandani, R. et al. (2020). Federated learning meets human emotions: a decentralized framework for human–computer interaction for IoT applications. *IEEE Internet of Things Journal* 8 (8): 6949–6962.

48 Yu, L., Albelaihi, R., Sun, X. et al. (2021). Jointly optimizing client selection and resource management in wireless federated learning for internet of things. *IEEE Internet of Things Journal* 9 (6): 4385–4395.

49 Li, J., Lyu, L., Liu, X. et al. (2021). FLEAM: a federated learning empowered architecture to mitigate DDoS in industrial IoT. *IEEE Transactions on Industrial Informatics* 18 (6): 4059–4068.

50 Sun, W., Lei, S., Wang, L. et al. (2020). Adaptive federated learning and digital twin for industrial internet of things. *IEEE Transactions on Industrial Informatics* 17 (8): 5605–5614.

51 Chen, Z., Liao, W., Hua, K. et al. (2021). Towards asynchronous federated learning for heterogeneous edge-powered internet of things. *Digital Communications and Networks* 7 (3): 317–326.

52 Chen, M., Poor, H.V., Saad, W., and Cui, S. (2020). Wireless communications for collaborative federated learning. *IEEE Communications Magazine* 58 (12): 48–54.

53 Issa, W., Moustafa, N., Turnbull, B. et al. (2023). Blockchain-based federated learning for securing internet of things: a comprehensive survey. *ACM Computing Surveys* 55 (9): 1–43.

54 Feraudo, A., Poonam Yadav, P., Vadim Safronov, V. et al. (2020). CoLearn: enabling federated learning in MUD-compliant IoT edge networks. *Proceedings of the Third ACM International Workshop on Edge Systems, Analytics and Networking*, Heraklion, Greece.

55 Tahir, M. and Ali, M.I. (2022). On the performance of federated learning algorithms for IoT. *IoT* 3 (2): 273–284.

56 Kishor, K. (2022). Personalized federated learning. In: *Federated Learning for IoT Applications* (ed. S.P. Yadav, B.S. Bhati, D.P. Mahato, and S. Kumar), 31–52. Cham: Springer International Publishing.

57 Wu, D., Ullah, R., Harvey, P. et al. (2022). Fedadapt: adaptive offloading for iot devices in federated learning. *IEEE Internet of Things Journal* 9 (21): 20889–20901.

58 Khan, L.U., Saad, W., Han, Z., and Hong, C.S. et al. (2021). Dispersed federated learning: vision, taxonomy, and future directions. *IEEE Wireless Communications* 28 (5): 192–198.

59 Rahman, A., Hasan, K., Kundu, D. et al. (2023). On the ICN-IoT with federated learning integration of communication: concepts, security-privacy issues, applications, and future perspectives. *Future Generation Computer Systems* 138: 61–88.

60 Ni, W., Liu, Y., Yang, Z. et al. (2021). Federated learning in multi-RIS-aided systems. *IEEE Internet of Things Journal* 9 (12): 9608–9624.

61 Boobalan, P., Ramu, S.P., Pham, Q.-V. et al. (2021). Fusion of federated learning and industrial internet of things: a survey. arXiv preprint arXiv:2101.00798.

62 Lu, Y., Huang, X., Zhang, K. et al. (2020). Communication-efficient federated learning for digital twin edge networks in industrial IoT. *IEEE Transactions on Industrial Informatics* 17 (8): 5709–5718.

63 Jiang, J.C., Kantarci, B., Oktug, S., and Soyata, T. (2020). Federated learning in smart city sensing: challenges and opportunities. *Sensors* 20 (21): 6230.

64 Otoum, S., Al Ridhawi, I., and Mouftah, H. (2021). Securing critical IoT infrastructures with blockchain-supported federated learning. *IEEE Internet of Things Journal* 9 (4): 2592–2601.

65 Saadat, H. et al. (2021). Hierarchical federated learning for collaborative IDS in IoT applications. *2021 10th Mediterranean Conference on Embedded Computing (MECO)*, Budva, Montenegro. IEEE.

66 (a) Balasubramanian, V., Aloqaily, M., Reisslein, M., and Scaglione, A. (2021). Intelligent resource management at the edge for ubiquitous IoT: an SDN-based federated learning approach. *IEEE Network* 35 (5): 114–121. (b) Li, C., Li, G., and Varshney, P.K. (2021). Decentralized federated learning via mutual knowledge transfer. *IEEE Internet of Things Journal* 9 (2): 1136–1147.

67 Saha, R., Misra, S., and Deb, P.K. (2020). FogFL: fog-assisted federated learning for resource-constrained IoT devices. *IEEE Internet of Things Journal* 8 (10): 8456–8463.

68 Guo, Y., Zhao, Z., He, K. et al. (2021). Efficient and flexible management for industrial internet of things: a federated learning approach. *Computer Networks* 192: 108122.

69 Briggs, C., Fan, Z., and Andras, P. (2021). A review of privacy-preserving federated learning for the internet-of-things. In: *Federated Learning Systems: Towards Next-Generation AI* (ed. M. Habib ur Rehman and M.M. Gaber), 21–50.

70 Wassan, S., Suhail, B., Mubeen, R. et al. (2022). Gradient boosting for health IoT federated learning. *Sustainability* 14 (24): 16842.

71 Shen, Y., Gou, F., and Jia, W. (2022). Node screening method based on federated learning with IoT in opportunistic social networks. *Mathematics* 10 (10): 1669.

72 Tonellotto, N., Gotta, A., Nardini, F.M. et al. (2021). Neural network quantization in federated learning at the edge. *Information Sciences* 575: 417–436.

73 Yang, W., Xiang, W., Yang, Y., and Cheng, P. (2022). Optimizing federated learning with deep reinforcement learning for digital twin empowered industrial IoT. *IEEE Transactions on Industrial Informatics* 19 (2): 1884–1893.

74 Yazdinejad, A., Parizi, R.M., Dehghantanha, A., and Karimipour, H. (2021). Federated learning for drone authentication. *Ad Hoc Networks* 120: 102574.

75 da Silva, L., Ferreira, G., Sadok, D.F.H., and Endo, P.T. (2023). Resource optimizing federated learning for use with IoT: a systematic review. *Journal of Parallel and Distributed Computing* 175: 92.

76 Arisdakessian, S., Wahab, O.A., Mourad, A. et al. (2022). A survey on IoT intrusion detection: federated learning, game theory, social psychology, and explainable AI as future directions. *IEEE Internet of Things Journal* 10 (5): 4059–4092.

77 Imteaj, A., Ahmed, K.M., Thakker, U. et al. (2022). Federated learning for resource-constrained IoT devices: panoramas and state of the art. In: *Adaptation, Learning, and Optimization* (ed. R. Razavi-Far, B. Wang, M.E. Taylor, and Q. Yang), 7–27. Springer.

78 Belenguer, A., Navaridas, J., and Pascual, J.A. (2022). A review of federated learning in intrusion detection systems for IoT. arXiv preprint arXiv:2204.12443.

79 Elayan, H., Aloqaily, M., and Guizani, M. (2021). Deep federated learning for IoT-based decentralized healthcare systems. *2021 International Wireless Communications and Mobile Computing (IWCMC)*, Harbin, China. IEEE.

80 Qolomany, B., Ahmad, K., Al-Fuqaha, A., and Qadir, J. (2020). Particle swarm optimized federated learning for industrial IoT and smart city services. *GLOBECOM 2020-2020 IEEE Global Communications Conference*. IEEE.

81 Yao, J. and Ansari, N. (2020). Enhancing federated learning in fog-aided IoT by CPU frequency and wireless power control. *IEEE Internet of Things Journal* 8 (5): 3438–3445.

82 AbdulRahman, S., Tout, H., Ould-Slimane, H. et al. (2020). A survey on federated learning: the journey from centralized to distributed on-site learning and beyond. *IEEE Internet of Things Journal* 8 (7): 5476–5497.

83 Imteaj, A. and Hadi Amini, M. (2019). Distributed sensing using smart end-user devices: pathway to federated learning for autonomous IoT. *2019 International conference on computational science and computational intelligence (CSCI)*, Las Vegas, Nevada, USA. IEEE.

84 Nguyen, D.C., Ding, M., Pathirana, P.N. et al. (2021). Federated learning for industrial internet of things in future industries. *IEEE Wireless Communications* 28 (6): 192–199.

85 Zhou, C., Fu, A., Yu, S. et al. (2020). Privacy-preserving federated learning in fog computing. *IEEE Internet of Things Journal* 7 (11): 10782–10793.

86 Ma, X., Jiang, Q., Shojafar, M. et al. (2022). DisBezant: secure and robust federated learning against byzantine attack in IoT-enabled MTS. *IEEE Transactions on Intelligent Transportation Systems* 24 (2): 2492–2502.

87 Mian, A.N. et al. (2022). A value-added IoT service for cellular networks using federated learning. *Computer Networks* 213: 109094.

88 Xu, Y., Lu, Z., Gai, K. et al. (2021). Besifl: blockchain empowered secure and incentive federated learning paradigm in IoT. *IEEE Internet of Things Journal*.

89 Ruzafa-Alcázar, P., Fernández-Saura, P., Mármol-Campos, E. et al. (2021). Intrusion detection based on privacy-preserving federated learning for the industrial IoT. *IEEE Transactions on Industrial Informatics* 19 (2): 1145–1154.

90 Zhang, C., Xu, Y., Elahi, H. et al. (2022). A blockchain-based model migration approach for secure and sustainable federated learning in iot systems. *IEEE Internet of Things Journal* 10 (8): 6574–6585.

91 Cui, Y., Cao, K., Cao, G. et al. (2021). Client scheduling and resource management for efficient training in heterogeneous IoT-edge federated learning. *IEEE Transactions on Computer-Aided Design of Integrated Circuits and Systems* 41 (8): 2407–2420.

92 Wei, Z., Pei, Q., Zhang, N. et al. (2021). Lightweight federated learning for large-scale IoT devices with privacy guarantee. *IEEE Internet of Things Journal* 10: 3179.

93 Xu, W., Fang, W., Ding, Y. et al. (2021). Accelerating federated learning for IoT in big data analytics with pruning, quantization and selective updating. *IEEE Access* 9: 38457–38466.

94 Zhang, L., Lei, X., Shi, Y. et al. (2023). Federated learning for IoT devices with domain generalization. *IEEE Internet of Things Journal* 10: 9622.

95 Wazzeh, M., Ould-Slimane, H., Talhi, C. et al. (2022). Privacy-preserving continuous authentication for mobile and IoT systems using warmup-based federated learning. *IEEE Network* 37: 224.

96 Han, Y., Li, D., Qi, H. et al. (2019). Federated learning-based computation offloading optimization in edge computing-supported internet of things. *Proceedings of the ACM Turing Celebration Conference-China*, 17–19 May 2019, China.

97 Zhou, J., Zhang, S., Lu, Q. et al. (2021). A survey on federated learning and its applications for accelerating industrial internet of things. arXiv preprint arXiv:2104.10501.

98 TensorFlow federated. https://www.tensorflow.org/federated (accessed 17 September 2023).

99 PySyft https://github.com/OpenMined/PySyft (accessed 17 September 2023).

100 FATE https://github.com/FederatedAI/FATE (accessed 17 September 2023).

101 IBM federated learning. https://www.ibm.com/docs/en/cloud-paks/cp-data/4.0?topic=models-federated-learning (accessed 17 September 2023).

102 Federated Learning Toolkit (FLTK). https://github.com/bacox/fltk (accessed 17 September 2023).

103 PaddleFL https://github.com/PaddlePaddle/PaddleFL (accessed 17 September 2023).

104 Flower https://flower.dev/ (accessed 17 September 2023).

105 FedML https://www.fedml.ai/v.

106 Microsoft Azure IoT edge. https://azure.microsoft.com/en-in/products/iot-edge (accessed 17 September 2023).

107 OpenMined https://www.openmined.org/ (accessed 17 September 2023).

108 TinyML Federated Learning (TEFL). https://arxiv.org/abs/2304.05201 (accessed 17 September 2023).

109 Sherpa.ai federated learning and differential privacy framework. https://www.sherpa.ai/federated-learning-platform/ (accessed 17 September 2023).

110 Leaf https://leaf.cmu.edu/ (accessed 17 September 2023).

111 Yarradoddi, S. and Gadekallu, T.R. (2022). Federated learning role in big data, IoT services and applications security, privacy and trust in IoT. In: *Trust, Security and Privacy for Big Data*, 28. Boca Raton, FL: CRC Press.

112 Abdellatif, A.A., Mhaisen, N., Mohamed, A. et al. (2022). Communication-efficient hierarchical federated learning for IoT heterogeneous systems with imbalanced data. *Future Generation Computer Systems* 128: 406–419.

113 Farahani, B. and Monsefi, A.K. (2023). Smart and collaborative industrial IoT: a federated learning and data space approach. *Digital Communications and Networks* 9 (2): 436–447.

114 Bao, W., Wu, C., Guleng, S. et al. (2021). Edge computing-based joint client selection and networking scheme for federated learning in vehicular IoT. *China Communications* 18 (6): 39–52.

115 Wang, X., Garg, S., Lin, H. et al. (2021). Toward accurate anomaly detection in industrial internet of things using hierarchical federated learning. *IEEE Internet of Things Journal* 9 (10): 7110–7119.

116 AbdulRahman, S., Ould-Slimane, H., Chowdhury, R. et al. (2022). Adaptive upgrade of client resources for improving the quality of federated learning model. *IEEE Internet of Things Journal* 10 (5): 4677–4687.

117 Zheng, Z., Zhou, Y., Sun, Y. et al. (2022). Applications of federated learning in smart cities: recent advances, taxonomy, and open challenges. *Connection Science* 34 (1): 1–28.

118 Chen, M., Shlezinger, N., Poor, H.V. et al. (2021). Communication-efficient federated learning. *Proceedings of the National Academy of Sciences* 118 (17): e2024789118.

119 Rothchild, D., Panda, A., Ullah, E. et al. (2020). Fetchsgd: communication-efficient federated learning with sketching. *International Conference on Machine Learning*. PMLR.

120 Shah, S.M. and Lau, V.K.N. (2021). Model compression for communication efficient federated learning. *IEEE Transactions on Neural Networks and Learning Systems* 34: 5937.

121 Sattler, F., Wiedemann, S., Müller, K.-R., and Samek, W. et al. (2019). Robust and communication-efficient federated learning from non-iid data. *IEEE Transactions on Neural Networks and Learning Systems* 31 (9): 3400–3413.

122 Pillutla, K., Kakade, S.M., and Harchaoui, Z. (2022). Robust aggregation for federated learning. *IEEE Transactions on Signal Processing* 70: 1142–1154.

123 McMahan, H., Brendan, F.X. Yu, P. et al. (2016). Federated learning: Strategies for improving communication efficiency. *Proceedings of the 29th Conference on Neural Information Processing Systems (NIPS)*, Barcelona, Spain, pp. 5–10.

124 Reisizadeh, A., Mokhtari, A., Hassani, H. et al. (2020). Fedpaq: a communication-efficient federated learning method with periodic averaging and quantization. *International Conference on Artificial Intelligence and Statistics*. PMLR.

125 Amiri, M.M., Gunduz, D., Kulkarni, S.R., and Vincent Poor, H. (2020). Federated learning with quantized global model updates. arXiv preprint arXiv:2006.10672.

126 Jiang, Y., Wang, S., Valls, V. et al. (2022). Model pruning enables efficient federated learning on edge devices. *IEEE Transactions on Neural Networks and Learning Systems* 34: 1037.

127 Lin, R., Xiao, Y., Yang, T.-J. et al. (2022) Federated pruning: improving neural network efficiency with federated learning. arXiv preprint arXiv:2209.06359.

128 Zhu, Z., Hong, J., and Zhou, J. (2021). Data-free knowledge distillation for heterogeneous federated learning. *International Conference on Machine Learning*. PMLR.

129 Singh, A., Vepakomma, P., Gupta, O., and Raskar, R. (2019). Detailed comparison of communication efficiency of split learning and federated learning. arXiv preprint arXiv:1909.09145.

130 Chen, T., Jin, X., Sun, Y., and Yin, W. (2020). Vafl: a method of vertical asynchronous federated learning. arXiv preprint arXiv:2007.06081.

131 Zhu, L., Lin, H., Lu, Y. et al. (2021). Delayed gradient averaging: tolerate the communication latency for federated learning. *Advances in Neural Information Processing Systems* 34: 29995–30007.

132 Park, J., Han, D.-J., Choi, M., and Moon, J. (2021). Sageflow: robust federated learning against both stragglers and adversaries. *Advances in Neural Information Processing Systems* 34: 840–851.

133 Li, Q., Wen, Z., Wu, Z. et al. (2021). A survey on federated learning systems: vision, hype and reality for data privacy and protection. *IEEE Transactions on Knowledge and Data Engineering* 35: 3347.

134 Ouadrhiri, E. and Ahmed, and Ahmed Abdelhadi. (2022). Differential privacy for deep and federated learning: a survey. *IEEE Access* 10: 22359–22380.

135 Zhang, C., Li, S., Xia, J. et al. (2020). {BatchCrypt}: efficient homomorphic encryption for {Cross-Silo} federated learning. *2020 USENIX Annual Technical Conference (USENIX ATC 20)*.

136 Byrd, D. and Polychroniadou, A. (2020). Differentially private secure multi-party computation for federated learning in financial applications. *Proceedings of the First ACM International Conference on AI in Finance*.

137 Truex, S., Baracaldo, N., Anwar, A. et al. (2019). A hybrid approach to privacy-preserving federated learning. *Proceedings of the 12th ACM Workshop on Artificial Intelligence and Security*.

138 He, S., Yan, Q., Wu, F. et al. (2023). GlueFL: reconciling client sampling and model masking for bandwidth efficient federated learning. *Proceedings of Machine Learning and Systems* 5: 695.

139 Mothukuri, V., Parizi, R.M., Pouriyeh, S. et al. (2021). A survey on security and privacy of federated learning. *Future Generation Computer Systems* 115: 619–640.

140 Ma, C., Li, J., Ding, M. et al. (2020). On safeguarding privacy and security in the framework of federated learning. *IEEE Network* 34 (4): 242–248.

141 Bouacida, N. and Mohapatra, P. (2021). Vulnerabilities in federated learning. *IEEE Access* 9: 63229–63249.

142 Bonawitz, K., Ivanov, V., Kreuter, B. et al. (2016). Practical secure aggregation for federated learning on user-held data. arXiv preprint arXiv:1611.04482.

143 Fereidooni, H., Marchal, S., Miettinen, M. et al. (2021). SAFELearn: secure aggregation for private federated learning. *2021 IEEE Security and Privacy Workshops (SPW)*. IEEE.

144 Kadhe, S., Rajaraman, N., Ozan Koyluoglu, O. et al. (2020). Fastsecagg: scalable secure aggregation for privacy-preserving federated learning. arXiv preprint arXiv:2009.11248.

145 Wei, K., Li, J., Ding, M. et al. (2020). Federated learning with differential privacy: algorithms and performance analysis. *IEEE Transactions on Information Forensics and Security* 15: 3454–3469.

14

Enhancing Cybersecurity Through Federated Learning: A Critical Evaluation of Strategies and Implications

M. Ashok Kumar[1], *Aliyu Mohammed*[2], *S. Sumanth*[3], *and V. Sivanantham*[4]

[1]*Department of Computer Science and Software Engineering, Skyline University Nigeria, Kano, Nigeria*
[2]*Department of Management, Skyline University Nigeria, Kano, Nigeria*
[3]*Department of Computer Science and Computer Applications, Government College for Women (Affiliated to Bengaluru North University), Kolar, Karnataka, India*
[4]*Department of Computer Science, Periyar University Salem, Tamil Nadu, India*

14.1 Introduction

The rise of cyber risks offers an unprecedented challenge to both organizations and people in today's linked digital environment. Protecting our digital infrastructure has become a top priority due to the constantly improving sophistication of hostile actors and the rising volume of sensitive data shared online. The traditional techniques of bolstering cybersecurity have frequently failed in this age of digital vulnerability. Thus, the need for creative and flexible strategies to strengthen our defences against cyber attacks is urgent. Federated Learning is one recent example of a ground-breaking strategy. A decentralized machine learning paradigm called federated learning enables several participants to jointly train a single model while keeping their data local and safe. This paradigm change not only tackles the privacy issues related to centralized data aggregation but also creates new opportunities for dispersed intelligence to improve cybersecurity. Figure 14.1 provides a visual representation of the general framework of Federated Learning, illustrating how this decentralized approach facilitates local model training without centralized data storage. This research paper's main objective is to undertake an in-depth and critical assessment of the benefits and drawbacks of using Federated Learning as a cybersecurity solution. We explore the fundamental ideas, practices, and practical implementations of Federated Learning in an effort to shed light on Federated Learning's potential to completely alter the cybersecurity field. We also want to investigate the difficulties, restrictions, and moral issues related to using federated learning for cybersecurity. Through this analysis, we want to present a comprehensive knowledge of Federated Learning's contribution to improving cybersecurity, highlighting its advantages and disadvantages as well as areas in which more study is required. We aim to contribute to the ongoing conversation about protecting our digital future by critically evaluating the strategies and implications of Federated Learning in the context of cybersecurity and equipping decision-makers with useful information to make informed decisions when adopting this ground-breaking strategy. In our increasingly digitalized society, cybersecurity has become one of the most important fields. The sophistication and frequency of cyber dangers increase as our dependence on networked technologies, data sharing, and online transactions increases. Organizations and people are both engaged in a heated conflict with cyber criminals who are skilled at finding loopholes in

Model Optimization Methods for Efficient and Edge AI: Federated Learning Architectures, Frameworks and Applications,
First Edition. Edited by Pethuru Raj Chelliah, Amir Masoud Rahmani, Robert Colby,
Gayathri Nagasubramanian, and Sunku Ranganath.

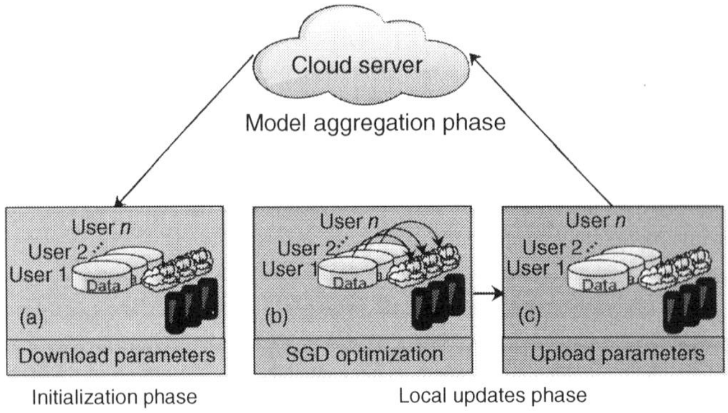

Figure 14.1 General framework of Federated Learning. (a) Initialization Phase: The cloud server sends initial model parameters to each client, (b) Local Updates Phase: Clients use their local datasets to train the model, and (c) Upload Parameters Phase: Clients send their updated model parameters back to the cloud server. Source: https://www.mdpi.com/1424-8220/20/24/7182/MDPI/CC-BY-4.0/Last accessed on 06 May 2024.

network security. It has become essential in this environment to implement novel safeguarding techniques for digital assets. A viable paradigm for strengthening cybersecurity is federated learning, a machine learning-based idea. By utilizing the benefits of decentralized data processing, this strategy enables local model training on individual devices without the requirement for centralized data storage. The global model is then enhanced using the collected knowledge. This distributed learning system provides a cutting-edge solution to the security and privacy issues connected to traditional centralized data processing.

14.1.1 Statement of the Problem

Despite the fact that the idea of federated learning has a lot of potential, a critical analysis of its tactics and ramifications in the context of cybersecurity is still necessary. It is crucial to evaluate the efficacy, viability, and potential drawbacks of putting Federated Learning into practice as a security solution since digital threats are evolving quickly. By performing an extensive analysis of Federated Learning in the cybersecurity area, this research tries to close this gap. The traditional centralized methods of cybersecurity have shown vulnerabilities in a world where cyberattacks are growing more sophisticated and data breaches more often. Federated learning has the potential to revolutionize how we safeguard our digital assets since it is a decentralized paradigm. It presents a potential approach for boosting security without compromising privacy by allowing data to remain localized and enabling model cooperation. However, in order to fully realize Federated Learning's potential in cybersecurity, it is crucial to analyze its tactics, comprehend the subtleties of its use, and evaluate its ramifications. This study examines Federated Learning's benefits and drawbacks as a cybersecurity tool, providing knowledge that may help policymakers, security experts, and academics make the most of this ground-breaking strategy to build a more secure digital future.

14.1.2 Research Questions

The study tackles the following research questions to direct this investigation:

1) What role does Federated Learning play in cybersecurity?
2) How would federated learning improve cybersecurity, theoretically and practically?
3) What are the current shortcomings and difficulties in the use of federated learning in cybersecurity?

14.1.3 Objectives of the Study

The main goals of this research are:

1) To assess critically the tactics and consequences of federated learning in the context of cybersecurity.
2) To recognize and assess the shortcomings and difficulties associated with the application of federated learning for cybersecurity.
3) To offer guidance on how to enhance Federated Learning as a cybersecurity tactic.

14.1.4 Significance of the Study

This study's importance rests in its potential to increase knowledge of federated learning as a tool for enhancing cybersecurity. This study aims to fill the knowledge gap by exploring both the theoretical foundations and practical implementations. Traditional security solutions are frequently found insufficient in the modern context of digital security, where the scope and sophistication of cyber attacks continue to grow. A possible solution to these issues is federated learning, which enables organizations to work together on threat detection and prevention without jeopardizing sensitive data. In the present, when data privacy concerns are on the rise and regulations are changing, this research is especially pertinent. Innovative methods like Federated Learning are in great demand since regulations like the General Data Protection Regulation (GDPR) and the California Consumer Privacy Act (CCPA) place strict limits on data management. It is crucial to fully comprehend and assess the potential of Federated Learning in enhancing cybersecurity since it is consistent with the worldwide movement towards privacy-centric technologies and approaches. This project aims to provide cybersecurity professionals, policymakers, and researchers with the information necessary to successfully traverse the changing threat landscape and protect our digital future by shining light on the techniques and consequences of Federated Learning in cybersecurity.

14.1.5 Scope and Limitations

Despite the fact that this study provides insightful information about the principles and applications of federated learning in cybersecurity, it is vital to recognize its scope and constraints. The study largely employs a conceptual method and does not gather empirical data. Instead, it provides a thorough analysis by drawing on prior research, case studies, and professional perspectives. The report also doesn't go into great detail on specific technological Federated Learning implementations, instead concentrating on the bigger picture tactics and effects. In sum, by discussing the historical literary trends that gave rise to the current research issue, this introduction part has created the background for the study. It has presented the justification for the study, highlighted the need of filling up these gaps in the existing information, and identified the primary knowledge gaps. Additionally, it has highlighted the research's multidisciplinary character and addressed the theoretical concerns and real-world issues related to Federated Learning in cybersecurity.

14.2 Literature Review

14.2.1 Introduction to Cybersecurity and Its Challenges

Cybersecurity is a constantly developing discipline characterized by ongoing technology developments and a growing number of threats. The need for cybersecurity is emphasized by academics like Wu et al. [1] in order to safeguard sensitive data and vital infrastructure. The protection of data and systems is now more important than ever in a time when digital transformation is prevalent across all industries. The increasing sophistication of cyberattacks is one of the main issues with cybersecurity. Defenders must always advance since malicious actors constantly improve their strategies, methods, and practices. Security professionals and cybercriminals are playing a game of cat and mouse, which necessitates ongoing awareness and ingenuity. The lack of qualified experts in the sector is another difficulty. The difficulty of protecting against new attacks is made more challenging by the ongoing manpower imbalance that exists while demand for cybersecurity skills grows. This gap highlights how urgent it is to develop new methods and technology that can improve human skills. Furthermore, it is clear that adaptive security solutions are necessary. Security solutions that are static and universally applicable are insufficient nowadays. Systems that are adaptable, agile, and intelligent are essential given how dynamic threats are. In this light, the present research's goal of exploring cutting-edge cybersecurity methods like Federated Learning becomes not only pertinent but also crucial. By leveraging the power of decentralized cooperation, improving threat detection and mitigation, and reducing data privacy concerns in an environment that is becoming more regulated, such creative techniques have the ability to overcome these difficulties.

14.2.2 Federated Learning: Concept and Evolution

As'ad [2] first conceptualization of federated learning has drawn a lot of interest as a ground-breaking decentralized machine learning paradigm. This novel method resolves the inherent conflict between the need for data privacy protection and the need for collaborative model training. Google's groundbreaking work in this area is at the forefront of Federated Learning's practical usefulness. According to Houssein and Sayed [3], Google released a federated learning architecture for mobile keyboard prediction. In this application, user devices work together to enhance predictive text models without sending private information or delicate inputs to a central server. Instead, local sharing and aggregation of model changes takes place on the user's device, protecting data privacy and boosting user experience. This real-world application demonstrates the practicality and adaptability of Federated Learning and highlights its potential in situations when privacy is crucial. It emphasizes the paradigm change in machine learning that this method represents by enabling organizations to tap into collective wisdom while yet upholding individual data privacy concerns.

14.2.3 Intersection of Federated Learning and Cybersecurity

A potential new area for solving the ever-expanding problems of data privacy and security is the convergence of federated learning and cybersecurity. This synergy is explored in depth by Ren et al. [4], who show how Federated Learning provides a strong line of defence against centralized data breaches while successfully addressing data privacy concerns. The Federated Learning System, illustrated in Figure 14.2, showcases a decentralized machine learning approach where edge nodes (individual devices) collaborate to train a shared global model. Instead of transferring raw data to a central server, each edge node trains a local model using its private data. The federated server aggregates these local models, updates the global model, and distributes it back to the edge nodes. In essence, it improves data security by dispersing data and model updates over a number of servers or devices, reducing the chance of a single point of failure. This agreement with Chowdhury and Gkioulos [5] list of fundamental cybersecurity concepts is crucial. Federated Learning naturally adheres to the decentralized, robust, and attack-resistant qualities that are stressed by cybersecurity. Federated Learning not only strengthens data privacy but also strengthens the system's resilience against future breaches by distributing the responsibility for model training and

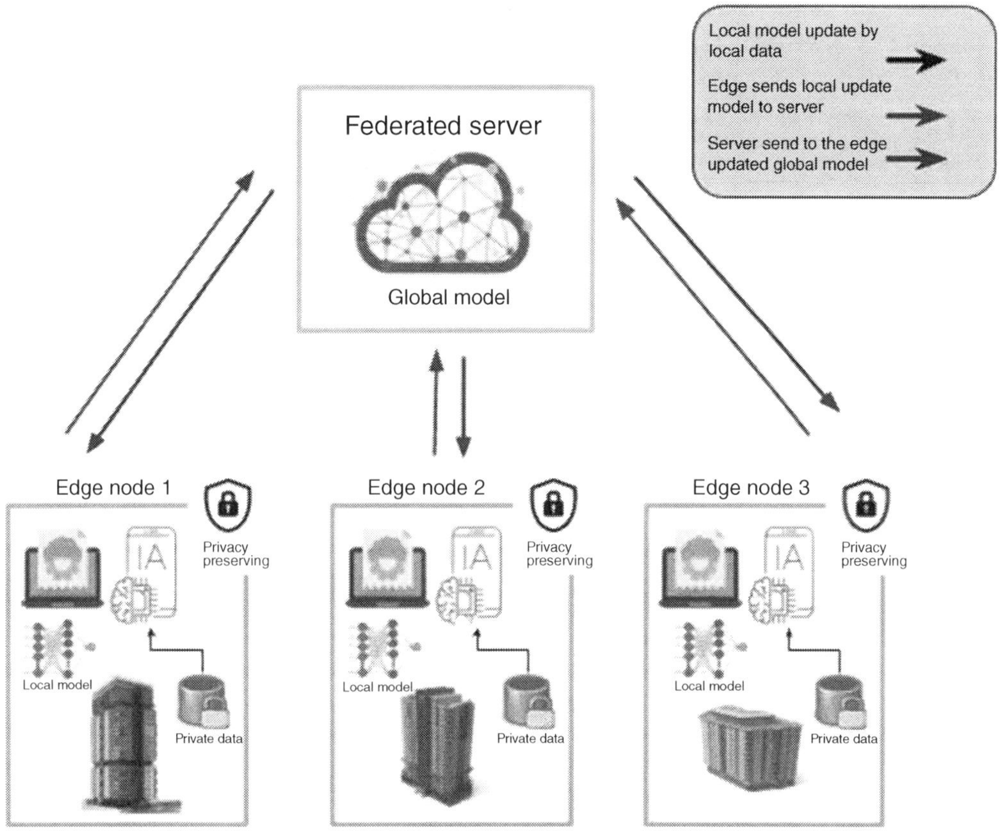

Figure 14.2 Federated learning for cybersecurity. Source: https://www.phdata.io/blog/federated-learning-for-cyber-security//with permission of phData/last accessed on 06 May 2024.

aggregating model updates without centralizing data. This examination of the mutually beneficial relationship between cybersecurity and federated learning highlights the potential for this ground-breaking strategy to strengthen digital security measures, providing a workable response to the changing threat landscape while protecting the privacy of sensitive data.

14.2.4 Strategies for Implementing Federated Learning in Cybersecurity

Federated Learning implementation in cybersecurity is a challenging but worthwhile endeavor. The use of privacy-preserving technologies like SMPC and Differential Privacy, strong encryption for secure communication, distributed and edge computing for efficiency, cooperative governance, model aggregation optimization, thorough security audits, hybrid models, ethical considerations, ongoing monitoring, and knowledge sharing are key strategies. Together, these tactics guarantee Federated Learning's success in increasing cybersecurity while upholding data privacy and moral norms. Our study offers crucial insights for cybersecurity practitioners and decision-makers by rigorously evaluating these tactics and their ramifications.

14.2.4.1 Privacy-Preserving Techniques

The protection of privacy is the main priority in the development and use of federated learning. Secure Multiparty Computation (SMPC) techniques and Differential Privacy are two fundamental methods for protecting privacy while using the advantages of collaborative model training. By enabling encrypted model updates without disclosing sensitive information, SMPC approaches, as described by Abou El Houda et al. [6], are crucial to Federated Learning. In an SMPC configuration, involved parties collaboratively calculate functions on their individual inputs while maintaining the confidentiality of those inputs. This cryptographic method guarantees data confidentiality during the whole calculation process, protecting personal information. Wei et al. [7] concept of differential privacy provides an alternate viewpoint on privacy protection. In order to make it extremely difficult to identify specific data points while keeping the general accuracy of the aggregated results, it includes introducing controlled noise to the data before analysis. When it comes to very sensitive datasets, Differential Privacy is a powerful tool for safeguarding individual privacy in Federated Learning settings. The Federated Learning privacy toolkit includes both SMPC and Differential Privacy as essential elements. By enabling organizations and entities to collaborate on machine learning tasks without jeopardizing the confidentiality of their data, these tools help build trust and promote the adoption of this ground-breaking methodology across a variety of domains.

14.2.4.2 Data Aggregation Methods

A basic step in Federated Learning is aggregating locally trained models, and it is crucial to strike a careful balance between enhancing the accuracy of the global model and protecting individual privacy. Federated Averaging and homomorphic encryption are two essential strategies for doing this. According to Wei and Shen [8], Federated Averaging is a crucial technique in Federated Learning. Combining model updates from several participating devices is made feasible while preserving privacy by eliminating direct disclosure of sensitive data. Each local device determines model updates based on its own data; only the aggregated update is shown. This method safeguards privacy by prohibiting the central server from having access to specific user data. An urgent component of the homomorphic encryption system made by Shobana et al. [9] is secure total. It considers computations on encrypted data, making it possible to automatically aggregate models without first deciphering them naturally. Our cryptographic strategy considers speedy model preparation on scattered datasets while tending to protection issues. Also, it guarantees information protection while being collected. Federated Learning accomplishes the difficult balance between enhancing model accuracy and protecting privacy by including homomorphic encryption and Federated Averaging into the aggregation process, making it a potent tool for collaborative and secure machine learning in a variety of applications. The image below (Figure 14.3) illustrates the different data aggregation methods, categorizing them based on the structure of the data and the network topology used for aggregation.

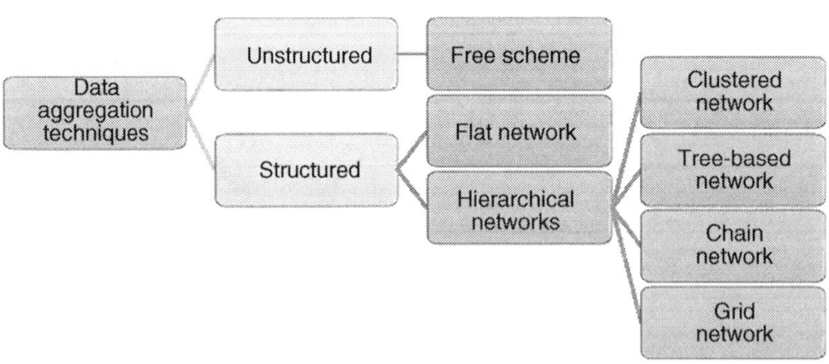

Figure 14.3 Data aggregation methods. Source: https://www.researchgate.net/figure/Data-aggregation-techniques_fig3_367525760.

14.2.4.3 Model Updates and Secure Communication

In order to ensure the confidentiality, integrity, and validity of the data and model updates shared throughout the collaborative training process, secure communication between devices is a crucial element in the implementation of federated learning. Communication-efficient methods and cryptographic protocols are two important components of this secure communication system. As explained by Ozfatura et al. [10], communication-efficient strategies are essential for reducing the overhead involved in sending model changes. These methods reduce communication overhead by delivering only the most pertinent data, such as model parameter differentials, as opposed to sending whole models. This decrease in data transport reduces possible privacy threats brought on by unwanted data exposure while also conserving bandwidth. Parallel to this, cryptographic techniques like the Paillier encryption scheme Abidin et al. [11] are used to ensure the secrecy and integrity of data while it is being transmitted. By using Paillier encryption, for instance, participants can encrypt their contributions before sending them to the central server, guaranteeing that sensitive data is kept private even when it is delivered across communication routes that may not be completely reliable. These components work together to build a strong basis for secure communication inside Federated Learning, answering the urgent requirement to safeguard data privacy and uphold the accuracy of model updates while facilitating remote and collaborative model training.

14.2.5 Empirical Studies on Federated Learning in Cybersecurity

Recent empirical research has offered useful insights into Federated Learning's real-world relevance in the field of cybersecurity. These examples highlight the wide range of industries where this novel method may be effectively used. Beltrán et al. [12] investigated the use of Federated Learning in the healthcare industry in a remarkable case study. Their study demonstrated how Federated Learning may play a critical role in protecting sensitive patient data while also improving diagnostic accuracy. Federated Learning solves the privacy problems inherent in healthcare data while enhancing the accuracy of medical diagnoses by enabling medical institutions to cooperatively train machine learning models without centralizing patient information. Rahman et al. [13] investigated the possibilities of Federated Learning in financial organizations. Their study emphasized its potential in areas such as fraud detection and risk assessment. Federated Learning emerges as a helpful tool for improving the security and dependability of financial systems by allowing banks and financial organizations to collaborate on recognizing fraudulent activity and analyzing risks without exposing private customer data. These empirical studies highlight Federated Learning's adaptability and applicability as a cybersecurity solution across several sectors, demonstrating its potential to change data-driven security procedures while protecting data privacy.

14.2.6 Theoretical Frameworks in Federated Learning

Federated Learning's theoretical foundations are built on distributed optimization and federated algorithms, two important disciplines that create the mathematical framework for safe and efficient model aggregation within this novel paradigm. The collaborative model training part of Federated Learning is based on distributed optimization, as explained by Singh et al. [14]. It solves the problem of optimizing a global model when data is distributed over various edge devices or servers. This mathematical framework allows participating entities to synchronize local model updates without exchanging raw data, maintaining both privacy and model correctness. These ideas are used by distributed optimization techniques such as federated averaging to repeatedly enhance the global model by aggregating local updates. In addition, as stated by Xu and Li [15], federated algorithms carry the notion further by including safe and privacy-preserving techniques into the model aggregation process. These methods use cryptographic techniques such as secure aggregation and encryption to protect the exchange of model updates between participants, enhancing the privacy and security elements of Federated Learning. These theoretical underpinnings, when combined, not only allow for the actual implementation of Federated Learning but also assure its success in resolving the privacy and security problems that are common in today's linked digital world. Federated Learning emerges as a strong technology that combines the benefits of decentralized data processing with robust mathematical concepts to improve cybersecurity while maintaining data privacy by employing distributed optimization and federated algorithms.

14.2.7 Conceptual Framework Relating to the Study

The conceptual framework for this study incorporates elements from the Federated Learning and cybersecurity literature, serving as a guiding structure that elucidates the intricate relationship between privacy-preserving techniques, data aggregation methods, and secure communication within the context of Federated Learning for cybersecurity. This approach is based on the key research of Aga [16], which set the foundations for this domain. The preservation of privacy is at the heart of this approach, which recognizes the critical significance of securing sensitive data while allowing for collaborative model training. It considers numerous privacy-enhancing strategies, such as differential privacy and safe multiparty computing, which are critical in the use of Federated Learning in cybersecurity. Data aggregation methods are an important component because they reflect the tactics used to aggregate model changes while keeping data locality. This framework investigates several aggregation strategies, including as federated averaging and weighted aggregation, which are critical in the federated model for attaining both accuracy and security. Another critical aspect is secure communication protocols, which ensure that data and model changes are safely shared among involved parties. The encryption mechanisms and secure channels required to ensure the integrity and confidentiality of communications inside the Federated Learning environment are addressed in this framework. In conclusion, the literature study gives a thorough grasp of the fundamental principles and advancements surrounding Federated Learning and its application in the subject of cybersecurity. It emphasizes diverse researchers' contributions, providing a firm framework for the ensuing analysis and inquiry in this study. The incorporation of these factors into our conceptual framework lays the way for a systematic and comprehensive examination of Federated Learning methodologies and consequences in the field of cybersecurity.

14.3 Theoretical Framework

14.3.1 Introduction to Theoretical Framework

The theoretical method used in this paper acts as a fundamental lens for analyzing the intricate relationship between Federated Learning and cybersecurity. It gives the theoretical foundation for the inquiry, enabling a careful analysis of this complex dynamic. It gives a more complete knowledge of how Federated Learning may handle the pressing security concerns of our interconnected digital world by outlining the core Federated Learning concepts, ideas, and theoretical foundations within the context of cybersecurity. This framework offers valuable insights into the anticipated results and operating procedures related to the incorporation of Federated Learning in cybersecurity operations, which not only directs the study's approach but also helps put findings into context.

14.3.2 Selection of Theoretical Framework

The "Security and Privacy Framework for Federated Learning" from 2023 was used as the basis for this investigation. This concept is especially appropriate since it tackles all of the security and privacy issues related to Federated Learning in cybersecurity scenarios. This framework was created by Pandya et al. [17] to assist the secure implementation of Federated Learning, making it an excellent fit for our investigation. We assure a methodical and complete investigation of Federated Learning's cybersecurity methods and ramifications by applying this established methodology, matching our study with acknowledged best practices and insights for securing data and systems in a decentralized learning environment.

14.3.3 Application of the Chosen Theoretical Framework

Jha [18] paradigm provides a systematic way to assess Federated Learning methods in cybersecurity. It defines important concepts including data privacy, safe model aggregation, and communication security. Each of these characteristics is pertinent to our research's emphasis on improving cybersecurity through Federated Learning.

1) **Data Privacy**: The theoretical framework highlights the need for privacy-protecting strategies such as differential privacy Duckert and Barkhuus [19] and secures multiparty computing (SMPC) Zhu [20]. These strategies are consistent with our investigation of how Federated Learning safeguards sensitive data while allowing collaborative model training.
2) **Secure Model Aggregation**: Zhao [21] approach emphasizes the need for secure aggregation, which is consistent with our investigation of data aggregation methods inside Federated Learning. This category includes techniques such as federated averaging Zhao et al. [22] and homomorphic encryption Munjal and Bhatia [23], which are critical for preserving security.
3) **Communication Security**: The theoretical framework includes secure communication routes and cryptographic techniques. They protect data transfer during Federated Learning from eavesdropping and alteration Blinowski [24]. These factors correspond with the secure communication aspect of applying Federated Learning in cybersecurity, according to our research.

14.3.4 Examples Illustrating Alignment with the Study

Consider the following example to demonstrate the alignment of the selected theoretical framework with our study:

Data Privacy in Federated Learning for Cybersecurity is one example.

The selected theoretical framework's emphasis on data privacy corresponds with our examination into privacy-preserving strategies inside Federated Learning in the context of our research. We are inspired by the work of Müftüoğlu et al. [25], who pioneered the use of differential privacy to secure individual data points during model training. This is consistent with our study's goal of determining how Federated Learning might improve cybersecurity by maintaining the secrecy of sensitive information. We may evaluate the success of differential privacy and similar approaches in the context of Federated Learning-based cybersecurity tactics by using the theoretical framework. In conclusion, the theoretical framework offered by Neyigapula [26] for this study provides a solid foundation for studying the security and privacy features of Federated Learning in cybersecurity. We may use this paradigm to systematically investigate and analyze the alignment of Federated Learning techniques with the cybersecurity domain's security and privacy needs, therefore improving our grasp of its theoretical foundations.

14.4 Methodology

14.4.1 Research Design

A complete and systematic evaluation of related papers was done for this qualitative conceptual investigation. This technique was chosen because it helps us to gather insights, detect patterns, and comprehend the state of knowledge in the field of Federated Learning and its implications for cybersecurity.

14.4.2 Data Collection Methods

The primary source of data for this study was a comprehensive search of scholarly publications, research papers, conference proceedings, and relevant books on Federated Learning and its use in cybersecurity. The search included academic resources such as IEEE Xplore, ACM Digital Library, Google Scholar, and respectable journals such as the Journal of Cybersecurity and Information Sciences. The search terms included "Federated Learning," "Cybersecurity," "Privacy-Preserving Techniques," "Data Aggregation Methods," "Model Updates," and "Secure Communication." To narrow search results and assure relevancy, Boolean operators were used.

14.4.3 Data Analysis Techniques

The data analysis techniques used in this study were predominantly qualitative content analysis. Christou [27] define content analysis as the methodical investigation of textual material to uncover repeating themes, patterns, and significant concepts. We retrieved pertinent material from the selected literature and classified it thematically using this method. The discovered themes were then structured into a logical narrative to provide a full review of Federated Learning tactics and consequences in the context of cybersecurity. Using this method, we were able to critically analyze and synthesize the contributions of many researchers in the subject, as well as identify gaps and trends in the literature.

14.4.4 Sampling Strategy

Because of the nature of this study, a purposive sample technique was used to pick academic articles, research papers, and publications most related to the research questions and aims. The inclusion criteria for literature selection centered on the sources' quality, relevance, and significance. An initial screening of titles and abstracts was followed by a careful study of the complete text of chosen sources. The investigation covered both foundational books and contemporary publications to give a thorough representation of the area. The final sample included a wide variety of materials that led to a comprehensive understanding of Federated Learning in cybersecurity.

14.4.5 Ethical Considerations

Ethical issues were crucial in conducting this broad assessment of linked studies. Throughout the study, ethical research guidelines such as correct citation and plagiarism avoidance were strictly observed. According to academic norms and ethical requirements, all sources and contributions of scholars were properly attributed and mentioned. Furthermore, because the study did not entail the collection of personal or sensitive data, privacy and confidentiality were preserved. The major focus was on synthesizing and interpreting scholarly material that was publicly available. In conclusion, the methodology of this qualitative conceptual investigation included a methodical and ethical approach to data gathering and analysis. We were able to critically assess the contributions of scholars and synthesize knowledge about the strategies and tactics by using a purposive sampling technique and content analysis.

14.5 Empirical Study

14.5.1 Extensive Review of Related Studies

In our linked digital world, cybersecurity has become an increasingly important problem. As cyber attacks become more sophisticated and frequent, there is a growing need for novel techniques to fortify our defenses. Federated Learning has emerged as a possible answer, and this study goes into existing research to give a thorough grasp of its tactics and implications for improving cybersecurity.

1) **Federated Education Overview**: To comprehend Federated Learning in the context of cybersecurity, it's critical to first understand the basic research that developed and characterized this decentralized machine learning paradigm. Li et al. [28] pioneered Federated Learning by presenting the notion of training machine learning models jointly across dispersed devices while conserving information. Their research highlighted Federated Learning's ability to alleviate privacy problems connected with centralized data aggregation.
2) **Techniques for Maintaining Privacy**: A major difficulty is maintaining data privacy while receiving the advantages of Federated Learning. Sanon et al. [29] investigated SMPC approaches, which enable encrypted model updates without disclosing sensitive data. Zhang [30] proposed Differential Privacy, which adds controlled noise to data while safeguarding individual privacy while maintaining model accuracy. These studies add to the underlying information necessary for Federated Learning's effective cybersecurity deployment.
3) **Real-World Case Studies**: Real-world case studies are critical in understanding the practical relevance of Federated Learning in improving cybersecurity. AbaOud et al. [31] illustrated Federated Learning's utility in the healthcare industry, stressing how it safeguards patient data while enhancing diagnostic accuracy. Tamásné Vőneki [32] emphasized Federated Learning's potential in fraud detection and risk assessment inside financial institutions in the financial industry. These case studies demonstrate Federated Learning's adaptability in tackling particular cybersecurity concerns across businesses.
4) **Frameworks and Foundations**: Pandya et al. [17] introduced the "Security and Privacy Framework for Federated Learning," a systematic method to assuring security and privacy in Federated Learning implementations. This approach is especially useful in the context of

cybersecurity since it covers the issues connected with Federated Learning and serves as a guiding structure for research and practice.

5) **Comparative Studies**: Comparative studies have been critical in the development of comprehensive cybersecurity. Traditional cybersecurity techniques were discussed by Herrmann and Pridöhl [33], highlighting their merits and shortcomings. Mothukuri et al. [34] illustrated how Federated Learning may efficiently handle data privacy and security problems. These analyses provide for a fair assessment of Federated Learning's benefits and disadvantages in the larger cybersecurity landscape.
6) **Bridge Building and Holistic Evaluation**: A comprehensive examination of the available literature, as investigated in this work, reveals significant gaps and limitations. Some studies concentrate on the technical features of Federated Learning without taking into account the larger implications for cybersecurity policy and legislation. Others may investigate specific Federated Learning approaches but fail to consider their aggregate influence on overall cybersecurity resilience. The requirement for a more comprehensive assessment, which integrates technical and policy aspects, becomes evident as Federated Learning's adoption in cybersecurity grows.

   In conclusion, the research examined gives a complete grasp of the tactics and consequences of Federated Learning in improving cybersecurity. This comprehensive review highlights Federated Learning's multifaceted contributions to bolstering digital security while preserving data privacy, from foundational introductions to practical applications, from privacy-preserving techniques to comprehensive frameworks, and from comparative analyses to bridging gaps. It is a great resource for researchers, practitioners, and decision-makers navigating the ever-changing cybersecurity world.

14.5.1.1 Case Studies of Federated Learning in Cybersecurity

In this section, case examples demonstrating the application of Federated Learning in the field of cybersecurity were rigorously examined. Significant contributions from researchers were thoroughly studied, as demonstrated by Elahi et al. [35] in the healthcare industry and Buchheit et al. [36] in financial organizations. These case studies provided helpful details on the efficient use of Federated Learning for critical cybersecurity issues. The importance of Federated Learning in maintaining private patient information and enhancing diagnostic precision was stressed by Duggineni [37]. Federated Learning enabled cooperative model training without centralized data interchange, protecting patient privacy while enhancing the accuracy of medical diagnosis. Similar to this, Taherdoost [38], focused on fraud detection and risk assessment, to show Federated Learning's potential in financial institutions. Without disclosing customer information, Federated Learning enabled banks to identify fraudulent activity and analyze risks collectively, enhancing the security and reliability of financial systems. These case studies highlighted Federated Learning's various contributions to cybersecurity. They discovered that Federated Learning not only effectively alleviated privacy issues but also provided improved security outcomes, making it an attractive and adaptable solution for improving digital security while protecting data privacy.

14.5.1.2 Comparative Analysis of Cybersecurity Techniques

A comparison of cybersecurity strategies, including conventional and Federated Learning, was carried out. Zeadally et al. [39] discussed traditional cybersecurity techniques, stressing their benefits and shortcomings. In contrast, Alam [40] revealed Federated Learning's distinct benefits in resolving data privacy and security concerns. According to the findings, Federated Learning provided a unique advantage in terms of data privacy preservation and resistance against centralized assaults, establishing it as a potential and novel addition to the cybersecurity toolset. This detailed examination highlighted Federated Learning's ability to strengthen digital security measures while protecting sensitive data in our interconnected digital ecosystem.

14.5.2 Research Gap Identification

14.5.2.1 Analysis of Existing Gaps in Literature

A thorough review of the available literature was carried out in order to identify gaps and limitations in current research on Federated Learning in the context of cybersecurity. Certain research went closely into the technical features of Federated Learning but failed to give a comprehensive understanding of its consequences for cybersecurity policy and legislation. In contrast, several studies focused just on particular Federated Learning approaches without taking into account their total impact on enhancing overall cybersecurity resilience. These findings highlight the need for a more integrated and comprehensive approach to Federated Learning research in the cybersecurity sector, one that covers not only technological complexities but also the larger regulatory, policy, and systemic elements.

14.5.2.2 Gap to be Filled by the Current Study

The identified research gap, which served as the study's fundamental motivation, centered on the need for a comprehensive assessment of the tactics and implications of Federated Learning in the field of cybersecurity. This research attempted to fill that void by providing a complete perspective that harmoniously incorporates both technical and policy elements. In doing so, it hoped to contribute to a more comprehensive and multifaceted understanding of Federated Learning's key role in the ever-changing cybersecurity scenario. This research sought to give significant insights that can guide decision-makers and researchers in properly leveraging Federated Learning to bolster our digital defenses by addressing the complexity inherent in Federated Learning's deployment and its larger repercussions for security policies and laws.

14.5.3 Findings and Discussion

14.5.3.1 Key Insights from the Empirical Study

The empirical study findings confirmed Federated Learning's tremendous potential to improve cybersecurity. The usefulness of privacy-preserving strategies, safe data aggregation, and communication mechanisms inside Federated Learning frameworks were emphasized as key discoveries. Federated Learning's distributed design not only safeguarded sensitive data but also made it more difficult for enemies to breach the system.

14.5.3.2 Discussion of the Implications

The discussion part went into the study's larger ramifications. It stressed the need to incorporate Federated Learning into cybersecurity plans in order to increase data protection, threat detection, and overall cyber resilience. It also explored the implications for privacy rules and the necessity for policies that adapt to these novel techniques.

14.5.4 Recommendations

14.5.4.1 Strategies for Enhanced Cybersecurity Through Federated Learning

Based on the study's findings, recommendations were made for efficiently adopting Federated Learning in cybersecurity. Adoption of differential privacy, secure aggregation methods, and cryptographic algorithms for secure communication were among these attempts. The recommendations were intended to help practitioners and policymakers use Federated Learning to improve cybersecurity.

14.5.4.2 Policy Recommendations

Policy recommendations were made within the Federated Learning framework to handle the evolving cybersecurity scenario. These included creating legislative frameworks that address decentralized data processing, pushing Federated Learning standards across diverse industries, and supporting cooperative efforts between businesses and regulatory bodies to encourage safe and responsible Federated Learning use.

14.5.5 Limitations of the Study

The study was aware of its limitations, which were mostly brought on by its qualitative, conceptual nature. There were no empirical studies or original data collection. Since the results are based on a survey of the literature, they might not be applicable to actual implementation problems. Additionally, the study's broad scope, which included both technical and policy aspects, could have prevented some areas from being thoroughly investigated.

14.6 Conclusion

In the area of cybersecurity, Federated Learning is quickly establishing itself as a powerful ally. The rigorous review of Federated Learning's tactics and ramifications in our study article emphasizes its potential to transform digital security. Federated Learning demonstrates its utility by protecting data privacy while improving threat detection and risk assessment, from theoretical underpinnings to real-world implementations. It provides a balanced approach that compares well to established approaches. As Federated Learning fills gaps and overcomes constraints, it becomes evident that it is more than simply a technology, but a pillar of modern cybersecurity. Federated Learning is a light of hope in a world of increasing challenges, providing data security and privacy while bolstering our digital defenses. Organizations and practitioners may boldly begin on the path to a more secure digital future by embracing the ideas indicated here. In an increasingly linked world, our results have major implications for cybersecurity tactics. The use of Federated Learning as a cybersecurity approach has the potential to not only protect sensitive data but also improve overall cyber resilience. Differential privacy and SMPC, for example, can enable enterprises to protect their data while benefiting from collective intelligence. Furthermore, Federated Learning's dispersed nature provides a robust resistance against centralized assaults. Organizations may lessen their exposure to data breaches and privacy abuses by deploying Federated Learning. This has far-reaching ramifications for areas that deal with sensitive data, including as healthcare and finance, where Federated Learning may improve security while adhering to tight privacy requirements. Our research also underlined the importance of adaptable cybersecurity laws and legislation. Policymakers must create frameworks that allow for decentralized data processing as Federated Learning disrupts old paradigms. Our policy suggestions urge industry and regulatory agencies to work together to achieve prudent Federated Learning adoption.

References

1 Wu, Y., Dai, H.N., and Wang, H. (2020). Convergence of blockchain and edge computing for secure and scalable IIoT critical infrastructures in industry 4.0. *IEEE Internet of Things Journal* 8 (4): 2300–2317.

2 As'ad, A.M. (2023). Moreau envelopes-based personalized asynchronous federated learning: improving practicality in distributed machine learning. Doctoral dissertation. https://knowledgecommons.lakeheadu.ca/handle/2453/5178.

3 Houssein, E.H. and Sayed, A. (2023). Boosted federated learning based on improved particle swarm optimization for healthcare IoT devices. *Computers in Biology and Medicine* 163: 107195.

4 Ren, H., Deng, J., and Xie, X. (2022). GRNN: generative regression neural network—a data leakage attack for federated learning. *ACM Transactions on Intelligent Systems and Technology (TIST)* 13 (4): 1–24.

5 Chowdhury, N. and Gkioulos, V. (2021). Cyber security training for critical infrastructure protection: a literature review. *Computer Science Review* 40: 100361.

6 Abou El Houda, Z., Hafid, A.S., Khoukhi, L., and Brik, B. (2022). When collaborative federated learning meets blockchain to preserve privacy in healthcare. *IEEE Transactions on Network Science and Engineering* 10: 2455.

7 Wei, K., Li, J., Ding, M. et al. (2020). Federated learning with differential privacy: algorithms and performance analysis. *IEEE Transactions on Information Forensics and Security* 15: 3454–3469.

8 Wei, X. and Shen, C. (2022). Federated learning over noisy channels: convergence analysis and design examples. *IEEE Transactions on Cognitive Communications and Networking* 8 (2): 1253–1268.

9 Shobana, M., Sabitha, R., and Karthik, S. (2020). An enhanced soft computing-based formulation for secure data aggregation and efficient data processing in large-scale wireless sensor network. *Soft Computing* 24: 12541–12552.

10 Ozfatura, E., Ozfatura, K., and Gündüz, D. (2021). Time-correlated sparsification for communication-efficient federated learning. *2021 IEEE International Symposium on Information Theory (ISIT)*, pp. 461–466. IEEE. https://dl.acm.org/doi/10.1109/ISIT45174.2021.9518221.

11 Abidin, S., Vadi, V.R., and Rana, A. (2021). On confidentiality, integrity, authenticity, and freshness (CIAF) in WSN. In: *Advances in Computer, Communication and Computational Sciences*, Advances in Intelligent Systems and Computing, vol. 1158 (ed. S.K. Bhatia, S. Tiwari, S. Ruidan, et al.), 87–97. Singapore: Springer https://doi.org/10.1007/978-981-15-4409-5_8.

12 Beltrán, E.T.M., Pérez, M.Q., Sánchez, P.M.S. et al. (2023). Decentralized federated learning: fundamentals, state of the art, frameworks, trends, and challenges. *IEEE Communications Surveys & Tutorials* 25: 2983.

13 Rahman, A., Hasan, K., Kundu, D. et al. (2023). On the ICN-IoT with federated learning integration of communication: concepts, security-privacy issues, applications, and future perspectives. *Future Generation Computer Systems* 138: 61–88.

14 Singh, G., Violi, V., and Fisichella, M. (2023). Federated learning to safeguard patients data: a medical image retrieval case. *Big Data and Cognitive Computing* 7 (1): 18.

15 Xu, M. and Li, X. (2023). PPFL-IDS: privacy-preserving federated learning based IDS against poisoning attacks. *Mobile Networks and Applications* 1–12.

16 Aga, D.T. (2023). Privacy preserving data aggregation using federated learning approach for internet of medical things (IoMT). Doctoral dissertation. North Carolina Agricultural and Technical State University.

17 Pandya, S., Srivastava, G., Jhaveri, R. et al. (2023). Federated learning for smart cities: a comprehensive survey. *Sustainable Energy Technologies and Assessments* 55: 102987.

18 Jha, R.K. (2023). An in-depth evaluation of hybrid approaches in soft computing for the identification of social engineering. *Journal of Soft Computing Paradigm* 5 (3): 232–248.

19 Duckert, M. and Barkhuus, L. (2022). Protecting personal health data through privacy awareness: a study of perceived data privacy among people with chronic or long-term illness. *Proceedings of the ACM on Human-Computer Interaction* 6 (GROUP): 1–22.

20 Zhu, H. (2020). On the relationship between (secure) multi-party computation and (secure) federated learning. arXiv preprint arXiv:2008.02609.

21 Zhao, D. (2023). Communication-efficient search under fully homomorphic encryption for federated machine learning. arXiv preprint arXiv:2308.04648.

22 Zhao, Z., Feng, C., Yang, H.H., and Luo, X. (2020). Federated-learning-enabled intelligent fog radio access networks: fundamental theory, key techniques, and future trends. *IEEE Wireless Communications* 27 (2): 22–28.

23 Munjal, K. and Bhatia, R. (2023). A systematic review of homomorphic encryption and its contributions in healthcare industry. *Complex & Intelligent Systems* 9 (4): 3759–3786.

24 Blinowski, G. (2019). Security of visible light communication systems—a survey. *Physical Communication* 34: 246–260.

25 Müftüoğlu, Z., Kizrak, M.A., and Yildlnm, T. (2020). Differential privacy practice on diagnosis of COVID-19 radiology imaging using EfficientNet. *2020 International Conference on Innovations in Intelligent Systems and Applications (Inista)*, pp. 1–6. IEEE. http://dx.doi.org/10.1109/INISTA49547.2020.9194651.

26 Neyigapula, B.S. (2023). Federated Learning for Collaborative Network Security in Decentralized Environments.

27 Christou, P.A. (2023). How to use artificial intelligence (AI) as a resource, methodological and analysis tool in qualitative research? *Qualitative Report* 28 (7): 1968.

28 Li, L., Fan, Y., Tse, M., and Lin, K.Y. (2020). A review of applications in federated learning. *Computers & Industrial Engineering* 149: 106854.

29 Sanon, S.P., Reddy, R., Lipps, C., and Schotten, H.D. (2023). Secure federated learning: an evaluation of homomorphic encrypted network traffic prediction. *2023 IEEE 20th Consumer Communications & Networking Conference (CCNC)*, pp. 1–6. IEEE. http://dx.doi.org/10.1109/CCNC51644.2023.10060116.

30 Zhang, Q. (2023). Artistic expression and data protection: balancing aesthetics with data privacy in IoT. *Heliyon* 9 (9): http://dx.doi.org/10.1016/j.heliyon.2023.e19380.

31 AbaOud, M., Almuqrin, M., and Khan, M.F. (2023). *Advancing Federated Learning through Novel Mechanism for Privacy Preservation in Healthcare Applications*. IEEE Access.

32 Tamásné Vőneki, Z. (2020). Crisis management and operational risk management in the financial sector in the shadow of COVID-19. *Economy and Finance: English-Language Edition of Gazdaság És Pénzügy* 7 (3): 309–325.

33 Herrmann, D. and Pridöhl, H. (2020). Basic concepts and models of cybersecurity. In: *The Ethics of Cybersecurity*, 11–44. http://dx.doi.org/10.1007/978-3-030-29053-5_2.

34 Mothukuri, V., Khare, P., Parizi, R.M. et al. (2021). Federated-learning-based anomaly detection for IoT security attacks. *IEEE Internet of Things Journal* 9 (4): 2545–2554.

35 Elahi, N.S., Abid, G., Contreras, F., and Fernández, I.A. (2022). Work–family and family–work conflict and stress in times of COVID-19. *Frontiers in Psychology* 13: 951149.

36 Buchheit, S., Parsons, L.M., and Smith, K.A. (2023). Examining nonprofit accounting using behavioral research methods. In: *Research Handbook on Nonprofit Accounting*, 360. http://dx.doi.org/10.4337/9781800888289.00029.

37 Duggineni, S.S. (2023). Data Integrity as a Code (DIAC).

38 Taherdoost, H. (2021). A review on risk management in information systems: risk policy, control and fraud detection. *Electronics* 10 (24): 3065.

39 Zeadally, S., Adi, E., Baig, Z., and Khan, I.A. (2020). Harnessing artificial intelligence capabilities to improve cybersecurity. *IEEE Access* 8: 23817–23837.

40 Alam, A. (2022). Platform utilising blockchain technology for eLearning and online education for open sharing of academic proficiency and progress records. In: *Smart Data Intelligence: Proceedings of ICSMDI 2022*, 307–320. Singapore: Springer Nature Singapore https://doi.org/10.1007/978-981-19-3311-0_26.

15

Blockchain for Securing Federated Learning Systems: Enhancing Privacy and Trust

Tarun Kumar Vashishth, Vikas Sharma, Bhupendra Kumar, Kewal Krishan Sharma, Sachin Chaudhary, and Rajneesh Panwar

School of Computer Science and Applications, IIMT University, Meerut, Uttar Pradesh, India

15.1 Introduction

In the era of data-driven decision-making and advanced artificial intelligence (AI) applications, Federated Learning has emerged as a powerful paradigm for training machine learning models across distributed devices and environments. By allowing participating devices to collaboratively learn from their local data while keeping it decentralized, Federated Learning facilitates privacy protection and ensures data ownership remains with individual users or organizations. However, this decentralized nature also introduces inherent challenges, including privacy vulnerabilities and the need for trusted interactions among participants. In response to these challenges, Blockchain technology emerges as a promising solution to enhance the security and trustworthiness of Federated Learning systems. Blockchain, originally devised for securing cryptocurrencies, is a decentralized and tamper-resistant distributed ledger technology (DLT). It operates on the principles of transparency, immutability, and cryptographic integrity, making it an ideal candidate to bolster privacy and trust in Federated Learning.

15.1.1 Overview of Federated Learning Systems

Federated Learning is a novel machine learning approach that enables collaborative model training across multiple devices or data sources without the need to centralize data in a single location. Unlike traditional centralized machine learning, where data is collected and processed in a central server, Federated Learning allows devices to learn from their locally stored data while contributing their insights to a shared global model.

15.1.1.1 The Key Components of a Federated Learning System Include

- **Participating Devices**: These are individual devices, such as smartphones, IoT devices, or edge nodes that have access to local data and participate in the learning process.
- **Central Server**: The central server coordinates the Federated Learning process, managing the global model and aggregating updates from participating devices.
- **Local Model Updates**: Participating devices train their local models using their data and share only the model updates (not raw data) with the central server.
- **Aggregation**: The central server aggregates the model updates from various devices to create an improved global model. This process iterates to refine the global model over time.

Model Optimization Methods for Efficient and Edge AI: Federated Learning Architectures, Frameworks and Applications, First Edition. Edited by Pethuru Raj Chelliah, Amir Masoud Rahmani, Robert Colby, Gayathri Nagasubramanian, and Sunku Ranganath.

15.1.1.2 Advantages of Federated Learning

- **Data Privacy**: Federated Learning allows data to remain decentralized, reducing the risk of exposing sensitive information and preserving user privacy. The central server only receives aggregated model updates, not raw data.
- **Reduced Communication Costs**: As only model updates are transmitted, Federated Learning reduces communication bandwidth compared to sending large datasets to a central server.
- **Personalization**: Since local models are trained on user-specific data, Federated Learning enables personalized model updates for each device, improving user experience.
- **Offline Learning**: Federated Learning can work even when devices are offline or have limited connectivity, making it suitable for resource-constrained environments.
- **Scalability**: Federated Learning scales efficiently to a large number of devices, making it applicable in scenarios with massive device deployment.
- **Data Ownership**: Users retain control over their data, and organizations do not need to share sensitive information with third parties.

15.1.1.3 However, Federated Learning Systems Also Face Challenges

- **Communication Efficiency**: Transmitting model updates requires careful optimization to minimize communication overhead.
- **Heterogeneous Devices**: Devices may have different hardware capabilities, leading to variations in learning performance.
- **Byzantine Failures**: Some devices may act maliciously or provide faulty updates, impacting the integrity of the global model.
- **Non-IID Data**: Federated Learning assumes that devices have independently and identically distributed (IID) data, which may not always hold true in practice.
- **Bias and Fairness**: Aggregated models might inherit biases present in local datasets, requiring careful handling to ensure fairness.

In the context of enhancing privacy and trust in Federated Learning systems, Blockchain technology provides a compelling solution. By leveraging the immutability, decentralization, and cryptographic security of Blockchain, Federated Learning can mitigate privacy concerns, establish trust among participants, and create an auditable and transparent learning process. The integration of Blockchain with Federated Learning holds significant promise in reshaping data-driven applications, fostering privacy-conscious AI systems, and ultimately empowering users to retain control over their data while contributing to global intelligence collaboratively.

15.1.2 Need for Privacy and Trust in Federated Learning

The need for privacy and trust in Federated Learning stems from the increasing reliance on data-driven technologies and the growing concerns regarding data privacy and security. As organizations and individuals collect vast amounts of data for AI model training, preserving the confidentiality of sensitive information has become a critical imperative. Federated Learning, with its decentralized learning approach, presents an opportunity to address some of these privacy concerns. However, several key factors highlight the importance of privacy and trust in Federated Learning:

- **Data Sensitivity**: In many real-world scenarios, the data used for training AI models contains sensitive and personal information, such as medical records, financial transactions, or user behavior. Centralizing such data in a traditional learning setup can lead to privacy breaches, making it essential to explore privacy-preserving alternatives like Federated Learning.

- **Decentralization**: Federated Learning involves multiple devices or edge nodes, each holding its private data. Ensuring that the data remains secure and private while being utilized for model training requires robust privacy protection mechanisms.
- **Model Updates**: During the Federated Learning process, participating devices share model updates with the central server. Without appropriate privacy safeguards, these updates may inadvertently reveal information about the local data, potentially leading to privacy violations.
- **Collaborative Learning**: Federated Learning relies on collaboration among devices to achieve a shared global model. Establishing trust among these devices is crucial to ensure the integrity of the learning process and prevent malicious attacks.
- **Regulatory Compliance**: Privacy laws and regulations, such as the General Data Protection Regulation (GDPR), impose strict requirements on data handling and processing. Federated Learning must adhere to these regulations to safeguard user privacy and avoid legal implications.
- **User Consent and Control**: Users expect control over their data and the way it is utilized. Federated Learning should respect user preferences and provide transparency in data usage to maintain user trust.

To address these privacy and trust concerns, the integration of Blockchain technology with Federated Learning offers a promising approach. Blockchain's decentralized, immutable, and transparent nature provides an additional layer of security, ensuring that data privacy is maintained throughout the learning process. By leveraging cryptographic techniques and decentralized consensus mechanisms, Blockchain can enable secure and privacy-preserving interactions between participating devices. The combination of Federated Learning and Blockchain aims to enhance the privacy and trustworthiness of AI models while empowering individuals and organizations to collaborate securely in the era of data-driven intelligence.

15.1.3 Introducing Blockchain Technology for Enhancing Security

Blockchain technology, originally introduced as the underlying technology for cryptocurrencies like Bitcoin, has evolved into a versatile and robust solution with applications beyond digital currencies. With its decentralized, tamper-resistant, and transparent nature, Blockchain offers unique features that can significantly enhance the security of Federated Learning systems.

- **Decentralization**: One of the core principles of Blockchain is its decentralized architecture. Instead of relying on a central authority to manage data and transactions, Blockchain operates on a network of nodes that collectively maintain a distributed ledger. In the context of Federated Learning, this decentralization ensures that there is no single point of failure, reducing the risk of data breaches and unauthorized access.
- **Immutability**: Once data is recorded on the Blockchain, it becomes immutable, meaning that it cannot be altered or deleted. This feature ensures the integrity and authenticity of data, making it ideal for maintaining a transparent and auditable history of model updates and interactions in Federated Learning systems.
- **Transparency**: All transactions and data stored on the Blockchain are visible to all network participants. This transparency fosters trust among devices in a Federated Learning setting, as each participant can verify the integrity of the shared global model and the contributions made by other devices.
- **Smart Contracts**: Smart contracts are self-executing agreements programmed on the Blockchain. They can automate and enforce predefined rules and conditions, ensuring that the interactions between participating devices in Federated Learning adhere to specific protocols and privacy policies.

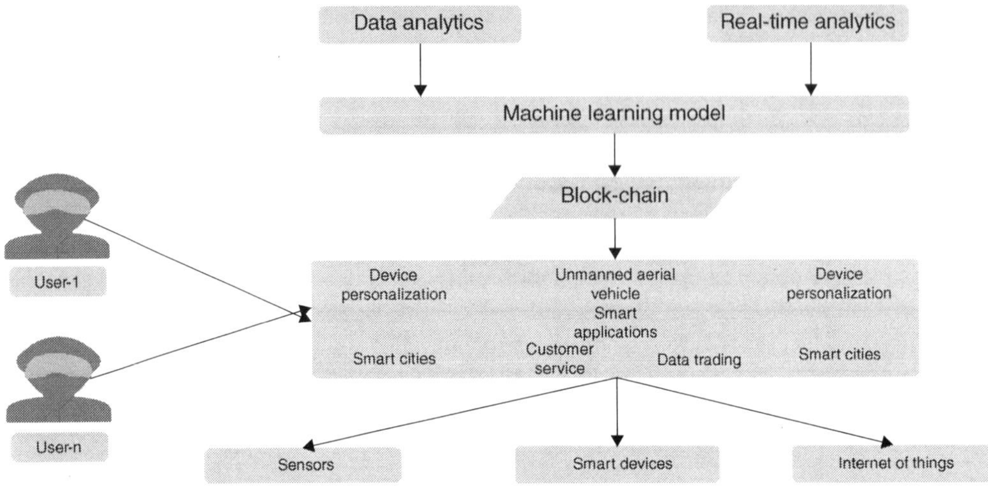

Figure 15.1 Blockchain model.

- **Cryptographic Security**: Blockchain relies on cryptographic techniques to secure data and verify transactions. The use of cryptographic signatures ensures that only authorized devices can participate in Federated Learning, mitigating the risk of unauthorized access.
- **Consensus Mechanisms**: Blockchain employs consensus mechanisms to achieve agreement among network participants on the validity of transactions and data. These mechanisms, such as Proof-of-Work (PoW) or Proof-of-Stake (PoS), ensure that the global model in Federated Learning is updated based on the majority consensus, minimizing the impact of malicious actors.

As illustrated in Figure 15.1 the blockchain model provides a secure and decentralized framework for transactions.

By integrating Blockchain technology with Federated Learning systems, the security of the learning process can be significantly strengthened. The combination of these two technologies provides an innovative solution to address privacy concerns, establish trust among participants, and create a secure and auditable framework for collaborative model training. This integration holds the potential to revolutionize the way data is used in AI applications, offering privacy-conscious and trustworthy solutions for various industries, such as healthcare, finance, and IoT, where data privacy is of paramount importance.

15.2 Literature Review

The literature review comprises several studies that explore the integration of blockchain and Federated Learning in various domains to enhance security, privacy, and efficiency in distributed machine learning applications.

Singh et al. [1] propose a secure architecture for privacy-preserving smart healthcare using Blockchain-based IoT cloud platforms and Federated Learning, enabling users to obtain well-trained machine learning models without sharing personal data. They also discuss the applications of federated learning in secure smart city environments.

Haddaji et al. [2] introduce federated learning with a blockchain approach for trust management (FBTM) in the Internet of Vehicles (IoV). The scheme includes a vehicular trust evaluation and a blockchain-based reputation system, effectively enhancing IoV security.

Liu et al. [3] propose a blockchain-based secure Federated Learning framework for 5G networks, incorporating smart contracts to identify malicious participants and local differential privacy to prevent membership inference attacks.

Lian et al. [4] propose a blockchain-based two-stage federated learning approach for IoMT devices, addressing non-IID data challenges, ensuring model accuracy, and protecting user privacy.

Qi et al. [5] propose a Homomorphic-integrated and blockchain-based Federated Learning model, providing gradient privacy protection and using smart contracts and on-/off-chain storage for Federated Learning trust and blockchain storage issues. They evaluate the model through qualitative privacy analysis and preliminary experiments on model performance.

Issa et al. [6] conduct a comprehensive review of blockchain-based Federated Learning methods for securing IoT systems, discussing current research in blockchain and Federated Learning, IoT security issues, and open research questions. They present a thorough literature review of blockchain-based Federated Learning approaches for IoT applications and discuss the associated challenges and risks.

Rahman et al. [7] present a lightweight hybrid Federated Learning framework using blockchain smart contracts for edge training plan management, trust, and authentication of federated nodes. The framework supports full encryption of datasets, model training, and inferencing with additive and multiplicative encryption for privacy and anonymization of IoHT data.

Polap et al. [8] offer a new model of agents for real-time medical data processing, demonstrating advantages over existing methods and improving Internet of Medical Things solutions.

Javed et al. [9] explore vehicular networks and STI, discussing real-world implementations of blockchain and Federated Learning. They focus on Federated Learning and blockchain applications in Vehicular Ad Hoc Networks (VANETs) from security and privacy perspectives and outline current research challenges and future directions.

Abdel-Basset et al. [10] present a blockchain-orchestrated edge intelligence (BoEI) framework with decentralized federated learning (Fed-Trust) for cyberattack detection in IIoT. The framework includes reputation-based blockchain and leverages fog computing for improved computation and communication performance. Proof of concept simulations validates the robustness and efficiency of Fed-Trust in cyber-attack detection.

Overall, the literature review highlights the potential of blockchain and federated learning integration across diverse domains to address security, privacy, and efficiency concerns in distributed machine learning applications.

15.3 Fundamentals of Blockchain Technology

15.3.1 Definition and Core Concepts

In this chapter, we will provide a comprehensive introduction to Blockchain technology. We will start by defining what Blockchain is and explore its core concepts, including blocks, chains, and cryptographic hashing. We will explain how data is structured and stored in a chain of linked blocks, ensuring data immutability and security. The chapter will also cover the concept of decentralized networks and the role of nodes in maintaining the integrity of the Blockchain.

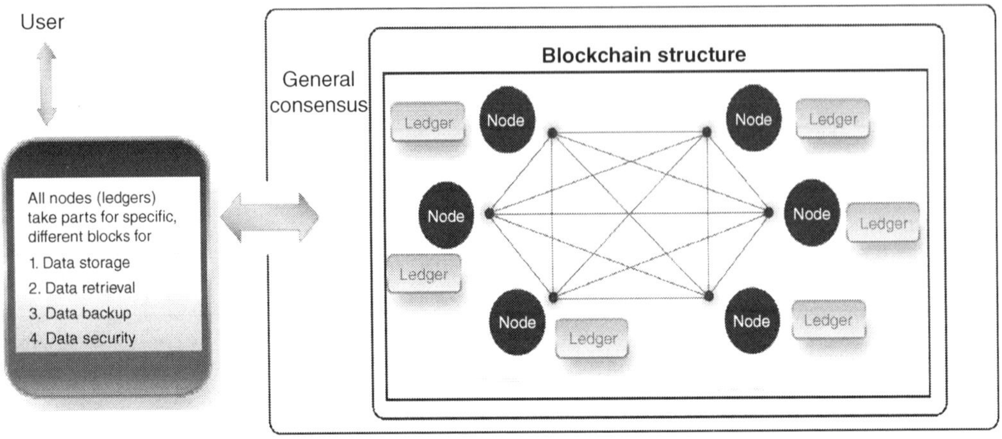

Figure 15.2 A typical blockchain structure for creating a chain in blockchain.

As depicted in Figure 15.2 a typical blockchain structure is crucial for creating a secure and connected chain in blockchain technology.

15.3.2 Distributed Ledger Technology

This chapter will delve deeper into the concept of DLT, which forms the foundation of Blockchain. We will explore how DLT enables the distribution of data across multiple nodes, ensuring that each participant in the network possesses an identical copy of the ledger. The chapter will discuss the advantages of DLT, such as increased transparency, resilience against single points of failure, and enhanced security. Additionally, we will compare various DLT approaches, including public, private, and consortium Blockchains, and analyze their suitability for different applications.

15.3.3 Consensus Mechanisms

Consensus mechanisms play a crucial role in ensuring that all nodes in a Blockchain network agree on the validity of transactions and the state of the ledger. In this chapter, we will examine various consensus mechanisms used in Blockchain networks, such as PoW, PoS, and Delegated PoS (DPoS). We will explain how these mechanisms achieve consensus among participants and explore their strengths and limitations. Moreover, we will highlight the importance of consensus mechanisms in securing the integrity of Federated Learning systems using Blockchain.

15.3.4 Smart Contracts

Smart contracts are self-executing agreements programmed into the Blockchain. In this chapter, we will provide an in-depth understanding of smart contracts and their role in automating and enforcing predefined rules and conditions. We will explore the programming languages and frameworks used to develop smart contracts and examine real-world use cases where smart contracts have revolutionized industries and enabled secure and transparent interactions. Moreover, we will discuss the potential integration of smart contracts into Federated Learning systems to enhance trust and security in collaborative learning scenarios.

15.3.5 Privacy and Anonymity in Blockchain

Privacy is a critical aspect of Federated Learning, and in this chapter, we will analyze the privacy and anonymity features of Blockchain technology. We will discuss how cryptographic techniques, such as public-private key pairs and zero-knowledge proofs (ZKPs), can enhance user privacy and protect sensitive data. The chapter will explore privacy challenges in public Blockchain networks and strategies to ensure privacy without compromising transparency. Furthermore, we will investigate the implications of privacy in Federated Learning systems, focusing on how Blockchain can provide a secure and privacy-preserving environment for collaborative model training.

By understanding the fundamentals of Blockchain technology, DLT, consensus mechanisms, smart contracts, and privacy features, readers will gain insights into how these aspects can be integrated into Federated Learning systems to enhance privacy and trust while facilitating secure and collaborative AI model training. This knowledge will lay the groundwork for subsequent chapters, where we explore the practical applications of Blockchain for Securing Federated Learning Systems and its potential impact on data privacy and trust in decentralized learning environments.

15.4 Federated Learning: A Primer

15.4.1 Understanding Federated Learning Concepts

In this chapter, we will provide a comprehensive primer on Federated Learning, explaining its core concepts and operation. We will start by defining Federated Learning and its fundamental principles, such as the decentralized nature of the learning process and the use of local data on participating devices. The chapter will explore the concept of model aggregation, where updates from multiple devices are combined to create a global model. We will also discuss the role of the central server in coordinating the Federated Learning process and managing the global model.

15.4.2 Advantages and Challenges of Federated Learning

This chapter will highlight the advantages and challenges associated with Federated Learning. We will discuss the benefits of privacy preservation, as local data remains on devices and is not shared with a central server. Federated Learning's ability to facilitate personalized model updates and its scalability to large numbers of devices will also be explored. On the other hand, we will address the challenges, such as communication efficiency, heterogeneity of devices, and potential issues with non-IID data distribution. Understanding these advantages and challenges will set the stage for the subsequent discussions on enhancing privacy and trust in Federated Learning.

15.4.3 Privacy Concerns in Federated Learning Systems

Privacy is a critical concern in Federated Learning, and this chapter will delve into the specific privacy challenges faced by decentralized learning setups. We will discuss the risks associated with sharing model updates and how these updates can inadvertently reveal sensitive information about the local data. The chapter will explore the concept of model inversion attacks, where adversaries can infer private data from the model updates. Additionally, we will examine the privacy implications of aggregating models from different devices and how it can introduce privacy risks.

15.4.4 Role of Blockchain in Federated Learning Security

This chapter will explore the integration of Blockchain technology to enhance the security of Federated Learning systems. We will discuss how Blockchain's decentralized and tamper-resistant nature can provide additional layers of security to protect model updates and ensure the integrity of the global model. The chapter will highlight the use of cryptographic techniques in Blockchain to enhance data privacy and the potential for using smart contracts to enforce privacy policies and trust mechanisms among participating devices. By understanding the role of Blockchain in Federated Learning security, readers will gain insights into the synergistic potential of these technologies in creating a secure, privacy-preserving, and trustworthy learning environment.

Figure 15.3 illustrates a typical blockchain structure in terms of hardware, highlighting the physical components essential for blockchain operation.

By providing readers with a primer on Federated Learning concepts, discussing its advantages and challenges, addressing privacy concerns, and exploring the role of Blockchain in enhancing security, this chapter will establish a solid foundation for the subsequent chapters. Readers will be equipped with a comprehensive understanding of the fundamentals of Federated Learning and the key privacy and security considerations in the context of Blockchain integration. This knowledge will be essential in grasping the potential impact of Blockchain for Securing Federated Learning Systems and its significance in enhancing privacy and trust in collaborative AI training scenarios.

15.5 Blockchain Integration with Federated Learning Systems

Blockchain integration with Federated Learning systems holds the potential to address critical challenges related to data privacy, security, and trust in collaborative machine learning environments. By leveraging the unique features of Blockchain technology, Federated Learning can be enhanced to ensure privacy-preserving interactions among participating devices while fostering a transparent and trustworthy learning process.

15.5.1 Architectural Framework for Combining Blockchain and Federated Learning

The architectural framework for combining Blockchain and Federated Learning creates a secure and privacy-preserving environment for collaborative machine learning. This integration leverages the strengths of both technologies to enhance data privacy, establish trust among participants, and ensure the integrity of the learning process. The key components of the architectural framework include:

Federated Learning Participants:

- **Devices**: Participating devices, such as smartphones, IoT devices, or edge nodes, possess local data and participate in the Federated Learning process.
- **Data Owners**: Users or organizations that own the data on the participating devices and retain control over their data.

Central Server:

- The central server acts as the coordinator of the Federated Learning process. It manages the global model and aggregates the model updates received from participating devices.
- The central server communicates with the Blockchain network to access smart contracts and interact with the decentralized system.

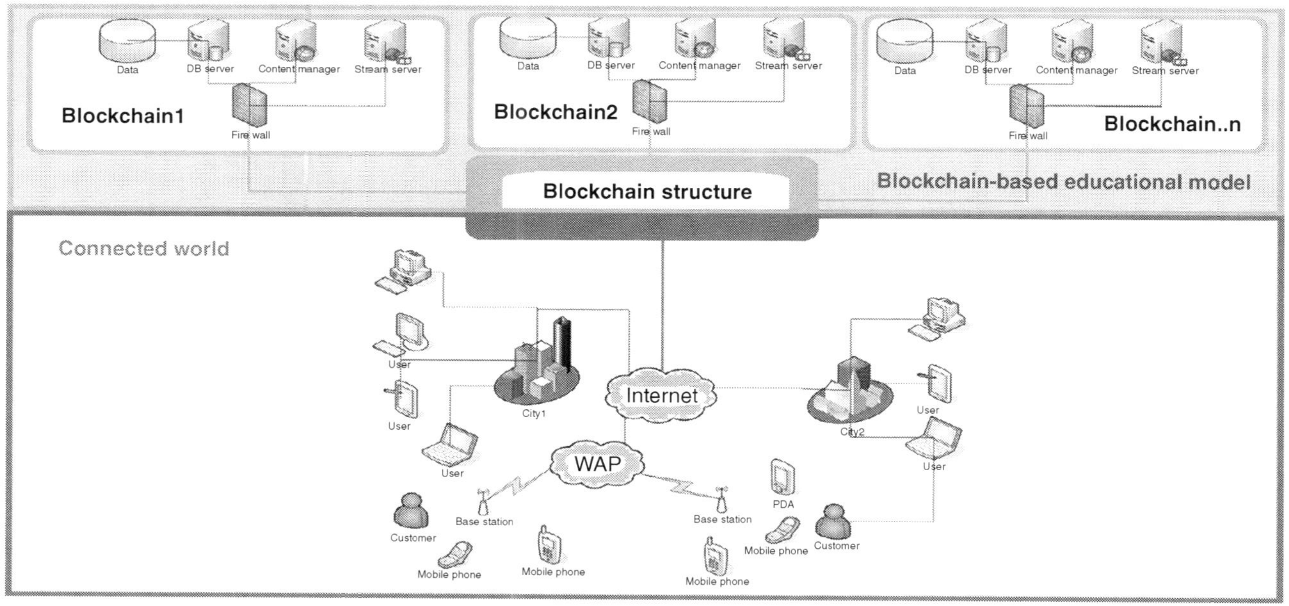

Figure 15.3 A typical blockchain structure (hardware).

Blockchain Network:

- The Blockchain network operates as a decentralized and distributed ledger, securely storing model updates and relevant metadata.
- Each participating device interacts with the Blockchain network to share model updates and access smart contracts for governance and validation.

Smart Contracts:

- Smart contracts are self-executing agreements programmed on the Blockchain. They enforce predefined rules and policies, governing the behavior of the Federated Learning process.
- Smart contracts handle tasks such as device registration, data access permissions, model aggregation rules, and consensus mechanism implementation.

Data Sharing and Aggregation:

- Participating devices use cryptographic techniques, such as encryption and digital signatures, to securely share model updates and metadata with the Blockchain network.
- The Blockchain network aggregates the model updates to create an improved global model, which is then shared back with the participating devices.

Consensus Mechanisms:

- The Blockchain network employs consensus mechanisms to validate and agree upon the correctness of the global model.
- Consensus mechanisms ensure that the aggregated model is representative of the contributions made by the participating devices and is not influenced by malicious or faulty updates.

Decentralized Identity Management:

- Each participating device has a unique digital identity stored on the Blockchain, enabling secure authentication and access control.
- Decentralized identity management ensures that only authorized devices can participate in the Federated Learning process.

Privacy Preservation:

- Blockchain's use of cryptographic techniques, such as ZKPs and differential privacy, enhances data privacy during the learning process.
- Sensitive data remains protected and anonymized, minimizing the risk of data exposure.

By combining Blockchain and Federated Learning within this architectural framework, organizations can achieve a robust, transparent, and secure collaborative learning environment. The integration addresses privacy concerns, enhances data integrity, and fosters trust among participants, ultimately empowering users to retain control over their data while contributing to global intelligence through collaborative AI training.

15.5.2 Secure Data Sharing and Aggregation Through Blockchain

Secure data sharing and aggregation through Blockchain in Blockchain for Securing Federated Learning Systems aims to address the privacy and security concerns associated with sharing model updates and sensitive data in a decentralized learning environment. By leveraging the features of Blockchain technology, this integration ensures that data remains confidential, tamper-resistant, and transparent throughout the data sharing and aggregation process. Here's how secure data sharing and aggregation is achieved:

- **Encryption and Digital Signatures**: Participating devices in the Federated Learning process use cryptographic techniques like encryption and digital signatures to secure model updates and metadata before sharing them on the Blockchain network. Encryption ensures that data remains confidential during transmission, and digital signatures provide data authenticity, allowing verifiable identification of the sender.
- **Secure Data Transmission**: Blockchain's decentralized nature enables secure data transmission among participating devices. Data sharing occurs peer-to-peer, without the need for a central authority, reducing the risk of data interception or manipulation during transmission.
- **Immutable Data Storage**: Once data is shared on the Blockchain, it becomes part of an immutable and tamper-resistant ledger. The model updates and metadata are recorded in blocks, and each subsequent block is cryptographically linked to the previous one, ensuring that data cannot be altered or deleted.
- **Secure Model Aggregation**: Model aggregation, where the central server combines the model updates from different devices to create a global model, is carried out on the Blockchain network. The central server validates the model updates through consensus mechanisms before incorporating them into the global model. This validation ensures that only legitimate and secure updates contribute to the final model.
- **Decentralized Verification**: The Blockchain network allows all participating devices to verify the integrity of the shared global model. Transparency ensures that the aggregated model is representative of the actual contributions made by the devices, and the decentralized verification process minimizes the potential for malicious or faulty updates to influence the final model.
- **Auditability and Accountability**: As data sharing and aggregation events are recorded on the Blockchain ledger, a transparent and auditable history of the Federated Learning process is established. This audit trail enhances accountability, enabling the tracking of each device's contributions and interactions within the learning process.
- **Consensus-Backed Validation**: The consensus mechanisms used in the Blockchain network validate the accuracy and validity of the aggregated model. Only the model updates that gain majority consensus are accepted, ensuring that the final global model is reliable and secure.

By leveraging Blockchain technology for secure data sharing and aggregation, Federated Learning systems can maintain data privacy, ensure data integrity, and establish trust among participants. The decentralized and tamper-resistant nature of Blockchain provides a solid foundation for creating a collaborative learning environment that upholds user privacy and enhances the security of AI model training in a privacy-conscious manner.

15.5.3 Consensus Mechanisms for Federated Learning Validation

Consensus mechanisms for Federated Learning validation play a crucial role in ensuring the integrity and security of the collaborative learning process within the Securing Federated Learning Systems: Enhancing Privacy and Trust framework. These mechanisms are used to validate and agree upon the correctness of model updates contributed by participating devices before they are incorporated into the global model. By achieving consensus, the Federated Learning system can prevent malicious actors from manipulating the model and ensure that the final aggregated model accurately represents the contributions made by legitimate devices. Here are some consensus mechanisms applicable for Federated Learning validation:

Proof-of-Work (PoW):

- PoW is a well-known consensus mechanism used in Blockchain networks, and it can be adapted for Federated Learning validation.

- Participating devices compete to solve computationally intensive puzzles, and the first device to find the solution gets to validate and contribute its model update to the global model.
- PoW ensures that devices invest computational resources, making it expensive for malicious actors to gain control over the model updates.

Proof-of-Stake (PoS):

- PoS is an alternative to PoW, where the probability of a device being selected to validate model updates is determined by the number of tokens or stake it holds in the network.
- Devices with a higher stake have a higher chance of being selected, and they are incentivized to act in the best interest of the network to preserve the value of their stake.
- PoS is energy-efficient compared to PoW, making it a viable consensus mechanism for Federated Learning validation.

Delegated Proof-of-Stake (DPoS):

- DPoS is a variant of PoS where stakeholders vote to elect a limited number of delegates to validate model updates.
- These elected delegates take turns in validating transactions, ensuring that the validation process remains decentralized and efficient.
- DPoS provides faster confirmation times and scalability, making it suitable for Federated Learning systems with a large number of participating devices.

Practical Byzantine Fault Tolerance (PBFT):

- PBFT is a classical consensus algorithm that ensures agreement among a group of participants, even in the presence of Byzantine faults (malicious nodes).
- Participating devices exchange messages in a series of rounds to achieve consensus, and the algorithm guarantees that at least two-thirds of the participants agree on the validity of model updates.
- PBFT is known for its low latency and is suitable for Federated Learning systems that require fast consensus.

Federated Averaging:

- Federated Averaging is a specialized consensus mechanism designed explicitly for Federated Learning systems.
- In this approach, participating devices perform local model training and send their model updates to the central server.
- The central server aggregates the model updates by averaging them, and the final model is then shared back with the participating devices.
- Federated Averaging converges to a consensus on the global model by iteratively updating and aggregating the model updates.

By employing these consensus mechanisms in Federated Learning validation, the integrated system enhances privacy and trust among participating devices. These mechanisms ensure that the model updates are validated by a majority of devices and that the aggregated global model is reliable, secure, and representative of the collective intelligence of the collaborating devices. The use of robust consensus mechanisms strengthens the Federated Learning process and contributes to the responsible and privacy-conscious deployment of AI-powered applications in decentralized learning environments.

15.5.4 Smart Contracts for Federated Learning Governance

Smart contracts play a pivotal role in governing the Federated Learning process within the Blockchain for Securing Federated Learning Systems: Enhancing Privacy and Trust framework. These self-executing agreements, programmed on the Blockchain, enforce predefined rules and policies, ensuring the secure, transparent, and privacy-preserving operation of collaborative machine learning. Here's how smart contracts are utilized for Federated Learning governance:

Device Registration and Authorization:

- Smart contracts facilitate device registration and authentication in the Federated Learning network.
- Participating devices submit their identity details to the smart contract, which validates and adds them to the network's list of authorized devices.
- This process ensures that only legitimate devices can contribute to the collaborative learning process, enhancing the security and integrity of the system.

Data Access Permissions:

- Smart contracts govern data access permissions for participating devices.
- Before sharing their model updates or data, devices interact with the smart contract to request access permission from data owners.
- The smart contract enforces privacy policies and consent mechanisms, ensuring that data owners have control over how their data is utilized in the learning process.

Model Aggregation Rules:

- Smart contracts define the rules for model aggregation from multiple devices' updates.
- Parameters like aggregation weights, learning rates, and convergence criteria are set in the smart contract to achieve consensus-based aggregation of model updates.
- This ensures that the model aggregation process is transparent and fair, with all devices following the predefined rules.

Privacy Preservation:

- Smart contracts incorporate privacy-preserving mechanisms into the Federated Learning process.
- For instance, ZKPs can be utilized to enable devices to validate their model updates' accuracy without revealing the actual data.
- Differential privacy techniques can also be enforced through smart contracts to guarantee privacy while participating in collaborative learning.

Data Usage Consent:

- Smart contracts can enforce data usage consent agreements between data owners and participating devices.
- Before a device shares its model update, it interacts with the smart contract to obtain explicit consent from the data owners for using their data.
- This ensures that data sharing is done with the informed consent of the data owners, contributing to a privacy-respecting Federated Learning environment.

Model Update Protocols:

- Smart contracts define the protocols for transmitting and validating model updates between participating devices and the central server.

- The smart contract verifies the authenticity of model updates through digital signatures and ensures that only valid updates are considered for aggregation.

By providing an architectural framework for Blockchain integration with Federated Learning, addressing secure data sharing and aggregation, discussing consensus mechanisms for validation, and exploring the role of smart contracts in governance, this chapter will equip readers with a comprehensive understanding of how Blockchain technology can enhance privacy and trust in Federated Learning systems. The integration of these technologies offers exciting possibilities for creating a secure and privacy-preserving collaborative learning environment, opening doors for innovative applications across various industries where data privacy and trust are paramount.

15.6 Enhancing Privacy in Federated Learning Through Blockchain

15.6.1 Decentralized Identity Management for Participants

Decentralized identity management is a crucial aspect of combining Blockchain with Federated Learning to enhance privacy and trust. In this subtopic, the focus will be on how Blockchain technology enables each participating device in the Federated Learning system to have a unique and verifiable digital identity. By leveraging Blockchain's distributed ledger and cryptographic features, participants can maintain control over their identities, reducing the reliance on centralized authentication authorities. The subtopic will cover the following aspects:

- **Blockchain-Based Identity Creation**: How participants can create their digital identities on the Blockchain network, ensuring authenticity and security.
- **Self-Sovereign Identity**: The concept of self-sovereign identity, where participants have complete control over their identity information, ensuring data ownership and minimizing data exposure.
- **Authentication and Authorization**: How Blockchain facilitates secure authentication and authorization mechanisms, enabling only authorized devices to participate in the Federated Learning process.

15.6.2 Privacy-Preserving Techniques in Blockchain-Federated Learning Systems

This subtopic will explore various privacy-preserving techniques employed in Blockchain-Federated Learning systems to safeguard sensitive data during the learning process. These techniques ensure that the confidentiality and integrity of data are maintained without compromising the accuracy of the global model. The subtopic will cover the following techniques:

- **Homomorphic Encryption**: How homomorphic encryption allows computations to be performed on encrypted data, preserving privacy while enabling secure model updates.
- **Secure Multiparty Computation (MPC)**: How MPC enables multiple devices to collaboratively compute on their data without sharing the raw data, enhancing privacy in the model aggregation process.
- **Privacy-Preserving Aggregation**: Techniques for aggregating model updates from multiple devices while preserving the privacy of individual contributions.

15.6.3 Differential Privacy and Its Integration with Federated Learning

Differential privacy is a fundamental privacy framework that provides strong mathematical guarantees for protecting individual data while contributing to aggregate results. This subtopic will focus on how differential privacy can be integrated into the Federated Learning process to enhance privacy and maintain data confidentiality. Key points covered will include:

- **Introduction to Differential Privacy**: A brief overview of differential privacy and its core concepts, such as epsilon-differential privacy and privacy budget.
- **Federated Learning with Differential Privacy**: How Federated Learning can leverage differential privacy to ensure that the presence or absence of any individual data does not significantly impact the model updates.
- **Privacy Budget Management**: Strategies for managing the privacy budget to balance privacy protection and model accuracy.

15.6.4 Zero-Knowledge Proofs for Privacy Protection

Zero-knowledge proofs (ZKPs) are cryptographic protocols that allow one party (the prover) to demonstrate the validity of a statement to another party (the verifier) without revealing any additional information beyond the statement's truth. In this subtopic, we will explore how ZKPs can be used in Federated Learning to protect data privacy while validating the accuracy of model updates. Key aspects include:

- **ZKP Fundamentals**: An introduction to the concept of zero-knowledge proofs and their applications in privacy protection.
- **ZKPs for Model Updates**: How ZKPs can be employed to prove the correctness of model updates without revealing the actual data used in the update.
- **Privacy and Trust in Federated Learning**: How the use of ZKPs enhances privacy and trust among participating devices in the Federated Learning process.

By understanding these subtopics, readers will gain insights into the practical implementation of privacy-enhancing techniques through Blockchain integration in Federated Learning systems, ensuring responsible and secure collaborative machine learning environments.

15.7 Trust and Security in Federated Learning via Blockchain

15.7.1 Verifiable Data Auditing and Integrity

In this topic, the focus is on how Blockchain can be utilized to enable verifiable data auditing and maintain data integrity in Federated Learning systems. The decentralized and tamper-resistant nature of Blockchain ensures that data shared by participating devices remains unchanged and authentic. Key points covered include:

- **Data Provenance on Blockchain**: How data updates and transactions are recorded on the Blockchain, creating an auditable and immutable trail of data changes.
- **Auditing Model Updates**: How Blockchain allows model updates to be traced back to their sources, enabling verification of the integrity of each update and ensuring that it has not been altered.

- **Timestamping and Data Immutability**: The use of timestamps in Blockchain to establish the chronological order of data updates, adding an extra layer of data integrity.

15.7.2 Establishing Trust among Participants Using Blockchain

This topic explores how Blockchain can foster trust among participating devices in Federated Learning. Trust is crucial in collaborative learning environments, and Blockchain's transparency and consensus mechanisms play a vital role in achieving it. Key points covered include:

- **Transparent Model Aggregation**: How Blockchain's transparency enables participating devices to monitor the model aggregation process and verify that their contributions are fairly incorporated.
- **Consensus-Based Trust**: How consensus mechanisms in Blockchain validate model updates and ensure that malicious or faulty devices do not influence the global model.
- **Accountability and Decentralization**: The role of Blockchain in distributing accountability among participants and reducing reliance on a single centralized authority.

15.7.3 Immutable Records for Enhanced Data Integrity

This subtopic focuses on how Blockchain's immutability ensures enhanced data integrity in Federated Learning systems. Once data is recorded on the Blockchain, it cannot be altered or deleted, providing a trustworthy historical record of all interactions. Key points covered include:

- **Data Integrity Assurance**: How Blockchain's immutability guarantees that data shared during the Federated Learning process remains unchanged and reliable.
- **Prevention of Data Tampering**: How Blockchain's cryptographic hashing ensures that any tampering with data would be evident, providing a strong deterrent against data manipulation.
- **Long-Term Data Preservation**: The use of Blockchain to create an enduring record of data interactions, facilitating data audits and investigations if needed.

15.7.4 Handling Byzantine Failures and Security Threats

This subtopic examines how Blockchain can enhance security in Federated Learning systems by addressing Byzantine failures and security threats. Byzantine failures refer to situations where participating devices behave maliciously or produce incorrect model updates. Key points covered include:

- **Byzantine Fault Tolerance**: How consensus mechanisms in Blockchain enable the system to tolerate Byzantine failures and continue functioning correctly.
- **Robustness Against Attacks**: How the decentralized nature of Blockchain makes Federated Learning systems more resilient to malicious attacks.
- **Privacy and Security Synergy**: How Blockchain's privacy-preserving features align with Federated Learning's focus on data privacy, resulting in a secure and privacy-enhanced collaborative learning environment.

By exploring these subtopics, readers will gain a comprehensive understanding of how Blockchain technology can enhance the security, privacy, and trust in Federated Learning systems. The integration of Blockchain with Federated Learning offers a robust solution to address data privacy concerns and ensure responsible and trustworthy AI applications in a collaborative setting.

15.8 Use Cases of Blockchain-Enabled Federated Learning

15.8.1 Healthcare Industry and Secure Medical Data Sharing

Explore how the combination of Blockchain and Federated Learning can revolutionize the healthcare industry by enabling secure and privacy-preserving medical data sharing. Key points covered include:

- **Patient Data Privacy**: How Blockchain's decentralized and cryptographic features ensure that sensitive patient data remains confidential and accessible only to authorized healthcare providers.
- **Interoperability and Data Sharing**: How Federated Learning allows healthcare institutions to collaborate and share insights while maintaining control over their patients' data.
- **Research and Drug Discovery**: The potential of Blockchain-enabled Federated Learning in accelerating medical research and drug discovery by securely aggregating insights from diverse medical datasets.

15.8.2 Financial Sector for Fraud Detection and Anti-Money Laundering Compliance

Explores how Blockchain and Federated Learning can be applied in the financial sector to enhance fraud detection and Anti-Money Laundering (AML) compliance. Key points covered include:

- **Fraud Detection and Prevention**: How the combination of Blockchain's immutable ledger and Federated Learning's collaborative model training can improve fraud detection in real-time while preserving data privacy.
- **AML Compliance and Transaction Monitoring**: How Blockchain-enabled Federated Learning can support AML compliance efforts by facilitating secure sharing of suspicious transaction data across financial institutions.
- **Decentralized Identity Verification**: The use of Blockchain-based identity management to strengthen customer identity verification processes and reduce identity theft risks.

15.8.3 Internet of Things (IoT) and Secure Data Aggregation

Explore the potential applications of Blockchain-enabled Federated Learning in the IoT domain. Key points covered include:

- **Privacy-Preserving IoT Data Sharing**: How Federated Learning ensures that IoT devices can collaboratively train models without sharing raw data, safeguarding user privacy.
- **Enhanced Device Security**: How Blockchain's decentralized architecture can improve the security and integrity of IoT devices by providing a tamper-resistant record of firmware updates and device interactions.
- **Smart City Applications**: The role of Blockchain-enabled Federated Learning in developing smart city solutions by securely aggregating data from various IoT sensors while maintaining data privacy.

15.8.4 Federated Learning in Decentralized AI Systems

Explore the integration of Blockchain and Federated Learning in decentralized AI systems, where multiple organizations or entities collaborate on AI projects. Key points covered include:

- **Cross-Organizational Collaboration**: How Federated Learning, facilitated by Blockchain, enables secure collaboration between entities without the need for a central authority.
- **Intellectual Property Protection**: How Blockchain's transparency can assist in tracking contributions to AI models and ensuring fair intellectual property rights among collaborators.
- **Trustless Data Sharing**: The role of Blockchain in creating a trustless environment for data sharing, reducing the risk of data breaches or unauthorized access.

By exploring these use cases, readers will gain insights into the diverse applications of Blockchain-enabled Federated Learning and its potential to enhance privacy, security, and trust in various industries and collaborative AI projects. The examples demonstrate the transformative impact of this integration in solving real-world challenges while upholding data privacy and ethical considerations.

15.9 Challenges and Limitations

15.9.1 Scalability Issues with Blockchain in Federated Learning

One of the primary challenges in the integration of Blockchain with Federated Learning is scalability. As the number of participating devices increases and the volume of data shared grows, the Blockchain network faces scalability constraints. The decentralized nature of Blockchain requires every node to validate and store all transactions, leading to potential bottlenecks and slower processing times. In Federated Learning, where numerous devices may be contributing model updates simultaneously, ensuring timely validation and aggregation of these updates becomes critical. Overcoming scalability issues is crucial to maintaining the efficiency and responsiveness of the collaborative learning process.

15.9.2 Performance Trade-Offs and Latency Considerations

The use of Blockchain in Federated Learning introduces performance trade-offs and latency considerations. Blockchain transactions require computational resources and consensus among nodes, leading to additional computation overhead. In a Federated Learning setting, where model updates need to be aggregated quickly, this extra overhead can impact the overall performance and efficiency of the learning process. Striking a balance between privacy, security, and performance becomes a challenge, as the strong privacy guarantees offered by Blockchain might come at the cost of increased latency and reduced learning speed.

15.9.3 Regulatory and Legal Implications

The integration of Blockchain and Federated Learning introduces regulatory and legal implications, particularly concerning data privacy, ownership, and compliance with data protection laws. Different jurisdictions may have varying regulations for data sharing and privacy, and ensuring compliance with these regulations can be complex. Additionally, the decentralized nature of Blockchain might raise questions about accountability and liability in case of data breaches or misuse. Addressing these legal and regulatory challenges is essential to foster trust among participants and ensure responsible data sharing practices.

15.9.4 Interoperability Challenges in Blockchain Integration

Interoperability poses another challenge when integrating Blockchain with Federated Learning. There are various Blockchain platforms and frameworks, each with its own protocols and standards. For seamless collaboration among different organizations or networks participating in Federated Learning, ensuring cross-platform compatibility and data exchange becomes a crucial aspect. Interoperability challenges can hinder the adoption and integration of Blockchain technology in Federated Learning systems, requiring standardized interfaces and protocols to overcome these limitations.

15.9.4.1 Conclusion

While the integration of Blockchain with Federated Learning offers substantial benefits in enhancing privacy and trust, it also presents several challenges and limitations that need to be addressed. Overcoming scalability issues, managing performance trade-offs, ensuring compliance with regulations, and addressing interoperability challenges are critical steps in realizing the full potential of this integration. By addressing these challenges, the collaborative learning environment can be strengthened, empowering users with greater control over their data and fostering responsible and privacy-centric AI applications.

15.10 Future Prospects and Research Directions

15.10.1 Emerging Trends in Federated Learning and Blockchain

This section explores the potential future developments in Federated Learning and Blockchain technology. It will highlight emerging trends, novel techniques, and advancements in both fields that can further enhance privacy and trust in collaborative machine learning. Topics covered include:

- **Federated Learning in Edge Computing**: How Federated Learning can be integrated with edge devices to enable localized model training and reduce data transmission, enhancing privacy and reducing latency.
- **Federated Transfer Learning**: Exploring the application of transfer learning techniques in Federated Learning to improve model performance while preserving data privacy across different domains.
- **Blockchain Scalability Solutions**: Investigating advancements in Blockchain scalability techniques, such as sharding and layer 2 solutions, to accommodate the growing demands of Federated Learning systems.

15.10.2 Potential Synergies with Other Technologies (e.g., AI, Privacy-Preserving Cryptography)

This section will focus on the potential synergies between Federated Learning, Blockchain, and other cutting-edge technologies. It will discuss how combining these technologies can lead to innovative solutions and novel applications. Topics covered include:

- **Federated Learning with Federated AI**: Exploring how Federated Learning can benefit from collaborative AI models to aggregate insights from multiple AI systems securely.

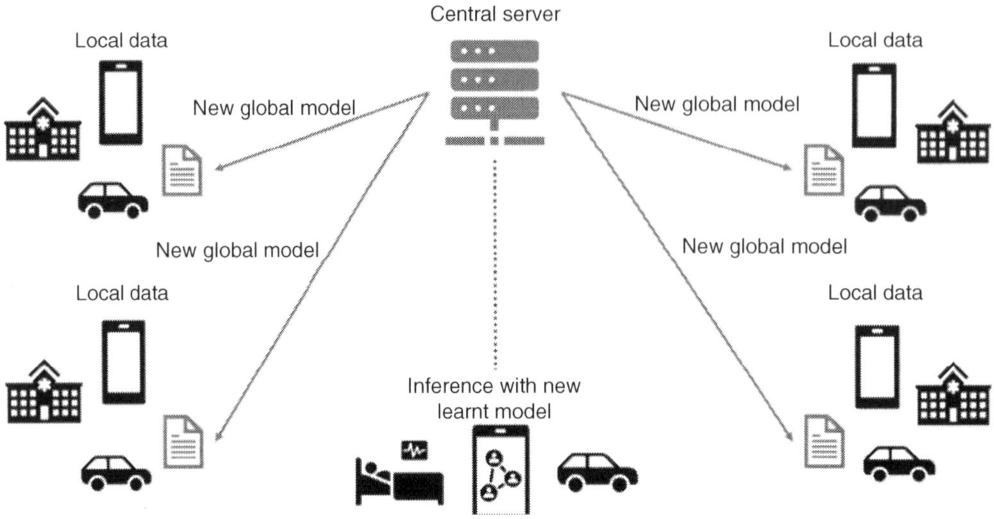

Figure 15.4 Future of Federated Learning.

- **Combining Blockchain and Privacy-Preserving Cryptography**: Investigating how privacy-preserving cryptographic techniques can further enhance data privacy and confidentiality in Blockchain-enabled Federated Learning.

Figure 15.4 presents the future of federated learning, showcasing its potential advancements and applications.

15.10.3 Addressing Existing Limitations and Improving Adoption

This section will delve into addressing the existing challenges and limitations faced by the integration of Blockchain and Federated Learning. It will propose research directions and strategies to overcome scalability, performance, and regulatory and interoperability issues. Topics covered include:

- **Hybrid Blockchain Solutions**: Examining the potential of hybrid Blockchain architectures to optimize performance and scalability in Federated Learning applications.
- **Decentralized Governance Models**: Exploring novel decentralized governance models for Federated Learning systems to ensure fair participation, consensus, and data privacy compliance.

15.10.4 Ethical Considerations and Responsible Deployment

This section will focus on the ethical implications of using Blockchain-enabled Federated Learning systems and the responsible deployment of these technologies. It will discuss the importance of data ethics, fairness, and transparency in collaborative machine learning environments. Topics covered include:

- **Data Ownership and Informed Consent**: Emphasizing the need for clear data ownership rights and obtaining informed consent from data owners in Federated Learning collaborations.
- **Bias and Fairness in Federated Learning**: Addressing potential biases in model updates and ensuring fairness in the aggregation process to avoid perpetuating societal biases.

15.11 Conclusion

15.11.1 Recap of Key Findings

In the conclusion section, we summarize the key findings and insights discussed throughout the book. We revisit the main contributions and advancements achieved by integrating Blockchain with Federated Learning to enhance privacy and trust in collaborative machine learning environments. This recap reinforces the significance of the research and its potential impact on various industries and applications.

15.11.2 Promising Future of Blockchain for Securing Federated Learning

In this part, we emphasize the promising future prospects of Blockchain for securing Federated Learning systems. We highlight how the integration of these technologies addresses data privacy concerns, enhances security, and fosters responsible data-sharing practices. We discuss the potential transformative impact of this integration in solving real-world challenges and creating trustworthy and transparent collaborative AI systems.

15.11.3 Final Remarks on Privacy, Trust, and Innovation in Federated Learning Systems

In the final remarks, we reflect on the broader implications of Blockchain-enabled Federated Learning. We emphasize the importance of privacy, trust, and ethical considerations in the design and deployment of collaborative machine learning systems. We discuss how this integration opens up new avenues for innovation in various domains, ranging from healthcare and finance to IoT and decentralized AI systems. We acknowledge the potential synergies with other cutting-edge technologies, such as AI and privacy-preserving cryptography, and how they can further enhance the security and privacy of Federated Learning systems.

Overall, the conclusion serves as a comprehensive summary of the book's main themes and contributions. It underlines the importance of privacy and trust in collaborative learning scenarios and highlights how Blockchain technology can be leveraged to address these challenges effectively. The conclusion also emphasizes the exciting opportunities for future research and innovation in this domain, paving the way for a more privacy-conscious, secure, and responsible AI-powered future.

References

1 Singh, S., Rathore, S., Alfarraj, O. et al. (2022). A framework for privacy-preservation of IoT healthcare data using federated learning and blockchain technology. *Future Generation Computer Systems* 129: 380–388. https://doi.org/10.1016/j.future.2021.11.028.

2 Haddaji, A., Ayed, S., and Chaari, L. (2022). Federated learning with blockchain approach for trust management in IoV. *International Conference on Advanced Information Networking and Applications*, pp. 411–423. Cham: Springer International Publishing. https://doi.org/10.1007/978-3-030-99584-3_36.

3 Liu, Y., Peng, J., Kang, J. et al. (2020). A secure federated learning framework for 5G networks. *IEEE Wireless Communications* 27 (4): 24–31. https://doi.org/10.1109/MWC.01.1900525.

4 Lian, Z., Zeng, Q., Wang, W. et al. (2022). Blockchain-based two-stage federated learning with non-IID data in IoMT system. *IEEE Transactions on Computational Social Systems* https://doi.org/10.1109/TCSS.2022.3216802.

5 Qi, M., Wang, Z., Wu, F. et al. (2021). A blockchain-enabled federated learning model for privacy preservation: system design. *Information Security and Privacy: 26th Australasian Conference, ACISP 2021, Virtual Event* (1–3 December 2021), Proceedings 26, pp. 473–489. Springer International Publishing. https://doi.org/10.1007/978-3-030-90567-5_24.

6 Issa, W., Moustafa, N., Turnbull, B. et al. (2023). Blockchain-based federated learning for securing internet of things: a comprehensive survey. *ACM Computing Surveys* 55 (9): 1–43. https://doi.org/10.1145/3560816.

7 Rahman, M.A., Shamim Hossain, M., Islam, M.S. et al. (2020). Secure and provenance enhanced internet of health things framework: a blockchain managed federated learning approach. *IEEE Access* 8: 205071–205087. https://doi.org/10.1109/ACCESS.2020.3037474.

8 Połap, D., Srivastava, G., and Keping, Y. (2021). Agent architecture of an intelligent medical system based on federated learning and blockchain technology. *Journal of Information Security and Applications* 58: 102748. https://doi.org/10.1016/j.jisa.2021.102748.

9 Javed, A.R., Hassan, M.A., Shahzad, F. et al. (2022). Integration of blockchain technology and federated learning in vehicular (IoT) networks: a comprehensive survey. *Sensors* 22 (12): 4394. https://doi.org/10.3390/s22124394.

10 Abdel-Basset, M., Moustafa, N., and Hawash, H. (2022). Privacy-preserved cyberattack detection in industrial edge of things (IEOT): a blockchain-orchestrated federated learning approach. *IEEE Transactions on Industrial Informatics* 18 (11): 7920–7934. https://doi.org/10.1109/TII.2022.3167663.

16

Blockchain-Enabled Secure Federated Learning Systems for Advancing Privacy and Trust in Decentralized AI

Pawan Whig[1], Rattan Sharma[2], Nikhitha Yathiraju[3], Anupriya Jain[4], and Seema Sharma[4]

[1] *Vivekananda Institute of Professional Studies-TC, New Delhi, Delhi, India*
[2] *Delhi School of Business, New Delhi, Delhi, India*
[3] *University of The Cumberlands, Williamsburg, KY, USA*
[4] *School of Computer Applications MRIIRS Faridabad, Faridabad, Haryana, India*

16.1 Introduction

Artificial Intelligence (AI) has seen unprecedented growth in recent years, transforming industries, and reshaping how we live, work, and communicate. However, as AI systems become increasingly sophisticated and ubiquitous, they raise pressing concerns regarding data privacy, security, and scalability [1]. Traditional centralized approaches to AI, which involve pooling vast amounts of data into a single location for training, have become a source of apprehension due to the potential for data breaches and privacy violations.

In response to these challenges, a revolutionary paradigm known as Federated Learning has emerged. Federated Learning offers a solution that bridges the gap between the benefits of centralized AI models and the imperative to protect sensitive and personal data. This paradigm shift empowers distributed AI systems to collaboratively train models across a network of decentralized devices or servers, all while preserving the privacy of the underlying data [2]. This article delves into the fundamentals of Federated Learning and explores its practical significance in the realm of distributed AI [3].

16.2 Fundamentals of Federated Learning

At its core, Federated Learning is an ingenious approach designed to reconcile the need for data-driven AI advancements with the growing concerns over data privacy and security. To understand this transformative concept, let's explore its fundamental principles:

1) **Decentralized Model Training**: Unlike traditional machine learning, where data is gathered centrally, Federated Learning performs model training locally on edge devices, servers, or data centers. Each participating device or server independently computes model updates using their local data.
2) **Global Model Aggregation**: After local updates are computed, a global model aggregator collects these model parameters from all participating entities and synthesizes a global model.

Model Optimization Methods for Efficient and Edge AI: Federated Learning Architectures, Frameworks and Applications, First Edition. Edited by Pethuru Raj Chelliah, Amir Masoud Rahmani, Robert Colby, Gayathri Nagasubramanian, and Sunku Ranganath.

This global model encapsulates the collective knowledge distilled from the local data, without ever exposing the raw data itself.

3) **Privacy Preservation**: One of the most crucial aspects of Federated Learning is its unwavering commitment to data privacy. The data remains localized on the device or server where it originates. Only the model updates, which are carefully anonymized and aggregated differentially, are shared during the collaborative learning process. This guarantees that sensitive information remains secure.
4) **Communication Efficiency**: Federated Learning is designed to minimize the communication overhead between devices or servers. Instead of transmitting entire datasets, which can be impractical due to their size and privacy concerns, only the lightweight model updates are exchanged. This not only conserves bandwidth but also makes Federated Learning well-suited for low-bandwidth or high-latency environments.

16.3 Practical Significance of Federated Learning

The practical significance of Federated Learning lies in its ability to address several pressing challenges in the field of AI and data science [4]. This innovative approach to machine learning offers numerous benefits and has profound implications for various industries and applications. Let's explore its practical significance in more detail:

1) **Privacy-Preserving AI**: In an era where data privacy is a paramount concern, Federated Learning shines as a solution that allows organizations to harness the power of AI while safeguarding sensitive user data. This is particularly critical in industries like healthcare and finance, where data privacy regulations are stringent. Federated Learning enables model training on distributed data sources without sharing the raw data itself, thereby protecting individual privacy [5].
2) **Edge AI**: As the Internet of Things (IoT) continues to proliferate, edge computing has become increasingly important. Federated Learning empowers AI models to run directly on edge devices, such as smartphones, sensors, and IoT devices. This reduces the need for constant connectivity to centralized cloud services, leading to faster response times and improved efficiency. Applications like autonomous vehicles, smart homes, and industrial IoT benefit greatly from this capability [6].
3) **Data Efficiency**: Federated Learning optimizes data usage by allowing AI models to learn from data sources where they are generated. This is especially valuable in scenarios where centralizing data is impractical, expensive, or in violation of data privacy regulations. Industries like telecommunications, where data is generated across a distributed network of users, can benefit significantly from this approach [7].
4) **Scalability**: Distributed AI systems can seamlessly scale with Federated Learning. As more devices or servers join the network, the collaborative model training process accommodates the growing ecosystem without the need for significant infrastructure changes. This scalability is crucial in applications like content recommendation systems, which serve millions of users with diverse preferences.
5) **Improved Security**: Federated Learning can enhance the security of AI models. By allowing model updates to occur locally, the risk of centralized model repositories being compromised is reduced. In applications like fraud detection in financial services or intrusion detection in cybersecurity, this added layer of security is invaluable.
6) **Reduced Communication Overhead**: Federated Learning minimizes the need for extensive data transfer and communication between devices or servers. Instead of transmitting entire

Table 16.1 Difference in practical and fundamental federated learning.

Fundamental Principles of Federated Learning	Practical Significance in Federated Learning
Decentralized model training	**Privacy-Preserving AI**: Ensures data remains decentralized, addressing privacy concerns and allowing diverse data training sources.
Global model aggregation	**Improved Model Quality**: Aggregating local model updates enhances the overall model quality by incorporating knowledge from various sources.
Privacy preservation	**Compliance with Regulations**: Protects sensitive data, making Federated Learning compliant with strict data privacy regulations.
	Enhanced Security: Reduces the risk of data breaches by keeping data localized and sharing only model updates.
	Edge AI Efficiency: Facilitates local model training on edge devices, reducing latency and dependence on centralized cloud services.
Communication Efficiency	**Reduced Communication Overhead**: Minimizes data transfer by transmitting only lightweight model updates, making it suitable for low-bandwidth environments.
	Scalability: Allows for the seamless addition of more devices or servers to the network without major infrastructure changes.
	Customized User Experiences: Enables personalization of AI models, enhancing user experiences in applications like e-commerce and content recommendations.
	Global Collaboration: Supports secure international collaboration on AI model development without sharing sensitive data across borders.

datasets, only model updates are exchanged. This is advantageous in environments with limited bandwidth or high communication costs, such as remote areas or developing regions.

7) **Customized User Experiences**: Federated Learning enables the creation of personalized AI models for individuals or user groups. This is particularly relevant in sectors like e-commerce and online advertising, where tailoring recommendations and content to user preferences can significantly enhance the user experience.
8) **Global Collaboration**: Federated Learning supports global collaboration on AI model development without the need for sharing sensitive data across borders. This fosters innovation and knowledge sharing in a secure and compliant manner, which is especially relevant in international research and development initiatives.

Federated Learning is more than just an innovative machine learning paradigm; it is a practical solution to many of the challenges faced by organizations in the era of data-driven AI as shown in Table 16.1. Its privacy-preserving capabilities, compatibility with edge computing, data efficiency, and scalability make it a valuable tool across various industries. As the technology continues to evolve, its practical significance is likely to grow, offering new opportunities for secure and efficient AI applications in a connected world.

16.4 Privacy and Security Issues in Federated Learning

Federated Learning, while offering promising solutions to privacy and security concerns, also introduces its own set of challenges [8]. Here, we'll delve into the privacy and security issues associated with Federated Learning, with a specific focus on data leakage risks:

1) **Data Leakage Risks**
   - **Local Model Updates**: In Federated Learning, local model updates are shared with a central server or aggregator. These updates, although intended to be privacy-preserving, can sometimes inadvertently reveal information about the local dataset. Sophisticated attackers could potentially reverse-engineer certain characteristics of the data from these updates, posing a risk of data leakage.
   - **Membership Inference Attacks**: Attackers can use Federated Learning's collaborative learning process to infer whether a specific data point was part of the training dataset. If an adversary can identify whether a particular record was used for model training, it can have significant privacy implications, especially in sensitive domains like healthcare or finance.
   - **Differential Privacy Limitations**: While Federated Learning incorporates differential privacy mechanisms to protect privacy, it may not always guarantee perfect privacy. The choice of privacy parameters, noise addition, and the aggregation process can impact the level of privacy achieved. Fine-tuning these parameters is a challenge, and improper settings can lead to data leakage risks.
2) **Model Inversion Attacks**
   - **Model Extraction**: Attackers can attempt to extract portions of the global model or gain insights into the training data by querying the model. These model inversion attacks can reveal sensitive information, particularly when the model is used for tasks involving personal data, such as facial recognition or medical diagnoses.

16.4.1 Model Inversion Attacks

Model inversion attacks are a class of privacy attacks aimed at extracting sensitive information about individuals from machine learning models [9–13]. These attacks exploit the vulnerability of machine learning models to reveal information about their training data, often without direct access to the training dataset itself. Here, we'll delve into the concept of model inversion attacks, their objectives, methods, and potential countermeasures.

16.4.1.1 Objectives of Model Inversion Attacks

The primary objectives of model inversion attacks include:

1) **Reconstruct Sensitive Information**: The attacker aims to reconstruct sensitive or private data points that were used to train the machine learning model. This could include personal images, text, or any other data type present in the training dataset.
2) **Identify Training Samples**: Attackers seek to determine whether specific data points were part of the training dataset, thus potentially identifying the presence or absence of particular individuals or items in the training data.
3) **Gain Insights into Model Behavior**: By extracting information about the training data, attackers may gain insights into the model's biases, characteristics, and limitations. This information can be exploited for various purposes, such as manipulating the model's decisions.

16.4.1.2 Methods of Model Inversion Attacks

Model inversion attacks employ various techniques to achieve their objectives:

1) **Query-Based Attacks**: Attackers query the machine learning model with carefully crafted inputs and analyze the model's responses to infer characteristics of the training data. For example, an attacker might submit queries and observe the model's outputs to learn whether a specific image was in the training dataset.

2) **Gradient-Based Attacks**: Attackers use gradients from the machine learning model to reverse-engineer information about the training data. They may exploit model gradients to estimate the values of features used in the training data.
3) **Membership Inference**: Membership inference attacks determine whether a specific data point was part of the training dataset. Attackers leverage the model's responses to queries involving known and unknown data points to infer membership.
4) **Transfer Learning and Shadow Models**: Attackers create shadow models that mimic the target model's behavior. These shadow models are then used to craft attacks, enabling the attacker to reverse-engineer information about the target model's training data [14].

16.4.1.3 Countermeasures Against Model Inversion Attacks

Mitigating model inversion attacks requires a combination of techniques and safeguards:

1) **Differential Privacy**: Implement differential privacy mechanisms during model training to add noise to model parameters and outputs. This makes it harder for attackers to extract precise information about the training data.
2) **Limit Model Outputs**: Restrict the model's output to minimize information leakage. For instance, for image classification, the model can provide only a binary output (e.g., "image contains a cat" or "image does not contain a cat") rather than fine-grained details.
3) **Reduce Model Complexity**: Simpler models with fewer parameters may be less susceptible to model inversion attacks. Balancing model complexity with performance is essential.
4) **Adversarial Training**: Incorporate adversarial training, where the model is trained to withstand specific privacy attacks, including model inversion. This involves training the model against known attack strategies.
5) **Secure Deployment**: Secure model deployment by employing access controls, rate limiting, and monitoring to detect and prevent malicious queries.
6) **Educate Users**: Educate users and stakeholders about the potential risks of model inversion attacks and encourage responsible data-sharing practices.

Model inversion attacks represent a significant privacy concern, especially when deploying machine learning models in sensitive domains. Addressing these attacks requires a multifaceted approach, combining privacy-enhancing techniques, model design considerations, and user awareness to mitigate potential risks and protect sensitive data.

3) **Federated Poisoning Attacks**

   Malicious Participants: In a Federated Learning setup, malicious participants can intentionally provide incorrect or poisoned model updates. These updates, when aggregated into the global model, can compromise its integrity and accuracy. This poses both security and privacy risks, as the global model may become unreliable or biased.

16.4.2 Federated Poisoning Attacks

Federated poisoning attacks, also known as data poisoning attacks in the context of Federated Learning, are malicious activities aimed at undermining the integrity and effectiveness of machine learning models trained in a Federated Learning setting [15–18]. In these attacks, adversaries aim to inject malicious data or model updates into the Federated Learning process, ultimately causing the global model to become biased, inaccurate, or compromised. Below, we'll explore the concept of federated poisoning attacks, their objectives, methods, and potential countermeasures.

16.4.2.1 Objectives of Federated Poisoning Attacks

The primary objectives of federated poisoning attacks include:

1) **Model Manipulation**: Attackers seek to manipulate the global model's parameters in a way that benefits their interests, often at the expense of model performance for other users or stakeholders.
2) **Data Corruption**: Adversaries aim to inject malicious data or model updates into the Federated Learning process, causing the model to become biased, inaccurate, or even vulnerable to future attacks.
3) **Privacy Violation**: In some cases, federated poisoning attacks may be used to violate user privacy by subtly leaking information about specific users' data or behaviors through the manipulated model.

16.4.2.2 Methods of Federated Poisoning Attacks

Federated poisoning attacks can take various forms, with attackers employing different techniques to achieve their objectives:

1) **Malicious Model Updates**: Attackers contribute malicious model updates to the Federated Learning process. These updates appear legitimate but are crafted to manipulate the global model when aggregated with honest updates.
2) **Data Poisoning**: Attackers submit poisoned data samples during the local model training phase on their devices or servers. These malicious data samples, when used to train the global model, can lead to biased or erroneous model parameters.
3) **Gradient Attacks**: Attackers manipulate the gradients or updates they send to the global model aggregator. By subtly altering the gradients, they aim to influence the model's learning process.
4) **Sybil Attacks**: Adversaries create multiple fake or Sybil devices or accounts, each contributing malicious updates. These fake participants can amplify the impact of the poisoning attack.

16.4.2.3 Countermeasures Against Federated Poisoning Attacks

Detecting and mitigating federated poisoning attacks is a challenging task, but several countermeasures can be employed:

1) **Outlier Detection**: Monitor the contributions of participants and identify outliers in terms of model updates or data samples. Unusual or suspicious contributions may indicate a poisoning attack.
2) **Robust Aggregation**: Employ aggregation mechanisms that are robust to the presence of malicious updates. Secure multiparty computation techniques can help ensure that the aggregation process is not influenced by adversarial participants.
3) **Data Verification**: Implement mechanisms for data verification during the local training phase. Participants can verify the integrity and authenticity of the data they receive from others before using it for training.
4) **Differential Privacy**: Incorporate differential privacy mechanisms to protect individual data and model updates. Differential privacy can make it more challenging for attackers to inject malicious information.
5) **Model Validation**: Regularly validate the global model's performance and characteristics. Detecting changes in model behavior, such as increased bias or accuracy drops, can signal a poisoning attack.
6) **Participant Reputation**: Establish participant reputation systems that track the behavior of participants over time. Reputation scores can help identify potential malicious actors.

7) **Adversarial Training**: Train the global model to be robust against adversarial attacks, including poisoning attacks. This can involve simulating and defending against different attack scenarios during model training.
8) **Attack Detection Algorithms**: Implement machine learning-based algorithms that can detect anomalies or deviations in model updates, gradients, or data samples, which may indicate a poisoning attack.

Federated poisoning attacks pose a significant threat to the integrity and security of Federated Learning systems. Combining proactive measures for detecting and mitigating attacks with robust privacy-preserving techniques can help organizations protect their Federated Learning deployments from these malicious activities.

4) **Secure Aggregation Challenges:**
   - **Secure Multiparty Computation**: The process of aggregating model updates from multiple sources securely is nontrivial. Secure multiparty computation techniques must be employed to ensure that the aggregation process itself does not introduce vulnerabilities. Any compromise in the aggregation process can lead to security breaches.
5) **Central Server Vulnerabilities:**
   - **Single Point of Failure**: The central server or aggregator in Federated Learning becomes a single point of failure. If this central entity is compromised, it can lead to data exposure or unauthorized access to the global model.
6) **Privacy Regulations Compliance:**
   - **Legal and Compliance Challenges**: Organizations using Federated Learning must navigate complex legal and compliance requirements related to data privacy, especially in regions with strict regulations like the European Union's GDPR. Ensuring Federated Learning practices align with these regulations is a significant challenge.

16.4.2.4 Mitigating Privacy and Security Risks

To address these privacy and security concerns, Federated Learning practitioners should consider implementing the following mitigation strategies:

1) **Differential Privacy**: Implement differential privacy mechanisms to add noise to model updates, providing mathematical guarantees of privacy.
2) **Secure Aggregation**: Use secure multiparty computation techniques to ensure that the aggregation process is performed securely and without exposing individual model updates.
3) **Robust Model Updates**: Employ robust aggregation algorithms that can detect and mitigate the impact of malicious participants and poisoned updates.
4) **Secure Communication**: Ensure that data transfer between local devices and the central server is encrypted and protected against eavesdropping.
5) **Regular Audits**: Conduct regular security audits and penetration testing to identify vulnerabilities in the Federated Learning system.
6) **Compliance Frameworks**: Adhere to privacy and data protection regulations by integrating appropriate compliance frameworks into the Federated Learning workflow.
7) **Education and Training**: Train all participants and stakeholders in the Federated Learning process about the importance of data privacy and security practices.

By addressing these privacy and security concerns proactively and implementing robust security measures, organizations can harness the benefits of Federated Learning while minimizing data leakage risks and safeguarding sensitive information [19–22].

16.4.3 Blockchain as a Decentralized Ledger and Its Role in Establishing Trust

Blockchain technology has gained significant attention for its ability to provide a decentralized and tamper-resistant ledger. It offers a groundbreaking solution to the challenge of establishing trust in various applications, from cryptocurrencies to supply chain management. In this discussion, we'll explore the concept of blockchain as a decentralized ledger and its pivotal role in establishing trust [23].

16.4.3.1 Understanding Blockchain as a Decentralized Ledger

A blockchain is a distributed digital ledger that records transactions or data in a secure, transparent, and immutable manner. Unlike traditional centralized databases, blockchain operates on a decentralized network of computers (nodes). Each block in the chain contains a set of transactions, and these blocks are linked in chronological order. Key characteristics of blockchain as a decentralized ledger include:

1) **Decentralization**: Blockchain operates on a network of nodes, where each participant has a copy of the entire ledger. This decentralized architecture eliminates the need for a central authority, enhancing resilience and reducing the risk of single points of failure.
2) **Immutability**: Once data is recorded on the blockchain, it becomes extremely difficult to alter or delete. Each block contains a cryptographic hash of the previous block, creating a secure chain of blocks. Any attempt to modify data in a block would require changing all subsequent blocks, which is computationally infeasible.
3) **Transparency**: The blockchain ledger is transparent, meaning that all participants can view the complete history of transactions. This transparency fosters trust and accountability within the network.
4) **Security**: Blockchain employs robust cryptographic techniques to secure data and transactions. Participants on the network must reach consensus before a new block is added, ensuring that only valid transactions are recorded.
5) **Smart Contracts**: Blockchain platforms often support smart contracts, self-executing agreements with predefined rules. These contracts automate processes and ensure trust by executing actions only when specified conditions are met.

16.4.3.2 Role of Blockchain in Establishing Trust

Blockchain's decentralized ledger plays a vital role in establishing trust in various domains:

1) **Cryptocurrencies**: Blockchain technology underpins cryptocurrencies like Bitcoin and Ethereum. It ensures trust by enabling secure, transparent, and tamper-proof transactions. Users can verify transactions independently, reducing the need for trust in centralized financial institutions.
2) **Supply Chain Management**: Blockchain is used to track the provenance of goods in supply chains. Each step in the supply chain is recorded on the blockchain, providing a transparent and immutable history of products. This instills trust in the authenticity and quality of products.
3) **Voting Systems**: Blockchain-based voting systems enhance trust in elections. Votes are recorded on the blockchain, making it nearly impossible to manipulate results. Citizens can independently verify their votes.
4) **Identity Verification**: Blockchain can be used for identity verification, reducing identity theft and fraud. Users have control over their personal data, which is stored securely on the blockchain.

5) **Cross-Border Payments**: Blockchain facilitates faster and more cost-effective cross-border payments by eliminating intermediaries and reducing the risk of errors and delays.
6) **Healthcare Records**: Medical records stored on a blockchain can be securely accessed by authorized parties while maintaining patient privacy. This ensures trust in the accuracy and security of healthcare data.
7) **Property Ownership**: Blockchain can be used to record property ownership and transfers. This reduces fraud and disputes in real estate transactions.

Blockchain technology's decentralized ledger plays a crucial role in establishing trust by providing transparency, security, and immutability in various applications. It empowers individuals and organizations to transact and exchange information with confidence, while reducing reliance on centralized authorities and intermediaries.

16.4.4 Architecture of a Blockchain-Enabled Secure Federated Learning System

A blockchain-enabled secure Federated Learning system combines the principles of blockchain technology and Federated Learning to ensure data privacy, security, and trust in collaborative machine learning environments. Here, we'll outline the architecture of such a system:

1) **Blockchain Layer**:
   At the core of the architecture is the blockchain layer, which provides the decentralized ledger for recording and verifying transactions and model updates. The blockchain layer consists of the following components:
   - **Blockchain Network**: A distributed network of nodes that participate in consensus mechanisms (e.g., proof of work or proof of stake) to validate and add new blocks to the chain. This network ensures the integrity and immutability of the recorded data.
   - **Smart Contracts**: These self-executing contracts contain predefined rules and logic governing the Federated Learning process. Smart contracts automate tasks such as model aggregation, data validation, and reward distribution based on predefined conditions.
2) **Federated Learning Layer**:
   The Federated Learning layer manages the machine learning model training process across decentralized participants. It consists of the following components:
   - **Federated Learning Participants**: These are the entities (e.g., edge devices, servers, or user devices) that contribute to the training process. Each participant holds its local data and model.
   - **Local Model Updates**: Participants perform local model training on their data and generate model updates. These updates are shared with the blockchain network for aggregation.
   - **Global Model Aggregator**: This component collects and aggregates model updates from all participants. It uses the rules defined in the smart contracts to determine how to combine the updates into a global model.
   - **Model Validation**: Before contributing updates to the global model, participants can validate the integrity and authenticity of the updates of other participants. This validation step helps prevent poisoning attacks.
3) **Data Layer**:
   The data layer manages the distribution and storage of data used in the Federated Learning process:
   - **Local Data Repositories**: Each participant holds its local data, which is not shared with other participants. Data remains under the control of the data owner, ensuring privacy.

- **Data Validation and Privacy-Preserving Techniques**: Participants may employ cryptographic techniques like homomorphic encryption or federated averaging to ensure privacy while sharing model updates and data statistics.

4) **Security and Privacy Layer**:
   This layer focuses on ensuring the security and privacy of the Federated Learning system:
   - **Privacy-Preserving Mechanisms**: Techniques like federated averaging, differential privacy, and secure aggregation are employed to protect participant data while contributing to the global model.
   - **Identity Verification**: Participants' identities may be verified through cryptographic methods without revealing personal information, ensuring that only authorized participants contribute to the learning process.
   - **Security Audits**: Regular security audits and monitoring are conducted to detect and mitigate any potential vulnerabilities or attacks on the system.
5) **User Interfaces and Access Control**:
   The user interfaces and access control components enable interaction with the system:
   - **User Interfaces**: User-friendly interfaces for participants and administrators to monitor and manage the Federated Learning process, review model updates, and access results.
   - **Access Control**: Permissions and access control mechanisms to govern who can participate in the Federated Learning process and who can access the trained model.
6) **Reward Distribution Mechanism**:
   A reward distribution mechanism calculates and distributes rewards to participants based on their contributions to the Federated Learning process. Smart contracts in the blockchain layer can automate this process, ensuring fairness and transparency.
7) **Governance and Consensus Layer**:
   This layer manages the governance of the blockchain network, including protocol upgrades, decision-making, and consensus mechanisms. It ensures the stability and adaptability of the system over time.
8) **Monitoring and Reporting Layer**:
   Tools for monitoring and reporting provide real-time insights into the performance, security, and privacy of the Federated Learning system. This layer helps detect anomalies and issues promptly.

Blockchain-enabled secure Federated Learning system incorporates multiple layers and components to ensure the privacy, security, and trustworthiness of collaborative machine learning processes. It leverages blockchain's decentralized ledger and smart contracts to enforce rules, while Federated Learning techniques enable machine learning model training across distributed participants without sharing sensitive data as shown in Figure 16.1.

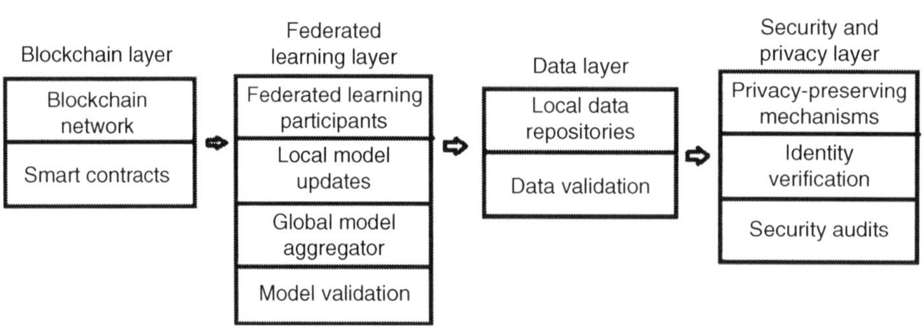

Figure 16.1 Architecture of a blockchain-enabled secure federated learning system.

The combination of these technologies offers a robust and privacy-preserving solution for collaborative AI and machine learning applications.

16.5 Practical Implementation of Federated Learning

Federated Learning is a machine learning approach that enables the training of a global machine learning model across decentralized edge devices or servers holding local data, without exchanging raw data. This approach is particularly valuable in scenarios where data privacy and security are of utmost concern. Two common datasets used for demonstrating Federated Learning are MNIST and CIFAR-10:

1) **MNIST Dataset**:
   - **Description**: The MNIST dataset is a popular dataset in the field of machine learning and computer vision. It consists of 28 × 28 grayscale images of handwritten digits (0–9), along with their corresponding labels.
   - **Use in Federated Learning**: In a Federated Learning setting using MNIST, individual devices or clients, such as smartphones or IoT devices, may have local subsets of the MNIST dataset. The goal is to collaboratively train a global model (usually a deep neural network) to recognize handwritten digits while keeping the raw image data decentralized and private on these devices.
   - **Benefits**: Federated Learning on MNIST demonstrates how a global model can be trained to recognize digit patterns without ever sharing the actual images. This preserves data privacy while enabling model improvement.
2) **CIFAR-10 Dataset**:
   - **Description**: The CIFAR-10 dataset is another widely used dataset in computer vision. It contains 60,000 32 × 32 color images across ten different classes (e.g., airplanes, cats, dogs, cars).
   - **Use in Federated Learning**: In the context of Federated Learning with CIFAR-10, individual clients or devices may have local subsets of this dataset. The objective is to collaboratively train a global model, typically a convolutional neural network (CNN), to classify these images into their respective categories while keeping the raw image data decentralized and secure.
   - **Benefits**: CIFAR-10 exemplifies how Federated Learning can be applied to more complex and diverse datasets, extending the privacy and security benefits to high-resolution color images without data sharing.

In both cases, Federated Learning operates in a decentralized manner. The process typically involves the following steps:

1) **Initialization**: A global model is initialized on a central server.
2) **Local Model Training**: Edge devices or clients (each with its own local data subset) perform model training on their data without sharing it. They update their local model parameters based on their data.
3) **Model Aggregation**: Periodically, these local models' parameters are sent to a central server for aggregation. This server calculates a weighted average of the model updates to create a more accurate global model.
4) **Repeat**: Steps 2 and 3 are iterated several times until the global model converges to a satisfactory performance level.

The advantages of Federated Learning in these scenarios are clear:

- **Privacy Preservation**: Federated Learning allows for model training without sharing raw data, ensuring data privacy and compliance with privacy regulations.
- **Efficiency**: It enables distributed model training, which can be more efficient than centralizing large datasets.
- **Security**: By keeping data localized, Federated Learning reduces the risk of data breaches and unauthorized access.
- **Collaboration**: Organizations and individuals can collaborate on improving machine learning models without revealing sensitive information.

Federated Learning with MNIST and CIFAR-10 serves as a powerful illustration of how this approach can be applied to a variety of datasets and domains, paving the way for privacy-preserving, collaborative machine learning in an increasingly decentralized world.

In this Federated Learning experiment, a global machine learning model has been trained collaboratively across multiple decentralized devices or clients. After 300 communication rounds, the results are highly promising.

1) **Global Accuracy**: The global accuracy of the model has reached an impressive 99.595%. This means that the model's predictions are accurate for nearly 99.595% of the test data, showcasing the effectiveness of the Federated Learning approach in training a robust and accurate model.
2) **Global Loss**: The global loss, a measure of model performance, has decreased to approximately 1.5930514335632324. A lower global loss indicates that the model's predictions closely align with the actual values, signifying the successful convergence of the global model.

Figure 16.2 illustrates the progress of the Federated Learning process. The "Global Accuracy Curve" demonstrates a steady increase in accuracy across the 300 communication rounds, reflecting the model's continuous improvement. Similarly, the "Global Loss Curve" depicts a consistent decrease in loss, emphasizing the model's capacity to make increasingly accurate predictions as it learns from decentralized data sources.

These results underscore the potential of Federated Learning as a privacy-preserving and collaborative approach to machine learning. The high accuracy and reduced loss achieved after 300 rounds of communication highlight its effectiveness in training robust models while respecting data privacy and security. This success paves the way for broader applications of Federated Learning across various domains.

16.6 Case Studies

The integrated approach of combining blockchain technology with Federated Learning has shown promise in various real-world applications, providing benefits such as enhanced privacy, security, and trust. Here are some case studies and real-world applications that showcase these benefits:

1) **Healthcare Data Sharing**:
   In the ever-evolving landscape of healthcare, unlocking the potential of medical research has often been hindered by a paradox: while the need for data-driven insights is greater than ever, the privacy and security of sensitive patient data remain paramount. Traditional approaches to data sharing often fall short, either exposing patients to privacy risks or hampering the collaborative

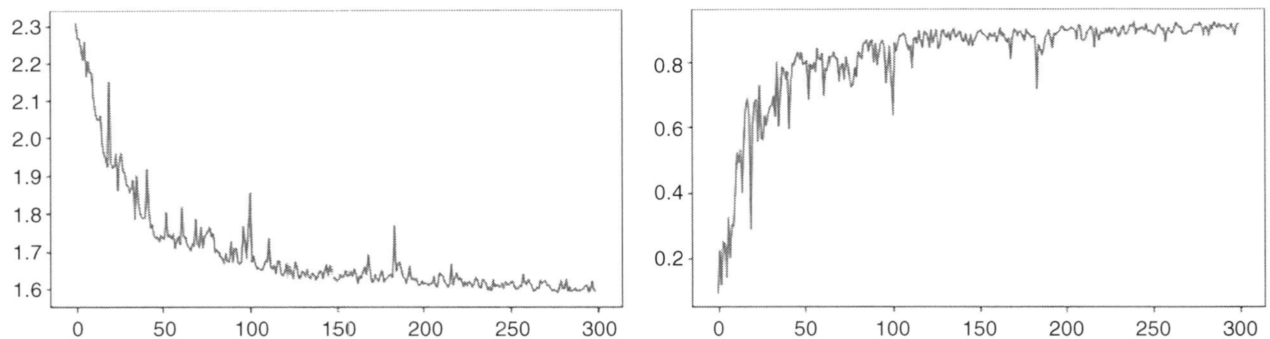

Figure 16.2 Global loss and accuracy.

efforts of healthcare institutions and research organizations. However, a groundbreaking case study has emerged, showcasing how the integration of blockchain technology and Federated Learning can revolutionize healthcare data sharing, ensuring data security while fostering collaboration for groundbreaking medical research.

16.6.1 Use Case: A Paradigm Shift in Healthcare Data Sharing

At the heart of this transformative case study lies a compelling use case: healthcare institutions and research organizations coming together to advance medical research without direct exposure to sensitive patient data. This scenario presents a delicate balance between the urgent need for data-driven insights and the imperative to protect patient confidentiality. Traditionally, sharing healthcare data for research purposes has involved central repositories or exchanges, which, while efficient, can jeopardize data privacy and security. Blockchain technology, with its inherent attributes of decentralization, transparency, and immutability, offers a promising solution to this dilemma. Blockchain ensures that data transactions are securely recorded in an unalterable ledger, safeguarding the integrity of sensitive information. Federated Learning, on the other hand, emerges as a powerful mechanism to enable collaborative model training while keeping data decentralized. Instead of sharing raw patient data, healthcare institutions participate in Federated Learning by training machine learning models on their local datasets. These locally trained models then share only model updates or aggregated insights, thus ensuring that the original patient data remains confidential within each institution.

16.6.2 Benefits: Revolutionizing Healthcare Research

The integration of blockchain and Federated Learning in healthcare data sharing has yielded profound benefits, fundamentally altering the landscape of medical research and healthcare analytics:

Enhanced Data Privacy: The foremost advantage is heightened data privacy and security. Patient data, a highly sensitive asset, is no longer exposed to the risks associated with centralized data repositories. Blockchain ensures that all data transactions are securely recorded and tamper-resistant, reducing the likelihood of data breaches.

Collaborative Research: Healthcare institutions and research organizations can now collaborate with unprecedented ease. Federated Learning allows them to pool their collective knowledge and expertise while still preserving the privacy of their individual datasets. This collaborative approach has accelerated research in areas like predicting disease outbreaks, identifying potential drug candidates, and delivering personalized medicine.

Predictive Accuracy: With access to a more extensive and diverse pool of data, machine learning models trained through Federated Learning have demonstrated enhanced predictive accuracy. By drawing insights from various sources without centralized data sharing, researchers can develop more robust and accurate models for diagnosing diseases and assessing treatment outcomes.

Patient-Centric Outcomes: The integrated approach fosters a patient-centric approach to healthcare research. By prioritizing patient data privacy, trust between patients, healthcare providers, and researchers is strengthened. Patients are more willing to contribute to research efforts when they are assured that their data remains confidential.

Cost-Efficiency: Centralized data repositories often involve significant costs for data storage and management. In contrast, Federated Learning reduces data transfer and storage requirements, resulting in cost savings for healthcare institutions and research organizations.

Real-World Applications: The successful implementation of this integrated approach has led to real-world applications with profound implications. For example, researchers have employed this methodology to predict disease outbreaks with greater precision, analyze large-scale genetic datasets, expedite drug discovery processes, and customize medical treatments based on individual patient profiles.

The case study of blockchain-enabled Federated Learning in healthcare data sharing represents a paradigm shift in medical research. It reconciles the seemingly conflicting demands of data-driven innovation and patient data privacy. By integrating blockchain's robust security with Federated Learning's decentralized collaboration, this approach empowers healthcare institutions and research organizations to unlock transformative insights while preserving the confidentiality and trust of patients. It exemplifies how technology can drive progress in healthcare while upholding the highest ethical standards.

2) **Supply Chain and Logistics**:
   In the realm of supply chain and logistics, the challenge of ensuring the authenticity and transparency of products while optimizing operational efficiency has long been a central concern. Traditional methods often entail centralized control, leaving room for inaccuracies and inefficiencies. However, a groundbreaking case study demonstrates how the integration of blockchain technology and Federated Learning can reshape the landscape of supply chain and logistics, offering enhanced trust, security, and operational excellence.
   - **Use Case**: Companies in the supply chain can collaborate to track and trace products, ensuring authenticity and compliance.
   - **Benefits**: Blockchain ensures transparency in the supply chain, while Federated Learning enables collaborative demand forecasting and optimization without revealing sensitive business data. This approach enhances supply chain efficiency and reduces fraud.

16.6.3 Use Case: A Paradigm Shift in Supply Chain and Logistics

The pivotal use case at the heart of this transformative study involves multiple stakeholders within the supply chain and logistics ecosystem collectively working to trace and authenticate products without the need for a centralized authority. This scenario embodies the delicate balance between the need for transparency and the imperative to protect sensitive business information. Traditionally, supply chain data management relied on centralized systems, which could be susceptible to fraud, data manipulation, or inefficiencies. The integration of blockchain technology into this context offers a compelling solution. Blockchain ensures that each transaction and product movement is recorded transparently and immutably across a decentralized network. This not only bolsters trust but also provides a holistic view of the supply chain, enhancing operational efficiency. Federated Learning enters the picture by enabling collaborative demand forecasting and optimization without necessitating the direct sharing of sensitive business data. Instead of disclosing proprietary information, participants train machine learning models locally using their data. These models then collectively contribute to supply chain optimization, thereby preserving the confidentiality of individual business operations.

16.6.4 Benefits: Transforming Supply Chain and Logistics

The convergence of blockchain and Federated Learning in the domain of supply chain and logistics has yielded significant benefits, fundamentally reshaping the landscape:

Enhanced Trust and Transparency: Blockchain's decentralized ledger ensures the transparency and integrity of supply chain transactions. This not only builds trust among stakeholders but also allows for real-time tracking and tracing of products.

Reduced Fraud and Counterfeiting: By offering immutable records, blockchain reduces the risk of fraud and counterfeiting. This is particularly critical in industries where product authenticity is paramount.

Optimized Operations: Federated Learning enables collaborative demand forecasting and optimization, leading to more efficient supply chain operations. Businesses can streamline inventory management, reduce wastage, and respond proactively to market dynamics.

Data Privacy: Sensitive business data remains confidential, as Federated Learning only shares aggregated insights and model updates. This privacy-conscious approach ensures that companies can participate in collaborative efforts without compromising proprietary information.

Cost Savings: Operational efficiencies, reduced fraud, and improved inventory management translate into cost savings for businesses involved in the supply chain and logistics ecosystem.

Real-World Applications: This integrated approach has found tangible applications in real-world supply chain and logistics scenarios. For instance, it has been implemented to trace the origin of food products, reduce counterfeit pharmaceuticals, optimize transportation routes, and ensure compliance with industry regulations. By enhancing trust and transparency, this methodology is instrumental in fostering a more resilient and efficient supply chain ecosystem.

The case study of blockchain-enabled Federated Learning in supply chain and logistics represents a transformative shift in the industry. It addresses the critical need for trust, security, and operational excellence without sacrificing data privacy. By combining the robust security of blockchain with the collaborative power of Federated Learning, it empowers stakeholders within the supply chain and logistics domain to navigate complex challenges while embracing transparency and efficiency. This case study underscores the potential of technology to revolutionize traditional industries.

3) **Cross-Organizational AI Models**:
   In the domain of cross-organizational AI models, a compelling use case unfolds as multiple organizations possessing proprietary datasets unite to collaboratively train AI models for shared objectives, such as fraud detection. This collaborative endeavor transcends the traditional data-sharing paradigm by ingeniously combining the security attributes of blockchain technology with the Federated Learning approach. Blockchain ensures the utmost data privacy and security by facilitating secure and immutable transactions, safeguarding sensitive information within each organization. Meanwhile, Federated Learning enables model training across the participating organizations without the necessity of sharing raw data, preserving data confidentiality. The result is a transformative approach that yields AI models of heightened accuracy and robustness, as they are enriched by diverse and decentralized data sources while respecting the paramount importance of data privacy and security among collaborating entities.
4) **Finance and Fraud Detection**:
   In the realm of finance and fraud detection, a compelling use case arises as financial institutions face the imperative to collectively enhance their fraud detection systems while upholding the utmost standards of customer data privacy. This intricate challenge is met through the fusion of blockchain technology and Federated Learning. Blockchain ensures the security and immutability of transaction records, creating a tamper-resistant ledger that underpins the financial ecosystem. Simultaneously, Federated Learning empowers banks to train and refine their fraud detection models using localized customer data, without necessitating the exchange of

sensitive information. The outcome is a revolutionary approach that significantly augments the accuracy of fraud detection systems, as they benefit from the diverse insights derived from various institutions' data, all while preserving the confidentiality and trust surrounding customers' personal and financial information. This harmonious synergy between blockchain and Federated Learning exemplifies a cutting-edge solution to a critical challenge faced by the financial industry.

5) **IoT and Smart Cities**:
In the context of IoT and smart cities, a compelling use case emerges as smart city infrastructure integrates data from a myriad of IoT devices to enhance services such as traffic management, urban planning, and resource allocation. This innovative endeavor is empowered by the harmonious amalgamation of blockchain technology and Federated Learning. Blockchain serves as the bedrock for securing IoT data, offering an immutable ledger that safeguards the integrity and confidentiality of the vast data generated by IoT devices. In parallel, Federated Learning takes centerstage by enabling the collaborative prediction and optimization of traffic patterns and urban services. Importantly, it achieves this without necessitating the centralization or sharing of sensitive data from individual devices. The resultant synergy between blockchain and Federated Learning redefines smart cities, bolstering the quality and efficiency of urban services, all while respecting and preserving the paramount importance of data privacy, an essential consideration in the era of interconnected urban environments. This integrated approach represents a paradigm shift in urban management, revolutionizing how cities harness IoT data for the betterment of their citizens while safeguarding their privacy and security.
    - **Use Case**: Smart city infrastructure collects data from IoT devices to optimize services like traffic management.
    - **Benefits**: Blockchain secures IoT data, while Federated Learning allows for collaborative traffic prediction and optimization. This approach improves city services while preserving data privacy.

These case studies and real-world applications demonstrate the benefits of combining blockchain and Federated Learning in various domains. This integrated approach enables organizations to harness the power of collaborative machine learning while addressing privacy, security, and trust concerns, ultimately leading to more effective and responsible data-driven solutions.

16.7 Conclusion

In conclusion, the exploration of blockchain-enabled Federated Learning across various domains has unveiled a transformative approach to data collaboration and innovation. This integrated methodology harmoniously merges the robust security, transparency, and immutability of blockchain technology with the decentralized, privacy-preserving capabilities of Federated Learning. Through a series of use cases and real-world applications, it becomes evident that this integrated approach offers a multitude of benefits. Firstly, it addresses the critical challenge of data privacy and security, ensuring that sensitive information remains confidential and tamper-proof, whether in healthcare, finance, supply chain, or smart cities. Secondly, it fosters collaboration among multiple stakeholders, allowing them to collectively train AI models and optimize processes without the need for centralized data sharing, thus unlocking new realms of innovation and accuracy. Moreover, the use cases highlighted demonstrate tangible applications that enhance predictive accuracy, reduce fraud, improve operational efficiency, and ultimately empower organizations to make more informed, data-driven decisions. This integrated approach is not only a testament to the potential of technology in reshaping industries but also a manifestation of the commitment to ethical data handling and privacy preservation. As we stand on the cusp of an era driven by data, blockchain-enabled Federated Learning emerges as a beacon, guiding industries toward responsible, collaborative, and secure data-driven transformations. It signifies a future where innovation and privacy coexist, allowing us to harness the power of data while safeguarding the trust and confidentiality of individuals and organizations alike.

16.8 Future Scope

Future work in the realm of blockchain-enabled Federated Learning holds promising avenues for further innovation and development. Here are several areas where researchers and practitioners can focus their efforts:

1) **Scalability Solutions**: Addressing scalability challenges is essential to accommodate larger and more diverse datasets in Federated Learning. Future research can explore methods to improve the scalability of Federated Learning algorithms, allowing for efficient model training across a broader range of participants.
2) **Interoperability Standards**: Developing interoperability standards and protocols is crucial for enabling seamless collaboration across different blockchain networks and Federated Learning platforms. This would facilitate cross-organizational initiatives and data sharing.
3) **Privacy-Preserving Techniques**: Continuous research into advanced privacy-preserving techniques, such as secure multiparty computation and differential privacy, can enhance the privacy guarantees of Federated Learning systems, making them even more robust against potential attacks.
4) **Security Auditing and Assurance**: As blockchain-enabled Federated Learning systems become integral to critical applications, rigorous security auditing and assurance mechanisms are needed to identify and mitigate vulnerabilities and threats effectively.
5) **Regulatory Compliance**: Staying abreast of evolving data privacy regulations and ensuring that blockchain-enabled Federated Learning systems comply with these regulations will be a top priority. Future work should focus on developing compliance frameworks and tools.
6) **Cross-Domain Applications**: Expanding the use cases of blockchain-enabled federated learning into new domains, such as energy, agriculture, and environmental monitoring, can unlock novel solutions for pressing global challenges.
7) **Decentralized Identity**: Research into decentralized identity systems, which leverage blockchain and federated learning for secure and privacy-preserving authentication, could revolutionize how individuals prove their identities without revealing sensitive information.
8) **Robustness and Fairness**: Ensuring that federated learning models are robust to adversarial attacks and that they produce fair and unbiased results is an ongoing area of research. Future work can delve into techniques for model fairness and robustness in federated learning settings.
9) **Education and Skill Development**: With the evolving landscape of blockchain and federated learning, there is a need for educational programs and skill development initiatives to train professionals in these emerging technologies.
10) **Real-Time Analytics**: Advancements in real-time analytics and processing of federated learning updates can further improve the agility and responsiveness of federated learning systems.

11) **Hybrid Models**: Combining federated learning with other machine learning approaches, such as transfer learning and ensemble methods, can lead to hybrid models that offer enhanced performance and versatility.
12) **Open-Source Tools**: The development and maintenance of open-source tools and libraries for blockchain-enabled federated learning can democratize access to these technologies and foster innovation.

The future of blockchain-enabled federated learning is rich with possibilities. By addressing scalability, privacy, security, and regulatory challenges while expanding into new domains and enhancing the robustness and fairness of models, researchers, and practitioners can continue to push the boundaries of what is achievable in the realm of decentralized, privacy-preserving machine learning. As these technologies mature, they hold the potential to drive significant positive change across industries and sectors.

References

1 Singh, S., Rathore, S., Alfarraj, O. et al. (2022). A framework for privacy-preservation of IoT healthcare data using federated learning and blockchain technology. *Future Generation Computer Systems* 129: 380–388.

2 Billah, M., Mehedi, S. T., Anwar, A. et al. (2022). A systematic literature review on blockchain enabled federated learning framework for internet of vehicles. arXiv preprint arXiv:2203.05192.

3 Qu, Y., Uddin, M.P., Gan, C. et al. (2022). Blockchain-enabled federated learning: a survey. *ACM Computing Surveys* 55 (4): 1–35.

4 Javed, A.R., Hassan, M.A., Shahzad, F. et al. (2022). Integration of blockchain technology and federated learning in vehicular (iot) networks: a comprehensive survey. *Sensors* 22 (12): 4394.

5 Ali, M., Karimipour, H., and Tariq, M. (2021). Integration of blockchain and federated learning for internet of things: recent advances and future challenges. *Computers & Security* 108: 102355.

6 Javed, L., Anjum, A., Yakubu, B.M. et al. (2023). ShareChain: blockchain-enabled model for sharing patient data using federated learning and differential privacy. *Expert Systems* 40 (5): e13131.

7 Ali, A., Al-Rimy, B.A.S., Tin, T.T. et al. (2023). Empowering precision medicine: unlocking revolutionary insights through blockchain-enabled federated learning and electronic medical records. *Sensors* 23 (17): 7476.

8 Qi, M., Wang, Z., Wu, F. et al. (2021). A blockchain-enabled federated learning model for privacy preservation: system design. *Information Security and Privacy: 26th Australasian Conference, ACISP 2021, Virtual Event* (1–3 December 2021), Proceedings 26, USA, pp. 473–489. Springer International Publishing.

9 Alsamhi, S.H., Almalki, F.A., Afghah, F. et al. (2021). Drones' edge intelligence over smart environments in B5G: blockchain and federated learning synergy. *IEEE Transactions on Green Communications and Networking* 6 (1): 295–312.

10 Wilhelmi, F., Guerra, E., and Dini, P. (2023, June). On the decentralization of blockchain-enabled asynchronous federated learning. *2023 IEEE 9th International Conference on Network Softwarization (NetSoft)*, 19–23 June 2023, Madrid, Spain, pp. 408–413. IEEE.

11 Sezer, B.B., Turkmen, H., and Nuriyev, U. (2023). PPFchain: a novel framework privacy-preserving blockchain-based federated learning method for sensor networks. *Internet of Things* 22: 100781.

12 Li, C., Yuan, Y., and Wang, F.Y. (2022, October). A novel framework for data trading markets based on blockchain-enabled federated learning. *2022 IEEE 25th International Conference on Intelligent Transportation Systems (ITSC)*, October 2022, Macau, China, pp. 3392–3397. IEEE.

13 Xu, R. and Chen, Y. (2022). μDFL: a secure microchained decentralized federated learning fabric atop IoT networks. *IEEE Transactions on Network and Service Management* 19 (3): 2677–2688.

14 Qi, Y., Hossain, M.S., Nie, J., and Li, X. (2021). Privacy-preserving blockchain-based federated learning for traffic flow prediction. *Future Generation Computer Systems* 117: 328–337.

15 Li, H., Li, C., Wang, J. et al. (2023). Review on security of federated learning and its application in healthcare. *Future Generation Computer Systems* 144: 271–290.

16 Chai, H., Leng, S., Chen, Y., and Zhang, K. (2020). A hierarchical blockchain-enabled federated learning algorithm for knowledge sharing in internet of vehicles. *IEEE Transactions on Intelligent Transportation Systems* 22 (7): 3975–3986.

17 Ayaz, F., Sheng, Z., Tian, D. et al. (2022). Blockchain-empowered AI for 6G-enabled internet of vehicles. *Electronics* 11 (20): 3339.

18 Cui, L., Qu, Y., Xie, G. et al. (2021). Security and privacy-enhanced federated learning for anomaly detection in IoT infrastructures. *IEEE Transactions on Industrial Informatics* 18 (5): 3492–3500.

19 Zhao, Y., Zhao, J., Jiang, L. et al. (2020). Privacy-preserving blockchain-based federated learning for IoT devices. *IEEE Internet of Things Journal* 8 (3): 1817–1829.

20 WHIG, P. (2023). Blockchain revolution: innovations, challenges, and future directions. *International Journal of Machine Learning for Sustainable Development* 5 (3): 16–25.

21 Whig, P., Kouser, S., Bhatia, A.B. et al. (2023). Explainable Machine Learning in Healthcare. In: *Explainable Machine Learning for Multimedia Based Healthcare Applications* (ed. M.S. Hossain, U. Kose, and D. Gupta), 77–98. Cham: Springer International Publishing.

22 Whig, P., Velu, A., Nadikattu, R. R., & Alkali, Y. J. (2023). Computational science role in medical and healthcare-related approach Ahmed A. Elngar, Vigneshwar M., Krishna Kant Singh, Zdzislaw Polkowski. *Handbook of Computational Sciences: A Multi and Interdisciplinary Approach*, 245-272 Wiley.

23 WHIG, P. (2023). A Comprehensive Review of Mask Detection Using Artificial Intelligence: Methods, Challenges, and Applications. *International Journal of Sustainable Development in Computing Science* 5 (2): 11–20.

17

An Edge Artificial Intelligence Federated Recommender System for Virtual Classrooms

M. Sirish Kumar[1], *T. Rupa Rani*[2], *U. Rakesh*[3], *Dyavarashetty Sunitha*[2], *and G. Sunil Kumar*[2]

[1] *School of Computing, Mohan Babu University, Tirupati, Andhra Pradesh, India*
[2] *Department of CSE, Narsimha Reddy Engineering College, Hyderabad, Telangana, India*
[3] *Department of Computational Intelligence, MRCET, Secunderabad, India*

17.1 Introduction

The information-active method of information discovery has replaced the user-active mode of search engines, in which intriguing items are routed to the user without (usually) their explicit demand [1]. Recommender systems are growing in popularity across industries, from social networking sites like Facebook and LinkedIn to e-commerce sites like Amazon. They can be used to suggest contacts, news to read, websites to visit, or things to Purchase.

Learning is no longer confined to traditional classroom settings because of the advancement of digital technology. Technology-Enhanced Learning (TEL) attempts to design, develop, and test sociotechnical solutions to improve learning activities both on an individual and organizational basis [2]. It allows pupils to study whenever and wherever they want.

Users may access, exchange, and systematize knowledge materials of all types and administer their own Individual Learning Environments (ILEs) thanks to an increasing number of online platforms, like Symbaloo4 and Graasp5.

The efficiency and effectiveness of learning are increased by ILEs because they provide learners the chance to make, coordinate, reuse, and bundle their own learning materials and devices [3]. ILEs are client-focused, as opposed to instructor- or course-focused, typical LMS (Learning Management Systems).

When it comes to educational resources, moving from LMS to ILEs introduces an overabundance and a distortion: with course-centered LMSs, students are directly given a small amount of carefully chosen and dedicated resources, whereas, with learner-centered ILEs, students must deal with a large number of resources for which they must make their own selection. Therefore, recommender systems are essential in ILEs and other open environments to assist students access materials that fit their interests from a pool of options that aren't preplanned and confined.

Students at Swiss institutions frequently encounter more diverse cultures during their studies and more than one platform for online learning. In actuality, there is a growing trend of student movement among universities. Swiss institutions do not, however, exclusively rely on even if they have all used a single sign-on method in common, the same online learning platform. There may also be more than one platform used inside a single university [4]. In order for students to create

Model Optimization Methods for Efficient and Edge AI: Federated Learning Architectures, Frameworks and Applications,
First Edition. Edited by Pethuru Raj Chelliah, Amir Masoud Rahmani, Robert Colby,
Gayathri Nagasubramanian, and Sunku Ranganath.

a richer learning environment and have access to more distributed materials, it becomes crucial to design a system that integrates resources from many platforms.

In this article, we suggest a federated recommender framework that uses users' interactions with dispersed resources to provide rich and pertinent recommendations.

By bringing data from other institutions to stages utilized locally (inside one foundation), these ideas can act as a focal point of access for learners, platforms that enrich already in use, and encourage the establishment of knowledge networks throughout Switzerland. The platforms themselves are made up of LMSs and ILEs that cater to various universities.

Segment 2 talks about akin work and Segment 3 portrays the principal parts of the framework for unified recommenders. The paper is concluded in Section 4, where future work is discussed.

17.2 Work Related To

The organisational strategy and architecture are the two essential elements of unified forage or proposition in educational contexts, and we examine current testing in this section. The league approach worries about how to get viable information, channel furthermore, and blend information, while the organization design focuses closely on the construction, whereby the structure can be productively carried out. The study provided a summary [5] concerned with integrating and sorting search results from several sources that can be found on ObjectSpot8, a vertical federated search engine. A number of algorithms are tested and compared to be able to effectively consolidate different indexed lists and improve execution. In [6], list items from various stages are gathered and once again positioned by a combined search administration. Positioned indexed lists are shipped off a gadget and shown to the client. Re-positioning has proceeded as follows: URLs, or Uniform Resource Locators, from search results, are sent to the recommendation service by the widget. After analyzing the database's attention metadata, this service provides customized recommendations [6]. Consideration metadata comprises recorded clients' cooperation with the suggested things (for example, review, similar to, disdain) to use client's advantage and rank things' worldwide and neighborhood ubiquity (inside the objective clients' organization). By mining this metadata, suggested records for comparative questions are further developed in resulting search demands.

The engineering utilized for combined search or data recovery from appropriated storehouses is the subject of several papers. A multi-specialist design wished for in [7] has an expert specialist that communicates through clients then searches are sent out solicitations to support specialists of various storehouses. There is a service agent in each repository. It responds to inquiries provided by the master agent and collects and updates metadata. Once the help specialists return their outcomes, the expert specialist consolidates the outcomes and presents them to clients. The middle person correspondence point is liable for sending client solicitations to various archives and afterward consolidating the reverted positioned outcomes. Within Ariadne [8], an avid collector acquires metadata having a place with new assets from various learning vaults consistently.

Furthermore, gradually. The effectiveness of handling search demands is improved by really taking a look at metadata. The OpenScout framework [9] follows a similar methodology.

A harvester is also used by our federated engine to collect data from various platforms. Nonetheless, we not just store metadata connected with learning items (or assets), but additionally, association information connecting clients and learning objects. Subsequently, individual client interests can be interpreted in a customized proposition that the unified can present. Furthermore, prescribed things are not restricted to learning objects, but rather likewise incorporate movement spaces and clients.

As in [6], we fostered a versatile gadget to show proposal results. A gadget is a little Web application that is embeddable in a Page and permits simple mashups [9]. This decision prompts a lightweight interoperability and a simple combination of the proposal shows administration in the web-based learning stages involved.

The created gadget permits students to arrange suggested things by type (for example, assets, clients, or action spaces) and gives an outline of the suggested things while giving a connection highlighting its actual area in one of the reaped learning stages.

17.3 Federated Recommender-Proposed Framework

17.3.1 Normal Information Model

The united promoter framework takes advantage of collaboration information put away in various web-learning stages. "Taking into account that information models, naming shows, and configurations change from one client to another and organization to organization" [10], it becomes significant to take on a standard information portrayal and trade design across various stages.

Graasp is a member of the web-based synchronized learning effort stages considered. It was based on top of the 3A communication model shown in Figure 17.1. We appreciate this compact design, as a source of perspective model to trade information spanning the various learning phases involved. The 3A association design depends on the accompanying fundamental developments [11]:

Actors: allude to clients or individuals. At first, the idea of entertainers coordinated additionally the thought of uses (and all the more, for the most part, any substance equipped for starting an occasion). Afterward, a reasonable division between human entertainers and applications was embraced.

Resources or assets: allude to various types of assets or computerized ancient rarities (documents, introductions, recordings, wikis, and so forth) made and divided between entertainers.

Movement spaces: allude to online settings or compartments where at least one entertainers shares within the scope of assets and applications of a serving action space, an expressly or verifiably expressed reason. For instance, in a formal learning environment, for each course,

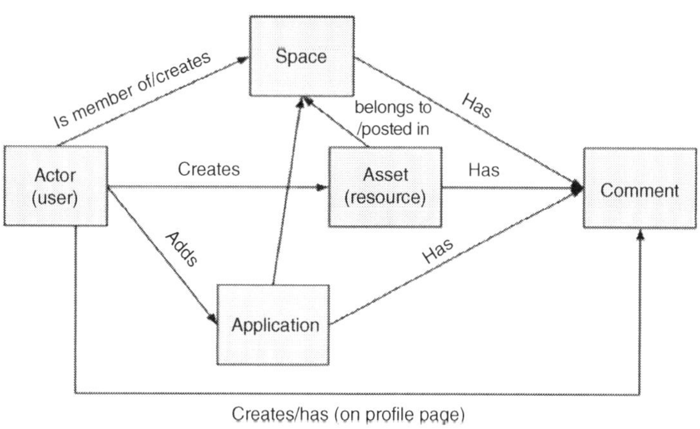

Figure 17.1

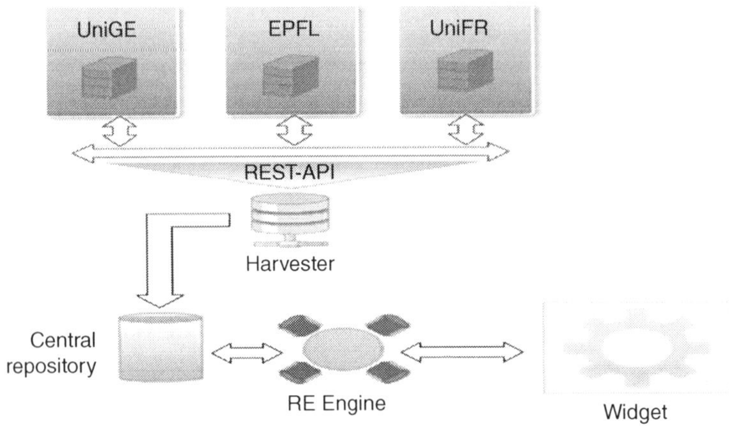

Figure 17.2 Engineering of the combined proposal framework.

a primary space is created, which all understudies, mentors, also, showing aides are welcome to join. Similarly, a local area could make a movement space to which every one of its individuals is welcomed. Each learning stage included (for instance, Moodle and Mahara) translates its information hooked on the 3A model and provides REST (Illustrative State Transfer)9. APIs are summoned gatherer to collect information in XML or JSON (JavaScript Item Notation)10 designs.

17.3.2 Design

The embraced engineering in support of the unified recommender framework is delineated in Figure 17.2. Its fundamental structure blocks are:

A gatherer which occasionally recovers the data about data from various stores parses using REST API's recovered information, and keeps it in the focal archive;

- A proposal motor (or RE)- which controls the suggestion calculation routinely and afterward the outcomes are saved in the central vault;

A gadget which inquiries the focal storehouse utilizing REST APIs and showcases prescribed things towards goal clients. This gadget can be implanted in any stage that executes the embraced REST APIs.

17.3.3 Information Assortment: The Reaper

The reaper fills in as a bridge among phases of learning and the focus store. Utilizing normal REST APIs executed by knowledge stages, the collector routinely recovers metadata, parses it in light of the 3A model scientific classification, as well as refreshes the focal archive. For security, just open and shut things are recovered. Public things comprise entertainers, spaces, assets and applications available to everybody with practically no limitations. The substance of shut spaces then again is simply open to approved individuals, while everybody else can see fundamental data (for example, portrayal, title, and watchwords).

During the collecting activity, two tables are utilized to accumulate the assembled information.

- An ordering table that contains the URLs of the gathered things;
- A line table that contains the URLs that are in the holding-up rundown to be reaped;

We depict from this point forward the means embraced as a result of the collector and work of above-mentioned tables in the information recovery and capacity process.

Step 1: Invoke the REST APIs and obtain the metadata and really look at the thing status

Each collected thing is related to an exceptional URL. The reaper computes the hash worth of the URL and the Programming interface reaction, and contains them in file stand. At the time the gatherer calls the REST APIs, it resolves initial check assuming the thing's URL to be available in the record table.

Assuming URL subsists and worth of the hash Programming interface reaction is as yet the matching, nonentity is finished; assuming that the URL exists, yet the hash esteem has altered, the thing is refreshed; in any case, it is another thing and it is incorporated into the table. On the off chance that vocation of the characterized APIs proceeds 404 mistakes for a current thing, it implies that the thing has been erased from the stage and, subsequently, it is likewise erased from the record board.

This check is significant, as in certain stages, for example, Moodle, it is preposterous to expect to question the gatherer each time a thing is refreshed or erased.

Step 2: Parse the reaction and hoard data in focal archive

For every URL obscured in the list table, the response is dissected in view of the 3A design, and the item's information and classification by science (for example, name, miniature image, depiction) are put away into the focal vault.

Step 3: Really take a look at thing's kid components

Spaces could contain individuals, assets, and subspaces as youngster components. Figure 17.3 displays the XML reaction information about an interplanetary thing


```
▼<rdf:RDF xmlns:rdf="http://www.w3.org/1999/02/22-rdf-syntax-ns#"
  xmlns:aaa="http://ple.unige.ch/re/1#"
  xmlns:foaf="http://xmlns.com/foaf/0.1/"
  xmlns:rdfs="http://www.w3.org/2000/01/rdf-schema#"
  xmlns:dc="http://purl.org/dc/elements/1.1/"
  xml:base="http://localhost/mahara/artefact/graaasp/service/data/">
 ▼<rdf:Description rdf:about="36e7e81b9d12dababcbe2febaba8b521-2">
    <aaa:id>36e7e81b9d12dababcbe2febaba8b521-2</aaa:id>
    <aaa:dataset>http://localhost/mahara/</aaa:dataset>
    <aaa:url>http://localhost/mahara/group/view.php?id=2</aaa:url>
    <aaa:name>test for creating space</aaa:name>
  ▶<aaa:description rdf:parseType="Literal">...</aaa:description>
    <aaa:type>space</aaa:type>
    <aaa:spaces rdf:resource="7e14ed175ce447742753df686bdcdaa0-2"/>
    <aaa:assets rdf:resource="63bdeaa8f804a116dd038e61b1452b1c-6"/>
    <aaa:actors rdf:resource="857c9a58e87a91b1a6044fe1e6dbaf26-1"/>
    <aaa:kind>group</aaa:kind>
    <rdf:type rdf:resource="http://ple.unige.ch/re/1#space"/>
    <rdfs:label>test for creating space</rdfs:label>
  ▶<rdfs:comment>...</rdfs:comment>
  ▶<dc:description rdf:parseType="Literal">...</dc:description>
  </rdf:Description>
</rdf:RDF>
```


Figure 17.3 A sample of information linked to a space.

including a subspace, an asset entertainer, as well as their metadata. For every kid component, the URL is put away from line table.

Step 4: Investigate things put away in the line table
For everything put away in the line table, The gatherer progresses from stage 1 to stage 3. The hierarchical cycle of recursion closes when the line has run out of URLs.

17.3.4 Information Handling: Proposal Motor

The RE utilizes the 3A positioning calculation to process the general significance of existing things for an objective client. The 3A positioning calculation is gotten as of first-page rank calculation, which Page and Brin intended for Google's placement of hypertext records [12]. If the owner of page j links to page I, they are demonstrating how important it is to check out page I. The additional approaching connections a page has, the more significant or legitimate it is. Additionally, pages linked from genuine pages are given more weight and are thus given priority status than those linked from invalid ones. As a result, thing positions are registered by applying an irregular stroll calculation on a nearness chart interfacing pages together, with two primary boundaries: one choosing the likelihood to pick an irregular sheet and the additional choosing of the likelihood to track a page connect.

The 3A positioning calculation pursues the very thought aside from that its chart comprises hubs addressing heterogeneous things, associated by border addressing, unlike connection types with equivalent or diverse significance loads. For example, A portion of a space edge that has the "enrollment" Weight-related the client hub to the freedom hub as an indication of awareness of that client in the space. The calculation's subtleties and likelihood condition can originate here [13].

To be relate to the calculation to the gathered information, the RE builds a diagram where every public or shut thing got is addressed with a hub. Any two items can have both unidirectional and bidirectional relationships planned in the diagram by a solitary boundary or two boundaries in inverse bearings interfacing the related hubs.

Figure 17.4 represents a straightforward illustration of a 3A proposal chart. In the diagram, ovals address client hubs even as square shapes address space hubs and parallelograms address assets. Bolts address both directions at once and connections between things, while Leona remarked on the "Introduction to HCI," she should be keen on HCI, when the creator of the asset "Introduction to HCI" and the maker of the "HCI course" space, Lion plainly shares similar awareness. So do the "HCI course" space part, Cindy. Successively, the 3A proposal calculation with Leona, as the objective client, force assisted her with finding spaces and individuals with shared notice by subsequent their between associations.

17.3.5 Information Perception: Widget

The gadget is utilized to show the suggested things and can be applied across various stages. Its connection point is displayed in Figure 17.5. It is executed among JavaScript. For every client

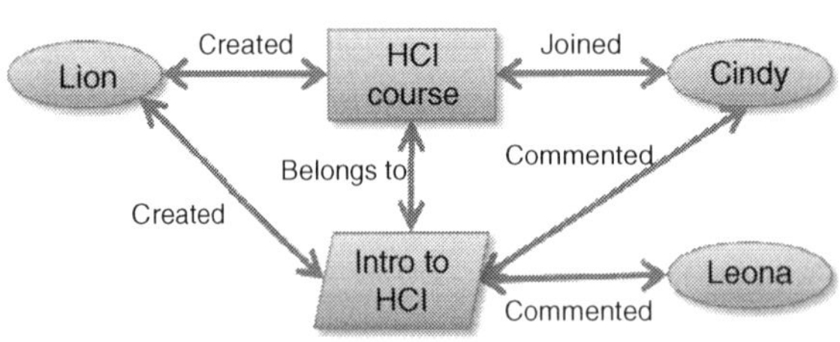

Figure 17.4 Model for suggestion diagram.

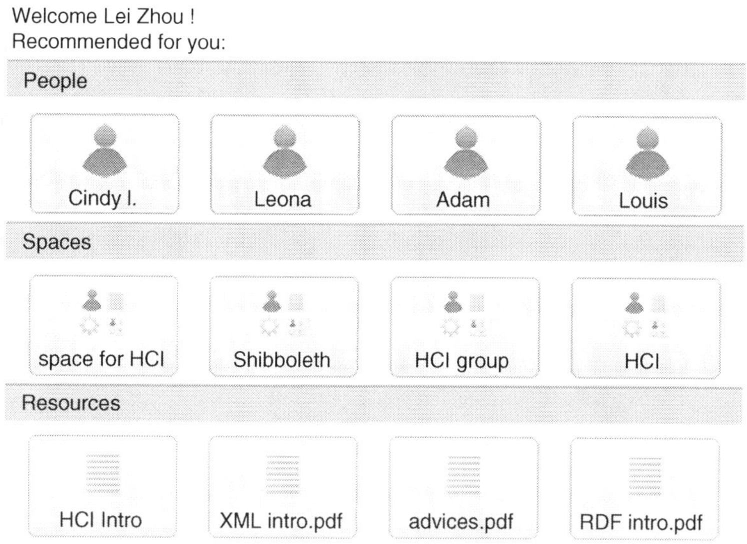

Figure 17.5 The point of interaction of the gadget.

(understudy or coach), the Programming interface force arrival of a customized rundown of suggested things, showing various lines as indicated by their sort.

At the point when the client taps on a suggested thing, it is the obligation of the stage where it was really presented to check whether the client has an adequate number of freedoms to get to the current thing. Swiss colleges have executed Custom [14] as a solitary sign-on (SSO) component. Custom permits clients to login once and have an extraordinary character across various stages and the proposal gadget. Thus, the route between the different stages of appreciation of gadgets can be delivered simply. The client can do the tasks naturally after being given the benefit of doubt throughout the response phase.

17.4 Final Thoughts and Future Work

A recommender that relies on connections is presented in this paper framework, which conveys a united proposal. As indicated by the normal information model. The collector compiles and analyzes metadata from a variety of sources. Then, at that point, the suggestion motor runs the 3A suggestion calculation and processes the overall significance of everything in light of the metadata put away in the focal vault. Each target client gets the suggestion naturally in the gadget and can get to suggested things from the reaped learning stages.

Sooner rather than later, we intend to take on a half-breed approach for the RE. For new clients, the RE can barely give helpful suggestion without adequate client activities. To address the cool beginning issue, content-based examination will be used to integrate the 3A connection-based proposition technique. Content-based investigation comprises registering content comparability between things (by looking at changed fields, for example, thing title, depiction, and catchphrases). Then, at that point, the delayed consequences of content-based isolation including thing-to-thing closeness measures are dealt with in the 3A proposition computation as an additional association.

An assessment by understudies commencing various Swiss colleges be additionally arranged. The assessment will zero in on the gadget's connection point case of use and the effect of unified search and suggestions when contrasted with nearby ones. The handiness of suggesting things from different Internet-based stages involved inside one college also as across various colleges will be surveyed.

References

1 Dabbagh, N. and Kitsantas, A. (2012). Personal learning environments, social media, and self regulated learning: a natural formula for connecting formal and informal learning. *The Internet and Higher Education* 15: 3–8.

2 Tang, T.Y. and Mccalla, G. Smart recommendation for an evolving E-learning system: architecture and experiment. *International Journal on E-Learning* 4: 105–129. http://www.editlib.org/p/5822 (retrieved March 2012).

3 LMS Installations (2010) At Swiss Institutions of Higher Education. http://eduhub.ch/info/lms-installations10.html (accessed April 2012).

4 Guy, I., Jaimes, A., Agulló, P. et al. (2010). Will Recommenders Kill Search? Recommender Systems - an Industry Perspective. *Proceedings of the 4th ACM Conference on Recommender Systems* (26–30 September 2010), pp. 7–12.

5 Manouselis, N., Drachsler, H., Vuorikari, R. et al. (2011). Recommender systems in technology enhanced learning. In: *Recommender Systems Handbook, Part 2* (ed. P. B. Kantor, F. Ricci, L. Rokach, et al.), 387–415. Springer.

6 Govaerts, S., El Helou, S., Duval, E., and Gillet, D. (2011). A federated search and social recommendation widget. *Proceedings of the 2nd International Workshop on Social Recommender Systems in conjunction with the 2011 ACM Conference on Computer Supported Cooperative Work* (04–05 September 2011), Sinaia, Romania, pp. 1–8.

7 Kaushik, S. and Kollipalli, D. (2011). Multi-agent based architecture for querying disjoint data repositories. *International Conference on Machine and Web Intelligence* (03–05 October 2010), Algiers, Algeria, pp. 28–34.

8 Ternier, S., Verbert, K., Parra, G. et al. (2009). The ariadne infrastructure for managing and storing metadata. *IEEE Internet Computing* 13: 18–25.

9 Ha, K.-H., Niemann, K., Schwertel, U. et al. (2011). A novel approach towards skill-based search and services of open educational resources. In: *MTSR 2011*, CCIS, vol. 240 (ed. E. García-Barriocanal, Z. Cebeci, M.C. Okur, and A. Öztürk), 312–323. Heidelberg: Springer.

10 El Helou, S., Salzmann, C., and Gillet, D. (2010). The 3A personalized, contextual and relation based recommender system. *Journal of Universal Computer Science* 16 (16): 2179–2195.

11 Tom, Scavo and Scott, Cantor. Shibboleth architecture technical overview. http://shibboleth.internet2.edu/docs/draft-mace-shibboleth-tech-overview-latest.pdf (accessed March 2012).

12 Grewe, L. and Pandey, S. (2011). Quantization of social data for friend advertisement recommendation system. In: *PDCTA 2011*, CCIS, vol. 203 (ed. D. Nagamalai), 596–614. Heidelberg: Springer.

13 El Helou, S., Li, N., and Gillet, D. (2010). The 3A interaction model: towards bridging the gap between formal and informal learning. *Proceedings of the Third International Conferences on Advances in Computer-Human Interactions* (10–16 February 2010), pp. 179–184.

14 Page, L., Brin, S., Motwani, R., and Winograd, T. (1999). The PageRank Citation Ranking: Bringing order to the Web. Technical Report. Stanford: Stanford Digital Library Technologies Project.

18

Federated Learning in Smart Cities

Seyedeh Yasaman Hosseini Mirmahaleh[1] *and Amir Masoud Rahmani*[2]

[1]*Department of Electrical Engineering, Science and Technology, Lille University, Lille, France*
[2]*Future Technology Research Center, National Yunlin University of Science and Technology, Douliou, Taiwan*

18.1 Introduction

To address the limitations of decentralized and centralized learning approaches, federated learning (FL) has emerged as a solution. FL eliminates the need for local data-sharing and supports heterogeneous data by following the principles of game theory to train local edge nodes based on their unique characteristics and features. FL involves creating a federal model incorporating local inference models and facilitates communication and data association between different groups of edge nodes. These selected nodes act as clients for the federal model and share the current state of the models. This approach ensures data privacy by preventing data-sharing between adjacent nodes and clusters, making it a secure Internet of Things (IoT) platform. The metaverse concept has gained prominence in the context of current technology and future modern life. The metaverse, which encompasses virtual tasks and combines real and virtual realities through digital twins, requires a robust model for learning heterogeneous data and informatics. The metaverse defines the virtual dimension of a smart city, which integrates augmented reality-based applications and devices with the communication infrastructure. This integration addresses the traffic problem of communication between digital twins and the variability of data processing and decision-making. FL offers a solution to tackle this issue by employing a modern decentralized technique for local training and inference, considering the statistical data dependency. Additionally, it necessitates a secure communication platform for data transport between agents within the smart city. FL offers a robust learning model for the smart city, ensuring data privacy through a global model for various applications. One critical application within the metaverse smart city and smart city technologies is addressing modern lifestyles' medical and healthcare requirements. This application can be described as the Internet of Medical Things (IoMT) and encompasses e-healthcare applications within IoT infrastructures. Due to the nature of healthcare services, data privacy must be prioritized when preparing an IoMT infrastructure and introducing medical-based applications to protect private records. FL emerges as a promising solution to meet the demands of IoMT and healthcare-based systems. It provides a secure platform for learning heterogeneous data and ensures privacy in healthcare settings.

In the context of industrial applications in smart cities, the mobility issue faced by electric vehicle (EV) devices poses a significant challenge in delivering services. This challenge arises from the need to connect various smart city components to diverse data sources. Fortunately, FL offers a

Model Optimization Methods for Efficient and Edge AI: Federated Learning Architectures, Frameworks and Applications,
First Edition. Edited by Pethuru Raj Chelliah, Amir Masoud Rahmani, Robert Colby,
Gayathri Nagasubramanian, and Sunku Ranganath.

solution by enabling the integration of these components and addressing the diverse demands of individuals, such as water consumption and energy sources, through robust learning techniques [1–3]. Additionally, traffic estimation plays a crucial role in determining the efficiency of a smart city, as it provides valuable insights into the condition of urban roads and offers optimal routes to users [4–6]. However, the scarcity of traffic data remains a hurdle for this technology. FL overcomes this obstacle by supporting local training and global control, ensuring effective resource management and secure communication between devices within a smart city.

Wireless health-monitoring devices and remote healthcare services become crucial in a smart city during risky situations such as dangerous virus epidemics and natural disasters [7, 8]. In these scenarios, FL provides a private framework for healthcare service delivery to patients. By employing game theory to define communications between demanders and servers, FL ensures effective iteration stages during the outbreak of COVID-19 [7]. Two significant challenges in IoMT infrastructures and e-healthcare services are alternative client and data imbalance, which need to be addressed to protect the records of individuals in health wireless communication and service providers like remote hospitals [8]. Trust, security, fault tolerance, and dependability are critical demands for medical-based applications in IoMT, IoT, smart cities, and metaverse communication platforms. FL can successfully address these demands by creating dependable communities between servers and clients and presenting a federal model for supporting local training (FL in trust, security, fault tolerance, and dependability). Another critical aspect of FL in healthcare applications is its role in drug discovery and medicine recommendation deployment on an industry scale [9]. Timely treatment and therapy actions are essential for efficient service delivery to requesters, and FL provides a model for learning and supporting these issues [10].

Digital twins of smart cities consist of various IoT, IoMT technologies, and independent deep learning approaches to communication between devices in physical and perception layers [11]. Since a strong and applicant-distributed system aids a smart city-based platform to support its operations in managing data variation, secure framework, and fault-tolerant structure, FL is an appropriate technique to satisfy the named purposes. According to the angels of FL acts in smart city, this study provides the following definitions of its role:

FL in Industrial Applications (FLIA): includes nonhealthcare applications and techniques to support them in a smart city.

FL in Healthcare Systems (FHLS): defined as the manners and facts related to healthcare-based applications and management by a smart city.

FL in Medicine Recommendation (FLMR): specializes in identifying the proper drugs and introducing them to doctors for prescribing medicine to patients on a smart city platform.

FL in Drug Discovery (FLDC): describes the role of FL's features in recommending medicine for joined people to medical-based infrastructure for treatment purposes and critical conditions in a smart city.

FL in Fault Tolerance (FLFT): attends to different approaches to prevent sharing malicious information between components of a smart city in classes of FLIA, FHLS, FLMR, and FLDC.

The chapter examines the role of FL in intelligent cities, following the five upper definitions. It analyzes the challenges faced by medical and industrial-based services and devices in wireless communications, which a federated model can address. The chapter classifies these problems in the first level of a purpose taxonomy tree and then explores FL solutions to overcome them. It defines several real-world applications and medical scenarios to illustrate these solutions. Figure 18.1 provides an overview of the chapter's structure, which includes detailed implementations and solutions divided into five defined acts: FLIA, FHLS, FLMR, FLDC, and FLFT.

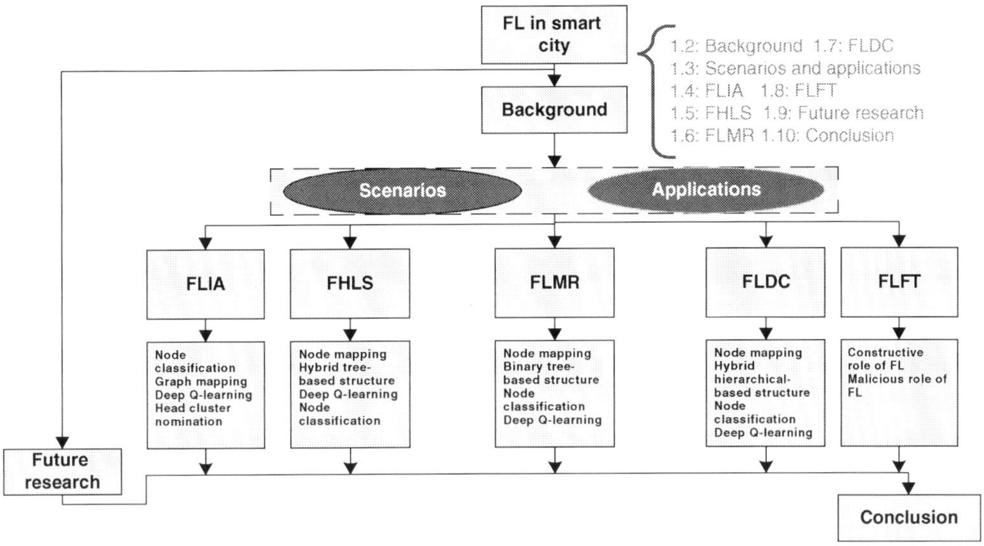

Figure 18.1 A taxonomy tree of the chapter purposes. FL; Smart city; Industrial application; Healthcare application; Drug discovery; Medicine recommendation; Fault tolerance.

18.2 Background

The modern lifestyle demands have led cities to adopt intelligent devices and wireless applications, resulting in the development of smart cities and metaverse smart city technology. These technologies enable remote control and access to various industrial, governmental, educational, and medical applications, simplifying human life. Smart cities have diverse data centers that can respond to human requirements and manage various aspects of urban life through intelligent control. However, due to the nature of these systems, it is crucial to have a secure framework and a fault-tolerant platform to ensure successful service delivery. Numerous recent studies have focused on smart cities and proposed various approaches and methodologies to address these challenges. This section provides a brief overview of these approaches.

Smart city technology faces the challenge of centralized learning to support communication between devices and agents and establish a secure framework for allocating heterogeneous data centers. Some studies have attempted to utilize a centralized approach by introducing novel learning approaches. For example, the recurrent neural network (RNN) has been targeted to overcome the challenges of smart cities in predicting the initial conditions of networked devices [12]. The Hony Boger algorithm was developed as an optimized RNN algorithm to predict the locations of vehicle components, reducing energy consumption and communication delay in a smart city [12]. However, centralized algorithms have struggled to support smart city technology successfully due to its fundamental essence, constructed by heterogeneous and asynchronously distributed nodes and data centers.

Some studies have classified smart city technology into IoT and metaverse-based techniques to address the security challenges of smart cities in more detail. Researchers have proposed a distributed Hash function algorithm for IoT-based smart cities to establish a secure platform for industrial communication and deploy Buffalo learning models to ensure security in fog computing [13]. This approach aims to reduce latency and energy consumption in smart cities' edge layer of IoT applications. Additionally, quantum computing has been used to protect private information and secure data transmission through quantum mechanics and the uncertainty principle to define public and private keys, known as quantum cryptography. This technique represents a future research direction for arming metaverse and IoT-based smart cities with a reliable and secure framework through quantum-based blockchain [14, 15]. Another study utilized the convolutional neural network (CNN) for data encryption and quantum cryptography to establish secure communication platforms in smart cities. This study also combined CNN with long short-term memory algorithms to forecast traffic and manage flow in a smart city [15]. FL-based blockchain approaches have also provided accurate and stable cryptography in smart cities [16]. These approaches reduce the time required for aggregating FL compared to other methods and successfully handle synchronization in edge computing. Furthermore, integrating digital and data twins has been explored to establish a secure framework for metaverse smart cities by facilitating direct communication [17, 18].

In terms of fault tolerance, smart cities must be able to continue functioning correctly and maintain access to services in the event of a fault. One approach involves implementing a fault-tolerant blockchain model to prevent sharing malicious data from access points to the cloud layer by inserting a hash table of validation keys [19]. This approach helps overcome limitations in security degrees and eliminates bottlenecks. Another study proposed a hierarchical structure for FL, known as Hierarchical FL (HFSA), which includes an edge middle layer between clients and servers. HFSA aims to increase fault tolerance in the edge layer and prevent instability of edge clients by allowing lazy clients to share their models asynchronously [20]. A harvesting system has also been developed to create a fault-tolerant platform for industrial IoT in smart cities. This system focuses on energy management and prevents hot spot points in head clusters through an energy-aware clustering algorithm for IoT nodes [21]. The concept of fault tolerance is also essential in medical-based applications of smart cities, where post-quantum cryptography has shown promise in establishing a fault-tolerant framework [22].

One important application of smart cities is in the field of medical and healthcare services. These services include remote monitoring of vital signs and remote medical advice, which require a reliable framework for communication and interaction. Tareen et al. [23] utilized the balanced load method to reduce the computing time of mining nodes in the blockchain for healthcare applications in smart cities. This approach helped establish a secure platform for the timely processing of patients' records and responding to their needs. Resource allocation was also considered crucial for efficient healthcare services in smart cities, ensuring timely allocation and utilization of resources [24]. A reward-based technique was implemented to improve the performance of healthcare systems in smart cities by scoring the treatment staff and resources based on their delivery of duties [25]. Telemedicine systems provide opportunities for remote disease diagnosis and treatment, requiring a secure framework for implementation [26]. Kalapaaking et al. [27] employed an encrypted inference method to protect healthcare applications from poisoning attacks, using a blockchain-based FL approach to detect illegible treatment staff and prevent sharing malicious information.

One of the strong points of FL is its role in drug discovery, which is crucial for medical-based applications in smart city technology. MELLODDY [10] is a robust drug discovery platform established on the Amazon web server to respond to the medical demands of citizens and other participants. It utilizes a secure framework inspired by the FL model, which provides a secure platform for drug discovery by updating the aggregated model using local gradients. In addition, Malik et al. [28] have demonstrated the effectiveness of blockchain-based communications in protecting private information and data in smart cities and IoT technologies for drug supply and healthcare services. FL has also successfully developed blockchain platforms to support all types of communication in smart cities.

Another aspect of medical-based applications of smart city technology involves medicine recommendation, which primarily targets doctors to prescribe appropriate drugs quickly. In this regard, electronic healthcare records and the background of patients play a crucial role in accurately defining medicine based on medical records. Saxena et al. [29] utilized a neural memory to store the history of doctor recommendations, vital signs, and e-records of patients to facilitate their drug prescriptions, especially in critical situations. The study also considered the medical history and medical documents of other patients with similar diseases to leverage their experiences for drug recommendation. Given the high sensitivity, variety, and sparsity of the data involved in such applications, a robust distributed technique is needed to ensure cybersecurity and supply management. FL is a suitable choice for addressing these challenges.

The studies above highlight the significance of FL in various smart city applications, including smart industry and smart healthcare, for supply management and data privacy protection. A smart city encompasses smart applications in healthcare, transportation, agriculture, chain management, grid, and homes, which our chapter focuses on regarding FL. The chapter discusses five definitions of FL in the context of smart cities: FLIA, FLHs, FLMR, FLDC, and FLFT [30]. FL plays a prominent role in modern electronic and intelligent technology, as demonstrated by developing an FL framework in the TensorFlow library [31]. This section provides clear definitions of key concepts related to smart cities and FL that will be used to explain the details of FL for smart city technology.

Federated Learning: is a contemporary technique that enables the adaptation of distributed systems to leverage the capabilities of emerging technologies such as smart cities and metaverse. It involves a federative model that allows local systems to upload their current networked device states.

IoT Smart City: refers to a city where smart devices are interconnected through an IoT infrastructure, enabling seamless communication between various applications.

Metaverse Smart City: describes a city incorporating a meta space of interconnected intelligent devices. It combines elements of virtuality and reality to define its applications.

Figure 18.2 illustrates the concepts of IoT and metaverse smart cities to represent their contents visually.

After presenting a schematic representation of a smart city incorporating metaverse and IoT technologies, Figure 18.3 provides an overview of the FL architecture that supports smart city applications.

18.3 Application and Scenario

The study aims to provide a detailed description of how FL supports the infrastructure of smart cities. The previous section classified the concepts of FL's role in technology. The chapter presents various scenarios and applications demonstrating FL processes to clarify this issue further. These scenarios are designed to resemble real-world applications and cover multiple situations closely. Section 18.1 defines five main classifications to explain FL's role in smart cities and proposes five applications based on these classifications. Each application is accompanied by a dedicated scenario that illustrates its operations.

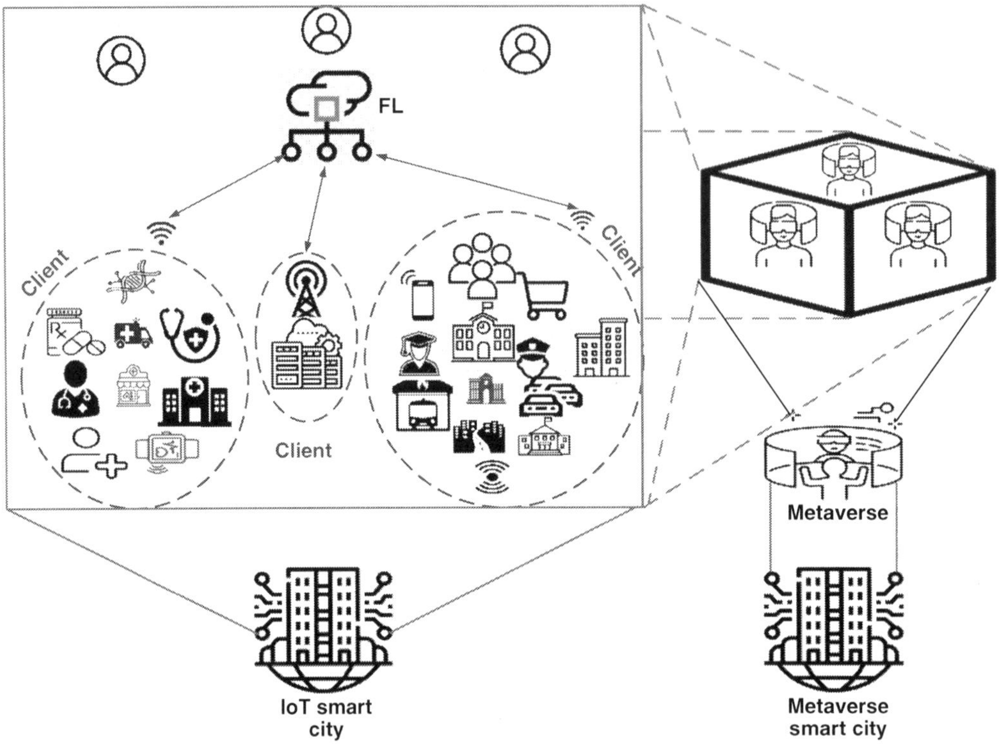

Figure 18.2 A schematic of IoT smart city and metaverse smart city.

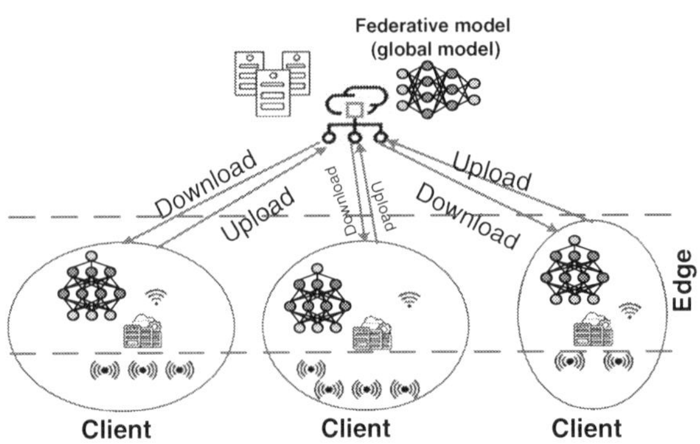

Figure 18.3 A schematic of IoT smart city and metaverse smart city.

Application 1 focuses on traffic management for industrial operations in a smart city. It involves a set of interconnected nodes such as traffic lights, temperature and air sensors, RFID, and AIDC tags. Figure 18.4a depicts Application 1 and its nodes for traffic management using FLIA. Scenario 1 describes a plan for communication between the nodes of Application 1 to determine the appropriate services for users in the smart city based on FLIA. This scenario assumes that Application 1 is distributed across different locations in a metaverse smart city. The only difference between a metaverse smart city and an IoT smart city, as shown in Figure 18.2, is the use of devices for augmented reality and the combination of virtual and real concepts. Scenario 1 aims to manage street traffic by gathering data from various locations and utilizing the outputs of sensors and tags.

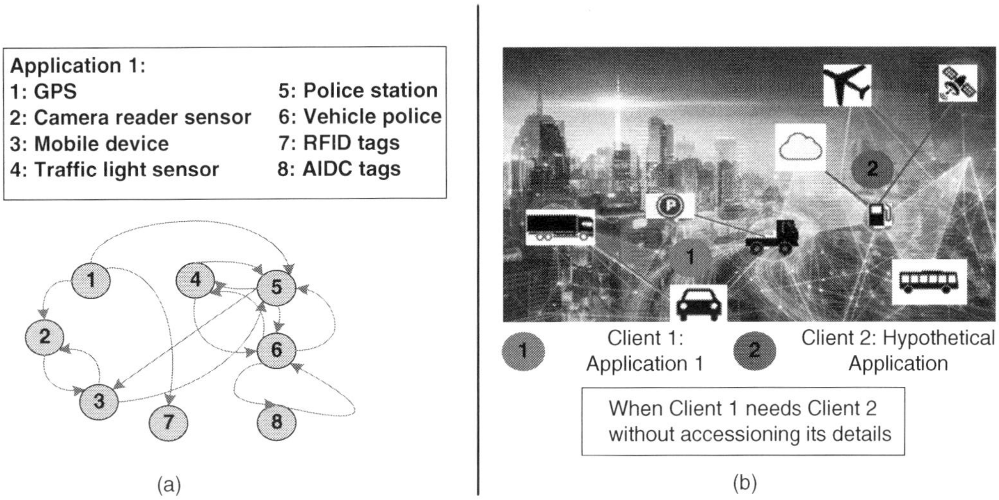

Figure 18.4 (a) Application 1 based on FLIA definition, (b) scenario 1 based on application 1. Source: jamesteohart/Adobe Stock Photos.

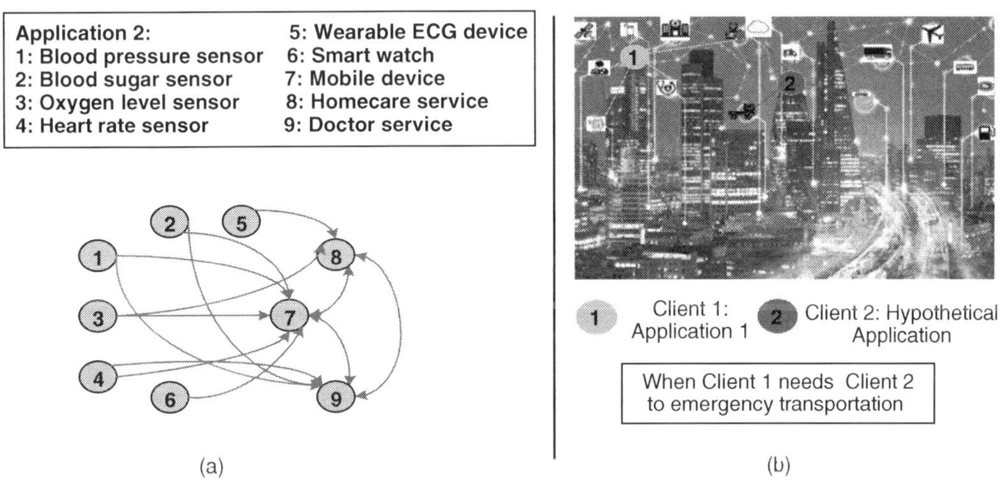

Figure 18.5 (a) Application 2 based on FHLS definition, (b) scenario 2 based on application 2. Source: IR Stone/Adobe Stock Photos.

Application 2 defines a set of e-healthcare nodes that are directly and indirectly connected to provide healthcare services to the users of a smart city. The application aims to depict the operations of FL for the FHLS definition. Application 2 assists patients with background diseases such as diabetes mellitus (DM), chronic obstructive pulmonary disease, congestive heart failure, and blood pressure connected to the IoMT platform of a metaverse smart city. Figure 18.5a demonstrates Application 2 and its nodes that provide healthcare services to patients with background diseases. Scenario 2 defines a risky situation based on emulating a medical-based accident for the mentioned risky group of people. Application 2 must successfully address this issue, as Figure 18.5b depicts.

Application 3 focused on supporting medicine recommendation services through the definition of FLMR (FL for Medicine Recommendation). It caters to two types of audiences, namely

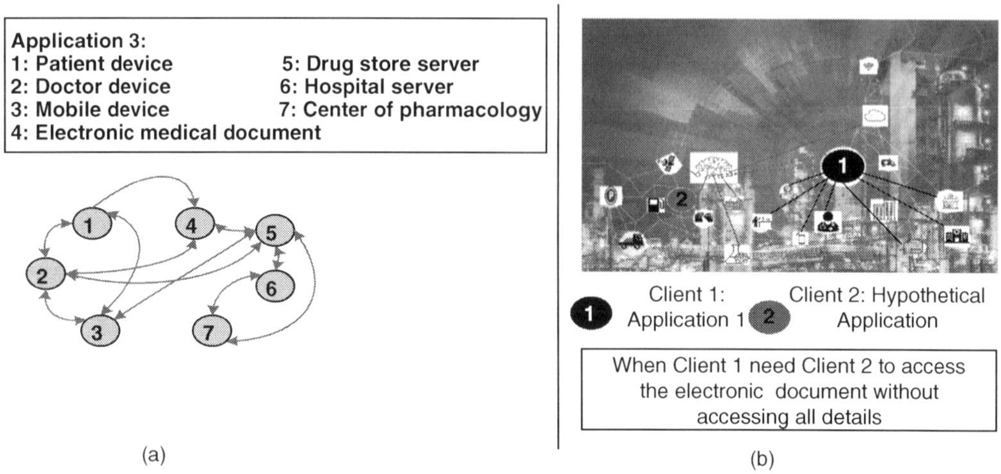

Figure 18.6 (a) Application 3 based on FLMR definition, (b) scenario 3 based on application 3. Source: metamorworks/Adobe Stock Photos.

patients and doctors. Patients have undergone recovery processes for breast cancer and acute depression, along with their corresponding digital documents. Figure 18.6a provides an overview of Application 3, its components, and their associations. The application assists doctors in prescribing suitable drugs for patients based on the latest advancements in pharmacy science, emphasizing the role of medication in inpatient treatment. The selection of diseases is based on reported statistics, with cancer and psychological distress being the most common in recent years [32, 33]. Scenario 3 outlines the actions of smart city nodes when a patient with breast cancer undergoing chemotherapy experiences symptoms of depression, highlighting the support provided by Application 3, as depicted in Figure 18.6b.

Application 4 focuses on providing drug discovery services in smart cities, explicitly targeting patients who suffer from the side effects of stroke and Alzheimer's neural attacks. This application utilizes wearable and biomarker sensors to monitor these patients' vital signs and symptoms. Drug discovery is crucial in supporting high-risk individuals facing dangerous conditions such as road accidents, natural disasters, and specific diseases with unpredictable side effects. The primary objective of Application 4 is to provide services to these patients and demonstrate the role of FLDC definition in smart cities. The details and communications of this application are depicted in Figure 18.7a. Scenario 4 outlines the tasks and operations of Application 4, highlighting the role of FL in drug discovery-based applications in smart cities, as shown in Figure 18.7b. This scenario focuses on the demand for drug discovery in prescribing and updating medication orders for patients with stroke or Alzheimer's, who experience brief psychotic disorders and require remote drug recommendations.

Application 5 focuses on smart transportation in a smart city, utilizing RFID and AIDC tags, as well as car sensors, to support the definition of FLFT. All components of various smart city applications need to connect to a secure platform for data transfer and communication purposes. The main objective of this application is to demonstrate the role of FL in creating a fault tolerance platform for communication between different devices in a smart city. Smart transportation is a fundamental and crucial aspect of a smart city, encompassing traffic management, emergency healthcare, fuel, and food transportation services. Therefore, this section aims to define Application 5 and its support for FLFT. Scenario 5 presents assumptions of faulty

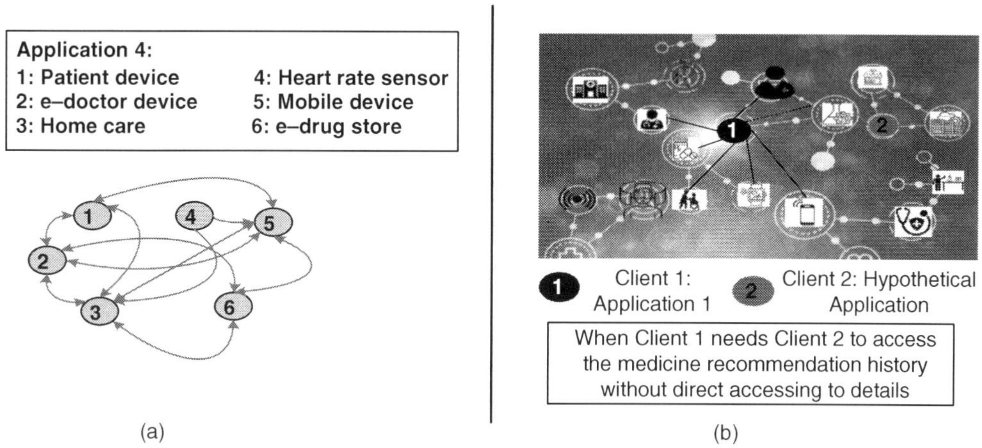

Figure 18.7 (a) Application 4 based on FLDC definition, (b) scenario 2 based on application 4.

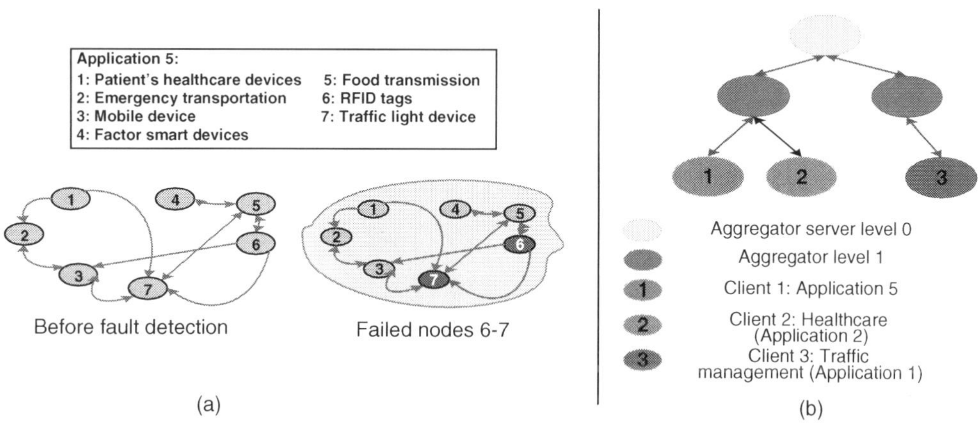

Figure 18.8 (a) Application 5 based on FLFT definition, (b) scenario 2 based on application 5.

devices in the actions of Application 5, illustrating the role of FL in satisfying fault tolerance demands. The scenario describes a situation where a smart city needs to support food and emergency healthcare transportation despite faults occurring in some devices involved in traffic monitoring. Figures 18.8a, b provide details of Application 5 and the related scenario depicting fault occurrences.

This section describes the five applications based on five task graphs, illustrating the relationship between their nodes and clarifying their purposes. By providing definitions and scenarios for each application, the subsequent sections of this chapter will build upon these concepts and explain the intended operations based on their respective tasks.

18.4 FL in Industrial Application

This section focuses on the role of FL in industrial applications supported by smart city technology, as defined by the FLIA. It aims to explain how FL can be utilized in traffic management scenarios to develop a global model for local training and data distribution. To illustrate the processes involved in FL, Application 1 is used as an example. Smart cities consist of distributed systems of diverse, interconnected devices that cater to various needs of modern lifestyles. FL supports these distributed systems by utilizing an adaptive global learning model, eliminating the need for local access between clients. To further clarify the role of FL in smart city industrial applications, Figure 18.8b is utilized to visually represent the association between Application 1 nodes and the global model.

Graph Mapping for FLIA: To define the mapping of the distributed graph nodes, we assume that the graph $G(G_g)$ consists of V vertices and E edges, with the initial vertex defining the global node as the federative model. The matrix $[G]_{i\times j}$ is a matrix with i rows and j columns, which maps the Application 1 graph (GA) nodes with X vertices and Y edges onto the nodes of the G graph. Equation (18.1) describes the mapping of the nodes in the distributed graph, where the matrix $[A]_{n\times m}$ with n rows and m columns defines the association between the nodes of Application 1.

$$G_A = (X, Y)\ [A]_{n\times m} = \begin{cases} n = X \\ m = n \end{cases} \quad A = \begin{array}{c} \\ 1 \\ 2 \\ 3 \\ 4 \\ 5 \\ 6 \\ 7 \\ 8 \end{array}\begin{array}{c} \begin{matrix} 1 & 2 & 3 & 4 & 5 & 6 & 7 & 8 \end{matrix} \\ \begin{bmatrix} 0 & 1 & 0 & 0 & 1 & 0 & 1 & 0 \\ 0 & 0 & 0 & 0 & 0 & 0 & 0 & 0 \\ 0 & 1 & 0 & 0 & 1 & 0 & 0 & 0 \\ 0 & 0 & 0 & 0 & 1 & 1 & 0 & 0 \\ 0 & 0 & 0 & 0 & 0 & 1 & 0 & 0 \\ 0 & 0 & 0 & 0 & 1 & 0 & 0 & 1 \\ 0 & 0 & 0 & 0 & 0 & 0 & 0 & 0 \\ 0 & 0 & 0 & 0 & 0 & 0 & 0 & 0 \end{bmatrix} \end{array}$$

$$G_g = (V, E)\ [G]_{ixj} = \begin{cases} i = V \\ j = i \end{cases} \quad G = \begin{bmatrix} g_{0,0} & \cdots & g_{0,j} \\ \vdots & \vdots & \vdots \\ g_{i,0} & \cdots & g_{i,j} \end{bmatrix}$$

$$Graph\ mapping = \begin{cases} g_{i,j} = a_{n,m} & (n \le n_{max}) \wedge (m \le m_{max}),\ \ 1 \le i \le i_{max}, 1 \le j \le j_{max} \\ g_{i,j} = 0, & n_{max} \le i \le i_{max},\ \ m_{max} \le j \le j_{\max} \end{cases} \tag{18.1}$$

$n_{\max}$ and $m_{\max}$ describe maximum number of rows and columns of $[A]_{n\times m}$ matrix when $n_{\max}$ and $m_{\max}$ reflect the maximum number of rows and columns of $[G]_{i\times j}$ matrix. The following graph depicts G_g after mapping Application 1's nodes, in which the red-colored and green-colored vertices and edges demonstrate an assumption mapped application and unmapped nodes, respectively.

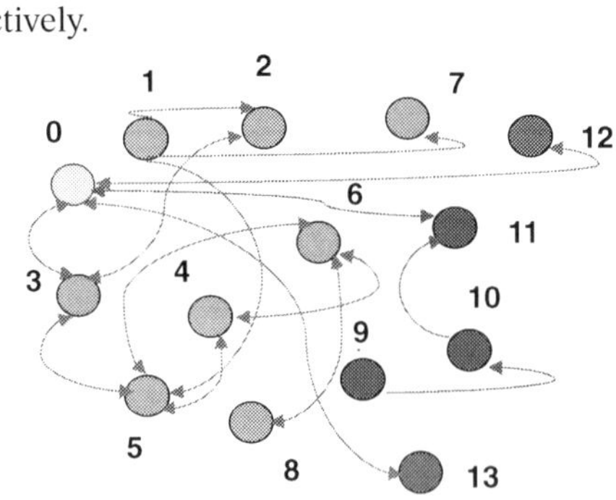

$$G = \begin{bmatrix} 0 & 0 & 0 & 1 & 0 & 0 & 1 & 0 & 0 & 0 & 0 & 1 & 1 & 1 \\ 0 & 0 & 1 & 0 & 0 & 1 & 0 & 1 & 0 & 0 & 0 & 0 & 0 & 0 \\ 0 & 0 & 0 & 0 & 0 & 0 & 0 & 0 & 0 & 0 & 0 & 0 & 0 & 0 \\ 1 & 0 & 1 & 0 & 0 & 1 & 0 & 0 & 0 & 0 & 0 & 0 & 0 & 0 \\ 0 & 0 & 0 & 0 & 0 & 1 & 1 & 0 & 0 & 0 & 0 & 0 & 0 & 0 \\ 0 & 0 & 0 & 0 & 1 & 0 & 1 & 0 & 0 & 0 & 0 & 0 & 0 & 0 \\ 1 & 0 & 0 & 0 & 1 & 1 & 0 & 0 & 1 & 0 & 0 & 0 & 0 & 0 \\ 0 & 0 & 0 & 0 & 0 & 0 & 0 & 0 & 0 & 0 & 0 & 0 & 0 & 0 \\ 0 & 0 & 0 & 0 & 0 & 0 & 1 & 0 & 0 & 0 & 0 & 0 & 0 & 0 \\ 0 & 0 & 0 & 0 & 0 & 0 & 0 & 0 & 0 & 0 & 1 & 0 & 0 & 0 \\ 0 & 0 & 0 & 0 & 0 & 0 & 0 & 0 & 0 & 0 & 0 & 1 & 0 & 0 \\ 1 & 0 & 0 & 0 & 0 & 0 & 0 & 0 & 0 & 0 & 0 & 0 & 0 & 0 \\ 1 & 0 & 0 & 0 & 0 & 0 & 0 & 0 & 0 & 0 & 0 & 0 & 0 & 0 \\ 1 & 0 & 0 & 0 & 0 & 0 & 0 & 0 & 0 & 0 & 0 & 0 & 0 & 0 \end{bmatrix}$$

Node Clustering for FLIA: There are two main approaches for implementing an FL-based model in a smart city: aggregation server and peer-to-peer. The complexity of a smart city makes the aggregation server approach suitable for handling different applications. This approach uses an aggregation server (represented by Node 0 in the mapped graph) to aggregate the global model from local training algorithms. Node clustering is essential in creating a federative model for a smart city, especially when dealing with complex technologies like the innovative city communication platform. There are two approaches to node clustering: clustering based on task definitions and clustering based on nodes' communications. Clustering based on task definitions can lead to limitations in terms of fixed clusters and head clusters. This can negatively impact the efficiency of a smart city system when tasks with similar concepts are classified in the same cluster but have no communication with each other. On the other hand, clustering based on nodes' communications allows for more flexibility in classifying components. Figure 18.9a, b illustrates the peer-to-peer steps in providing an aggregation model and clustering the nodes based on their tasks' concepts. This figure also highlights the challenges associated with these two approaches.

After node classification, the mapped graph (G_g) nodes are clustered to implement an FL-based aggregation model. The clustering is based on the communications between the nodes. The blue-colored cluster represents Application 1, while the black-colored node is a head cluster for each depicted partition.

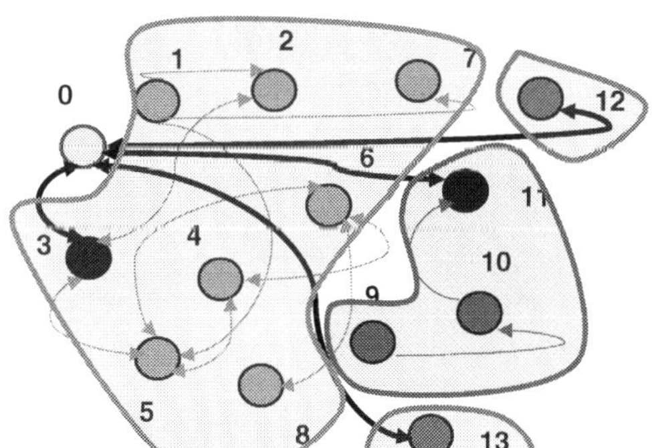

After node classification, the mapped graph (G_g) nodes are clustered based on their communication patterns to implement an FL-based aggregation model. The clusters are formed by dedicating the vertices with more edges to the head cluster role, which serves as an initial condition for the

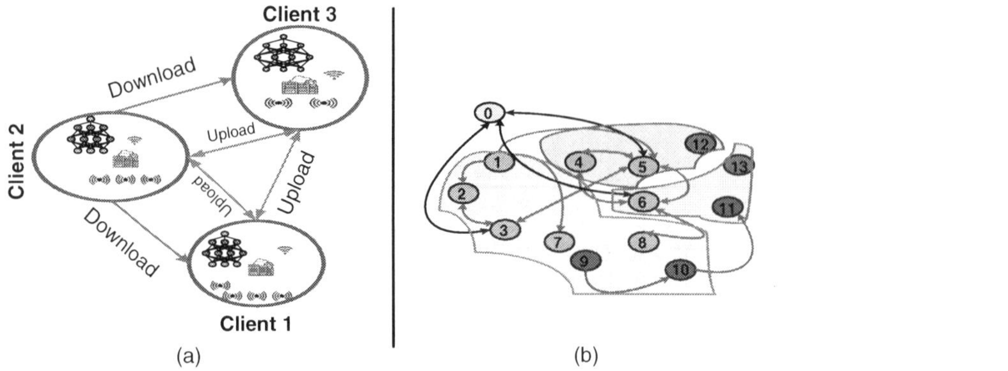

Figure 18.9 (a) Peer-to-Peer topology for FL, (b) task-based node classification.

FL model. The head clusters change based on a flexible pattern to prevent hotspot points and bottleneck problems during the upload and download of local and global models. Equation (18.2) describes a model for node classification and defines clusters based on the relationship between smart city components. Initially, classification is based on the nodes of each application, but it is not permanent, as a device failure may cause a node to be shared between two classes. Each cluster is denoted as $Cluster_c$ where c represents the cluster number, and the *Start node* is the first node of each cluster that initiates the classification process. The nodes of G_g join a cluster when the condition $(g_{i,j} = 1) \vee (g_{j,i} = 1)$ is satisfied, indicating communication with other clusters, and the start node is registered when the condition $(g_{i,j} = 0) \wedge (g_{j,i} = 0)$ is met, indicating no communication with other clusters. This model of node classification can be generalized to cluster the vertices of different distributed graphs.

$$Node\ clustering = \begin{cases} Cluster_c = node_{(i \times j_{max})+j} & (g_{i,j} = 1) \vee (g_{j,i} = 1), 1 \leq i \leq i_{max}, 1 \leq j \leq j_{max} \\ Start\ node = node_{(i \times j_{max})+j} & (g_{i,j} = 0) \wedge (g_{j,i} = 0), 1 \leq i \leq i_{max}, 1 \leq j \leq j_{max} \end{cases} \tag{18.2}$$

Figure 18.10 illustrates the steps involved in aggregating the model of local training methods based on FL in a smart city.

This approach is flexible and allows for changes in the cluster size and its nodes through variations in communication between the components of a smart city. As shown in Figure 18.10, the

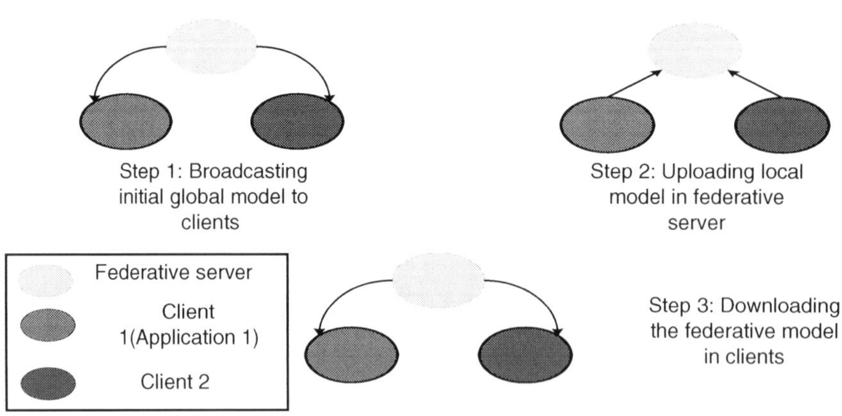

Figure 18.10 The three main steps of aggregating the global model based on FL.

members and sizes of clusters can be changed by altering the association between nodes without affecting the applications.

$$G = \begin{bmatrix} 0 & 0 & 0 & 1 & 0 & 0 & 1 & 0 & 0 & 0 & 0 & 1 & 1 & 1 \\ 0 & 0 & 1 & 0 & 0 & 1 & 0 & 0 & 0 & 1 & 0 & 0 & 0 & 0 \\ 0 & 0 & 0 & 0 & 0 & 0 & 0 & 0 & 0 & 0 & 0 & 0 & 0 & 0 \\ 1 & 0 & 1 & 0 & 0 & 1 & 0 & 0 & 0 & 0 & 0 & 0 & 0 & 0 \\ 0 & 0 & 0 & 0 & 0 & 1 & 1 & 0 & 0 & 0 & 0 & 0 & 0 & 0 \\ 0 & 0 & 0 & 0 & 1 & 0 & 1 & 0 & 0 & 0 & 0 & 0 & 0 & 0 \\ 1 & 0 & 0 & 0 & 1 & 1 & 0 & 0 & 1 & 0 & 0 & 0 & 0 & 0 \\ 0 & 0 & 0 & 0 & 0 & 0 & 0 & 0 & 0 & 0 & 0 & 0 & 0 & 0 \\ 0 & 0 & 0 & 0 & 0 & 0 & 1 & 0 & 0 & 0 & 0 & 0 & 0 & 0 \\ 0 & 0 & 0 & 0 & 0 & 0 & 0 & 0 & 0 & 0 & 0 & 0 & 0 & 0 \\ 0 & 0 & 0 & 0 & 0 & 0 & 0 & 0 & 0 & 0 & 0 & 1 & 0 & 0 \\ 1 & 0 & 0 & 0 & 0 & 0 & 0 & 0 & 0 & 0 & 0 & 0 & 1 & 0 \\ 1 & 0 & 0 & 0 & 0 & 0 & 0 & 0 & 0 & 0 & 0 & 0 & 0 & 0 \\ 1 & 0 & 0 & 0 & 0 & 0 & 0 & 0 & 0 & 0 & 0 & 0 & 0 & 0 \end{bmatrix}$$

Deep Q-Learning in FLIA: Deep Q-learning is a popular machine learning algorithm that supports intelligent decision-making and devices in solving complex modern technology problems. This technique can be successfully applied to industrial applications in a smart city due to its implementation based on a backward chain of rewards and different states' actions. The initial setting involves defining a primary model to propose clients set their models for training. The first step is to analyze the local statistics model in data distribution between devices and upload it to the aggregation server to define a global model. The next step involves downloading the current state for local training. This section utilizes a Markov decision process to provide an initial definition of communication between local and global models, as follows:

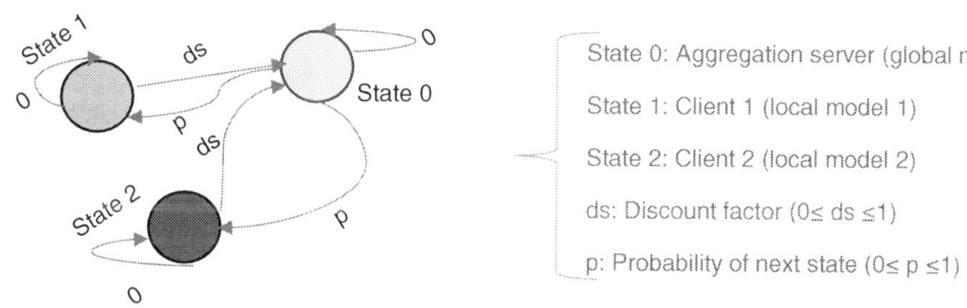

According to the initial definition of communication between local and global models based on the Markov decision process, states 0 change to states 1–2 by initializing a primary model for training the local node. States 1–2 determine a discount factor (*ds*) value to alert the global model of the efficiency of the initial definitions after inferring the local pattern of data distribution and dependency, defined as uploading the model. The ds value helps try states to obtain rewards (R) until they are close to expected services. The Markov model generalizes to a deep Q-learning method when the local node trains with a local dataset based on the technique. When k clients are candidates to contribute to the federative model, the Q matrix is defined as a matrix with k rows and columns, where $k_{\max}$ describes the maximum number of clients.

$$Q = \begin{array}{c} \\ State\,0 \\ State\,1 \\ State\,2 \\ \vdots \\ State\,k \end{array} \begin{array}{c} \begin{array}{ccccc} State\,0 & State\,1 & State\,2 & \cdots & State\,k \end{array} \\ \begin{bmatrix} 0 & p & p & \cdots & q_{0,k} \\ ds & 0 & 0 & \cdots & q_{1,k} \\ ds & 0 & 0 & \vdots & q_{2,k} \\ \vdots & \vdots & \vdots & \vdots & \vdots \\ q_{k,0} & q_{k,1} & q_{k,2} & \cdots & q_{k,k} \end{bmatrix} \end{array}$$

The p value is defined as the probability of migrating to the next state based on the feedback of the local model and the initial tables of the variables to the federative model. When the global model is fixed, all values of the p variable are set to 0, indicating that there is no need for iteration and initialization of the federated model and Q rate. The federated function $F(w)$ at time t in the r round of iteration can be defined with k client candidates by Equation (18.4). Before presenting a mathematical description of federative learning, Equation (18.3) represents the local function based on an initial aggregated model:

$$f(w)_k = \sum_{r=0}^{r=r_{\max}} (ds_{k_r} \times f(w)_r) \tag{18.3}$$

$$F(w)_{r_{t+1}} = \frac{1}{k}\left(\sum_{k=1}^{k=k_{\max}} (p_{r_k} \times f(w)_{k_t})\right) \tag{18.4}$$

In Equations (18.3–18.4), $r_{\max}$ and $k_{\max}$ define the maximum number of iterations and clients, respectively. To clarify the role of Q-learning in FL, this section presents three definitions as follows:

Discount Factor (*ds*): consists of the main concept of the parameter in Q-learning and aids in defining a valid impact factor based on ds for client priority in providing services in a smart city.

Probability (*p*): is based on the main concept of rewarding a state in Q-learning due to its success in satisfying an action, which indicates the efficiency of any updating federative model to support local training.

Iterative (*r*): describes the number of iterations of the steps of FL to aggregate a global model, combining with the concept of iteration of episodes in Q-learning.

Algorithm 18.1 (FL-based Q-learning): By defining these three key concepts, Algorithm 18.1 can effectively implement FL based on Q-learning to support a smart city.

In the first step of Algorithm 18.1, the candidate clients are mapped to the states of the Q-learning model with their private actions, and the initial value for the probability of migrating to the next state is set to 0. The initial parameters of the Q-table are listed to start aggregating a federative model of local training in clients. In step 2 of the algorithm, Algorithm 18.1 implements the relevant operations to FL based on the concept of Q-learning while setting the initial model and iteration by Q rate. This means the initial global model is transferred to clients and fixed when (Q rate $= 0$) is met. Step 2 continues to identify the proper global model until ($ds_k \neq 0$) and ($p_k \neq 0$) are satisfied by all states. The discount factor and probability of migrating from state 0 to other states reduce the computation and decision-making time to provide a secure service to smart city citizens when facing many demands in aspects of industrial, transportation, healthcare, and automated factories.

The model communication between the aggregator and client is performed by a head cluster of each partition of smart city devices. A flexible-based approach aids in selecting an appropriate head cluster for clients to prevent creating a hotspot point in data transfer between local devices and aggregator servers. To support the process, an intelligent and straightforward cost-aware algorithm can make a fair selection process to candidate a device as a head cluster. This role is periodically exchanged between the nodes of a client and transferred to a spare head cluster when the candidate fails or faces variations in local communication. The efficiency of FL and smart cities is enhanced by a flexible technique of head cluster nomination when it mitigates the time and cost of model transferring. The first characteristic of a device is its reliability, which is analyzed based on data transfer between cluster nodes. Other features can be determined based on application priority for service providing, such as computation time, response delay, and maximum transit in t time. The aggregated model of each client head cluster can be a candidate based on a separate simple

Algorithm 18.1 FL-based Q-learning (FLQ)

Inputs: 1- client: States whereas client ≥ 0
2- Q: Learning rate whereas $Q \geq 0$
Output: FederativeModel: The federative function
whereas $0 \leq$ FederativeModel

Parameter declaration:
s: *States whereas* $s \geq 0$
t: *time to represent state* whereas $0 \leq t \leq 1$
s_n^a: *state of client at time t*
a_n^a: *local training processes and operations = action*
ds_{n+1}^a: *Discount factor for action* s_n^a at time n and $(n + 1)$
Q: *Learning rate*
r: *round of iteration*

Start
  Procedure FLQ (Q_1, client)
    Start phase 1 *//Markov node mapping and Initialize Q based on primary lookup table*
      for each client in clients do
        $s_k \leftarrow client_k$
        $p_k \leftarrow 0$
      end for;
      for each episode in episodes do
        do in iteration
          Initialize primary Q table with Q
          Examine the state, execute movement prediction, and predict the next state
        end iteration;
      end for;
    end phase 1*;*
    Start phase 2 *// Initialize primary parameter of FL and aggregating a federative model*
      Defining the initial global model (w_0)
      if (Q ≠ 0) do
        do in iteration
          for each state do
            sent global model (w_0) to client k (s_k)
            set local model f $(w)_k$ based on Equation (18.3)
            $ds_n^a \leftarrow DS_w(s_n^a, a_n)$ *// Discount factor process*
          end for;
          $p_r \leftarrow P_r(s_0^a, s_{k+1}^a)$
          Updating the initial global model $F(w)_r$ based on Equation (18.4)
        end iteration;
      Updating initial parameters and set federative model
    end phase 2*;*
  end procedure;

training model and data dependency between local devices. In Application 1, a head cluster can be determined by a multilayer neural network and gradient-distant method. The activation function of the neural network is the Rectified Linear Unit (ReLU), which can prune weak weights and neurons based on imposing a penalty value through a backward chain process, as shown in Figure 18.11.

Head Cluster Nomination: Equation (18.5) describes a simple mathematical model to score each device based on application priority, where d reflects the number of client nodes. α, β, and μ

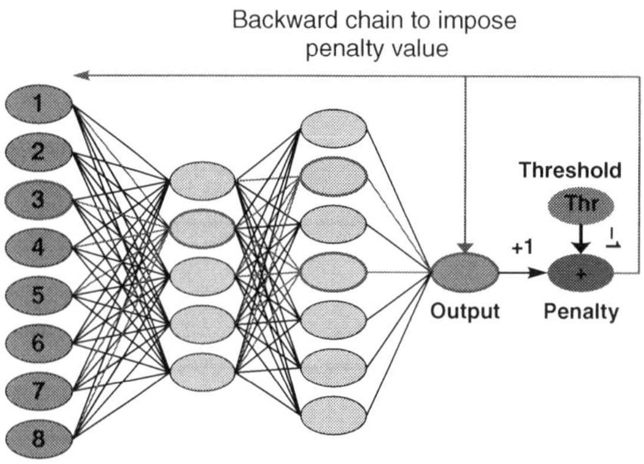

Figure 18.11 A neural network-based head cluster nomination.

describe the related impact factors for reliability, availability, and cost, respectively, which are determined based on application priority.

$$NodeCandidate_d = \alpha(reliability)_d + \beta(availability)_d + \mu(cost)_d \qquad (18.5)$$
$$\alpha + \beta + \mu = 1$$

Algorithm 18.2 implements the operations for head cluster nomination based on Equation (18.5) and application priority without employing a neural network. The main process of the algorithm iterates when t periodic time passes since the previous head cluster nomination or when a client

Algorithm 18.2 Head cluster nominator (HCN)

Inputs: 1- R: Reliability whereas $0 \leq R \leq 1$
2- A: Availability whereas $0 \leq A \leq 1$
3- C: Cost whereas $C \geq 0$
4- N: The nodes of application *// As an example: nodes for Application 1* $\{1, 2, 3, 4, 5, 6, 7, 8\}$
Output: HeadCluster: The federative function whereas HeadCluster ≥ 0

Parameter declaration:
d: *Nodes number whereas* $d \geq 0$
f_d: *Impact factor for condidate node d*
d_{max}: *Maximum number of nodes*
t: *time whereas* $t \geq 0$
ε: A permanent value whereas $0 \leq \varepsilon \leq 1$

Start
  Procedure HCN (R, A, C)
    Start *//Markov node mapping and Initialize Q based on primary lookup table*
      Defining application priority
      $f \leftarrow 1$ *// set the impact factor for all nodes*
      if (t = period) or (Situation = event) do
        do in iteration
          for each node in nodes do
            $(NodeCondidate)_d \leftarrow (\alpha(reliability)_d + \beta(availability)_d + \mu(cost)_d) \times f_d$
          end for;
          HeadCluster $\leftarrow$ Max$\{(NodeCondidate)_0, (NodeCondidate)_1, \ldots, (NodeCondidate)_{d_{max}}\}$
          $f_d \leftarrow \frac{(NodeCondidate)_d}{3 \times (\alpha + \beta + \mu)} - \varepsilon$ *// Normalizing the value of NodeCondidate to its impact factor*
        end iteration;
      end if;
    end;

faces an unpredictable event, as indicated by the IF instruction in Algorithm 18.2. The impact factor of each candidate node assists in achieving flexible and fair node dedication in the head cluster role, with its value being reduced after each selection.

FL technique provides a secure platform for communication among devices in a smart city. By employing FL, learning techniques involving indirect connections between clients can protect data confidentiality without additional measures to prevent unauthorized access to private data. Combining FL with the popular machine learning model, Q-learning enhances the efficiency of smart applications, cities, and the metaverse. FL introduces a positive constraint in the migration between states of a Markov decision process, enabling the definition of initial actions for each state based on the successful operation percentage in achieving their objectives. The aggregator server serves as the fixed starter state, while other states migrate based on feedback from their actions. This secure communication between nodes in an industrial application of a smart city is facilitated by flexible clustering methods that enable data transfer and distribution between components without compromising system security. The flexible pattern of head cluster nomination further enhances the efficiency of FL in smart city technology by reducing various types of delay and energy consumption at hotspot points. All approaches can effectively support industrial-based applications in a smart city by embracing the core concept of FL.

18.5 FL in the Healthcare System

The section focuses on the healthcare applications that analyze the efficiency of FL in smart cities. One crucial requirement for healthcare applications is protecting patients' private records and preventing unauthorized access by malicious individuals such as illegitimate doctors or intruders. FL helps address these demands in medical-based applications within a smart city by securing client communication and restricting global access to data centers. To explain how FL can address these issues in a smart city, this section outlines the steps involved in FL based on the FHLS definition and Application 2. We utilize Scenario 2 to highlight the importance of using information from other clients without direct local and global access. This scenario illustrates a common situation in healthcare services where a patient requires emergency transportation and information about traffic patterns on urban roads. To address this issue, two or more clients must collaborate to find an appropriate route for transporting the patient to a hospital, clinic, or other medical emergency center. Application 2 focuses on patients with background diseases due to the high sensitivity of their health situation when facing a virus epidemic, urban accidents, and natural disasters. The relief groups prioritize the infected people with specific diseases in healthcare services. These patients register in the medical data bank of a smart city. The concepts of smart healthcare and decentralizing services reflect that a distributed system definition constructs the architecture of the smart city platform. We utilize a distributed graph to depict the nodes of the application and the relationship between them by mapping the graph's vertices. In the previous section, we represented G_g with V vertices and E edges to describe the association between nodes of Application 1 and the aggregator. In this section, we describe the relationship between the nodes of Application 2 and the global model by G_{g2} graph with $V2$ vertices and $E2$ edges. We map the nodes of Application 2 onto vertices and define the edges of the graph based on the association between components, which G_{A2} demonstrates with $X2$ vertices and $Y2$ edges. Figure 18.12 depicts node mapping on the vertices of the graph.

Node Mapping for FLMR: $[A2]_{n2\times m2}$ is a binary matrix with $n2$ rows and $m2$ columns, where $n2_{\max}$ and $m2_{\max}$ represent the maximum number of rows and columns, respectively.

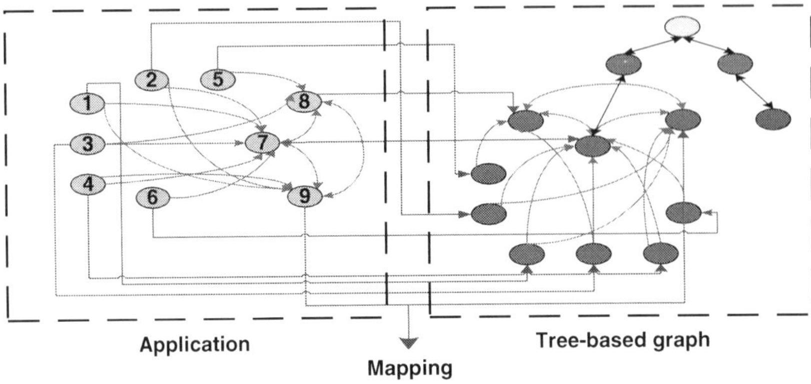

Figure 18.12 Graph mapping by application nodes.

Each element ($a2_{n2,\,m2}$) in the matrix represents the relationship between two nodes of G2. $[G2]_{i2\times j2}$ is a binary matrix that explains the association between the nodes of *the* G_{g2} graph, where $i2$ and $j2$ describe the row and column of the graph, respectively. The variables $i2_{\max}$ and $j2_{\max}$ reflect the maximum number of rows and columns in $[G2]_{i2\times j2}$.

$$
G_{A2} = (X2, Y2)\ \ [A2]_{n2\times m2} = \begin{cases} n2 = X2 \\ m2 = n2 \end{cases} \quad A2 = \begin{array}{c} \\ 1 \\ 2 \\ 3 \\ 4 \\ 5 \\ 6 \\ 7 \\ 8 \\ 9 \end{array} \begin{array}{c} 1\ 2\ 3\ 4\ 5\ 6\ 7\ 8\ 9 \\ \begin{bmatrix} 0 & 0 & 0 & 0 & 0 & 1 & 0 & 0 \\ 0 & 0 & 0 & 0 & 0 & 0 & 0 & 0 \\ 0 & 0 & 0 & 0 & 0 & 0 & 0 & 0 \\ 0 & 0 & 0 & 0 & 0 & 0 & 0 & 0 \\ 0 & 0 & 0 & 0 & 0 & 0 & 0 & 0 \\ 0 & 0 & 0 & 0 & 0 & 0 & 0 & 0 \\ 0 & 0 & 0 & 0 & 0 & 0 & 0 & 0 \\ 0 & 0 & 0 & 0 & 0 & 0 & 0 & 0 \end{bmatrix} \end{array}
$$

$$
G_{g2} = (V2, E2)\ \ [G2]_{i\times j} = \begin{cases} i2 = V2 \\ j2 = i2 \end{cases} \quad G2 = \begin{bmatrix} g2_{0,0} & \cdots & g2_{0,j} \\ \vdots & \vdots & \vdots \\ g2_{i,0} & \cdots & g2_{i2,j2} \end{bmatrix}
$$

The high sensitivity of smart healthcare applications necessitates using a different architecture for FL. This architecture conforms to a hierarchical structure to implement FL and define a global model. As shown in Figure 18.13, a hybrid tree-based structure can efficiently support Application 2 in accomplishing FL.

Equation (18.6): The node mapping process of a binary tree by application components and aggregator servers is described, where *TreeNode* and $Client_{\max}$ represent the node of the tree in level L and the maximum number of clients, respectively. TN denotes the total number of nodes in the tree, and $Node_{L1}$ represents the number of nodes in level 1 of the tree.

$$
\begin{aligned}
Node_{L1} &= \lceil \log_2 Client_{\max} \rceil \\
TN &= 1 + Node_{L2} + Client_{max} \\
TreeNode_{L_{NI}} &= \begin{cases} Aggregator_{L0_{NI}}\ , & L = 0, NI = 0 \\ Aggregator_{L1_{NI}}, L = 1, & 1 \le NI < Node_{L1} \\ Client_{L_{NI}}, L = 2, & Node_{L1} \le NI \le TN \end{cases}
\end{aligned} \tag{18.6}
$$

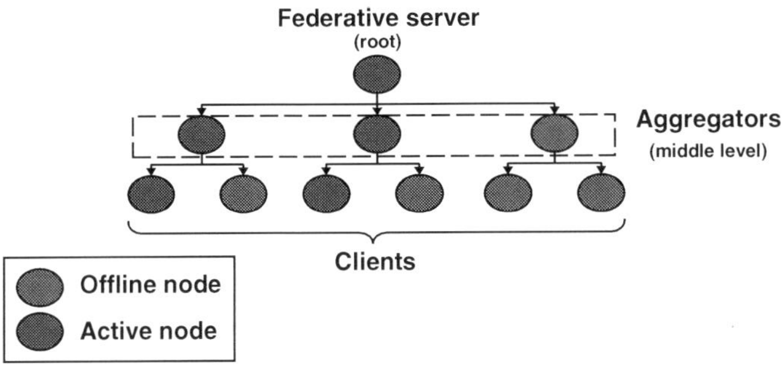

Figure 18.13 A hybrid hierarchal-based structure for implementing FL.

In Equation (18.6), $Aggregator_{L_{NI}}$ represents the aggregator *NI* (node number of tree) at level *L* of the tree, with the root of the tree (*L*0) being mapped by the global model. The nodes in level 1 are the first level of aggregating a training model based on local data dependency. The graph shows the green and yellow nodes representing the aggregators in levels 1 and 0 (root), respectively. The gray nodes indicate the mapped vertices of the graph corresponding to Application 2 tasks. The graph illustrates the relationship between the application nodes and the aggregators at two levels. The red vertices represent assumed unmapped nodes in the graph. The vertex indices are based on level and node numbers, such as node 1 in level 2, defined as 2-1.

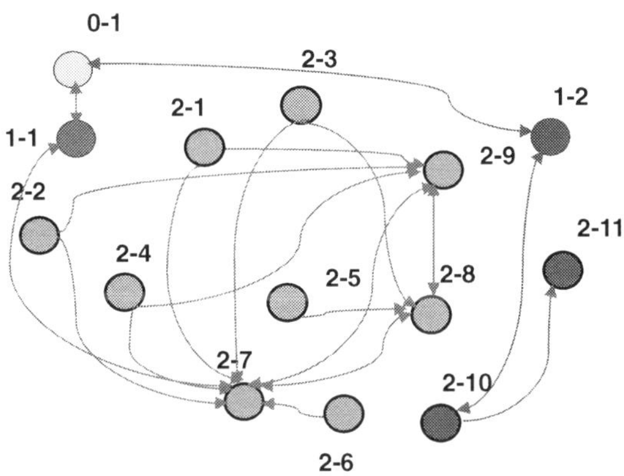

$$G2 = \begin{bmatrix} 0 & 1 & 1 & 0 & 0 & 0 & 0 & 0 & 0 & 0 & 0 & 0 & 0 & 0 \\ 1 & 0 & 0 & 0 & 0 & 0 & 0 & 0 & 0 & 1 & 0 & 0 & 0 & 0 \\ 1 & 0 & 0 & 0 & 0 & 0 & 0 & 0 & 0 & 0 & 0 & 0 & 1 & 0 \\ 0 & 0 & 0 & 0 & 0 & 0 & 0 & 0 & 0 & 1 & 0 & 1 & 0 & 0 \\ 0 & 0 & 0 & 0 & 0 & 0 & 0 & 0 & 0 & 1 & 0 & 1 & 0 & 0 \\ 0 & 0 & 0 & 0 & 0 & 0 & 0 & 0 & 0 & 1 & 1 & 0 & 0 & 0 \\ 0 & 0 & 0 & 0 & 0 & 0 & 0 & 0 & 0 & 1 & 0 & 1 & 0 & 0 \\ 0 & 0 & 0 & 0 & 0 & 0 & 0 & 0 & 0 & 0 & 1 & 0 & 0 & 0 \\ 0 & 0 & 0 & 0 & 0 & 0 & 0 & 0 & 0 & 1 & 0 & 0 & 0 & 0 \\ 0 & 1 & 0 & 0 & 0 & 0 & 0 & 0 & 0 & 0 & 1 & 1 & 0 & 0 \\ 0 & 0 & 0 & 0 & 0 & 0 & 0 & 0 & 0 & 1 & 0 & 1 & 0 & 0 \\ 0 & 0 & 0 & 0 & 0 & 0 & 0 & 0 & 0 & 1 & 1 & 0 & 0 & 0 \\ 0 & 0 & 1 & 0 & 0 & 0 & 0 & 0 & 0 & 0 & 0 & 0 & 0 & 1 \\ 0 & 0 & 0 & 0 & 0 & 0 & 0 & 0 & 0 & 0 & 0 & 0 & 0 & 0 \end{bmatrix}$$

A Hybrid Tree Structure and Node Classification: The $G2$ matrix demonstrates the association between vertices of G_g, presenting a mathematical description of node clustering for all connected components contributing to healthcare applications. The clustering approach simplifies the implementation of FL by using an aggregator to train a global model based on local data. In this approach, a head cluster is a candidate for each cluster to upload its local model and download the global model from the aggregator server. This section introduces a hybrid tree structure that supports the implementation of FL for healthcare applications in a smart city. The nodes in the following graph are classified based on their relationship with the tree structure.

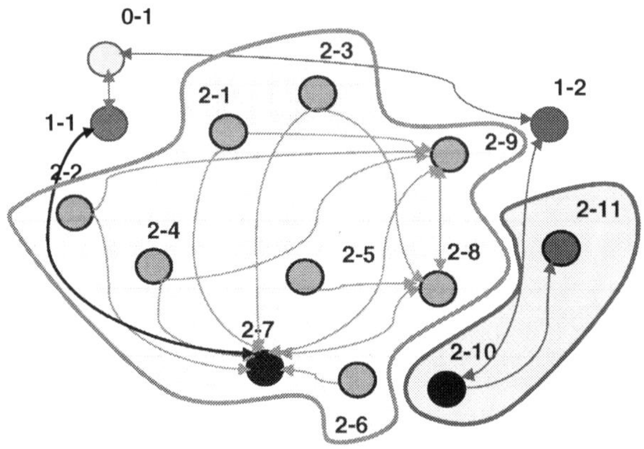

In node classification, each black-colored node serves as the head cluster for the existing blue- and red-based color partitions. The selection of head clusters is based on the characteristic of vertices having a higher degree than other nodes. The benchmark for this selection can be adjusted through application priority. As explained in the previous section, candidate nodes are modified to distribute the workload evenly across devices and prevent any hotspots in clients. The nodes with blue color partitions represent tasks related to their corresponding components in Application 2. The vertices connected to node 0-1 are the level 1 aggregators in the tree structure, where node 0-1 acts as the root. Based on the mapping of G_g and the tree structure, Equation (18.7) describes the clustering of nodes at level 1 of the tree using the $G2$ matrix. Equation (18.2) is a generalization used to classify local nodes whose partitions are attached to the tree's leaves as clients. Two adjacent clients contribute to a similar level 1 node (aggregator), depending on the number of available clients for fulfilling the application's purpose.

$$MiddleLayerCluster = deg\, g2(0-1) = d(0-1) \tag{18.7}$$

The *MiddleLayerCluster* represents the number of clusters at level 1, where each cluster consists of only one node. The $d(0-1)$ denotes the degree of the root node (node 0-1) in the tree structure or the $G2$ matrix. This chapter introduces two definitions of tree-based structures to distinguish between a pure tree and a hybrid tree, which are used to implement FL based on two specifications.

Pure Tree Structure: Consists of a binary tree-based topology for communication between clients and aggregators. The tree's root defines the global model in this structure, while the clients depict the leaves.

Hybrid Tree Structure: Depicts a tree based on a custom design. This structure is used when active and related clients need to communicate with aggregator nodes in level 1 of the tree. In this case, joining multiple clients into a middle aggregator is unnecessary.

To establish a suitable structure for implementing FL, this section draws inspiration from the hybrid tree definition to clarify the two concepts of tree topologies.**0-1**

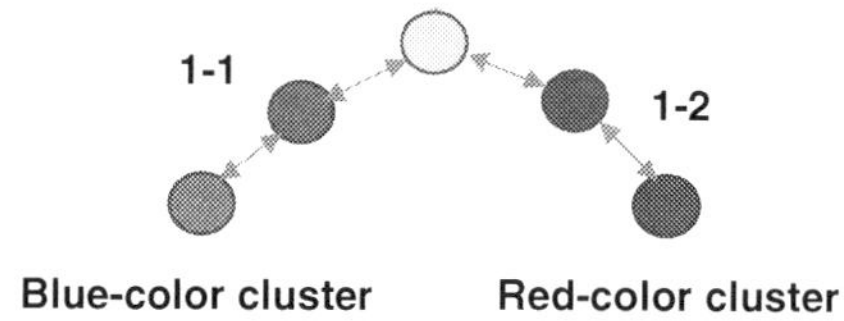

Deep Q-learning for FLMR: To effectively cater to the high data and diverse range of smart applications in a smart city, a deep learning model is required to provide services to its citizens. To address the specific demands related to healthcare in smart cities, reinforcement learning, a popular unsupervised learning approach, can be implemented using the FL concept. The accompanying graph illustrates the global, aggregative, and client aspects of this process, employing the concepts and rules of the Markov decision process.

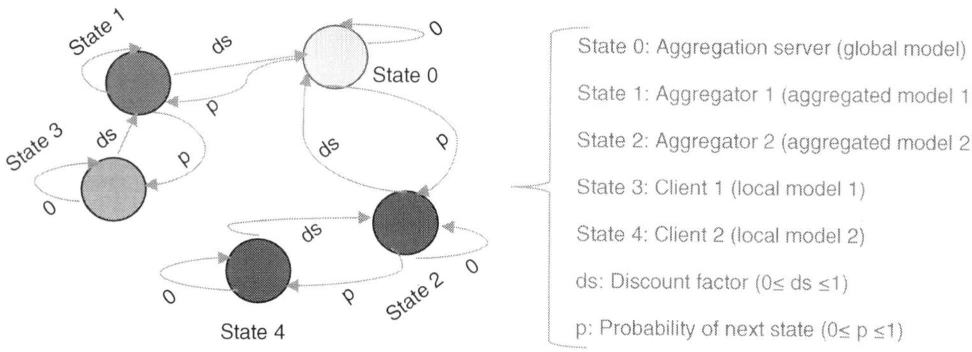

Discount factor aids in deciding about returning to the previous state for uploading the local model to an aggregator and the aggregated model to the aggregation server. Its value is determined by rewarding passing actions for each client and aggregator. The *p-value* reflects that the global and aggregated models satisfy the requirements of a smart city-based system or need more learning, which defines the probability of migrating to the next state. Therefore, Q table is initialized by an initial federative model and details of states and their actions. Q matrix describes the states and their association based on the value of p and ds between couple states, which are all variable, and parameter decelerations are the same as in the previous section (Section 1.4).

$$
\begin{array}{cc}
 & \begin{array}{ccccccc} State\,0 & State\,1 & State\,2 & State\,3 & State\,4 & \cdots & State\,k \end{array} \\
Q = \begin{array}{c} State\,0 \\ State\,1 \\ State\,2 \\ State\,3 \\ State\,4 \\ \vdots \\ State\,k \end{array} &
\begin{bmatrix}
0 & p & p & 0 & 0 & \cdots & q_{k,6} \\
ds & 0 & 0 & p & 0 & \cdots & q_{k,6} \\
ds & 0 & 0 & 0 & p & \cdots & q_{k,6} \\
0 & ds & 0 & 0 & 0 & \cdots & q_{k,6} \\
0 & 0 & ds & 0 & 0 & \cdots & q_{k,6} \\
\vdots & \vdots & \vdots & \vdots & \vdots & \cdots & q_{k,6} \\
q_{k,0} & q_{k,1} & q_{k,2} & q_{k,3} & q_{k,4} & \vdots & q_{k,6}
\end{bmatrix}
\end{array}
$$

Changes in the communication infrastructure between clients and aggregators necessitate the use of FL models with two levels of aggregator nodes, as described by Equation (18.10). Equation (18.9) outlines the function of the federative model at level 1, denoted as $f(w_{L1})$. The hybrid tree structure allows for two literation levels to aggregate a global model, with the first level identified by r_{L1} and the root level representing the global model denoted as r. Equation (18.8) provides a mathematical representation of updating local models using federative models from level 1.

$$f(w)_k = \sum_{r_{L1}=0}^{r_{L1}=r_{L1\max}} (ds_{k_{r_{L1}}} \times f(w_{L1})_{r_{L1}}) \tag{18.8}$$

$$f(w_{L1}) = \frac{1}{k}\left(\sum_{k=1}^{k=k_{\max}} (p_{r_{L1k}} \times f(w)_{k_t})\right) \tag{18.9}$$

$$F(w)_{r_{t+1}} = \frac{1}{k}\left(\sum_{k=1}^{k=k_{\max}} (p_{r_k} \times f(w_{L1})_{k_t})\right) \tag{18.10}$$

Algorithm 18.3 (Hybrid Tree Structure-Based FL): Algorithm 18.3 outlines the operations of FL in a hybrid tree structure, drawing inspiration from the Q-learning technique to define the roles of aggregators and clients. The process begins with the initial broadcast of the global model to aggregator nodes in level 1, triggering parallel processes where each branch of the tree uploads local models and downloads the global model. Once aggregators in level 1 receive models from clients, the aggregator server uploads the aggregated models and proceeds to iterate the steps. In the first step of Algorithm 18.3, the clients and aggregators are assigned to the leaf and middle levels of the hybrid tree, respectively. *StateNodeL1* represents the states of the nodes at level 1 of the hybrid tree (middle level), which correspond to the tasks of the aggregators. Step 2 focuses on implementing the necessary operations to initialize the global model and define parameters such as ds and p for the next state migration. The aggregator server, located at the tree's root, broadcasts the initial global model to the aggregator nodes, and the middle-level states begin downloading the federated model to the clients. Finally, the aggregators upload the aggregated model to the aggregator server to initiate the second round of iteration and obtain $F(w)_r$.

This section aims to demonstrate the effectiveness of FL in supporting smart healthcare applications in a smart city. We use Application 2 as an example to illustrate the communication between medical devices and the need for indirect connections in a smart city. FL successfully addresses this issue by utilizing a hybrid tree structure and combining FL with Q-learning. The hierarchical structure of the hybrid tree ensures privacy protection and enables indirect communication between clients without accessing all their details. The node selection process for the head cluster remains the same as in the previous section, utilizing a flexible-based approach to reduce latency and energy consumption. Implementing FL across three levels may introduce additional delay and energy consumption, which can be mitigated through pruning techniques and predicting the closest initial global model to the actual function.

18.6 FL in Medicine Recommendation

Medical-based applications play a significant role in modern and smart cities, catering to citizens' healthcare needs, drug discovery, and medicine recommendations. These applications also extend to smart surgeries based on augmented reality in the metaverse smart cities. The previous section focused on healthcare-based applications, and now we will analyze medicine recommendations. The critical difference between medicine recommendation and drug discovery lies in meeting the demands for recently discovered medicines and providing access to appropriate drugs for a specific disease, which is crucial for patient care. Due to the lack of direct data access between

Algorithm 18.3 FL-based Q-learning and hybrid tree structure (FLQT)

Inputs: 1- client: States whereas client ≥ 0
2- Q: Learning rate whereas Q ≥ 0
3- aggregator: Nodes of tree's level 1 whereas client ≥ 0
Output: FederativeModel: The federative function
whereas $0 \leq$ FederativeModel

Parameter declaration:
s: *States whereas* $s \geq 0$
t: *time to represent state* whereas $0 \leq t \leq 1$
s_n^a: *state of client's actions at time t*
a_n^a: *local training processes and operations (action)*
ds_{n+1}^a: *Discount factor for action* s_n^a *at time n and* $(n+1)$
Q: *Learning rate*
r: *round of iteration for federative server*
$p_{r_{L1}}$: *the probability of migrating to next state based on round* r_{L1} *of iteration for aggregators in level 1* whereas $0 \leq p \leq 1$

```
Start
  Procedure FLQT (Q_1, client, aggregator)
    Start phase 1 //Markov node mapping and Initialize Q based on primary lookup table
        for each client in clients do // mapping states with client
            s_k ← client_k
            p_k ← 0 // initializing the related p value to the corresponding states
        end for;
        for each aggregator in ⌈log_2 clients⌉ do //mapping aggregators on states
            StateNodeL1_{⌈log_2 clients⌉} ← Aggregator_{⌈log_2 clients⌉} // mapping aggregator on the states
          end for;
          for each episode in episodes do
            do in iteration
              Initialize primary Q table with Q
              Examine the state, execute movement prediction, and predict the next state
            end iteration;
          end for;
      end phase 1;
      Start phase 2 // Initialize primary parameter of FL and aggregating a federative model
          Defining the initial global model (w_0) and broadcasting to all aggregators in middle level of tree
          if (Q ≠ 0) do
            do in iteration
              do in iteration
                for each state in states do
                  sent global model (w_0) to client k (s_k)
                  set local model f (w)_k based on Equation (18.8)
                  ds_n^a ← DS_w (s_n^a, a_n) // Discount factor process
                end for;
                p_{r_{L1}} ← P_{r_{L1}} (s^a_{⌈log_2 k⌉}, s^a_{k+1}) // Determining the value of the probability of migrating to
                next state
                Updating the initial federative model in level 1 f(w_{L1}) based on Equation (18.9)
              end iteration;
              p_r ← P_r (s_0^a, s_{k+1}^a ) // Determining the value of the probability of migrating to next state
              for state 0
              Updating the initial federative model based on Equation (18.10)
          end iteration;
        Updating initial parameters and set federative model
      end phase 2;
    end procedure;
```

components in a smart city, FL emerges as an appropriate approach to support data processing for various applications. In this section, we will explain the role of FL in medicine recommendation, explicitly focusing on the definition of FLMR and highlighting the efficiency it brings to Application 3. Scenario 3 illustrates the importance of medicine recommendations and meeting citizen demands in a smart city. This scenario focuses on urgent medicine recommendations for patients with specific diseases, depression, and anxiety disorders. For instance, if a cancer patient experiences unpredictable side effects during chemotherapy that jeopardize their health and potentially their life, an oncologist needs to prescribe an appropriate alternative drug with minimal risk. This process can be time-consuming using routine and traditional approaches, but specialist doctors can expedite it by utilizing a smart medicine recommendation system in a smart city. Security also plays a critical role in ensuring the efficient support of this application, as any faults or attacks could endanger human lives.

The distributed graph can be an effective technique for modeling the interconnectedness of nodes within an application in a smart city. In this context, the edges ($X3$) of the G_{A3} graph represent the relationships between the nodes of Application 3, which are then mapped onto the vertices ($Y3$) of G_{A3}. The matrix $[A3]_{n3\times m3}$, with $n3$ rows and $m3$ columns, describes the associations between the vertices of G_{A3}, where $n3_{\max}$ and $m3_{\max}$ indicate the maximum number of rows and columns in the matrix.

$$G_{A3}=(X3,Y3)\qquad [A3]_{n3\times m3}=\begin{cases} n3=X3\\ m3=n3\end{cases}\qquad A3=\begin{matrix} & \begin{matrix}1&2&3&4&5&6&7\end{matrix}\\ \begin{matrix}1\\2\\3\\4\\5\\6\\7\end{matrix} & \begin{bmatrix}0&1&1&1&0&0&0\\1&0&1&1&1&0&0\\1&1&0&0&0&1&1\\1&0&0&0&0&0&0\\0&0&0&0&0&0&0\\0&0&1&1&0&0&1\\0&1&0&0&1&1&1\end{bmatrix}\end{matrix}$$

Node Mapping Technique in FLMR: The preceding section utilized a hybrid tree structure for implementing FL, where certain clients were assigned to leaf nodes that could be inactive, and some middle-level nodes could only accommodate one client. However, this setup can lead to an unstable situation for Application 3, especially when it requires a secure and robust platform to safeguard highly sensitive information regarding recently discovered drugs. To address this issue, we propose the use of a pure tree structure based on the provided definition, which offers a federative model and enables indirect communication between clients. G_{g3} represents the relationship between the root node, middle-level aggregators, and components of Application 3. This graph comprises $V3$ vertices and $E3$ edges. $[G3]_{i3\times j3}$ denotes a matrix with $i3$ rows and $j3$ columns, which describes the relationship between the vertices of G_{g3} that are assigned tasks corresponding to Application 3.

$$G_{g3}=(V3,E3)\qquad [G3]_{i3\times j3}=\begin{cases} i3=V3\\ j3=i3\end{cases}\qquad G3=\begin{bmatrix} g3_{0,0} & \cdots & g3_{0,j3}\\ \vdots & \vdots & \vdots\\ g3_{i3,0} & \cdots & g3_{i3,j3}\end{bmatrix}$$

In this section, Equation (18.6) maps the clients and aggregators onto the tree's leaf and middle-level nodes (level 1). The graph below shows the relationship between the nodes of G_{g3}, with gray vertices representing the nodes mapped by Application 3's tasks. Yellow and green nodes represent federative servers and aggregators, while red vertices indicate unmapped nodes.

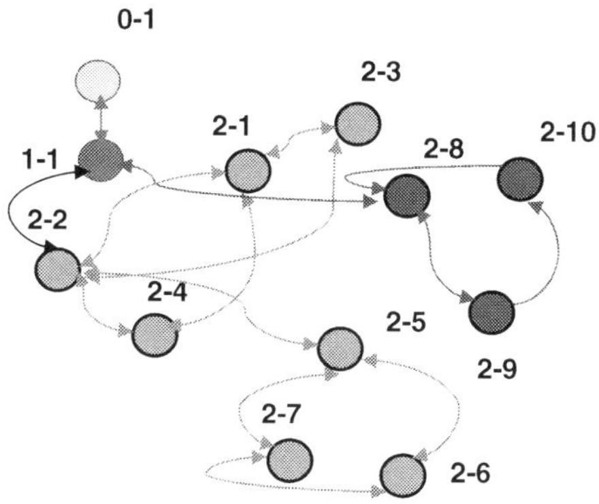

Pure Tree Structure and Node Classification for FLMR: The tree structure definition provides a clear classification method for the partitions of aggregators and their association with connected clients. It allows the classification of each client's existing nodes to specify their component, as shown in the following graph. The black-colored nodes are designated as the head cluster for each classification, with the collection process based on nodes with more edges (degree).

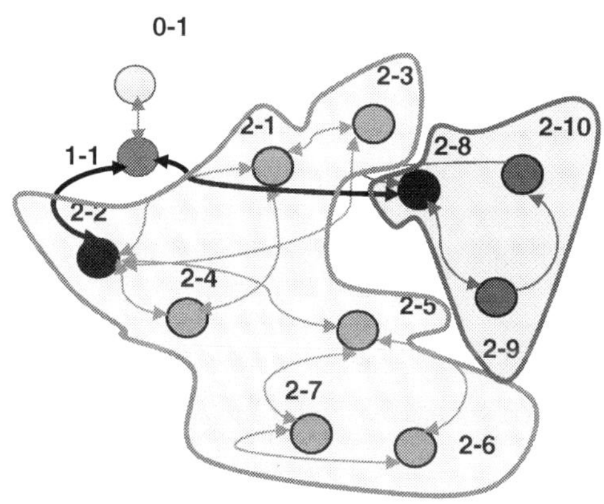

The classification of elements in the $[G3]_{i3\times j3}$ matrix, which represents the node classification of G_{g3}, also demonstrates the node clustering.

$$G3 = \begin{bmatrix} 0 & 1 & 0 & 0 & 0 & 0 & 0 & 0 & 0 & 0 & 0 & 0 \\ 1 & 0 & 0 & 1 & 0 & 0 & 0 & 0 & 0 & 1 & 0 & 0 \\ 0 & 0 & 0 & 1 & 1 & 1 & 0 & 0 & 0 & 0 & 0 & 0 \\ 0 & 1 & 1 & 0 & 1 & 1 & 0 & 0 & 0 & 0 & 0 & 0 \\ 0 & 0 & 1 & 0 & 0 & 1 & 0 & 0 & 0 & 0 & 0 & 0 \\ 0 & 0 & 1 & 1 & 0 & 0 & 0 & 0 & 0 & 0 & 0 & 0 \\ 0 & 0 & 0 & 0 & 0 & 1 & 0 & 1 & 1 & 0 & 0 & 0 \\ 0 & 0 & 0 & 0 & 0 & 0 & 1 & 0 & 1 & 0 & 0 & 0 \\ 0 & 0 & 0 & 0 & 0 & 0 & 1 & 1 & 0 & 0 & 0 & 0 \\ 0 & 0 & 0 & 0 & 0 & 0 & 0 & 0 & 0 & 0 & 1 & 0 \\ 0 & 0 & 0 & 0 & 0 & 0 & 0 & 0 & 0 & 1 & 0 & 1 \\ 0 & 0 & 0 & 0 & 0 & 0 & 0 & 0 & 0 & 1 & 1 & 0 \end{bmatrix}$$

As shown in the $G3$ matrix, the classification nodes of clients follow a pattern similar to the one presented in Section 1.4 and described by Equation (18.2). The middle-level nodes of the tree are classified using Equation (18.7).

Deep Q-Learning in FLMR: The process of recommending medicine involves a complex procedure to identify and prescribe a new drug for a specific disease. This process requires a robust, unsupervised learning approach. Q-learning can be a successful method to support this application in a smart city, especially when combined with the concept of FL. Utilizing the states and actions makes it possible to transition between different levels of the tree, allowing for the uploading of local models and downloading of global models based on the FL concept.

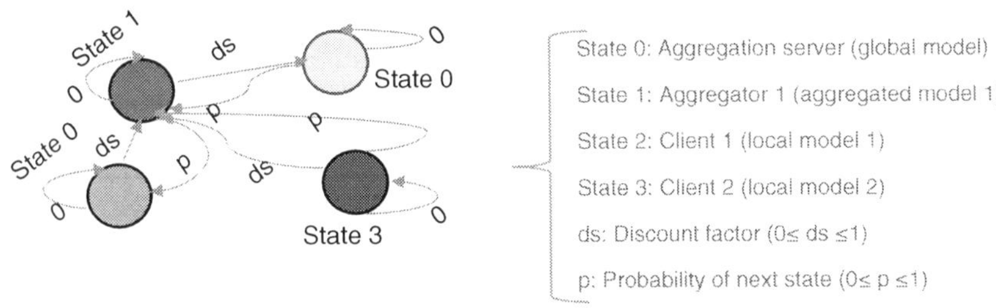

State 0 consists of the initial global model, which changes to state 1 based on the probability of migrating to the next state for downloading state 1 with the federated model. State 1 then changes to states 2-3 based on the value of p for downloading clients by the aggregated model. All states can transition back to the previous state by analyzing their actions and determining their necessity for defining the reward value. The discount factor determines the probability of iteration of the processes. Algorithm 18.3 encompasses all the mentioned operations to obtain a global model based on the local data dependency of clients. The federated function is obtained by Equations (18.8–18.10) due to the employment of a tree-based structure and the combination of FL ($F(w)_r$) with Q-learning.

In this section, we emphasized the role of FL in medicine recommendation applications to demonstrate the efficiency of FL in a smart city. The sensitive nature of medicine recommendations necessitates greater accuracy and security, prompting us to explore recent pharmacology-based research to assist specialist doctors in prescribing drugs. We proposed a pure tree-based structure for the hierarchical implementation of FL, creating a platform that enhances security and facilitates indirect accessibility between clients when seeking information outside their cluster. Implementing FL with a fixed structure like a pure binary tree reduces the probability of encountering attacks and data loss, and the federative model is established based on all participating clients connecting to the middle aggregator. However, this approach may result in increased delay and energy consumption. Additionally, aggregator nodes may absorb inactive clients, adding complexity to the implementation of FL. This issue can be resolved by developing a prediction model to identify active nodes and enable their contribution to the aggregators.

18.7 FL in Drug Discovery

A learning technique can be considered efficient in supporting a communication system when it can be generalized for different routines and commonly used applications. This chapter aims to depict and emphasize the role of FL in a smart city, including various smart applications. FL can be implemented using multiple structures such as peer-to-peer, aggregator server, tree, hybrid tree, and other decentralized and centralized communication infrastructures. These structures facilitate both direct and indirect connections between clients and aggregator nodes. The smart city concept is based on a distributed system with configurability characteristics that enable the inclusion of more citizens and various intelligent devices without compromising performance. We have explained how FL supports industrial, healthcare, and medicine recommendation applications, which are popular in a smart city. Additionally, FL plays a crucial role in drug discovery, particularly in identifying appropriate medicines for patients with emergencies such as brief psychotic disorders and sudden panic attacks. The application must discover suitable drugs to propose to patients and their caregivers when they face unpredictable emergencies to ensure the mental health of infected individuals. This section clarifies the role of FL in drug discovery based on the FLDC definition in a smart city, highlighting its importance in Application 4 and Scenario 4. In the previous sections, we explained how to implement FL for three different applications and presented mathematical descriptions and algorithms to support the steps involved in FL in a smart city.

Distributed Graph for Application 4: This section demonstrates the generalization of the addressed techniques for supporting various applications in a smart city using FL. It also explains the role of FL in drug discovery applications. To support this purpose, we utilize a distributed graph called G_{A4}, which consists of $X4$ vertices and $Y4$ edges, to illustrate the association between the components of Application 4. The relationship between the nodes of G_{A4} is determined by the elements of $[A4]_{n4 \times m4}$, where the matrix has $n4$ rows and $m4$ columns.

$$G_{A4} = (X4, Y4) \qquad [A4]_{n4 \times m4} = \begin{cases} n4 = X4 \\ m4 = n4 \end{cases} \qquad A3 = \begin{array}{c|cccccc} & 1 & 2 & 3 & 4 & 5 & 6 \\ \hline 1 & 0 & 1 & 1 & 0 & 1 & 0 \\ 2 & 1 & 0 & 1 & 0 & 1 & 1 \\ 3 & 0 & 1 & 0 & 0 & 0 & 1 \\ 4 & 0 & 0 & 1 & 0 & 1 & 0 \\ 5 & 0 & 0 & 0 & 0 & 0 & 1 \\ 6 & 0 & 0 & 0 & 0 & 1 & 0 \end{array}$$

According to the communication between the nodes of Application 4 and the priorities of the drug discovery-based application, a hybrid tree structure can efficiently support the related demands. This structure ensures security to protect private information about various drugs and allows indirect access to the medicine data bank. G_{g4} describes the association between located nodes at different levels of the hybrid tree, which consists of $V4$ vertices and $E4$ edges. The purpose of this structure is to provide a secure platform and access to information about patients' experiences with similar disorders and drug reactions. Scenario 4 demonstrates that the drug discovery-based application needs to generate safe medicine prescriptions without direct involvement of specialist doctors. In this case, a hybrid tree enables indirect access to active clients, eliminating the need for all clients to connect to located aggregators in the middle level of the tree. This section defines a fully customized structure based on the pattern of a binary tree for communication between clients and aggregators, similar to the hybrid tree described in Section 1.5.

$$G_{g4} = (V4, E4)) \; [G4]_{i \times j} = \begin{cases} i4 = V4 \\ j4 = i4 \end{cases} \quad G3 = \begin{bmatrix} g4_{0,0} & \cdots & g4_{0,j4} \\ \vdots & \vdots & \vdots \\ g4_{i4,0} & \cdots & g4_{i4,j4} \end{bmatrix}$$

Node Mapping Technique for a Hybrid Tree: The node mapping description is based on Equation (18.6) to transfer clients' roles onto the hybrid tree leaf. Aggregators are also mapped onto the nodes of the middle level of the hybrid tree, with the root mapped by the aggregator server. G_{g3} illustrates the association between the clients and aggregators on the tree, where the gray-colored and red-colored vertices represent the assumption nodes and mapped nodes of Application 4, respectively.

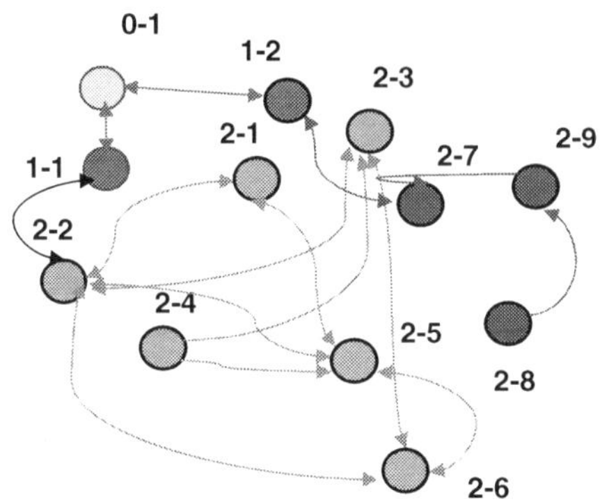

Node Classification of a Hybrid Tree: A clear plan is implemented using a tree structure to classify the nodes. The middle and root aggregator nodes are classified, while the client nodes are clustered based on Equation (18.2). Each client is created based on the association between nodes, which describes an application or joins together to ascertain a task. The node classification for G_{g3} is as follows: the black-colored nodes are candidates for the head cluster of each classification. The blue-colored cluster represents the client included in Application 4, and the red-colored classification demonstrates an assumption client, where their nodes are clustered in one partition based on their relationship.

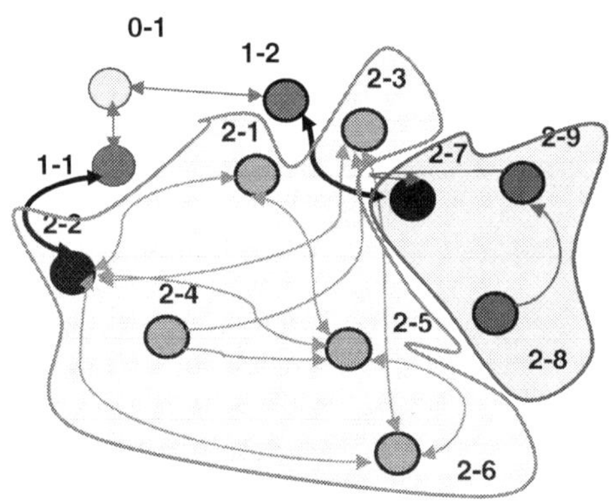

We aim to elucidate the classification of nodes by utilizing the association between their vertices, represented by the $[G4]_{i4 \times j4}$ matrix. The quadrilaterals colored in green, black, blue, and red signify the aggregator, head cluster, Application 4 cluster, and assumption application cluster, respectively.

$$G4 = \begin{bmatrix} 0 & 1 & 1 & 0 & 0 & 0 & 0 & 0 & 0 & 0 & 0 & 0 \\ 1 & 0 & 0 & 1 & 0 & 0 & 0 & 0 & 0 & 0 & 0 & 0 \\ 1 & 0 & 0 & 0 & 0 & 0 & 0 & 0 & 0 & 1 & 0 & 0 \\ 0 & 0 & 0 & 0 & 1 & 1 & 0 & 1 & 0 & 0 & 0 & 0 \\ 0 & 0 & 0 & 1 & 0 & 1 & 0 & 1 & 1 & 0 & 0 & 0 \\ 0 & 0 & 0 & 0 & 1 & 0 & 0 & 0 & 1 & 0 & 0 & 0 \\ 0 & 0 & 0 & 0 & 0 & 1 & 0 & 1 & 0 & 0 & 0 & 0 \\ 0 & 0 & 0 & 0 & 0 & 0 & 0 & 0 & 1 & 0 & 0 & 0 \\ 0 & 0 & 0 & 0 & 0 & 0 & 0 & 1 & 0 & 0 & 0 & 0 \\ 0 & 0 & 1 & 0 & 0 & 0 & 0 & 0 & 0 & 0 & 0 & 0 \\ 0 & 0 & 0 & 0 & 0 & 0 & 0 & 0 & 0 & 0 & 0 & 1 \\ 0 & 0 & 0 & 0 & 0 & 0 & 0 & 0 & 0 & 1 & 0 & 0 \end{bmatrix}$$

Aggregator Merging: During model discovery using an FL-based approach on a hybrid tree structure, it is possible to obtain similar models for certain aggregators. This presents an opportunity to simplify the structure of the hybrid tree by merging these similar cases. In order to support this method, we merge the adjacent aggregators with one connected client if they are similar in the aggregated model before finalizing the global model at the root of the tree. This approach helps to reduce the complexity of communication between various applications in a smart city and improves energy efficiency. To define adjacent aggregators, the structure must be considered a three-dimensional (3D) tree instead of a two-dimensional (2D) tree or a flat structure.

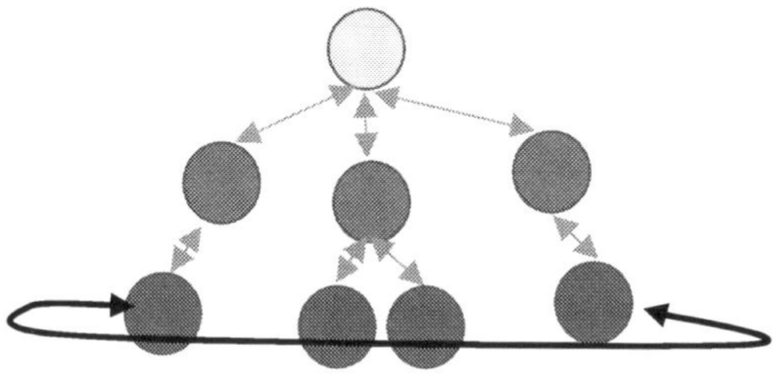

Equation (18.11) describes aggregator merging ($MergedNode_{NL1}$) where $Node_{NL1+1}$ demonstrates node $NL1$of middle level (level 1) of a hybrid tree and $Node_0$ reflects first node of the level. $NL1$ indicates node number for the located nodes in middle level.

$$MergedNode_{NL1} = \begin{cases} Node_{NL1+1}, 0 \leq NL1 < NL1_{max}, (d(Node_{NL1}) = 1) \wedge (d(Node_{NL1+1}) = 1) \\ Node_0, \quad NL1 = NL1_{max}, (d(Node_{NL1}) = 1) \wedge (d(Node_0) = 1) \end{cases} \tag{18.11}$$

The Deep Q-learning approach can be a good choice for combining with the FL technique to support smart drug discovery-based applications. This can be achieved by defining various actions of aggregators and clients. The discount factor and p-value aid in changing states based on the success of the initial global model in clients and their feedback. The Markov process chain is as follows when node 1-1 (aggregator 1-1) and node 1-2 (aggregator 1-2) obtain a similar pattern.

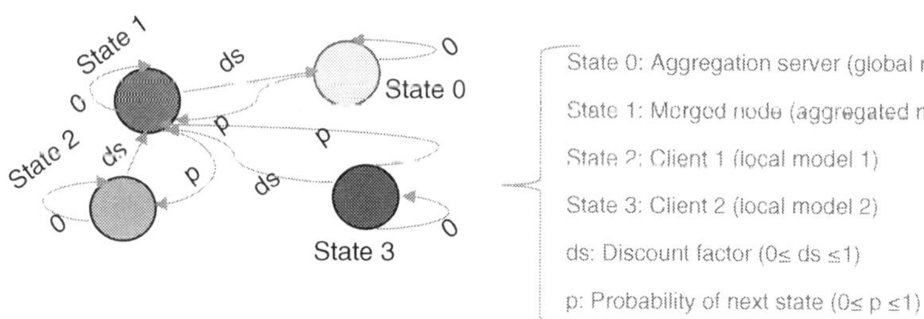

State 1 is formed by merging two aggregators, 1 and 2. The upper state diagram represents the Markov process chain that results from implementing the merging process. The Q matrix illustrates the relationships between states, based on the ds and p edges connecting them. The value of k represents the number of rows and columns in the matrix. State 0 transitions to state 1 when the probability of migration exceeds 0, which is dependent on the initial global model. Additionally, the role of migration also applies to shifting state 1 to states 2 and 3.

$$Q = \begin{array}{c} \\ \textit{State } 0 \\ \textit{State } 1 \\ \textit{State } 2 \\ \textit{State } 3 \\ \vdots \\ \textit{State } k \end{array} \begin{array}{c} \begin{array}{cccccc} \textit{State } 0 & \textit{State } 1 & \textit{State } 2 & \textit{State } 3 & \ldots & \textit{State } k \end{array} \\ \begin{bmatrix} 0 & p & 0 & 0 & \cdots & q_{0,k} \\ ds & 0 & p & p & \cdots & q_{1,k} \\ 0 & ds & 0 & 0 & \cdots & q_{2,k} \\ 0 & ds & 0 & 0 & \cdots & q_{3,k} \\ \vdots & \vdots & \vdots & \vdots & \cdots & \vdots \\ q_{k,0} & q_{k,1} & q_{k,2} & q_{k,3} & \cdots & q_{k,k} \end{bmatrix} \end{array}$$

Algorithm 18.4 (Aggregator Merging): This section utilizes a hybrid tree structure for communication between aggregator and client, combining Q-learning with FL. Algorithm 18.3 encompasses the implementation of FL for drug discovery-based applications. It also supports

---

Algorithm 18.4 Node merging (NM)

Inputs: 1- client: States whereas client ≥ 0
2- Q: Learning rate whereas $Q \geq 0$
3- aggregator: Nodes of tree's level 1 whereas client ≥ 0
Output: FederativeModel: The federative function
whereas $0 \leq$ FederativeModel

Parameter declaration:
s: *States whereas* $s \geq 0$
t: *time to represent state* whereas $0 \leq t \leq 1$
s_n^a: *state of client's actions at time t*
a_n^a: *local training processes and operations (action)*
ds_{n+1}^a: *Discount factor for action* s_n^a *at time n and* $(n + 1)$
Q: *Learning rate*
r: *round of iteration for federative server*
$p_{r_{L1}}$: *the probability of migrating to next state based on round* r_{L1} *of iteration for aggregators in level 1* whereas $0 \leq p \leq 1$
ag: *the number of aggregator nodes whereas* $ag \geq 0$

Start
Procedure NM (Q_1, client, aggregator)
Start phase 1 *//Markov node mapping and Initialize Q based on primary lookup table*
Call phase 1 of Algorithm 18.3
Performing all instructions of phase 1 of Algorithm 18.3
end phase 1 of Algorithm 18.3;
end phase 1;
Start phase 2 *// Initialize primary parameter of FL and aggregating a federative model*
Defining the initial global model (w_0) and broadcasting to all aggregators in middle level of tree
do in phase 2 of Algorithm 18.3
Updating the initial federative model in level 1 $f(w_{L1})$ based on Equation (18.9)
if $(f(w_{L1})_{ag} = f(w_{L1})_{ag+1})$ **do**
check the degree of aggregator nodes
if ($d(aggregator_{ag}) = 1$) **and** $d(aggregator_{ag}) = 1$(**do**
Merging the two aggregators
end if;
end if;
Updating the initial federative model $F(w)_r$ based on Equation (18.10)
end phase 2 of algorithm 18.3;
Updating initial parameters and set federative model
end phase 2;
end procedure;

---

aggregator merging operation by incorporating red color-based instructions into step 2 of Algorithm 18.3, thereby defining Algorithm 18.4.

Algorithm 18.4 is designed to implement the steps of FL-based on a hybrid tree structure while merging aggregators with similar patterns and combining them with Q-learning. It includes two red color-based IF instructions that analyze the situation of the federative function at level 1 of the tree and their degree when identifying similar patterns. In phase 1, Algorithm 18.4 calls phase 1 of Algorithm 18.3 to apply any variations in this phase using the CALL instruction. In phase 2, it starts the operations of phase 2 of Algorithm 18.3 and adds the red color-based instructions to merge the aggregators that are connected to one client by checking ($d(aggregator_{ag}) = 1$) and $d(aggregator_{ag}) = 1$).

This section explores the application of smart city technology in drug discovery, focusing on providing timely and accurate services to meet the demands of citizens. To illustrate this, Application 4 and Scenario 4 are introduced, which demonstrate the practical tasks that can be supported by smart city technology. It is evident from this application and scenario that FL plays a crucial role in establishing a secure and adaptable platform for communication between medical devices, medical-based sensors, and citizens within a distributed system in a smart city. The section also introduces the concept of FLDC to highlight the significance of FL in drug discovery. FLDC enables indirect and secure access between clients to facilitate their tasks. To implement FLDC, a hybrid tree structure is proposed, consisting of two levels of learning represented by the aggregator nodes located at the root and level 1 of the tree. To ensure transparency in implementing the global model in clients and analyzing local models in aggregators, the section defines the steps of FL based on the states of deep Q-learning. This approach provides a clear understanding of how the global model is implemented and how local models are analyzed by the aggregator nodes. Additionally, the section investigates the merging of nodes to create an efficient hybrid tree structure. This merging process helps reduce the cost overhead associated with having multiple connected clients with similar patterns. By combining aggregators with similar patterns, the overall efficiency of the system can be improved.

18.8 FL in Fault Tolerance

An essential characteristic of a smart city is defined based on its fault tolerance in supporting various smart applications to respond to the possible demands of citizens with maximum dependability. This means that the technology must be supported by a robust technique to provide solutions in responding to failures before impacting the performance of a system or delivering a defective service to the requester. FL is a modern technique used to define a global model for distributed systems that have numerous heterogeneous and asynchronous devices and data centers. This technique follows a multistage decision-making process to describe a federative model of the clients' local training, which locks direct access between clients and protects their private information. FL can create a fault-converging process to prevent system failure in responding to demands by employing similar patterns from correct clients to cover the tasks of failed nodes. However, the technique can also create a dangerous situation for clients and applications when the federative server or aggregators appear as malicious nodes that broadcast corrupted global models to clients. This section presents a short emphasis to depict the role of FL in the fault tolerance of a smart city based on two definitions:

Constructive Federative Model: describes the role of FL in preparing a fault-tolerant platform by locking direct access between clients and its role in fault coverage for failed nodes.

Malicious Federative Model: defines the risky aspect of FL when aggregators appear as malicious nodes that compromise clients by downloading corrupted global models.

Application 5 depicts a smart transportation application whose two nodes fail because of some static and dynamic faults. Scenario 5 describes a fault scenario for Application 5 when two related tags and sensors receive the information of traffic management and have to access some details of way situation to emergency transport. The section explains the role of FL based on two upper definitions by considering Application 5 and its scenario.

Constructive Federative Model: To demonstrate the constructive role of FL, we first map Application 5 onto the vertices of a distributed graph named G_{A5} with $X5$ vertices and $Y5$ edges to indicate the association between the application's nodes to clarify that fault occurrence makes a disturbance in the operations of an application. $[A5]_{n5\times m5}$ matrix demonstrates the association between the vertices of G_{A5} by its elements where $n4$ and $m4$ describe the rows and columns of the matrix. The red color-based elements demonstrate the failed communication induced by the failure of two nodes 6-7 of the application.

$$G_{A5} = (X5, Y5) \qquad [A5]_{n5\times m5} = \begin{cases} n5 = X5 \\ m5 = n5 \end{cases} \qquad A5 = \begin{array}{c} \\ 1 \\ 2 \\ 3 \\ 4 \\ 5 \\ 6 \\ 7 \end{array} \begin{array}{c} \begin{array}{ccccccc} 1 & 2 & 3 & 4 & 5 & 6 & 7 \end{array} \\ \begin{bmatrix} 0 & 1 & 0 & 0 & 0 & 0 & 1 \\ 0 & 0 & 1 & 0 & 0 & 0 & 0 \\ 0 & 1 & 0 & 0 & 0 & 1 & 1 \\ 0 & 0 & 0 & 0 & 1 & 0 & 0 \\ 0 & 0 & 0 & 1 & 0 & 1 & 0 \\ 0 & 0 & 1 & 0 & 1 & 0 & 1 \\ 0 & 0 & 1 & 0 & 1 & 0 & 0 \end{bmatrix} \end{array}$$

Equation (18.12) computes the total number of failed commutations (FC) induced by the failure of a node where $d(FaultyNode)_f$ describes the degree of faulty node f. The f and $f_{\max}$ demonstrate the number and total number of faulty nodes, respectively.

$$FC = \sum_{f=0}^{f=f_{\max}} d(FaultyNode)_f \tag{18.12}$$

As one of the efficient structures in implementing FL, the tree-based structure can help cover the tasks and information of failed nodes. This explanation highlights the role of FL in creating a fault-tolerant platform using the tree structure. The G_{g5} graph illustrates the relationship between the root, aggregators in the middle level of the tree, and clients based on a pure binary tree structure. The graph below showcases how FL locks an application and seamlessly continues with other clients, ensuring a reliable and uninterrupted service.

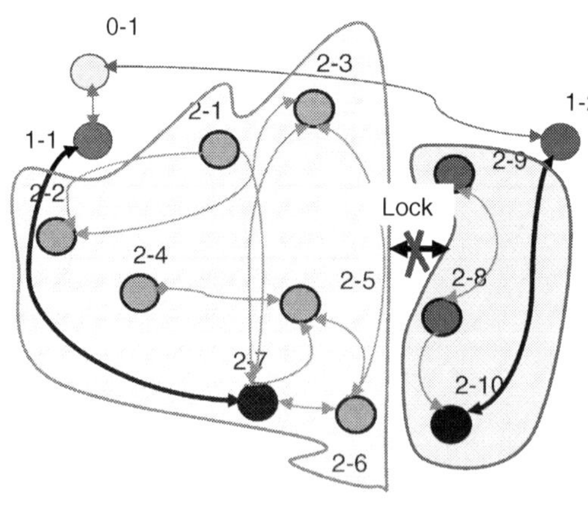

The blue color-based cluster consists of the nodes of Application 5, and the red color-based cluster, includes the assumption nodes of an application. These two clusters have their models but follow a global model. They have direct access to each other's details to obtain a coordinated model for handling a transportation system in Scenario 5 when FL does not contribute to covering the demands. A secure communication system is defined as one that is fault-tolerant, meaning it can handle failures and maintain system security. When nodes 6-7 fail, the blue color-based cluster can utilize other nodes with similar tasks located in other clients to cover the purposes of Application 5, such as responding to emergencies and transporting fuel. This ensures that even if the internal nodes related to traffic monitoring fail, the system can still function effectively. In a smart city, many smart applications and devices can handle demands beyond their client. Using FL techniques can increase the number of tolerable clients and devices, improving fault tolerance in the smart city. This means that intelligent applications can support each other's tasks and effectively cover citizens' demands. Equation (18.13) describes the probability of failure service (FS) when FL is employed to define a global model for a smart city with k clients. FL increases the repair rate (μ) over time, aiding in repairing failures (λ) when many clients are involved in covering faulty nodes' tasks. The variables h, and h_{max} represent the number and total number of failures at a given time.

$$FS = 1 - \left(\frac{\sum_{h=0}^{h=h_{max}} (1 - \mu_h) \times \lambda_h}{k} \right) \tag{18.13}$$

As shown in Equation (18.13), the probability of fault tolerance in a smart city increases as the number of clients enhances and the probability of client repairing rises. In a smart city with many clients, FL creates an opportunity to repair a service by employing the correct clients.

Malicious Federative Model: FL can sometimes have negative implications when aggregating a federative model for different clients in a system or communication platform, especially when the aggregator server is faulty. This can be detrimental to all clients connected to the defective aggregator server and can result in system failures and disruptions in the operations of a smart city. In this section, we depict the association between the aggregator server and clients' nodes in Application 5 using the G_{g5} graph, which consists of $V5$ vertices and $E5$ edges. The graph illustrates how a malicious federative node influenced the faulty nodes (nodes 6-7) of Application 5 and broadcasted a defective global model to other clients.

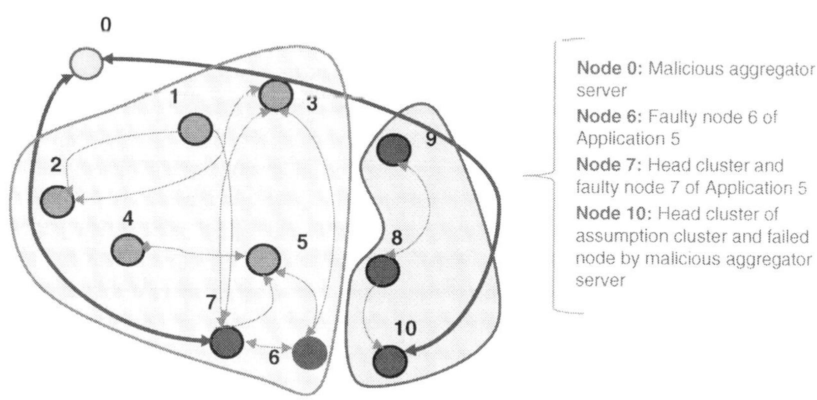

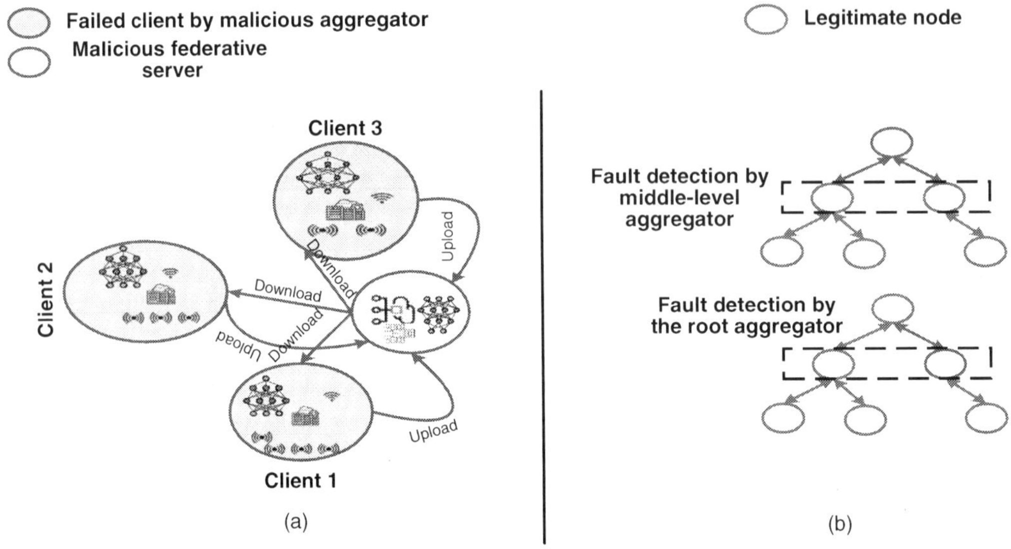

Figure 18.14 (a) Malicious role of FL for fault-tolerant, (b) constructive role of FL for fault-tolerant.

In addition to Application 5 (blue color-based partition), the upper graph shows that the assumption application fails when the aggregator server broadcasts a defective global model. This graph illustrates the role of the aggregator server in causing system failures based on Application 5 and Scenario 5. The aggregator server can endanger a system or multiple connected clients without any internal faults occurring when it broadcasts a malicious model. One possible solution to this issue is using a multistage aggregating model, such as tree-based structures or full custom designs. In a tree-based structure, the aggregating model is implemented in two stages before being broadcast to clients. The initial defined model of the federative server is first broadcasted to the located aggregator nodes in the middle level of the tree, and each aggregator downloads the model for two connected clients based on local time scheduling. Therefore, if the aggregator server located at the tree's root behaves maliciously, the middle-level aggregators can detect the fault and stop transferring the model to clients. The aggregator server can also identify the defective middle-level aggregators and temporarily lock them, preventing the broadcasting of the model to the faulty aggregators. Figure 18.14a, b depicts the malicious role of the federative server and the benefits of tree-based FL implementation in preventing the broadcasting of defective models to clients through multistage model aggregating.

Equation (18.14) calculates the fault tolerance value for aggregator nodes (FTA) when transferring the global model to clients. Herel represents, μ_{NL1} represents the probability of covering faults of an aggregator, and $NL1 \geq 1$ indicates the number of middle-level aggregators. The node number indicates the aggregator server when $NL1 = 0$ is satisfied. As shown in Equation (18.14), when FL is implemented using a basic structure of a federative server, the fault tolerance of a system depends on the probability of the federative server being correct without any faults being covered by another aggregator. The fault tolerance of the aggregator increases when FL is implemented using a tree-based structure, as each aggregator is more likely to be covered by another.

$$FTA = \frac{\sum_{NL1=0}^{NL1=NL1\max}(\mu_{NL1} + (1-\lambda_{NL1}))}{k}$$

$$\begin{cases} \text{When the total number of aggregators} = 1, \\ FTA = \dfrac{(0 + (1-\lambda_0))}{k} = \dfrac{1-\lambda_0}{k} \\ \text{When the total number of aggregators} = \text{NL1}_{\max}, \\ FTA = \dfrac{\sum_{NL1=0}^{NL1=NL1\max}(\mu_{NL1} + (1-\lambda_{NL1}))}{k} \end{cases} \tag{18.14}$$

This section analyzes the constructive and malicious roles of FL in fault-tolerant problems based on Application 5 and Scenario 5 in a smart city. The implementation structure of FL, specifically the broadcasting of initial global models and gathering of local models, significantly impacts satisfying or improving the fault-tolerant factor for a smart city with various heterogeneous applications. The proposed structure must handle multistage model aggregation to prevent the transfer of defective models. This can be achieved through tree and full costume-based designs. It is important to note that while the FL technique can be dangerous for a system, its productivity depends on the implementation method used to address the issue of defective aggregator nodes.

18.9 Future Aspects and Research Benefits

FL is a popular technique used to support distributed systems for data management and establish a secure platform for communication between heterogeneous devices. Due to its concepts, FL holds significant importance in future research and offers benefits in two aspects.

Implementation Structure: In a smart city, various intelligent applications have different priorities and characteristics. FL is crucial in aggregating a global model from clients' local training. The efficiency and cost of FL in a smart city depend on its implementation structure, including delay and energy consumption. By providing a generalized structure, the performance of FL can be improved, enabling timely model aggregation at minimal cost. This, in turn, enhances the efficiency of various applications in meeting citizens' demands in a smart city.

Fault-tolerant: FL addresses a major challenge in distributed systems by providing a global model for local training without direct client access. It establishes a secure platform for client communication, ensuring privacy protection and preventing cyber-attacks. However, there is a potential risk when the aggregative server or aggregator behaves as a malicious node by broadcasting a defective model to candidate clients. The implementation structure can offer a fault-tolerant approach to prevent the transfer of defective global models to clients. Nonetheless, this issue presents a challenge that requires further research and the development of state-of-the-art techniques for robust fault-tolerant FL, particularly when an aggregator acts as a malicious node.

18.10 Conclusion

A smart city has emerged as a modern technology encompassing various intelligent applications in healthcare, industry, transportation, education, homes, and government. This technology is built on the foundation of a distributed system with heterogeneous and asynchronous applications, where FL has proven to be a successful technique for supporting the system. This chapter focuses on the role of FL in five commonly used applications in a smart city: industry, healthcare, medicine recommendation, drug discovery, and fault tolerance. The research also explores different approaches to node mapping, clustering, and head cluster nomination in implementing FL in a smart city. The study highlights Deep Q-learning as a popular machine learning algorithm that can be combined with FL to support smart applications and intelligent devices. The analysis includes examining aggregator servers, trees, and hybrid hierarchical topologies for implementing FL and facilitating communication between aggregators and clients. It is found that fully connected and tree-based designs can offer more efficient solutions for FL. Furthermore, the study investigates the issue of fault tolerance and defines both constructive and malicious roles of FL in establishing a reliable platform for devices in a smart city. The implementation structure for communicating between clients and aggregators is crucial in addressing this issue.

18.11 Parameters' Declaration

$[A]_{n\times m}$: A Matrix with n rows and m columns

$[G]_{i\times j}$: G Matrix with i rows and j columns

$g_{i,j}$: Element of G matrix whereas $g_{i,j}=0$ or $g_{i,j}=1$

$a_{n,m}$: Element of A matrix whereas $a_{n,m}=0$ or $a_{n,m}=1$

i: The row number of G matrix whereas $0\le i\le i_{\max}$

j: The column number of G matrix whereas $0\le j\le j_{\max}$

$i_{\max}$: The maximum number of rows of G matrix whereas $i_{\max}\ge 0$

$j_{\max}$: The maximum number of columns of G matrix whereas $j_{\max}\ge 0$

$n_{\max}$: The maximum number of rows of A matrix whereas $n_{\max}\ge 0$

$Cluster_c$: The cluster with c number whereas $c\ge 0$

Start node: Starter node that a cluster identifies based on its edges

L: The level number of the tree whereas $0\le L\le 2$

NI: The node number of the middle level (level 1) of the tree whereas $NI\ge 0$

$TreeNode_{L_{NI}}$: The tree node located on middle level of level L

$Aggregator_{L1_{NI}}$: The aggregator node located on level 1 with NI number

$Client_{L_{NI}}$: The connected client to aggregator NI located on level L of tree

MiddleLayerCluster: The clusters of middle level of tree

$MergedNode_{NL1}$: The merged node $NL1$ whereas $NL1\ge 0$

$Node_{NL1+1}$: The located node $LN1+1$ (next node after node $NL1$) on level 1 of the tree

FC: The number of failed communications of faulty nodes

f: The faulty node number whereas $0\le f\le f_{\max}$

$f_{\max}$: The maximum number of faulty nodes whereas $f_{\max}\ge 0$

$f(w)_k$: Local function of client k whereas $k \geq 0$

r: The number of iterations round whereas $r \geq 0$
$r_{\max}$: The maximum number of iterations round whereas $r_{\max} \geq 0$

k: The number of clients whereas $k \geq 0$

$k_{\max}$: The maximum number of clients whereas $k_{\max} \geq 0$
ds_{k_r}: Discount factor whereas $0 \leq ds \leq 1$

t: Time whereas $t \geq 0$

$F(w)_{r_{i+1}}$: Global function at next time of last round iteration

$NodeCandidate_d$: The candidate node for head cluster with d number whereas $d \geq 0$

α: Impact factor for reliability whereas $0 \leq \alpha \leq 1$

β: Impact factor for availability whereas $0 \leq \beta \leq 1$

μ: Impact factor for cost whereas $0 \leq \mu \leq 1$

$(reliability)_d$: Reliability for candidate node with d index whereas $0 \leq reliability \leq 1$
$(availability)_d$: Availability for candidate node with d index number whereas $0 \leq availability \leq 1$
$(cost)_d$: Cost for candidate node with d index number whereas $0 \leq cost \leq 1$

$(FaultyNode)_f$: The faulty node with f index number

h: The faulty node number of FL components whereas $0 \leq h \leq h_{\max}$
$h_{\max}$: The maximum faulty number of FL components whereas $h_{\max} \geq 0$

μ_h: The repair rate of faulty node whereas $0 \leq \mu \leq 1$

λ_h: The fault rate of node h whereas $0 \leq \lambda \leq 1$

FTA: The fault-tolerant for aggregators

$NL1$: The node number of level1 of a purr binary tree whereas $0 \leq NL1 \leq 1$

$NL1_{\max}$: The maximum node number of level1 of a pure binary tree whereas $NL1_{\max} \geq 0$

μ_{NL1}: The repair rate of faulty aggregator $NL1$ whereas $0 \leq \mu \leq 1$

λ_{NL1}: The fault rate of aggregator $NL1$ whereas $0 \leq \lambda \leq 1$

$m_{\max}$: maximum number of columns of A matrix whereas $m_{\max} \geq 0$

References

1 Hanjri, M.E., Kabbaj, H., Kobbane, A., and Abouaomar, A. (January 30 2023). Federated learning for water consumption forecasting in smart cities. arXiv preprint arXiv:2301.13036.

2 Bousbiat, H., Bousselidj, R., Himeur, Y. et al. (17 April 2023) Crossing roads of federated learning and smart grids: overview, challenges, and perspectives. arXiv preprint arXiv:2304.08602.

3 Zekiye, A. and Özkasap, Ö. (1 May 2023). The internet of energy systems: blockchain and smart contracts meet federated learning. *2023 IEEE International Conference on Blockchain and Cryptocurrency (ICBC)*, pp. 1–3. IEEE, Dubai, United Arab Emirates.

4 da Silva, M.V., Bittencourt, L.F., and Rivera, A.R. (10 December 2020). Towards federated learning in edge computing for real-time traffic estimation in smart cities. *Anais do IV Workshop de Computação Urbana*, pp. 166–177. SBC, Rio de Janeiro, Brazil.

5 Musa, A.A., Malami, S.I., Alanazi, F. et al. (2023). Sustainable traffic management for smart cities using internet-of-things-oriented intelligent transportation systems (ITS): challenges and recommendations. *Sustainability* 15 (13): 9859.
6 Liu, G., Furth, N., Shi, H. et al. (8 May 2023). Federated learning aided deep convolutional neural network solution for smart traffic management. *NOMS 2023-2023 IEEE/IFIP Network Operations and Management Symposium*, pp. 1–8. IEEE.
7 Deng, H., Hu, J., Sharma, R. et al. (2023). NVAS: a non-interactive verifiable federated learning aggregation scheme for COVID-19 based on game theory. *Computer Communications* 206: 1–9.
8 Farooq, M.S., Tehseen, R., Qureshi, J.N. et al. (2023). FFM: flood forecasting model using federated learning. *IEEE Access* 11: 24472–24483.
9 Ullah, F., Srivastava, G., Xiao, H. et al. (2023). A scalable federated learning approach for collaborative smart healthcare systems with intermittent clients using medical imaging. *IEEE Journal of Biomedical and Health Informatics*. 1–13. doi: 10.1109/JBHI.2023.3282955.
10 Oldenhof, M., Ács, G., Pejo, B. et al. (2022). Industry-scale orchestrated federated learning for drug discovery. arXiv. arXiv preprint arXiv:2210.08871.
11 Houssein, E.H. and Sayed, A. (2023). Boosted federated learning based on improved particle swarm optimization for healthcare IoT devices. *Computers in Biology and Medicine* 22: 107195.
12 Wang, W., He, F., Li, Y. et al. (2023). Data information processing of traffic digital twins in smart cities using edge intelligent federation learning. *Information Processing and Management* 60 (2): 103171.
13 Asha, A., Arunachalam, R., Poonguzhali, I. et al. (2023). Optimized RNN-based performance prediction of IoT and WSN-oriented smart city application using improved honey badger algorithm. *Measurement* 210: 112505.
14 Irshad, R.R., Hussain, S., Hussain, I. et al. (2023). An intelligent buffalo-based secure edge-enabled computing platform for heterogeneous IoT network in smart cities. *IEEE Access* 11: 69282.
15 Ghosh, U., Das, D., Chatterjee, P., and Shetty, S. (12 June 2023). Quantum-enabled blockchain for data processing and management in smart cities. *2023 IEEE 24th International Symposium on a World of Wireless, Mobile and Multimedia Networks (WoWMoM)*, pp. 425–430. IEEE, Boston, Massachusetts, USA.
16 Mohammed, N.J. (2023). Quantum cryptography in convolution neural network approach in smart cities. *Journal of Survey in Fisheries Sciences* 10 (2S): 2043–2056.
17 Kuang, Z. and Chen, C. (2023). Research on smart city data encryption and communication efficiency improvement under federated learning framework. *Egyptian Informatics Journal* 24 (2): 217–227.
18 Devarajan, G.G., Thirunnavukkarasan, M., Amanullah, S.I. et al. (2023). An integrated security approach for vehicular networks in smart cities. *Transactions on Emerging Telecommunications Technologies* 15: e4757.
19 Li, G., Luan, T.H., Li, X. et al. (2023). Breaking down data sharing barrier of smart city: a digital twin approach. *IEEE Network* 38: 238.
20 Vivekanandan, M., Premkamal, P.K., Johnpaul, C.I., and Ebinazer, S.E. (11 February 2023). Blockchain based secure data storage verification algorithm for smart city environment. *2023 4th International Conference on Innovative Trends in Information Technology (ICITIIT)*, pp. 1–5. IEEE, Kottayam, India.
21 Li, Y., Yu, H., Zeng, Y., and Pan, Q. (2023). HFSA: a semi-asynchronous hierarchical federated recommendation system in smart city. *IEEE Internet of Things Journal* 10: 18808.

22 Singh, S., Rathore, S., Alfarraj, O. et al. (2022). A framework for privacy-preservation of IoT healthcare data using federated learning and blockchain technology. *Future Generation Computer Systems* 129: 380–388.

23 Jayaraman, R., Alshehri, M., Kumar, M. et al. (2023). Secure biomedical document protection framework to ensure privacy through blockchain. *Big Data* 11: 437.

24 Tareen, F.N., Alvi, A.N., Malik, A.A. et al. (2023). Efficient load balancing for blockchain-based healthcare system in smart cities. *Applied Sciences* 13 (4): 2411.

25 Oueida, S., Aloqaily, M., and Ionescu, S. (2019). A smart healthcare reward model for resource allocation in smart city. *Multimedia Tools and Applications* 78: 24573–24594.

26 Sahatiya, P. and Singh, D.K. (25 April 2023). Role of telemedicine in healthcare sector for betterment of smart city. *Soft Computing: Theories and Applications: Proceedings of SoCTA 2022*, pp. 925–935. Singapore: Springer Nature Singapore.

27 Kalapaaking, A.P., Khalil, I., and Yi, X. (2023). Blockchain-based federated learning with SMPC model verification against poisoning attack for healthcare systems. *IEEE Transactions on Emerging Topics in Computing* 12: 269.

28 Malik, H., Anees, T., Faheem, M. et al. (2023). Blockchain and internet of things in smart cities and drug supply management: open issues, opportunities, and future directions. *Internet of Things* 26: 100860.

29 Saxena, N., Saxena, P., and Veenadhari, S. (8 April 2023). Adaptive multi-hop deep learning based drug recommendation system with selective coverage mechanism. *2023 IEEE 12th International Conference on Communication Systems and Network Technologies (CSNT)*, pp. 133–138. IEEE, Bhopal, India.

30 Cheikhrouhou, O., Amdouni, I., Mershad, K. et al. (6 June 2022). Blockchain for the cybersecurity of smart city applications. arXiv preprint arXiv:2206.02760.

31 https://www.tensorflow.org/federated (accessed 19 September 2024).

32 https://seer.cancer.gov/statfacts/html/common.html (accessed 19 September 2024).

33 https://www.who.int/news-room/fact-sheets/detail/mental-disorders (accessed 19 September 2024).

Index

Model Optimization Methods for Efficient and Edge AI: Federated Learning Architectures, Frameworks and Applications, First Edition. Edited by Pethuru Raj Chelliah, Amir Masoud Rahmani, Robert Colby, Gayathri Nagasubramanian, and Sunku Ranganath.

Made in the USA
Las Vegas, NV
19 September 2026

e079952a-222d-4b8e-ba87-0359104a10ebR01